APPLIED MATHEMATICS IN CHEMICAL ENGINEERING

McGraw-Hill Chemical Engineering Series

THE SERIES

Anderson and Wenzel—*Introduction to Chemical Engineering*
Aries and Newton—*Chemical Engineering Cost Estimation*
Badger and Banchero—*Introduction to Chemical Engineering*
Bennett and Myers—*Momentum, Heat, and Mass Transfer*
Clarke and Davidson—*Manual for Process Engineering Calculations*
Comings—*High Pressure Technology*
Coughanowr and Koppel—*Process Systems Analysis and Control*
Dodge—*Chemical Engineering Thermodynamics*
Griswold—*Fuels, Combustion, and Furnaces*
Groggins—*Unit Processes in Organic Synthesis*
Harriott—*Process Control*
Henley and Bieber—*Chemical Engineering Calculations*
Huntington—*Natural Gas and Natural Gasoline*
Johnson—*Automatic Process Control*
Johnstone and Thring—*Pilot Plants, Models, and Scale-up Methods in Chemical Engineering*
Katz, Cornell, Kobayashi, Poettmann, Vary, Elenbaas, and Weinaug—*Handbook of Natural Gas Engineering*
Kirkbride—*Chemical Engineering Fundamentals*
Knudsen and Katz—*Fluid Dynamics and Heat Transfer*
Kohl and Riesenfeld—*Gas Purification*
Lapidus—*Digital Computation for Chemical Engineers*
Leva—*Fluidization*
Lewis, Radasch, and Lewis—*Industrial Stoichiometry*
Mantell—*Adsorption*
Mantell—*Electrochemical Engineering*
McAdams—*Heat Transmission*
McCabe and Smith, J. C.—*Unit Operations of Chemical Engineering*
Mickley, Sherwood, and Reed—*Applied Mathematics in Chemical Engineering*
Nelson—*Petroleum Refinery Engineering*
Perry (Editor)—*Chemical Business Handbook*
Perry (Editor)—*Chemical Engineers' Handbook*
Peters—*Elementary Chemical Engineering*
Peters—*Plant Design and Economics for Chemical Engineers*
Reid and Sherwood—*The Properties of Gases and Liquids*
Robinson and Gilliland—*Elements of Fractional Distillation*
Schechter—*The Variational Method in Engineering*
Schmidt and Marlies—*Principles of High-polymer Theory and Practice*
Schweyer—*Process Engineering Economics*
Sherwood and Pigford—*Absorption and Extraction*
Shreve—*Chemical Process Industries*
Smith, B. D.—*Design of Equilibrium Stage Processes*
Smith, J. M.—*Chemical Engineering Kinetics*
Smith, J. M., and Van Ness—*Introduction to Chemical Engineering Thermodynamics*
Treybal—*Liquid Extraction*
Treybal—*Mass-transfer Operations*
Van Winkle—*Distillation*
Vilbrandt and Dryden—*Chemical Engineering Plant Design*
Volk—*Applied Statistics for Engineers*
Walas—*Reaction Kinetics for Chemical Engineers*
Walker, Lewis, McAdams, and Gilliland—*Principles of Chemical Engineering*
Williams and Johnson—*Stoichiometry for Chemical Engineers*
Wilson and Ries—*Principles of Chemical Engineering Thermodynamics*

APPLIED MATHEMATICS IN CHEMICAL ENGINEERING

HAROLD S. MICKLEY

Professor of Chemical Engineering
Massachusetts Institute of Technology

THOMAS K. SHERWOOD

Professor of Chemical Engineering
Massachusetts Institute of Technology

CHARLES E. REED

Engineering Manager, Chemical Department
General Electric Company

SECOND EDITION

McGRAW-HILL BOOK COMPANY

New York Toronto London

1957

APPLIED MATHEMATICS IN CHEMICAL ENGINEERING

Library of Congress Catalog Card Number 56-10325

1415 FGBP 77654

ISBN 07-041800-4

PREFACE

Applied mathematics has assumed increasing importance to the chemical engineer. The authors of current professional literature freely use transform, vector, and finite-difference methods to attack a problem. The practicing engineer is finding modern mathematical techniques a valuable tool in the analysis of a variety of situations. Present trends in the chemical and process industries involve increased emphasis on automatic-control systems, high-speed machine computation, operational analysis, and the like. These developments directly depend upon the application of advanced mathematical procedures.

The purpose of this book is to consolidate the advanced methods of mathematics into a form that can be applied readily by both the student and the professional engineer. Emphasis is focused on the engineering applications of mathematics. Considerable attention is given to the problem of expressing a physical situation in mathematical language. Problems drawn from the engineering literature are used to illustrate the mathematical procedures. The material covered falls into three broad categories: (1) the treatment and interpretation of engineering data, (2) the analysis of situations involving only one independent variable, and (3) the analysis of situations involving two or more independent variables.

The mathematical background of the reader is presumed to be limited. The text material begins with a discussion of the process of differentiation, and the development of more advanced procedures follows in a step-by-step manner.

The book may be used as either an undergraduate or a graduate text. The material of Chaps. 1, 3, 4, and 5 can be handled readily by the undergraduate and will provide the background needed for the assimilation of the more advanced material in a subsequent study program.

The second edition represents an extensive revision of the original work. Chapter 8 (The Laplace Transform), Chap. 9 (Analysis of Stagewise Processes by the Calculus of Finite Differences), and Chap. 10 (The Numerical Solution of Partial Differential Equations) are completely new. The remaining chapters have been rewritten, and new material has been added. Such topics as the statistics of small samples, analysis of variance, factorial design of experiments, expansion in a series of orthogonal functions, vector notation, and others have been included.

The decision to add new material and to omit material formerly included has been difficult. The action taken has been based upon an appraisal of trends both in engineering education and in the engineering profession.

The treatment of the material used has been influenced by a number of sources. Primary impetus was supplied by the first edition of "Applied Mathematics in Chemical Engineering." Marshall and Pigford's "The Application of Differential Equations to Chemical Engineering Problems" focused attention on the utility of the Laplace transform and the calculus of finite differences in the analysis of the transient behavior of unit operations equipment. The advanced mathematics courses given at the Massachusetts Institute of Technology, and in particular those that have led to F. B. Hildebrand's books, "Advanced Calculus for Engineers" and "Methods of Applied Mathematics," have been of great help. The texts by R. V. Churchill, "Fourier Series and Boundary Value Problems," "Modern Operational Mathematics in Engineering," and "Introduction to Complex Variables and Applications," have been consulted frequently. Other major sources of material have been listed in the bibliography at the end of each chapter.

The authors are indebted to Professor Sir Ronald A. Fisher, Cambridge, and to Oliver & Boyd, Ltd., Edinburgh and London, for permission to reprint Table 2-1 from their book "Statistical Methods for Research Workers,"; to Professor P. C. Mahalanobis, F.R.S., Calcutta, and to the Indian Statistical Institute, Calcutta, for permission to reprint Table 2-4 from an article appearing in *Sankhyā;* to Professor G. W. Snedecor, Ames, Iowa, and to Collegiate Press, Inc., of Iowa State College, Ames, for permission to reprint Table 2-5 from their book "Statistical Methods Applied to Experiments in Agriculture and Biology"; to Dr. T. v. Kármán and Dr. M. A. Biot and to McGraw-Hill Book Company, Inc., New York, for permission to reprint Table 5-2 from their book "Mathematical Methods in Engineering"; to Professor R. V. Churchill, Ann Arbor, Mich., and to McGraw-Hill Book Company, Inc., New York, for permission to reprint Table 8-1 from their book "Modern Operational Mathematics in Engineering." The constructive suggestions and encouragement of the author's professional colleagues are gratefully acknowledged. In particular, the invaluable aid in the preparation of the second edition by the authors of the first edition, Professor T. K. Sherwood and Dr. C. E. Reed, is sincerely appreciated.

HAROLD S. MICKLEY

CONTENTS

CHAPTER 1

TREATMENT OF ENGINEERING DATA

1-1. Introduction. The engineer constantly utilizes experimental data. He tests theoretical predictions by comparison with experiment; he analyzes process performance by examination of experimental results; he makes critical decisions on the basis of his interpretation of experimental measurements. Consequently, the ability to extract maximum information from engineering data is important. This chapter discusses several useful techniques for the treatment of experimental results.

1-2. Graphical Representation. Graphical methods have proved invaluable in the analysis of the relatively complex processes with which the chemical engineer deals. Much of the basic physical and chemical data are best represented graphically, and graphical methods are introduced in this way into the analytical treatment of the process.

One or more of the many types of graphical representations may be employed for the following purposes: (1) as an aid in visualizing a process or the meaning of a computation, (2) for the representation of quantitative data or of a theoretical or empirical equation, (3) for the comparison of experimental data with a theoretical or empirical expression, and (4) as a means of computation.

The relation between two physical quantities p and x is commonly obtained as a tabulation of values of p for a number of different values of x. The relation between p and x is not easy to visualize by studying the tabulated results and is best seen by plotting p vs. x. If the conditions of the experiment are such that p is known to be a function of x only, the functional relation will be indicated by the fact that the points may be represented graphically by a smooth curve, and deviations of the points from a smooth curve indicate the reliability of the data. If p is a function of two variables x and y, a series of results of p in terms of x may be obtained for each of several values of y. When plotted, the data will be represented by a family of curves, each curve representing the relation between p and x for a definite constant value of y. If another variable z is involved, we may have separate graphs for constant values of z, each showing a family of curves of p vs. x. Extension of this method of representing data to relate more than four quantities is impractical unless general relations between two or more of the variables can be obtained. This point is further discussed below.

The ordinary graphical representation of experimental results is usually the first step in finding an empirical equation to represent the data, as described in Sec. 1-5. Even when the empirical equation is to be obtained directly from the tabulated data by a numerical process, it is usually desirable to plot the points in order that the nature of the function may be visualized. Where the calculated relation between two quantities is dependent wholly on sound theoretical or empirical equations, it may be desirable to plot the resulting function in order to visualize its properties.

The use to which the graph is to be adapted should be kept in mind in preparing a graphical representation either of data or of an equation. The coordinates should be chosen in such a way that the accuracy of reading the graph will be good for all ranges of the variables involved, with the resulting curve falling with a slope of roughly ± 1 on a square diagram. The scale should be arranged so that interpolation is readily accomplished.

Consider the relation

$$E = \frac{8}{\pi^2}\left\{\exp\left[-\left(\frac{\pi}{2}\right)^2\tau\right] + \frac{1}{9}\exp\left[-9\left(\frac{\pi}{2}\right)^2\tau\right] + \frac{1}{25}\exp\left[-25\left(\frac{\pi}{2}\right)^2\tau\right] + \cdots\right\} \qquad (1\text{-}1)$$

which is obtained as the analytical solution to the problem of the un-

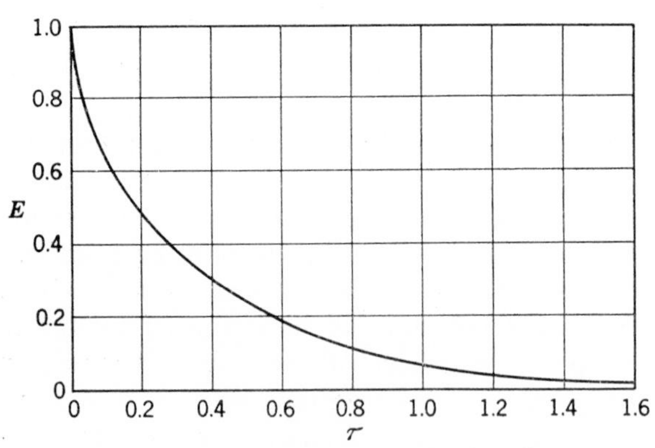

FIG. 1-1. Linear-scale plot of E-τ function.

steady-state cooling of an infinite solid slab. Values of E may be obtained directly for any value of τ, but a tedious trial-and-error calculation is necessary if τ is to be obtained for a given value of E. By means of a graph of E vs. τ, E may be read in terms of τ or τ in terms of E with equal facility.

Figure 1-1 shows E plotted vs. τ on ordinary rectangular coordinate paper. It is apparent that τ cannot be read accurately for large values of E, nor can E be obtained accurately at large values of τ. When τ is large, the first term of the series is the only one of significance, and the series reduces to an exponential relation, suggesting a graph of log E vs. τ

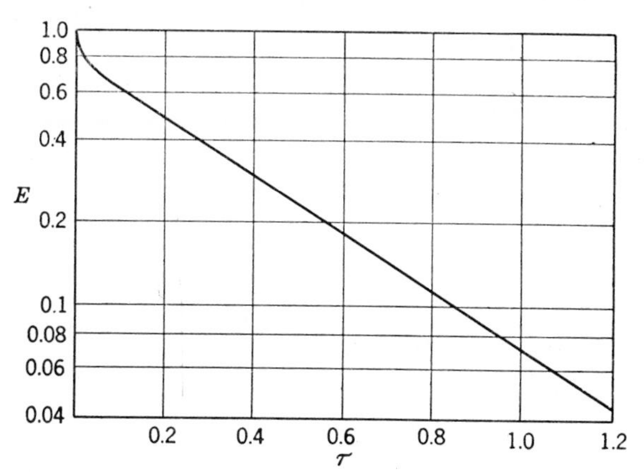

FIG. 1-2. Semilog plot of E-τ function.

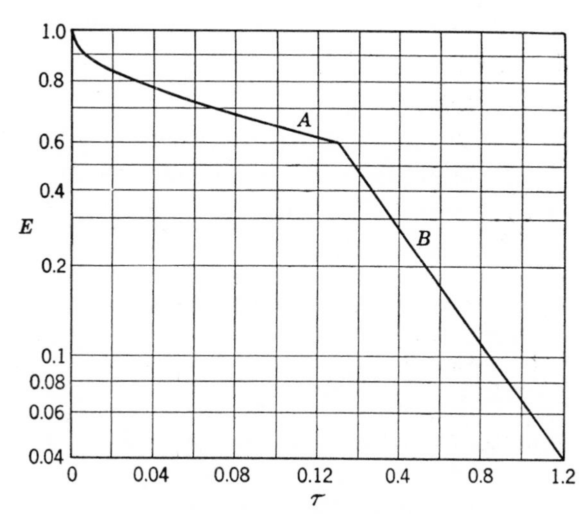

FIG. 1-3. Extended abscissa scale.

or a graph of E vs. τ on semilogarithmic coordinate paper. Figure 1-2 is such a graph of E vs. τ, with a logarithmic ordinate scale and a linear abscissa scale. It is apparent that this method of plotting represents a considerable improvement, as the graph may be read with good accuracy (in view of its size) except for small values of τ. Figure 1-3 represents a further modification of the same graph. The logarithmic scale from

$E = 0.6$ to $E = 1.0$ has been doubled and a larger abscissa scale used in this region. The result is a combination of two graphs similar to Fig. 1-2, with different ordinate and abscissa scales for values of E greater and less than 0.6. The two graphs are fitted together, giving a curve with two branches. The accuracy in reading values from branch A is considerably improved, although the accuracy at large values of τ is less than in Fig. 1-2.

1-3. Elimination of Trial and Error. Experimental data or correlations of experimental data should be plotted, if possible, so that the use of the graph should eliminate trial-and-error calculations. This is not

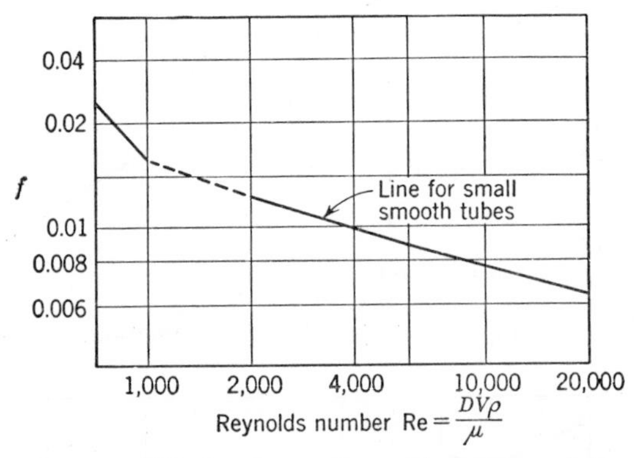

FIG. 1-4. Friction-factor–Reynolds-number plot.

always possible for all uses to which the graph may be adapted, but alternative methods of plotting may be used to advantage in employing the data in different ways. The familiar friction-factor graph for fluids in round pipes represents a correlation of a large amount of data, usually plotted as f vs. $DV\rho/\mu$ with logarithmic ordinate and abscissa scales (Fig. 1-4). This is used in connection with the Fanning equation

$$H = \frac{2fLV^2}{g_cD} \tag{1-2}$$

where H = head lost because of friction in length L of pipe having diameter D

g_c = "consistency factor"

V = superficial fluid velocity

ρ = fluid density

μ = fluid viscosity

f = dimensionless friction factor

When all quantities are known except the head loss H, which is to be calculated, the group Re is first obtained, f is read from the plot, and H is calculated directly from the equation. If it is desired to calculate the necessary pipe diameter for a specified flow and head, the problem becomes one of trial and error, since the Reynolds group is not immediately obtainable. Similarly, trial and error are involved in the calculation of the flow to be expected with a specified head, fluid, and pipe size. It is clear that the method of plotting is inconvenient for two of the three usual calculations in connection with which the correlation is of value. The following procedure will determine the manner in which the graphed data should be replotted in order to eliminate trial and error in a specific type of calculation.

Consider the case in which it is desired to calculate the pipe diameter corresponding to a specified flow and head. What is needed is a plot prepared by using Fig. 1-4, in which the abscissa contains only known quantities and the ordinate the unknown (and any necessary known) parameters.

Let Q represent the given volume rate of flow. Then the relations which are available are Eq. (1-2),

$$Q = \frac{\pi}{4} D^2 V \tag{1-3}$$

$$\text{Re} = \frac{DV\rho}{\mu} \tag{1-4}$$

and the relation between f and Re given by Fig. 1-4. The "unknowns" are four in number: D, V, Re, and f. There are four "equations" (one of them graphical) and four unknowns, and so a solution is possible. The algebraical relations are used to eliminate those unknowns which do not appear directly in the graph, in this case, D and V. Combination of (1-3) and (1-4) gives

$$V = \frac{\pi(\text{Re})^2\mu^2}{4Q\rho^2} \tag{1-5}$$

$$D = \frac{4Q\rho}{\pi(\text{Re})\mu} \tag{1-6}$$

Substitution of (1-5) and (1-6) in (1-2) gives

$$\text{Re}\,\sqrt[5]{f} = \left(\frac{32g_cH}{L}\right)^{1/5} \left(\frac{Q}{\pi}\right)^{3/5} \frac{\rho}{\mu} \tag{1-7}$$

The right-hand side of Eq. (1-7) contains known quantities only. Furthermore, Fig. 1-4 may be used to prepare a plot of Re $\sqrt[5]{f}$ vs. either f or Re. With the aid of such a plot, Re or f may be obtained, without trial and error, from the specified data. Equations (1-5) to (1-7) may then be used to calculate the remaining variables.

A graph of Re $\sqrt[5]{f}$ vs. Re is shown in Fig. 1-5. Since Re appears in both variables plotted, the calculated value of D is quite insensitive to f. This is true because of the relation between the variables and not because of the method of plotting, although it will be shown below that graphical representations of data should not, in general, involve the same variable in both ordinate and abscissa.

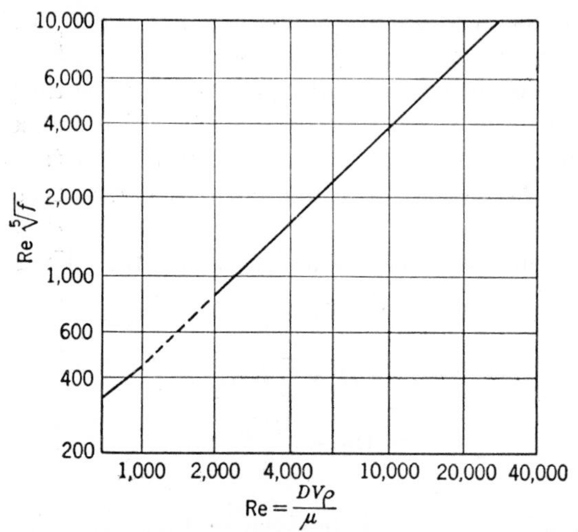

FIG. 1-5. Modified friction-factor plot.

Following the same principle of eliminating the variable to be determined from one of the quantities plotted, it may be shown that the third calculation to determine the flow for a specified head and pipe size is easily carried out, using a plot of Re $\sqrt{f}$ vs. Re or f.

1-4. Misleading Methods of Correlation. Any correlation of experimental data based on a graph in which the same variable appears in both ordinate and abscissa should be viewed with suspicion. When one of the less important variables is placed in both quantities plotted, it is possible to extend the scale and make the correlation appear to be much better than it really is. Such correlations are occasionally presented in the literature. The investigator, trying various methods of plotting his results, hits upon a method of plotting that brings his data together and presents a correlation that is unintentionally deceiving as to its generality. Such methods of plotting may be arrived at by fairly sound analysis of the physical problem involved and may be defended as being rational, although a poor test of the data. A rather subtle example of this process arises in the study of heat transfer to boiling liquids. The surface coefficients obtained are large and are relatively difficult to reproduce, so that the problem of correlating such data is difficult. Suppose a set of data to have been obtained, covering a range of temperature differences between steam and boiling water in an evaporator relying on liquid circulation by natural convection. The experimenter is confronted with a series of values of h and Δt, all the results having been obtained with a constant liquid composition and boiling

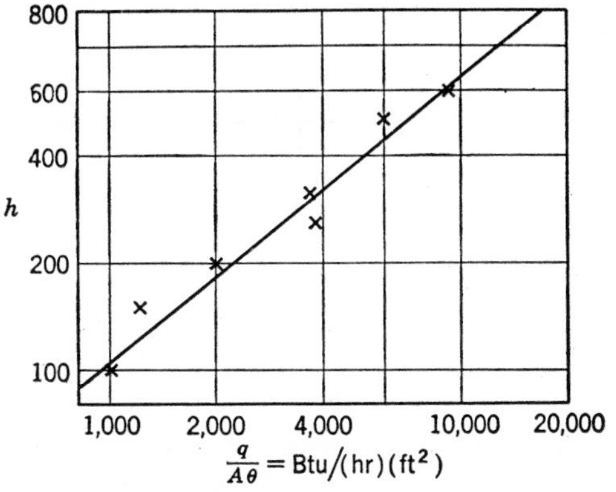

FIG. 1-6. Correlation of data on heat transfer.

temperature. He reasons that the surface coefficient of heat transfer is dependent on the effective thickness of some type of surface film and that this in turn should be a function of the degree of agitation of the liquid in the evaporator. The agitation should be a function of the rate of boiling and the rate of heat transfer, in turn. He reaches the conclusion, therefore, that h should depend on the heat "current density" $q/A\theta$ and prepares a graph to test this conclusion. The result is shown in Fig. 1-6, which indicates a better correlation than is often obtained for this case of heat transfer.

It should be noted, however, that the abscissa $q/A\theta$ is the calculated product of h and Δt, and the graph, therefore, involves h in both ordinate and abscissa. If h is plotted vs. Δt, the result indicates a poor correlation of the same data, as shown by Fig. 1-7. Fundamentally, the two graphs, Figs. 1-6 and 1-7, are equivalent, but the former appears to present a better correlation of the data because large variations in h overshadow small deviations of Δt. It may be argued that the quantities varied were the rate of heat flow and the steam temperature and that these

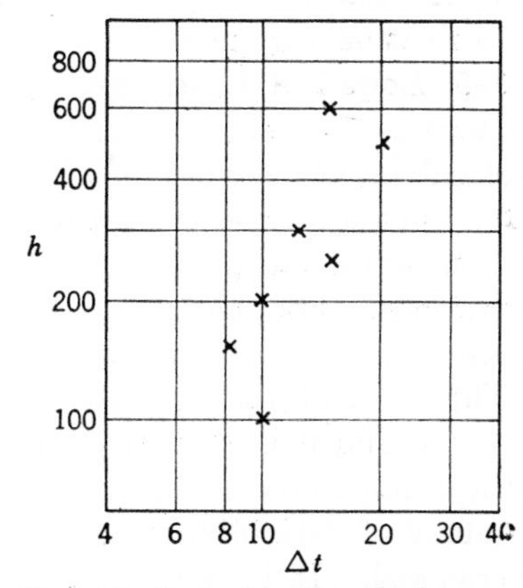

FIG. 1-7. Test of boiling heat-transfer data.

should be the variables plotted, but general correlations on such a basis would be inconvenient, and the graph of h vs. Δt is an excellent test of the experimental data.

From the example quoted above, it follows that if one variable plotted is *divided* by some function of the other, a graph of the resulting ratio may present a more severe test of the data. For example, fluid-flow data are frequently plotted as

$$f = \frac{\pi^2 H g_c D^5}{32 L Q^2}$$

vs.

$$\mathrm{Re} = \frac{4Q\rho}{\pi \mu D}$$

If the experimental variables are H and D, with L, Q, μ, and ρ held constant, the graph of f vs. Re is really a graph of HD^5 vs. $1/D$. If the measured value of D is 10 per cent too great but H is correct, the abscissa will be 9 per cent too small, and the ordinate will be 61 per cent high. The general slope of the curve is negative, and the experimental point will be considerably off. In this case, the method of plotting represents a severe test of the data. If the experimental variables are H and ρ, all others being held constant, then the usual friction-factor graph represents a direct plotting of the variables studied. When a number of variables are to be related, it is seldom possible to find a single method of plotting that will represent an equally critical test of the experimental data relating various pairs of the variables.

1-5. Empirical Equations. The representation of experimental data by means of algebraic equations is a practical necessity in engineering. Not only are such equations shorthand expressions for a large amount of data, but they serve as the necessary mathematical expressions by which the empirical information may be treated in subsequent mathematical operations. For the first use, the equation must be truly representative of the experimental data; for the second, it should be simple in form.

The form of the equation is frequently suggested by a theoretical analysis, and it is necessary only to evaluate certain constants. If the form is not known, dimensional analysis may be helpful in suggesting grouping of variables, and obvious practical considerations must not be overlooked. It is often evident that the curve must go through the origin or some fixed point or perhaps become asymptotic to some definite value of one variable. The form of the empirical equation chosen must be consistent with such considerations. The general problem of fitting data by an empirical equation may be divided into two parts: the determination of a suitable form of equation and the evaluation of the constants. The determination of the form proceeds largely by trial, although certain rules may be laid down as practical aids. If the data can be plotted in such a way as to give a straight line, either by choice of graph paper or by a proper arrangement of variables for plotting, the linear form leads immediately to an expression relating the original variables. Likely forms are tested by plotting in such a manner as to make the expression linear, and the constants are evaluated from the straight line obtained.

The experimental data may be assumed to be given in the form of a table of values of a variable y for corresponding values of another variable x. For example, let it be desired to represent the following data by means of an empirical equation:

x	0.2	0.5	1.0	2.0	3.0	4.0
y	3.2	3.7	4.1	8.1	13.7	22.6

To make the problem general, it will be assumed that there is no additional information as to the nature of the function.

The data may always be represented by a trigonometric series or by a polynomial having a sufficient number of terms; the six points tabulated may be represented by a polynomial of the fifth degree, containing six arbitrary constants. A form involving so many constants is undesirable for three reasons: The constants would be quite difficult to evaluate; the expression obtained would be relatively complicated to use; and the determinations presumably involve experimental errors of such magnitude that an approximate representation is all that is justified. The problem, therefore, is to represent the data by an expression of the simplest possible form that fits the tabulated results within the estimated experimental error.

The first thing to do is to plot the data on ordinary rectangular coordinates, as shown in Fig. 1-8. This serves two purposes: The nature of the function is easily visualized, so that the empirical form to be employed may be more readily selected; and the possibility of the simple linear form

$$y = a + bx \tag{1-8}$$

is tested at the outset. The points do not fall on a straight line in Fig. 1-8, and the simple linear form is not applicable. Usually, the next thing to do is to plot the data on logarithmic paper in order to test the form

$$y = ax^n \tag{1-9}$$

This form is very often applicable and is not too complicated for ordinary use. It is represented by a straight line of y vs. x on logarithmic paper, having a slope n and an "intercept" a at $x = 1$. If the ordinate and abscissa scales are equal, i.e., if the distances along the axes are the same for a tenfold change in the variable y as for a tenfold change in x, the actual slope of the line is the value of n. If the scales are unequal, n is best obtained by substituting coordinates of the straight line in (1-9). In the case of the illustrative example, it is apparent from Fig. 1-8 that (1-9) is not applicable, since the curve does not go through the origin. The power function (1-9) passes through the origin for all positive values of n; if n is negative, y decreases as x increases. The logarithmic plot may be omitted in this case.

The shape of the curve is typical of expressions of the form (1-9), a fact which suggests that (1-9) might be applicable if a suitable change in coordinates were

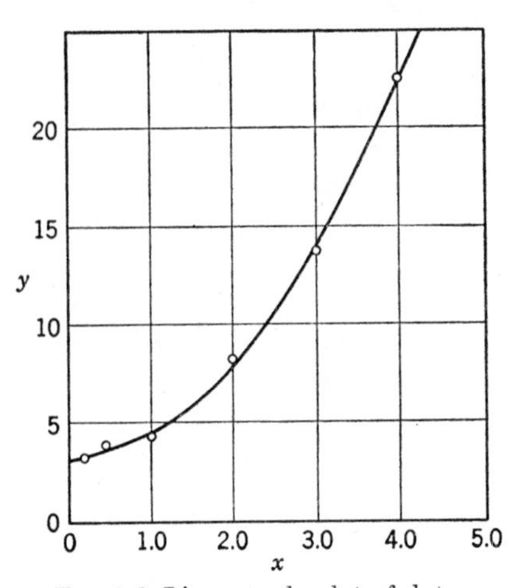

Fig. 1-8. Linear-scale plot of data.

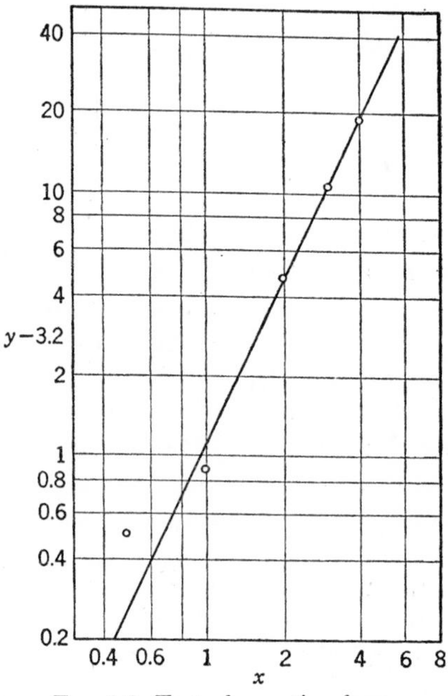

Fig. 1-9. Test of equation form.

made. Since (1-9) must pass through the origin, the change must place the new origin at some point on the curve. An obvious possibility is the intercept $y = 3.2$ at $x = 0$. The new variables $y - 3.2$ and x are plotted in Fig. 1-9 and may be approximated by a straight line. The equation of the line is

$$y = 3.2 + ax^n \tag{1-10}$$

which may be assumed to be a satisfactory representation of the data, provided that after evaluation of the constants the discrepancies between calculated and observed values are no greater than the assumed error in the experimental determinations. The procedures for evaluation of constants will be taken up in the next section.

The following is a useful summary of the simpler forms that may be used, with suggested methods of plotting to give straight lines:

1. $y = a + bx$ — Plot y vs. x.
2. $y = ax^n$ — Plot log y vs. log x or y vs. x on logarithmic coordinates.
3. $y = c + ax^n$ — First obtain c as intercept on plot of y vs. x; then plot log $(y - c)$ vs. log x, or y vs. x^n, or $(y - c)$ vs. x on logarithmic coordinates.
4. $y = ae^{bx}$ — Plot log y vs. x or y vs. x on semilogarithmic coordinates.
5. $y = ab^x$ — Plot log y vs. x or y vs. x on semilogarithmic coordinates.
6. $y = a + \dfrac{b}{x}$ — Plot y vs. $1/x$.
7. $y = \dfrac{x}{a + bx}$ — Plot x/y vs. x or $1/y$ vs. $1/x$.
8. $y = a + bx + cx^2$ — Plot $(y - y_n)/(x - x_n)$ vs. x, where y_n, x_n are the coordinates of any point on a smooth curve through the experimental points.
9. $y = \dfrac{x}{a + bx} + c$ — Plot $(x - x_n)/(y - y_n)$ vs. x, where y_n, x_n are the coordinates of any point on the smooth curve.

Various additional forms involving four constants are discussed by Lipka,[1]† who gives a detailed treatment of the subject of selecting the form of the equation. Equations involving more than two constants should not be tested unless the simpler forms fail or unless it is obvious from the first graph that the function is quite complicated. Although type 3 appeared to be satisfactory in the case of the illustrative problem, it may be found that some other form, such as type 8, will fit the data better.

Instead of attacking the problem of fitting the original data by an empirical equation, any simple expression that approximates the function may be taken and then an empirical equation fitted to the differences between the observed values and those computed by this simple expression. The data are plotted as y vs. x and the curve approximated by a function $f(x)$, which is preferably of some simple form such as $y = bx$, $y = a + bx$, or $y = ax^n$. An empirical equation for the differences $y - f(x)$, in terms of x, is then determined by the methods described in the preceding section. This procedure has the advantage that the errors involved in fitting the empirical equation apply to the differences, which are small quantities if the first approximation is good. The func-

† Superscript numerals refer to numbered references in the Bibliography at the end of each chapter.

tion $f(x)$ should not only be of a simple form but should be so chosen that the differences will not pass through a maximum or minimum. It is obvious that the method may be extended by writing a second approximation for the first differences and then finding an empirical expression for the second differences.

Figure 1-10 represents a graph of y vs. x on logarithmic paper to be fitted by an empirical equation. This is first approximated by the dotted straight line representing the function $y = x^{1/2}$. The differences $y - x^{1/2}$

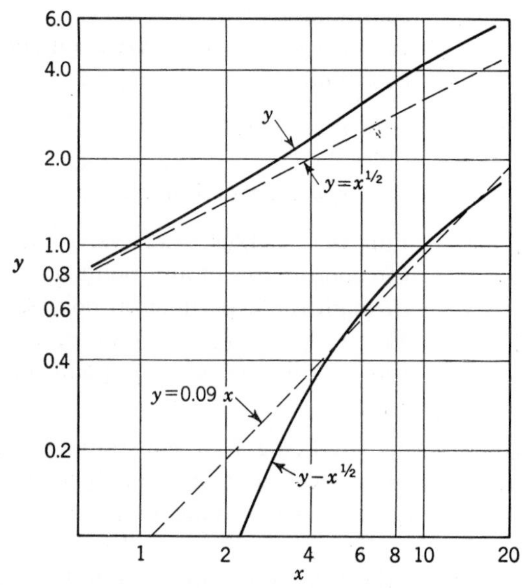

FIG. 1-10. Empirical equation by successive approximation.

are plotted vs. x, as shown by the lower solid curve. This is fitted approximately by the lower dotted straight line, representing the function $y = 0.09x$. The original function is then expressed as

$$y = x^{1/2} + 0.09x$$

It is apparent from a comparison of the two lower curves that this expression fits the function within a maximum error of about 0.10 in y. A still better result may be secured by writing an approximate expression for the first differences and then obtaining an empirical expression for the second differences. If this were done, the lower straight line $y = 0.09x$ would not be used, as this makes the second differences pass through a maximum. A straight line asymptotic to the upper end of the difference curve would be a better first approximation, as the equation for the second differences would then be of a simpler form. In obtaining the empirical equation for the last differences plotted, the larger differences should be fitted closely, as these represent the larger errors in the function.

This method has obvious utility in engineering work because of its directness and simplicity, although some practice is needed if final difference functions that are difficult to fit empirically are to be avoided. The error in the final result may be made as small as desirable by continuing this process. By differentiation of the result, values of the derivative may be obtained, and the method may be employed as an alternative procedure for numerical differentiation.

Evaluation of Constants. The form of the empirical equation having been determined, the remaining problem is the determination of the arbitrary constants. This may be done (1) by writing the equation for the best straight line placed by visual inspection, (2) by the method of averages, and (3) by the method of least squares. The first method is the simplest and is often satisfactory in the practical cases encountered in chemical engineering. The least-squares procedure involves the most numerical computation. On the other hand, *when certain requirements concerning the behavior of the data are met,* the least-squares method provides the most reliable set of constants and an estimate of the probable precision of the result.

In the illustrative example given above, a straight line on logarithmic coordinates was obtained in Fig. 1-9. This passes through the points $y - 3.2 = 0.4$ at $x = 0.6$, and $y - 3.2 = 25$ at $x = 4.5$, whence

$$0.4 = a(0.6)^n$$

and

$$25 = a(4.5)^n$$

which give $a = 1.15$ and $n = 2.05$. Equation (1-10) becomes

$$y = 3.2 + 1.15x^{2.05} \tag{1-11}$$

This result may be tested by comparing the calculated and observed values of y for each value of x.

x	y	y_{calc}	R	R^2
0.2	3.2	3.2	0	0
0.5	3.7	3.5	−0.2	0.04
1	4.1	4.4	0.3	0.09
2	8.1	8.0	−0.1	0.01
3	13.7	14.1	0.4	0.16
4	22.6	22.9	0.3	0.09
			$\Sigma R = 0.7$	$\Sigma R^2 = 0.39$

The last two columns give the values of R and R^2, where R is the difference between the calculated and observed values of y. The maximum error is 0.4, and ΣR is 0.7. Even though the equation were a poor representation of the data, ΣR might be zero because of cancellation of terms, and ΣR^2 is a better indication of how well the equation fits the data.

The method of averages involves the preliminary division of the data into groups corresponding in number to the number of the arbitrary constants to be evaluated. The constant 3.2 in (1-10) was obtained graphically as the intercept of the line of Fig. 1-8. Letting this be one of the constants to be evaluated, there results

$$y = c + ax^n \tag{1-12}$$

and the tabulated data will be divided into three groups to determine the three constants. Writing the six equations for R, there result

$$\begin{aligned} R_1 &= c + a(0.2)^n - 3.2 \\ R_2 &= c + a(0.5)^n - 3.7 && (a) \\ R_3 &= c + a(1)^n - 4.1 \\ R_4 &= c + a(2)^n - 8.1 && (b) \\ R_5 &= c + a(3)^n - 13.7 \\ R_6 &= c + a(4)^n - 22.6 && (c) \end{aligned}$$

which are divided into three pairs. Setting $\Sigma R = 0$ in each group, the three equations

$$\begin{aligned} 2c + [(0.2)^n + (0.5)^n]a - 6.9 &= 0 \\ 2c + [1 + (2)^n]a - 12.2 &= 0 \\ 2c + [(3)^n + (4)^n]a - 36.3 &= 0 \end{aligned}$$

are obtained. These may be solved simultaneously, giving $a = 1.07$, $c = 3.3$, and $n = 2.08$. When these values are used, ΣR is 0.2, and ΣR^2 is 0.22. It follows that the result

$$y = 3.3 + 1.07x^{2.08} \tag{1-13}$$

is somewhat better than (1-11). In general, the method of averages gives a better result than is obtained by placing the best straight line through the points on an ordinary sheet of graph paper, although the result is dependent on the method of grouping the points. Consecutive grouping, as followed above, is usually the most reliable. If the number of points is not evenly divisible by the number of constants, the groups are necessarily unequal in size. If there had been eleven points, for example, groups (*a*), (*b*), and (*c*) might have contained four, three, and four equations, respectively.

The method of least squares is commonly regarded as the most reliable method of fitting an empirical equation to a set of data. While this is often the case, it is not necessarily true. Like all techniques, it is subject to limitations which, unfortunately, are frequently disregarded. Consequently, the method is often misused. Inasmuch as the numerical calculations involved may be quite tedious, it is desirable to ensure that the technique is suitable before proceeding. The justification of the method of least squares is based upon an analysis of the factors causing variance in the data set. Such topics and the least-squares procedure are considered in Chap. 2.

1-6. Interpolation. Interpolation is of practical importance to the engineer because of the constant necessity of referring to sources of information expressed in the form of a table. Logarithms, trigonometric functions, properties of steam, vapor pressures, and other physical and chemical data are commonly given in the form of tables in the standard reference works. Although these tables are sometimes given in sufficient detail, so that interpolation is not necessary, it is important to be able to interpolate properly when the need arises.

It will be assumed that there is available a tabulation of a series of values of y for corresponding tabulated values of the independent variable x. The problem of interpolation is to obtain the correct value of y corresponding to some value of x lying between two tabulated values x_n and x_{n+1}. Extrapolation refers to the problem of finding the value of y corresponding to a value of x lying outside the range of tabulated values of x. Clearly, interpolation or extrapolation may be accomplished by using an empirical relation found by one of the methods of the preceding section. It is more customary, however, to employ one of several formulas, these expressions being based on fitting a polynomial to the function represented by the table and then using the polynomial to calculate the desired value of y. The simplest illustration of the general method would be to fit two points y_0, x_0, y_1, x_1 by means of

$$y = a + bx$$

and then employ this equation to calculate y for some x lying between x_0 and x_1. Most engineers do this mentally when reading values from the steam tables. If a number of points are used, the polynomial may be of a correspondingly high degree, and the method may be quite exact.

It may be noted that, if the nature of the function is known, it is foolish to apply an arbitrary polynomial, since the constants in the correct form may be obtained by substituting a sufficient number of values of the variables. An exception to this statement may be the case of a complicated function difficult to evaluate numerically. In interpolation of values of vapor pressures, it is sometimes better to use the fact that the logarithm of the vapor pressure is very nearly linear in the reciprocal of the absolute temperature than to employ an interpolation formula.

Of the various interpolation formulas that are in standard use, only Newton's and Lagrange's will be described. For a discussion of the so-called "central-difference formulas" of Bessel and Stirling, the reader is referred to Scarborough,[4] who summarizes the particular advantages and uses of the various forms. Newton's formula requires a table of values of y for equidistant values of the independent variable x. Lagrange's formula does not have this limitation.

1-7. Newton's Formula. In order to use Newton's interpolation formula, it is necessary to understand the construction of a "difference table." Table 1-1 is such a table, constructed from a series of values of y and x, where $y = x^3$.

TABLE 1-1. DIFFERENCE TABLE

x	$y = x^3$	Δ^1	Δ^2	Δ^3	x	y	Δ^1	Δ^2	Δ^3
0.50	0.125				x_0	y_0			
		0.091					$\Delta^1 y_0$		
0.60	0.216		0.036		x_1	y_1		$\Delta^2 y_0$	
		0.127		0.006			$\Delta^1 y_1$		$\Delta^3 y_0$
0.70	0.343		0.042		x_2	y_2		$\Delta^2 y_1$	
		0.169		0.006			$\Delta^1 y_2$		$\Delta^3 y_1$
0.80	0.512		0.048		x_3	y_3		$\Delta^2 y_2$	
		0.217		0.006			$\Delta^1 y_3$		$\Delta^3 y_2$
0.90	0.729		0.054		x_4	y_4		$\Delta^2 y_3$	
		0.271		0.006			$\Delta^1 y_4$		$\Delta^3 y_3$
1.00	1.000		0.060		x_5	y_5		$\Delta^2 y_4$	
		0.331		0.006			$\Delta^1 y_5$		$\Delta^3 y_4$
1.10	1.331		0.066		x_6	y_6		$\Delta^2 y_5$	
		0.397					$\Delta^1 y_6$		
1.20	1.728				x_7	y_7			

The left half of the table shows the numerical values for the illustrative example; the right half of the table indicates the nomenclature used in referring to the various quantities. $\Delta^1 y_0$ is $(y_1 - y_0)$, and $\Delta^1 y_n$ is $(y_{n+1} - y_n)$.

$$\Delta^2 y_0 = \Delta^1 y_1 - \Delta^1 y_0 = (y_2 - y_1) - (y_1 - y_0) = y_2 - 2y_1 + y_0$$

$$\Delta^2 y_n = \Delta^1 y_{n+1} - \Delta^1 y_n = (y_{n+2} - y_{n+1}) - (y_{n+1} - y_n)$$
$$= y_{n+2} - 2y_{n+1} + y_n$$

Similarly,

$$\Delta^3 y_0 = \Delta^2 y_1 - \Delta^2 y_0 = (y_3 - 2y_2 + y_1) - (y_2 - 2y_1 + y_0)$$
$$= y_3 - 3y_2 + 3y_1 - y_0$$

and
$$\Delta^3 y_n = y_{n+3} - 3y_{n+2} + 3y_{n+1} - y_n$$

It is apparent that in the case of the example the third differences are constant. It has already been seen that the first differences are constant for a linear equation, and it is not difficult to prove that the nth differences are constant for an nth-degree polynomial, provided that the values of x are equidistant. This suggests the use of a difference table to determine the complexity of a function and the degree of the empirical polynomial that must be used.

The tabulated function is to be fitted by an nth-degree polynomial, $f(x)$. This may be written

$$f(x) = a_0 + a_1(x - x_0) + a_2(x - x_0)(x - x_1) + a_3(x - x_0)(x - x_1)(x - x_2) + \cdots + a_n(x - x_0)(x - x_1) \cdots (x - x_{n-1}) \quad (1\text{-}14)$$

Substituting $f(x) = y_0$ at $x = x_0$, there results $a_0 = y_0$.
Substituting $f(x) = y_1$ at $x = x_1$ gives

$$a_1 = \frac{y_1 - a_0}{x_1 - x_0} = \frac{\Delta^1 y_0}{h} \quad (1\text{-}15)$$

where h represents the constant x increment. If this process is continued, the values of a_2, a_3, etc., are obtained as

$$a_2 = \frac{\Delta^2 y_0}{2h^2} \qquad a_3 = \frac{\Delta^3 y_0}{3!h^3} \qquad a_n = \frac{\Delta^n y_0}{n!h^n} \quad (1\text{-}16)$$

whence

$$f(x) = y_0 + (x - x_0)\frac{\Delta^1 y_0}{h} + (x - x_0)(x - x_1)\frac{\Delta^2 y_0}{2h^2} + (x - x_0)(x - x_1)(x - x_2)\frac{\Delta^3 y_0}{3!h^3} + \cdots + (x - x_0)(x - x_1) \cdots (x - x_n)\frac{\Delta^n y_0}{n!h^n} \quad (1\text{-}17)$$

Although this is the desired expression, it is usually rewritten in terms of another variable p defined by

$$p = \frac{x - x_0}{h} \quad (1\text{-}18)$$

whence

$$y = f(x) = y_0 + p\,\Delta^1 y_0 + \frac{p(p-1)}{2!}\Delta^2 y_0 + \frac{p(p-1)(p-2)}{3!}\Delta^3 y_0 + \frac{p(p-1)(p-2)(p-3)}{4!}\Delta^4 y_0 + \cdots + \frac{p(p-1)\cdots(p-n+1)}{n!}\Delta^n y_0 \quad (1\text{-}19)$$

Referring to Table 1.1, it will be noted that y_0, $\Delta^1 y_0$, $\Delta^2 y_0$, $\Delta^3 y_0$, etc., are the differences lying in the top diagonal column. It is apparent that these values are of significance primarily in the upper part of the table, and the formula (1-19), known as "Newton's formula for forward interpolation" (in the direction of increasing values of x), is best applied for interpolation near the first of the table.

The use of the formula may be illustrated by means of the numerical differences tabulated for x^3. Let it be desired to obtain the value of y for $x = 0.62$. Here $y_0 = 0.125$, $h = 0.10$, $p = 1.2$, $n = 3$, and the values of the differences are read from the table. Hence,

$$y = 0.125 + 1.2 \cdot 0.091 + \frac{1.2(1.2-1)}{2} 0.036 + \frac{1.2(1.2-1)(1.2-2)}{6} 0.006 = 0.238328$$

In this case, the result is $(0.62)^3$ exactly, since the function is of the third degree; the third differences are constant; and the interpolation formula included the term involving the third difference. Actually, the last term was quite small, and if only the second differences had been used the result would have been $y = 0.23852$. If the relation between y and x represents experimental data, the third or fourth differences may be only approximately constant and the interpolation accordingly inexact. In using the interpolation formula, it should be noted that the accuracy of the result depends on the constancy of the last differences used as well as on the number of terms employed.

Referring again to the difference table for $y = x^3$, Table 1-1, suppose that y is desired for some value of x near the end of the table. If x lies between x_n and x_{n+1} and if the kth differences are to be used, then $\Delta^1 y_0$, $\Delta^2 y_0$, etc., may be replaced by the differences lying on a diagonal

TABLE 1-2

x	y	Δ^1	Δ^2	Δ^3
1.20	1.728			
		−0.397		
1.10	1.331		0.066	
		−0.331		−0.006
1.00	1.000		0.060	
		−0.271		−0.006
0.90	0.729		0.054	
		−0.217		−0.006
0.80	0.512		0.048	
		−0.169		
0.70	0.343			
	etc.			

ending with the kth difference horizontally opposite x_n or x_{n+1}, depending on whether k is odd or even. For example, for $x = 1.06$, use 0.271, 0.060, and 0.006 in place of $\Delta^1 y_0$, $\Delta^2 y_0$, and $\Delta^3 y_0$. Accordingly, y_0 is replaced by y_4, which is 0.729, and p is $(1.06 - 0.90)/h = 1.6$. This simply amounts to the omission of the tabulated values for x less than 0.9 and the construction of the difference table on the basis of the last four values tabulated.

If y is desired at $x = 1.16$, only Δ^1 is available for substitution in the formula. In this case, an inversion of the table is used which appears as Table 1-2. Using this inverted table, the values of y and Δ in the top diagonal row are substituted in the formula. In terms of the original nomenclature, Eq. (1-19) becomes

$$y = y_7 + q\,\Delta^1 y_7 + \frac{q(q+1)}{2!}\,\Delta^2 y_6 + \frac{q(q+1)(q+2)}{3!}\,\Delta^3 y_5 + \cdots + \frac{q(q+1)(q+2)\cdots(q+n-1)}{n!}\,\Delta^n y_{8-n}$$

where

$$q = \frac{x - x_7}{h}$$

In general, where x_n, y_n are the last values given, interpolation near the end of the table may be carried out by substitution in the general form

$$y = y_n + q\,\Delta^1 y_n + \frac{q(q+1)}{2!}\,\Delta^2 y_{n-1} + \frac{q(q+1)(q+2)}{3!}\,\Delta^3 y_{n-2} + \cdots + \frac{q(q+1)(q+2)\cdots(q+n-1)}{n!}\,\Delta^k y_{n-k+1} \qquad (1\text{-}20)$$

where $q = (x - x_n)/h$. In the case of $x = 1.16$, this gives

$$y = 1.728 - 0.4 \cdot 0.397 - \frac{0.4 \cdot 0.6}{2}\,0.066 - \frac{0.4 \cdot 0.6 \cdot 1.6}{6}\,0.006$$
$$= 1.560896$$

1-8. Lagrange's Interpolation Formula. The construction of a difference table and the use of Newton's formula have the advantage that the relative constancy of the last differences is immediately apparent and the accuracy of the interpolation indicated qualitatively in this way. The quantitative error may be calculated quite accurately by equations expressing the error in terms of the differences tabulated. The method has the disadvantage, however, of being limited to tabulated values of x, which are equidistant. Lagrange's method is not restricted in this way.

As before, it is assumed that the function may be represented by a polynomial, now written in the form

$$\begin{aligned} y = {} & a_1(x - x_2)(x - x_3)(x - x_4)\cdots(x - x_n) \\ & + a_2(x - x_1)(x - x_3)(x - x_4)\cdots(x - x_n) \\ & + a_3(x - x_1)(x - x_2)(x - x_4)\cdots(x - x_n) \\ & + \cdots + a_n(x - x_1)(x - x_2)(x - x_3)\cdots(x - x_{n-1}) \end{aligned} \qquad (1\text{-}21)$$

Substituting y_1, x_1, there results

$$a_1 = \frac{y_1}{(x_1 - x_2)(x_1 - x_3)(x_1 - x_4) \cdots (x_1 - x_n)}$$

The other constants are obtained by similar substitutions, and

$$a_n = \frac{y_n}{(x_n - x_1)(x_n - x_2)(x_n - x_3) \cdots (x_n - x_{n-1})} \tag{1-22}$$

Rewriting the polynomial gives

$$\begin{aligned} y = y_1 & \frac{(x - x_2)(x - x_3)(x - x_4) \cdots (x - x_n)}{(x_1 - x_2)(x_1 - x_3)(x_1 - x_4) \cdots (x_1 - x_n)} \\ & + y_2 \frac{(x - x_1)(x - x_3)(x - x_4) \cdots (x - x_n)}{(x_2 - x_1)(x_2 - x_3)(x_2 - x_4) \cdots (x_2 - x_n)} + \cdots \\ & + y_n \frac{(x - x_1)(x - x_2)(x - x_3) \cdots (x - x_{n-1})}{(x_n - x_1)(x_n - x_2)(x_n - x_3) \cdots (x_n - x_{n-1})} \end{aligned} \tag{1-23}$$

As an example of the use of Lagrange's formula, let it be desired to obtain the density of a 26 per cent solution of phosphoric acid in water at 20°C. The following data are quoted by Perry:[2]

y, density	1.0764	1.1134	1.2160	1.3350
x, percentage H_3PO_4	14	20	35	50

It is obvious that the interpolation formula may be applied to the figures after the decimal point, replacing the 1 at the end of the calculation. Substituting in (1-23), there results

$$\begin{aligned} y = 1.0 + 0.0764 & \frac{(26 - 20)(26 - 35)(26 - 50)}{(14 - 20)(14 - 35)(14 - 50)} \\ + 0.1134 & \frac{(26 - 14)(26 - 35)(26 - 50)}{(20 - 14)(20 - 35)(20 - 50)} \\ + 0.2160 & \frac{(26 - 14)(26 - 20)(26 - 50)}{(35 - 14)(35 - 20)(35 - 50)} \\ + 0.3350 & \frac{(26 - 14)(26 - 20)(26 - 35)}{(50 - 14)(50 - 35)(50 - 20)} \\ = 1.0 - \tfrac{2}{7}0.0764 & + \tfrac{24}{25}0.1134 + \tfrac{64}{175}0.2160 - \tfrac{1}{25}0.3350 \\ = 1.1528 & \end{aligned}$$

This may be compared with the correct value 1.1529, given in the same table from which the data quoted above were obtained.

In the example given, four tabulated points were used, which means that the interpolation is based on fitting the function with a cubic equation. The accuracy of the result naturally depends on how well such an equation fits the data, and this should be kept in mind in using

Lagrange's formula. In the case of vapor-liquid equilibrium data for binary mixtures, the vapor composition is related to the liquid composition by a function frequently too complex to be interpolated by means of Lagrange's formula with only four points. This is particularly true in the case of azeotropic mixtures. A large number of points are required for interpolation of functions having points of inflection. When the formula is employed, the arithmetic may be tested if it is noted that the coefficients of y_1, y_2, etc., must add to unity, since the formula must hold true if all values of y are equal.

1-9. Extrapolation. Although the interpolation equations may be used for extrapolation, the answers cannot be considered reliable. In general, it is unwise to extrapolate any empirical relation significantly beyond the last data point. If, however, a certain form of equation is predicted by theory and substantiated by the available data, reasonable extrapolation is ordinarily justified.

1-10. Differentiation. The fundamental natural laws on which engineering is based are frequently expressed in the form of differential equations giving the rate of change of an important quantity with respect to an independent variable. Particularly in the interpretation of experimental data, therefore, it is necessary to be able to carry out the process of differentiation, analytically, graphically, or numerically. It is assumed that the reader is familiar with the general concepts and procedures of differentiation. Consequently, only a brief review will be given here.

A function $y = f(x)$ expresses the relation between y and x. The rate of change of y corresponding to a change in x, or the ratio of a differential change in y to the corresponding differential change in x, is called the "derivative" of the function. It is expressed as dy/dx or $f'(x)$. The formal process of differentiation is based upon the following concepts:

Consider a continuous function $y = f(x)$. Let x be changed by the infinitesimal amount Δx. Then the increment

$$\Delta y = f(x + \Delta x) - f(x)$$

is also an infinitesimal, since $f(x)$ is continuous. By definition,

$$f'(x) \equiv \lim_{\Delta x \to 0} \frac{\Delta y}{\Delta x} = \lim_{\Delta x \to 0} \frac{f(x + \Delta x) - f(x)}{\Delta x} \tag{1-24}$$

Then

$$\Delta y = f'(x)\,\Delta x + \epsilon\,\Delta x \tag{1-25}$$

where ϵ is an infinitesimal of order $(\Delta x)^n$, $n \geqq 1$. $f'(x)\,\Delta x$ is called the "principal part" of Δy or the differential of y:

$$f'(x)\,\Delta x \equiv dy \tag{1-26}$$

For the independent variable x, dx represents *all* of Δx; for the dependent variable y, dy represents only the *principal part* of Δy. These concepts may be shown graphically. In Fig. 1-11, $\Delta y/\Delta x$ represents the slope of the *chord* connecting the points (x,y) and $(x + \Delta x, y + \Delta y)$, whereas dy/dx represents the slope of the tangent to the curve $y = f(x)$ at the point (x,y).

Not all functions have derivatives, since for such to exist it is necessary that the ratio of the increment of y, Δy, to the increment of x, Δx, approach a definite limit as Δx approaches zero. This limit is defined as

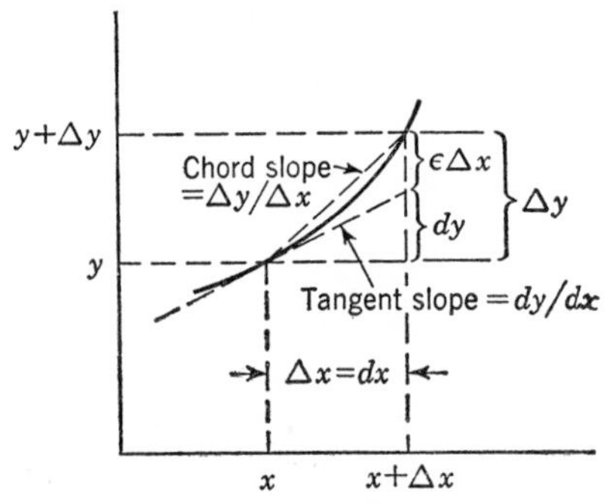

FIG. 1-11. Illustration of infinitesimals.

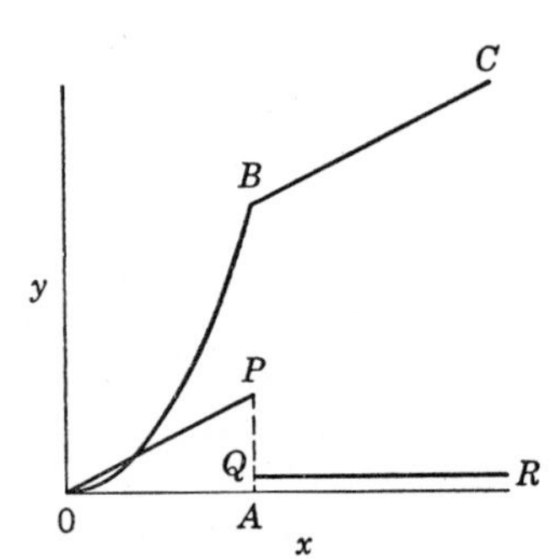

FIG. 1-12. Graph of function and derived curve.

the "derivative." In Fig. 1-12, for example, curve OBC represents a continuous function having no derivative at point B. The derivative, or "derived function," is plotted as shown by $OPQR$ and is seen to be discontinuous at B. Since the derivative has no *definite* value at B, the function has no derivative at B.

It is important to note that the *first* derivative, dy/dx, represents the ratio of two distinct quantities, dy and dx. Consequently, this ratio may be handled by ordinary algebraic procedures. For example,

$$dx \frac{dy}{dx} = dy \tag{1-27}$$

$$\frac{dy}{dx}\frac{dx}{dz} = \frac{dy}{dz} \tag{1-28}$$

On the other hand, in general, the higher derivatives cannot be treated in this manner. This statement is a consequence of the following considerations:

Suppose that $y = \phi(z)$ and $z = \gamma(x)$, where x is the independent variable. It may be shown that

$$dy = \phi'(z)\,dz \qquad \text{and} \qquad \frac{dy}{dz} = \phi'(z)$$

The second derivative is defined by the operations

$$\phi''(z) \equiv \frac{d}{dz}[\phi'(z)] \equiv \frac{d}{dz}\left(\frac{dy}{dz}\right) \tag{1-29}$$

An alternative, commonly used notation is

$$\phi''(z) \equiv \frac{d^2y}{dz^2} \tag{1-30}$$

In general, however,

$$\frac{d^2y}{dz^2} \neq \frac{d^2y}{(dz)^2}$$

By definition,

$$\begin{aligned} d^2y &= d(dy) = d[(\phi'z)\,dz] \\ &= \phi''(z)\,(dz)^2 + \phi'(z)\,d^2z \end{aligned} \tag{1-31}$$

Hence,

$$\frac{d^2y}{(dz)^2} = \phi''(z) + \phi'(z)\,\frac{d^2z}{(dz)^2} \tag{1-32}$$

and $d^2y/(dz)^2$ is not equal to $d^2y/dz^2 = \phi''(z)$ unless d^2z is zero. However,

$$\begin{aligned} d^2z &= d(dz) = d[\gamma'(x)\,dx] \\ &= \gamma''(x)\,(dx)^2 + \gamma'(x)\,d^2x \end{aligned} \tag{1-33}$$

Although (1-33) shows that, in general, d^2z is not zero, d^2z is not specified unless d^2x is defined. Since x is the independent variable, d^2x may be defined arbitrarily. The accepted definition is

$$d^2x = d^3x = \cdots d^nx = 0 \tag{1-34}$$

when x is the independent variable. It then follows that $d^2z, d^3z, \ldots, d^nz$ are, in general, *not* equal to zero. Consequently, the higher derivatives may not be treated as the ratio of two quantities unless the denominator represents the independent variable.

1-11. Change of Variable. In later work, it will be necessary to interchange variables in derivatives. For example, the solution to a differential equation involving the derivatives $dy/dx, d^2y/dx^2, \ldots, d^ny/dx^n$ is often simplified if either y or x or both are replaced by new variables. The necessary manipulations follow as a corollary to the discussion of Sec. 1-10.

Case I. y replaced by z, where $y = \phi(z)$

a.

$$\begin{aligned} \frac{dy}{dx} &= \frac{dy}{dz}\frac{dz}{dx} \\ &= \phi'(z)\,\frac{dz}{dx} \end{aligned}$$

b.
$$\frac{d^2y}{dx^2} = \frac{d}{dx}\left(\frac{dy}{dx}\right)$$
$$= \frac{d}{dx}\left[\phi'(z)\frac{dz}{dx}\right]$$
$$= \phi''(z)\left(\frac{dz}{dx}\right)^2 + \phi'(z)\frac{d^2z}{dx^2}$$

c.
$$\frac{d^ny}{dx^n} = \frac{d}{dx}\left(\frac{d^{n-1}y}{dx^{n-1}}\right)$$

Case II. x replaced by t, where $x = \alpha(t)$

a.
$$\frac{dy}{dx} = \frac{1}{\alpha'(t)}\frac{dy}{dt}$$

since $dx = \alpha'(t)\,dt$.

b.
$$\frac{d^2y}{dx^2} = \frac{d}{dx}\left(\frac{dy}{dx}\right)$$
$$= \frac{1}{\alpha'(t)}\frac{d}{dt}\left[\frac{1}{\alpha'(t)}\frac{dy}{dt}\right]$$
$$= \frac{\alpha'(t)(d^2y/dt^2) - \alpha''(t)(dy/dt)}{(\alpha't)^3}$$

c.
$$\frac{d^ny}{dx^n} = \frac{d}{dx}\left(\frac{d^{n-1}y}{dx^{n-1}}\right)$$

1-12. Graphical Differentiation. Since the derivative of a function at a particular value of x is the slope of the curve of y vs. x at that value of x, it follows that any means of measuring the slope of a curve may be employed to obtain the derivative.

One rather obvious method is to plot the function and obtain the slope with the aid of a straightedge. The latter is swung about the point on the curve until it is judged by eye best to represent the slope of the curve at the point, a straight line is drawn for record, and the slope of the straight line is calculated. In this last step, it is important to obtain the slope in terms of the variables, taking into account the scales used in ordinate and abscissa. Only where these are equal is the geometric slope the correct value.

A better way is to place a protractor or triangle on a fixed straightedge or T square at a definite angle to the x axis and slide it along until it touches the curve representing the function. This procedure locates the point on the curve where the slope is equal to the predetermined value. As an aid in locating the exact point of tangency, several chords, such as AB of Fig. 1-13, may be drawn parallel to the tangent. These are bisected, and a line QP is drawn through their mid-points to intersect the curve at the desired point P. Although the principle is quite similar to the direct measurement of the slope at a point, the results obtained are found to be somewhat better. With a graph of reasonable size, it is usually possible to obtain the derivative in this way with an error of about 3 per cent, although the accuracy varies considerably with the curvature of the line representing the function.

A very convenient device for measuring slopes of curves is a movable protractor fitted with a prism at the center. This is placed over the curve at the point in question and turned until the two branches of the curve seen through the two faces of the prism appear to join to form a smooth continuous curve. The slope of the curve with reference to the base of the instrument is read directly. Some of the instruments of this type employ a protractor of such small diameter that the accuracy obtainable is no better than, if as good as, that obtainable by the preceding methods.

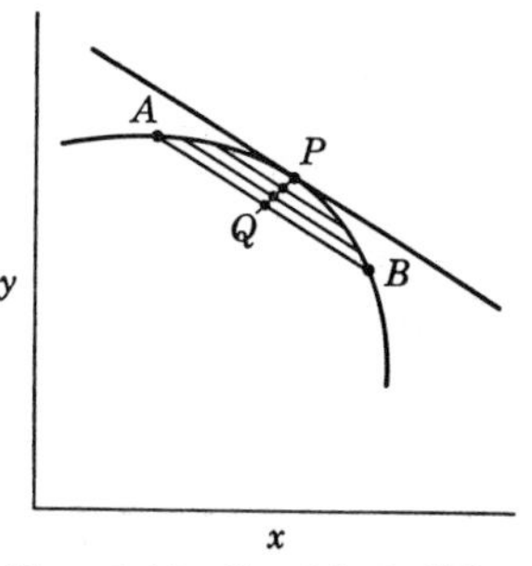

FIG. 1-13. Graphical differentiation.

1-13. Numerical Differentiation. Although graphical differentiation is customarily used when the relation between the variables is given in graphical or tabular form, on occasion numerical procedures are more convenient. The general procedure is to represent the function in the region where the derivative is desired by an interpolation formula or some other empirical formula. This relation is then differentiated analytically. For example, differentiation of Newton's interpolation formula (1.19) yields

$$\frac{dy}{dx} = \frac{dy}{dp}\frac{dp}{dx} = \frac{1}{h}\frac{dy}{dp} = \frac{1}{h}\left(\Delta^1 y_0 + \frac{2p-1}{2!}\Delta^2 y_0 + \frac{3p^2-6p+2}{3!}\Delta^3 y_0 + \frac{4p^3-18p^2+2p-6}{4!}\Delta^4 y_0 \cdots\right) \tag{1-35}$$

Douglass-Avakian Method. An alternative procedure, developed by Douglass and Avakian,[5] employs a fourth-degree polynomial, which is fitted to seven *equidistant* points by the method of least squares. The polynomial is

$$y = a + bx + cx^2 + dx^3 + ex^4 \tag{1-36}$$

The values of the independent variable must be spaced at equal intervals, h, and a translation of coordinates made so that $x = 0$ for the central point of the seven values of the independent variable. The seven values of the new independent variable x are then $-3h$, $-2h$, $-h$, 0, h, $2h$, and

$3h$. *Let k represent the coefficient of h in the values of x.* Thus at $x = -3h$, $k = -3$; at $x = -2h$, $k = -2$; etc. The values of the constants in (1-36) are then given by the expressions

$$a = \frac{524\Sigma y - 245\Sigma k^2y + 21\Sigma k^4y}{924} \tag{1-37a}$$

$$b = \frac{397\Sigma ky}{1,512h} - \frac{7\Sigma k^3y}{216h} \tag{1-37b}$$

$$c = \frac{-840\Sigma y + 679\Sigma k^2y - 67\Sigma k^4y}{3,168h^2} \tag{1-37c}$$

$$d = \frac{-7\Sigma ky + \Sigma k^3y}{216h^3} \tag{1-37d}$$

$$e = \frac{72\Sigma y - 67\Sigma k^2y + 7\Sigma k^4y}{3,168h^4} \tag{1-37e}$$

The slope of the polynomial (1-36) at the central point ($x = 0$) is

$$\frac{dy}{dx} = b = \frac{397\Sigma ky}{1,512h} - \frac{7\Sigma k^3y}{216h} \tag{1-38}$$

As an example of the use of this technique, let it be desired to find the constants in the polynomial approximation (1-36) and the slope dy/dz at $z = 4.5$ from the data of Table 1-3.

TABLE 1-3

y	2	3	2	−1	−2	−2	−1
z	0.0	1.5	3.0	4.5	6.0	7.5	9.0

The values of the independent variable z are evenly spaced. Hence only a change of variable is necessary. Move the origin along the independent variable axis to the mid-point at $z = 4.5$. Then

$$x = z - 4.5$$

Table 1-4 is prepared as an aid in substitution into Eqs. (1-37) for the constants.

TABLE 1-4

z	y	x	k	ky	k^2y	k^3y	k^4y
0.0	2	−4.5	−3	−6	18	−54	162
1.5	3	−3.0	−2	−6	12	−24	48
3.0	2	−1.5	−1	−2	2	−2	2
4.5	−1	0.0	0	0	0	0	0
6.0	−2	1.5	1	−2	−2	−2	−2
7.5	−2	3.0	2	−4	−8	−16	−32
9.0	−1	4.5	3	−3	−9	−27	−81
Σ.....	...	.....	...	−23	13	−125	97

Then use of Eq. (1-37*a*) gives

$$a = \frac{524 \cdot 1 - 245 \cdot 13 + 21 \cdot 97}{924} = -0.68$$

In a similar fashion,

$$b = -1.32 \qquad c = 0.21 \qquad d = 0.049 \qquad e = -0.0075$$

The fourth-order polynomial approximation to the data is then

$$y = -0.68 - 1.32x + 0.21x^2 + 0.049x^3 - 0.0075x^4$$

which may be converted to an equation in z, if desired, by the substitution $x = z - 4.5$. The correspondence of the polynomial with the data is shown in Fig. 1-14. The slope at $x = 0$ ($z = 4.5$) is

$$\frac{dy}{dx} = \frac{dy}{dz} = b = -1.32$$

Forcing a polynomial expression to fit each point of a set of experimental data does not necessarily result in the best representation of the physical relationship. The experimental errors inherent in the measurement can introduce anomolous results. This is particularly true when the empirical relationship is to be differentiated. In Fig. 1-15, the solid

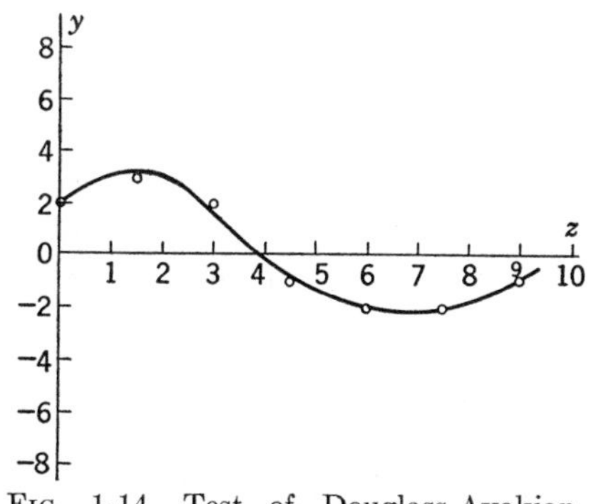

FIG. 1-14. Test of Douglass-Avakian method.

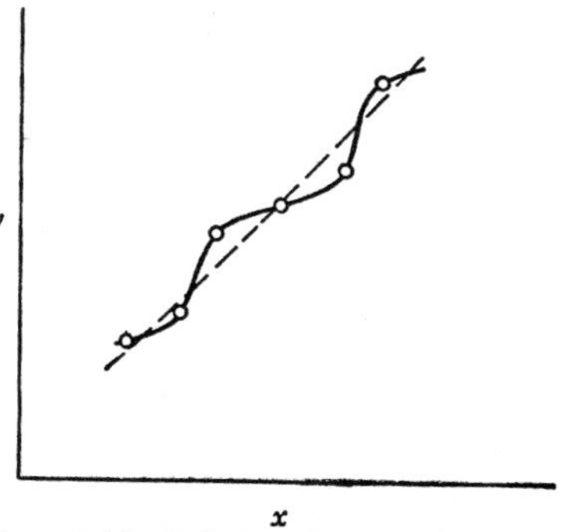

FIG. 1-15. Polynomial relation contrasted with smoothed curve.

line represents a polynomial whose constants were evaluated by forcing the polynomial to pass through each of the six data points. The dotted line is a "smoothed"-curve representation of the data. It is clear that, at the same value of x, differentiation of the solid curve ordinarily will give a significantly different value of the derivative than will differentiation of the smoothed (dotted) curve. In many cases, the smoothed relation is a more valid representation of the actual state of affairs.

In view of the situation discussed above, it is evident that the differentiation of experimental data involves the exercise of judgment. The statistical methods of Chap. 2 are frequently helpful when decisions of this kind are involved. A further discussion of curve fitting, differentiation, etc., is presented in Chap. 2.

1-14. Maxima and Minima. Since the derivative dy/dx, or $f'(x)$, represents the rate of change of y with change in x, it is evident that if the function passes through a maximum or minimum the derivative will be zero. If the occurrence of such a maximum or minimum is to be determined and located, the derivative is equated to zero, thus giving the condition for which the maximum or minimum exists. This procedure is of considerable value in engineering calculations, since the location of a maximum or minimum is frequently of great practical importance. This is particularly true in chemical-engineering plant design, since the optimum operating and design variables are to be determined in order that the maximum profit will result.

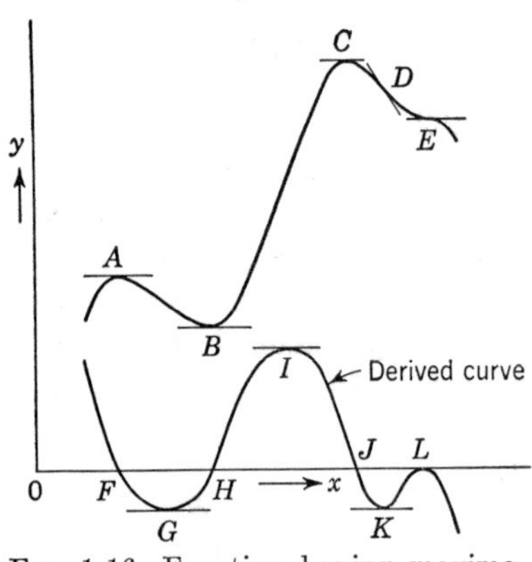

FIG. 1-16. Function having maxima and minima and derived curve.

In Fig. 1-16, the curve $ABCDE$ represents the function $y = f(x)$. The values of y in the neighborhood of points A and C are all less than the values of y at A and C, and the function is said to go through a maximum at these points. Similarly, because the values of y in the neighborhood of B are all greater than the value of B, the function is said to go through a minimum at B. This curve emphasizes that the terms "maximum" and "minimum" do not necessarily denote the greatest and the least possible values a function may assume.

It is evident geometrically that the slope of the curve is zero at the maximum and minimum points, and since the slope is given by the derivative dy/dx, or $f'(x)$, these points may be determined and located by solution of the equation $f'(x) = 0$. The roots of this equation merely locate the maximum and minimum points and cannot distinguish between them. Furthermore, the condition locates such points as E. One method of distinguishing among points A, C, and E is to calculate values of $f(x)$ in their immediate vicinity. An alternative method, often more convenient, is to evaluate the second derivative at these points. In Fig. 1-16, curve $FGHIJKL$ is a graph of dy/dx vs. x. The slope of this curve is d^2y/dx^2, or $f''(x)$. This slope is negative at those values of x for which $f(x)$ has a maximum and positive at those values of x for which $f(x)$ has a minimum. The second derivative is zero at point E. The slope of the derived curve is zero at those points on $y = f(x)$ where the tangent crosses the curve. These are known as "points of inflection." A convenient table of these relations follows:

Maximum	*Minimum*
$dy/dx = 0$	$dy/dx = 0$
$d^2y/dx^2 < 0$	$d^2y/dx^2 > 0$

If the function does not pass through either a maximum or minimum, this will be indicated in some way by the expressed condition, as by giving the square root of a negative number when solved for the dependent variable.

Calculations to determine the optimum conditions from the point of view of costs and monetary return are termed "economic balances." The basic criterion of the true optimum is maximum return on the investment, but the problem frequently simplifies to one of determining the minimum cost, the maximum production from a given piece of equipment, the minimum power, etc. The quantity considered is expressed as a function of the design or operating variables. On differentiating and equating to zero, we find the condition which determines the optimum value of the variable under control.

As an example, consider the two-stage reversible adiabatic compression of a gas from an initial pressure p_1 to a final pressure p_3. If the fixed charges on the compressors are assumed to be essentially independent of the interstage pressure employed, then the optimum operation involves the determination of the interstage pressure for which the total power requirement is a minimum. For a fixed gas flow, this corresponds to the minimum work for the two stages. If the gas enters at T_1 and is cooled to T_1 between stages, the total work is given by

$$W = NRT_1 \frac{k}{k-1}\left[\left(\frac{p_2}{p_1}\right)^{(k-1)/k} - 2 + \left(\frac{p_3}{p_2}\right)^{(k-1)/k}\right] \qquad (1\text{-}39)$$

where N = pound moles of gas compressed
R = molal gas constant (1,544)
k = ratio of specific heat at constant pressure to specific heat at constant volume for gas compressed
T_1 = inlet gas temperature, °R
W = total work, ft-lb

If this quantity is to be a minimum, then the derivative must be zero:

$$\frac{dW}{dp_2} = NRT_1 \frac{k}{k-1}\left(\frac{k-1}{k} p_1^{(1-k)/k} p_2^{-(1/k)} - \frac{k-1}{k} p_3^{(k-1)/k} p_2^{(1-2k)/k}\right) = 0$$

Solving for the value of p_2 that satisfies this condition gives

$$p_2^{(2k-2)/k} = p_3^{(k-1)/k} p_1^{(k-1)/k}$$

or

$$p_2 = \sqrt{p_1 p_3} \tag{1-40}$$

This is the relation found in handbooks for the optimum interstage pressure in a two-stage compression. Since k does not enter into the result, it is apparent that the equation will hold for any polytropic compression in which it is allowable to replace k by a constant exponent n.

The location of a maximum or minimum becomes obvious if the function is plotted, and this alternative procedure may be followed in economic-balance calculations. The analytical method has the advantage of directness and simplicity in many cases. The graphical method involves calculation of the function at a number of points and usually requires considerably more time to obtain the result. When the curve is obtained, however, its shape indicates the importance of operating at the exact optimum. If the curve is flat, a considerable variation in operating conditions will not affect the costs or profits appreciably. If the maximum or minimum is sharp, it may be quite important to operate at the optimum point. If the calculation of costs contains a graphical step, such as the use of a humidity chart, it may not be possible to obtain an analytical function for differentiation. An empirical equation may be substituted for the plot used in the graphical step and the analytical method followed if the resulting cost function is not too complicated. The graphical procedure of plotting the function may be followed in any case.

In chemical-engineering situations, the dependent variable is often a function of several independent variables. The determination of maximum or minimum value under such circumstances is carried out by the techniques of partial differentiation discussed in Chap. 5.

1-15. Evaluation of Indeterminate Fractions. Not infrequently, the function $f_1(x)/f_2(x)$ will give rise to one of the indeterminate forms $0/0$ or ∞/∞ at some particular value of x, $x = a$. In such a case, the value of the function at $x = a$ is defined as the limit of the value approached by the function as x approaches a.

$$\frac{f_1(a)}{f_2(a)} = \lim_{x \to a} \frac{f_1(x)}{f_2(x)} \tag{1-41}$$

This limit may sometimes become evident upon algebraic rearrangement. Consider the fraction $(x - a)/(x^2 - a^2)$, which reduces to $1/(x + a)$ after division of numerator and denominator by $x - a$. When $x = a$, the fraction becomes $0/0$. The value at $x = a$ is obtained from definition (1-41) as $\frac{1}{2a}$ because, as $x \to a$,

$$\lim_{x \to a} \frac{x - a}{x^2 - a^2} = \lim_{x \to a} \frac{1}{x + a} = \frac{1}{2a}$$

A more general procedure for the evaluation of such fractions is indicated by L'Hôpital's rule, which is derived in most texts on calculus. When the rule is followed, both numerator and denominator are differentiated with respect to x, and the desired value of the original fraction is obtained as the ratio of the derivatives, evaluated at $x = a$.

Expressed mathematically,

$$\lim_{x \to a} \frac{f_1(x)}{f_2(x)} = \frac{f_1'(a)}{f_2'(a)} \text{ when } \frac{f_1(x)}{f_2(x)} = \frac{0}{0} \text{ or } \frac{\infty}{\infty} \text{ at } x = a$$

If the fraction obtained is again indeterminate, the rule may be applied again. Sometimes the process must be repeated two or three times before a finite value is obtained. It is essential to note that the limiting process involves *independent* differentiation of numerator and denominator.

An equation encountered in absorption, distillation, extraction, and leaching is of the form

$$y = \frac{x^{n+1} - x}{x^{n+1} - 1} \tag{1-42}$$

In gas absorption, with systems where Henry's law applies to the gas-liquid equilibrium, y represents the fraction of solute in the inlet gas that is absorbed in a tower of n perfect plates, and x represents the ratio of solute-free liquor rate to gas rate, divided by the Henry's-law constant. In practice, this equation is used rather widely, with x varying from zero to very large numbers. No trouble is experienced except when x is unity, resulting in the fraction $0/0$. On application of L'Hôpital's rule, the ratio of the derivative of the numerator to the derivative of the denominator becomes

$$\frac{(n + 1)x^n - 1}{(n + 1)x^n}$$

which reduces to $n/(n + 1)$, the desired value of y for $x = 1$.

An expression giving rise to an indeterminate form such as $0 \cdot \infty$, $\infty - \infty$, 0^0, ∞^0, or 1^∞ may usually be rearranged to give a fraction taking the form $0/0$ or ∞/∞ and evaluated by L'Hôpital's rule.

1-16. Integration. Integration is the reversal of the process of differentiation. In general, it is the process whereby the function $y = f(x)$ is determined when the function $dy/dx = f'(x)$ which defines the derivative is known. It is presumed that the reader is familiar with the elementary processes of integration. A table of integrals such as that of

Peirce[3] will prove helpful in later work and should be a part of the library of any engineer. Subsequent chapters in this book deal with the problem of integrating differential equations.

1-17. Graphical Integration. The graphical evaluation of definite integrals is of great utility in the many cases where the algebraic form of the integral is difficult to evaluate or impossible to obtain. In general, the problem is to evaluate an integral of the form $\int_{x_1}^{x_2} f(x)\,dx$, where $f(x)$ is a function of x. The procedure is to plot $f(x)$ vs. x and measure the area under the curve, between the vertical lines at $x = x_1$ and $x = x_2$.

The area is most conveniently obtained by the use of a planimeter, although it may also be obtained by counting squares and fractions of squares if good cross-sectional paper is used. The result is obtained either as a planimeter reading or as a number of squares. It is important to be able to convert these values to the correct numerical value of the desired integral. The planimeter should be used not only to trace the figure to be measured but also to trace any convenient square or rectangle of approximately the same area. The lengths of the two sides of the rectangle (not in inches or centimeters but in units of the scales employed for ordinate and abscissa) are multiplied to obtain the area of the rectangle in the proper units. The desired integral is then obtained as the area of the rectangle multiplied by the ratio of the two planimeter readings.

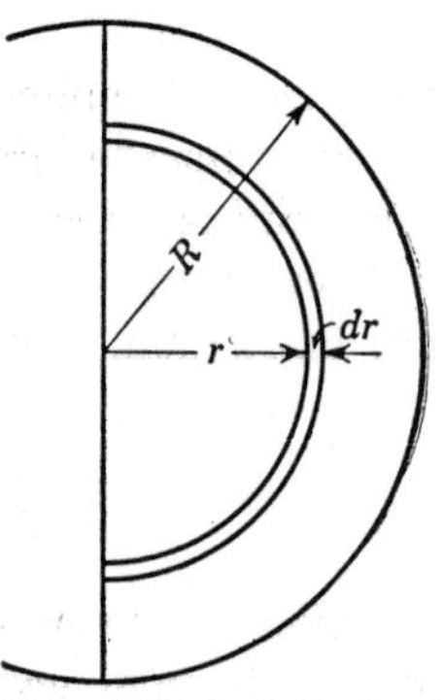

FIG. 1-17. Infinitesimal area in cylindrical coordinates.

Consider the problem of evaluating the quantity of oil flowing in a round pipe, under conditions such that the flow-velocity distribution across any pipe diameter is given by an expression of the form $V = f(r)$. The radius of the pipe is R, and V is the velocity at the distance r from the center line. In the face of the pipe cross section, consider any semicircular ring of width dr and length πr (see Fig. 1-17). The area of this differential ring is $\pi r\,dr$, and the flow through this element is $V\pi r\,dr$ cfs. The total flow for one-half the cross section of the pipe is the summation, or integration, of the flows through all such elements having radii between 0 and R. Since the flow in both halves of the pipe is the same, the total flow Q is given by

$$Q = 2\pi \int_0^R Vr\,dr \tag{1-43}$$

When the known relation between V and r is substituted,

$$Q = 2\pi \int_0^R rf(r)\,dr \tag{1-44}$$

This may or may not be integrable in the ordinary way, depending on the complexity of the function $f(r)$. It may be integrated quite easily by the graphical procedure, however. When the method described is followed, a plot is prepared of $rf(r)$ vs. r and the area under the curve between the limits 0 and R multiplied by 2π to give Q.

Since $r\,dr = \frac{1}{2}\,d(r^2)$, (1-44) may be written

$$Q = \pi \int_0^{R^2} f(r)\,d(r^2) \tag{1-45}$$

The integration also may be performed graphically by preparing a plot of $f(r)$ vs. r^2 and measuring the area between $r^2 = 0$ and $r^2 = R^2$.

1-18. Graphical Construction of Integral Curves. The function $y = f(x)$ has associated with it an integral function $F(x)$ such that

$$\frac{dF(x)}{dx} = f(x) \tag{1-46}$$

and

$$F(x) = \int f(x)\,dx + C \tag{1-47}$$

The definite integral $\int_{x=a}^{x=b} f(x)\,dx$ is interpreted as the area bounded by the ordinates at $x = a$ and $x = b$ under the curve of $f(x)$ vs. x. An obvious method of plotting $F(x)$ is to write first a definite integral with variable upper limit

$$F(x) = \int_{x=x_0}^{x=x} f(x)\,dx \tag{1-48}$$

By measuring the area under the curve $f(x)$ vs. x between the fixed ordinate at $x = x_0$ and the variable ordinate at $x = x$ and plotting this area against x, the curve for $F(x)$ is determined. The arbitrary selection of an initial point (x_0, y_0) on the integral curve is equivalent to assigning a definite value to the arbitrary constant C in (1-47). The integral function $F(x)$ is a function of an arbitrary constant C and therefore represents a family of curves, one of which may be plotted when a value is assigned to C.

The measurement of areas is frequently inconvenient and in any event unnecessary to the development of an accurate integral curve, in view of the relation expressed by (1-46). To illustrate the method, suppose that $f(x)$ is the stepwise curve $ABCDEFGH$ of Fig. 1-18. The initial point A' on the integral curve is selected at any convenient distance above A, and, since $f(x)$ has a constant value of 0.5 over the interval AB, by (1-46) $F(x)$ must proceed to C' with a constant slope of 0.5. Similarly, $C'D'$, $D'F'$, and $F'H'$ have slopes of 1.0, 0.25, and -0.5, respectively.

A convenient method of construction for determining the slopes is to lay off to the left of the origin a distance OP equal to one unit on the x axis. If the ordinate of a point C on the curve to be integrated is projected onto the y axis at V, the slope of the line PV is the same as the slope of the integral curve corresponding to point C, and $C'D'$ is parallel to PV. In many cases, the range of variation of x and y may be such that it would be impractical to use the same scale on the x and y axes. Under these circumstances, a simple relation may be derived between the length of the units on the x and y axes, the distance OP (called "the polar distance"), and the length of the unit that must be used to measure

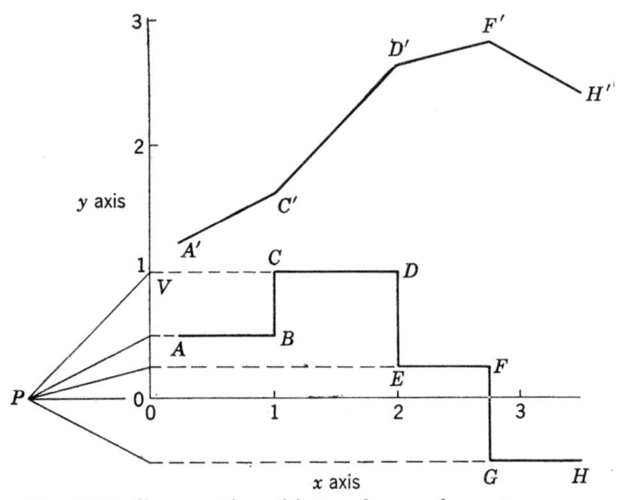

FIG. 1-18. Construction of integral curve from step curve.

the ordinates of the integral curve. Let u_x and u_y be the lengths in terms of any unit (inches, centimeters, or millimeters) on the x and y axes, respectively; p and u_i will be taken as the lengths of the polar distance OP and the length of the unit used to measure integral-curve ordinates. Consider $f(x)$ over the interval CD. If the preceding construction is to be used and $C'D'$ taken parallel to PV,

$$\frac{dF(x)}{dx}\frac{u_i}{u_x} = \frac{f(x)u_y}{p} \tag{1-49}$$

Since $dF(x)/dx = f(x)$, irrespective of the lengths of the various scales, (1-49) reduces to

$$u_i = \frac{u_y u_x}{p} \tag{1-50}$$

This is a necessary relation connecting the length of the polar distance p with the lengths of the units on the x and y axes and the length of the unit used to measure ordinates of the integral curve. In Fig. 1-18, $u_x = u_y = p = u_i$.

The procedure whereby this method of plotting integral curves may be extended to any function is illustrated in Fig. 1-19. With point A as a starting point, a series of triangles is constructed, by means of a step curve, about the curve to be integrated. The step curve is constructed so that the areas of successive triangles below and above the curve are equal. The polar distance may now be laid off and the step curve integrated, as in Fig. 1-18. The accuracy of the integral curve so obtained will depend upon how

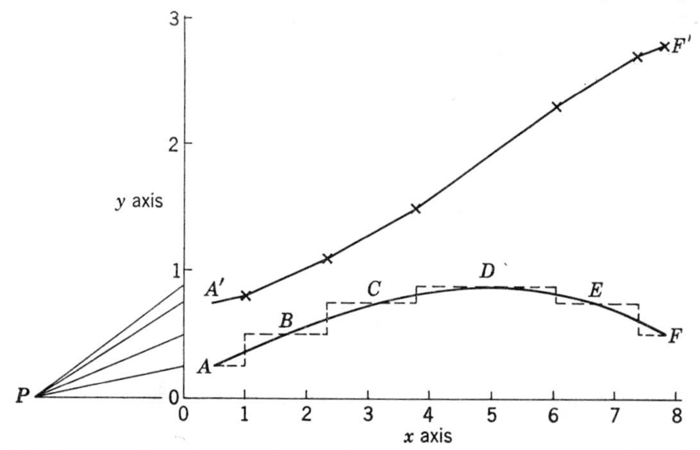

FIG. 1-19. Construction of integral curve from step curve approximating a function.

many triangles are used and how closely the condition of equality of areas is approached. This method can be quite accurate in practice. The series of straight lines resulting from the construction may be smoothed into a curve when it is seen that at the abscissas B, C, D, and E the integral curve must be tangent to the straight lines. In Fig. 1-19, for purposes of illustration, if the length of the unit on the x axis is considered as 1, the length of the unit on the y axis will be 2, and the polar distance will be 2.5. Hence, by (1-50), the length of the unit used to measure ordinates of the integral curve must be $u_i = 2 \cdot 1/2.5 = 0.8$ times as long as the length of the unit on the x axis.

1-19. Numerical Integration. Numerical-integration procedures are available. An obvious method is to fit the data with an empirical expression and use formal analytical methods.

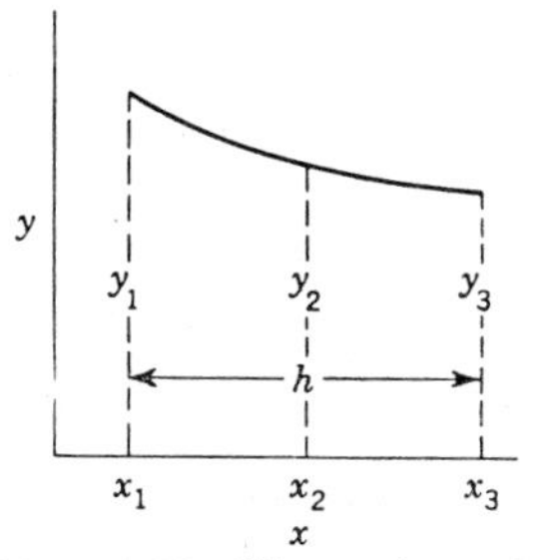

FIG. 1-20. Illustration of Simpson's rule.

If the area to be measured is visualized, it is usually apparent how it may be divided into a number of narrow trapezoids, the area of each approximated numerically, and the summation obtained. If the base of the figure (Fig. 1-20), having a length $x_3 - x_1$, is divided into n equal increments, and the $n + 1$ corresponding ordinates are $y_0, y_1, \ldots, y_n$, then the area is given by

$$\int_{x_1}^{x_3} y\,dx = \frac{x_3 - x_1}{n}\left(\frac{y_0}{2} + y_1 + y_2 + \cdots + y_{n-1} + \frac{y_n}{2}\right) \quad (1\text{-}51)$$

This method of finding the integral is known as the "trapezoidal rule."

A somewhat better method, much used by engineers, is Simpson's rule. Referring to Fig. 1-20, assume that the curve between x_1 and x_3 may be represented exactly by a cubic† or simpler equation; i.e., assume

$$y = a_0 + a_1x + a_2x^2 + a_3x^3 \quad (1\text{-}52)$$

In Fig. 1-20, y_1 is the value of the function at $x = x_1$, and y_3 is the value of the function at $x = x_1 + h$, where h represents the interval $x_3 - x_1$. By direct integration,

$$\int_{x_1}^{x_3} y\,dx = a_0h + \frac{a_1}{2}(2x_1h + h^2) + \frac{a_2}{3}(3x_1^2h + 3x_1h^2 + h^3) + \frac{a_3}{4}(4x_1^3h + 6x_1^2h^2 + 4x_1h^3 + h^4) \quad (1\text{-}53)$$

By direct substitution in (1-52), y_1, y_2, and y_3 are obtained as follows:

$$y_1 = a_0 + a_1x_1 + a_2x_1^2 + a_3x_1^3$$

$$y_2 = a_0 + a_1\left(x_1 + \frac{h}{2}\right) + a_2\left(x_1 + \frac{h}{2}\right)^2 + a_3\left(x_1 + \frac{h}{2}\right)^3$$

$$y_3 = a_0 + a_1(x_1 + h) + a_2(x_1 + h)^2 + a_3(x_1 + h)^3$$

where x_2 is placed midway between x_1 and x_3, that is, at

$$x = x_1 + \frac{h}{2}$$

and y_2 is defined as the ordinate at that point. If the expressions for the values of y_1, y_2, and y_3 are expanded and compared with (1-53), it may be readily shown that

$$\int_{x_1}^{x_3} y\,dx = \int_{x_1}^{x_1+h} y\,dx = \frac{h}{6}(y_1 + 4y_2 + y_3) \quad (1\text{-}54)$$

This is Simpson's rule.

† It is frequently stated that Simpson's rule amounts to replacing the actual curve by a second-degree parabola over the interval x, $x + h$. As a matter of fact, it can be shown that the area under such a parabola passing through three equidistant points on any third-degree curve is the same as the area under the third-degree curve itself. Although it is true that the curve is replaced by the arc of a second-degree parabola for the interval involved, the integral obtained is actually accurate if the curve is a cubic.

The procedure in carrying out an actual integration is to note the ordinates (or values of the function) at the two limits and at a point midway between. The sum of the end ordinates plus four times the middle ordinate, all multiplied by one-sixth the increment of x, represents the value of the integral.

It is seen that the procedure outlined is exact if the function can be represented by a linear, quadratic, or cubic equation in the interval x_1, x_3. If the curvature is great and the function complicated, the interval may be divided into a number of small intervals and the rule applied to each.

A somewhat more efficient method is due to Gauss. By the use of three ordinates, it provides a means of obtaining the exact integral for a fifth-degree function, where Simpson's rule is exact only for a cubic. Values of the ordinate y are noted at n predetermined values of x. Each is multiplied by a predetermined constant, and the sum of the products so obtained is multiplied by the interval of x to give the desired integral. If n points are used, the method is exact if the function can be represented by a polynomial of the $(2n - 1)$st degree.

The derivation for the special case of a fifth-degree function will be developed briefly. The function is of the form

$$y = a_0 + a_1x + a_2x^2 + a_3x^3 + a_4x^4 + a_5x^5$$

When the substitution is made,

$$x = \frac{a + b}{2} + \frac{b - a}{2} u \tag{1-55}$$

the desired integral becomes

$$\int_a^b y\, dx = \frac{b - a}{2} \int_{-1}^{+1} y\, du = N(b - a) \tag{1-56}$$

where N is the average value of y over the interval a, b. Since y can be represented by a fifth-degree polynomial in x, it can also be represented by a similar function of u, such as

$$y = a_0' + a_1'u + a_2'u^2 + a_3'u^3 + a_4'u^4 + a_5'u^5 \tag{1-57}$$

Integrating (1-57) and substituting in (1-56) result in

$$N = a_0' + \frac{a_2'}{3} + \frac{a_4'}{5} \tag{1-58}$$

When the rule outlined is followed, this result may be obtained from three ordinates y_1, y_2, and y_3 and three constants K_1, K_2, and K_3, as follows:

$$N = K_1y_1 + K_2y_2 + K_3y_3 \tag{1-59}$$

where y_1, y_2, and y_3 correspond to values of x or u yet to be determined. From (1-59),

$$\begin{aligned} N &= K_1(a_0' + a_1'u_1 + a_2'u_1^2 + a_3'u_1^3 + a_4'u_1^4 + a_5'u_1^5) \\ &\quad + K_2(a_0' + a_1'u_2 + a_2'u_2^2 + a_3'u_2^3 + a_4'u_2^4 + a_5'u_2^5) \\ &\quad + K_3(a_0' + a_1'u_3 + a_2'u_3^2 + a_3'u_3^3 + a_4'u_3^4 + a_5'u_3^5) \\ &= a_0'(K_1 + K_2 + K_3) + a_1'(K_1u_1 + K_2u_2 + K_3u_3) \\ &\quad + a_2'(K_1u_1^2 + K_2u_2^2 + K_3u_3^2) + a_3'(K_1u_1^3 + K_2u_2^3 + K_3u_3^3) \\ &\quad + a_4'(K_1u_1^4 + K_2u_2^4 + K_3u_3^4) + a_5'(K_1u_1^5 + K_2u_2^5 + K_3u_3^5) \end{aligned} \tag{1-60}$$

If this is to be identical with the correct result (1-58), it is clearly necessary that the following relations hold:

$$\begin{aligned} K_1 + K_2 + K_3 &= 1 \\ K_1u_1 + K_2u_2 + K_3u_3 &= 0 \\ K_1u_1^2 + K_2u_2^2 + K_3u_3^2 &= 1/3 \\ K_1u_1^3 + K_2u_2^3 + K_3u_3^3 &= 0 \\ K_1u_1^4 + K_2u_2^4 + K_3u_3^4 &= 1/5 \\ K_1u_1^5 + K_2u_2^5 + K_3u_3^5 &= 0 \end{aligned}$$

These equations may be solved to give values for K_1, K_2, K_3, u_1, u_2, and u_3 as follows: $K_1 = K_3 = 5/18$; $K_2 = 4/9$; $u_1 = -\sqrt{3/5} = -0.7746$; $u_2 = 0$; $u_3 = \sqrt{3/5} = 0.7746$. Values of y corresponding to these values of u are multiplied by their respective K's and added together to give N, as shown by (1-59). The integral is $N(b - a)$, since N is the average value of y.

In general, in the procedure of Gauss, a sufficient number of simultaneous equations are obtained to evaluate nK's and n values of u to be employed, where the polynomial is of the $(2n - 1)$st degree. Thus four points are needed for an exact integration of a seventh-degree equation, etc. Although any number of seventh-degree curves could be drawn through the four points on the curve, they would necessarily all have the same area under them between $u = -1$ and $u = +1$.

The application of the method will appear much clearer if the simplicity of the change of variables is realized. All that is done is to change x to u so that the limits will be -1 and $+1$ instead of a and b.

As an example of the procedure, consider a simple case of a function which, for comparison, is easily integrated by the usual procedure. Let it be desired to obtain the integral

$$\int_0^2 y\,dx = \int_0^2 (10 + x - x^2 + x^3 - x^4 + x^5)\,dx \tag{1-61}$$

On changing variables, it is apparent that $x = 0$ at $u = -1$; $x = 2$ at $u = 1$; $x = 1 - 0.7746 = 0.2254$ at $u = -\sqrt{3/5}$; $x = 1$ at $u = 0$;

$x = 1.7746$ at $u = +\sqrt{3/5}$. The value of y at u_1 ($x = 0.2254$) is 10.27; y at u_2 ($x = 1$) is 11; and y at u_3 ($x = 1.7746$) is 31.9. The average value of the function is

$$N = 5/18 \cdot 10.27 + 4/9 \cdot 11.0 + 5/18 \cdot 31.9 = 13.8$$

and the integral $(b - a)N$ is $2 \cdot 13.8 = 27.6$. This is identical with the value obtained by direct integration of (1-61).

Numerical values of u and of K to be used in applying the method are tabulated below for $n = 2$, 3, and 4, i.e., when the integral is to be obtained from two, three, or four values of the function.

For $n = 2$,

$$K_1 = K_2 = 1/2 \qquad u_1 = -u_2 = \sqrt{1/3}$$

For $n = 3$,

$$K_1 = K_3 = 5/18 \qquad K_2 = 4/9$$

$$u_1 = -\sqrt{3/5} = -0.7746 = -u_3 \qquad u_2 = 0$$

For $n = 4$,

$$K_1 = K_4 = 0.1739 \qquad K_2 = K_3 = 0.3261$$

$$u_1 = -0.8611 \qquad u_2 = -0.34 \qquad u_3 = 0.34 \qquad u_4 = 0.8611$$

For $n = 5$,

$$K_1 = K_6 = 0.118463 \qquad K_2 = K_4 = 0.239314 \qquad K_3 = 0.284444$$

$$u_1 = -0.906180 \qquad u_2 = -0.538469 \qquad u_3 = 0 \qquad u_4 = 0.538469$$

$$u_5 = 0.906180$$

As pointed out above, these will give exact results for a cubic-, fifth-, seventh-, and ninth-degree function, respectively, but may be used as approximations in other cases. The four-point method gives sufficient accuracy for most engineering work, and the three-point method may often be used to advantage.

Gauss's method of integration should be readily adaptable to engineering test work, where an observation is to be made frequently to obtain the average value of an important variable over the period of the test. As an example, it may be applied to flow measurements by pitot tubes, to accomplish a simplification of the necessary traverse. In the usual case of a single traverse across one diameter of a round duct, the total fluid flow is given by

$$Q = \pi \int_0^{R^2} V d(r^2) = \frac{\pi R^2}{2} \int_{-1}^{+1} V d\left(\frac{r}{R}\right)^2 \tag{1-62}$$

As suggested earlier, the integration may be obtained by plotting the velocities as obtained by the pitot tube vs. r^2, or r^2/R^2, and integrating graphically. An alternative procedure is to measure the velocity at 10 points representing half rings of equal areas, in which case the average velocity is one-tenth the sum of the 10 measured velocities. The readings are taken at values of r^2/R^2 of 0.1, 0.3, 0.5, 0.7, and 0.9, which correspond to 2.6, 8.2, 14.6, 22.6, 34.2, 65.8, 77.4, 85.4, 91.8, and 97.4 per cent of the distance from one wall to the other across a diameter.

If the three- or four-point Gauss methods are to be used, it must be possible to represent the velocity distribution in terms of position across the diameter by a fifth- or seventh-degree function. If the flow conditions result in a velocity distribution of such complexity that this is not possible, more points are needed for an exact result, and the simpler methods should be looked upon as approximations although they are usually sufficiently good for most plant test work. This is shown by the results of a number of pitot-tube traverses obtained under widely different conditions in a steel plant, a paper mill, and a heavy chemical plant. The results of the three- and four-point Gauss integrations are compared in Table 1-5 with those of the 10-point method described above and with accurate graphical integrations. It will be noted that the comparison made is of the methods of integration only and that the result will not be the true average velocity in the duct unless the flow in each half of the duct is symmetrical radially.

In making a simplified traverse as described, the problem is the integration from $-R$ to $+R$ of the area under a curve of V vs. r^2/R^2. If three points are used, the velocity is observed at the center ($u = 0$) and at $r^2/R^2 = -\sqrt{3/5}$ and $+\sqrt{3/5}$, i.e., at the center and at two points at a radius of $0.880R$. The velocity at the center is multiplied by $4/9$, and the two other observed velocities are multiplied by $5/18$. The sum of the three products is the desired average velocity.

If four points are used, two readings are taken at $0.928R$ and two at $0.583R$. The two nearest the wall are multiplied by 0.1739, and the two nearest the center are multiplied by 0.3261. The sum of the four products is the average velocity in the duct.

It must be remembered that any method of numerical integration is necessarily based on an adequate and representative *sampling* of the function to be integrated. The general nature of the function should be known, so that special attention may be paid to any region in which y is suspected of varying rapidly or in a peculiar manner. If Simpson's rule is used, large increments of x may be taken in regions where the function is smooth and approximately linear, and small increments, with small values of h, employed in regions where the function changes rapidly.

Fortunately, in the natural processes encountered in most engineering work, the functions are relatively simple, and the danger is not great of failing to take a "sample" in a region where the function jumps rapidly.

TABLE 1-5. COMPARISON OF AVERAGE VELOCITIES CALCULATED BY VARIOUS METHODS

Test	Fluid	Duct diameter, ft	Fluid density, pcf	Average velocity, fps				Location of traverse
				By graphical integration	10-point	3-point Gauss	4-point Gauss	
A	Air	2.17	0.071	51.2	52.1	50.5	51.9	Air to preheater in billet reheater
B	Air	2.66	0.077	21.7	21.7	21.6	21.5	6 in. downstream from open inlet
C	Blast-furnace gas	6.33	0.078	62.0	62.3	63.9	61.7	19 diam downstream from 45° bend
D	Air	2.0	0.075	43.5	43.7	44.0	43.4	5 diam downstream from right-angle bend
E	Air	1.05	0.073	25.1	25.2	25.1	25.3	4 diam downstream from blower
F	Air	1.05	0.073	26.7	26.7	26.6	26.8	20 diam downstream from right-angle bend
G	Water	0.33	62.3	10.9	10.9	10.9	10.8	25 diam downstream from bend
H	Water	0.33	62.3	11.2	11.2	11.4	11.1	25 diam downstream from bend
I	Sulfur-burner gas	1.46	0.076	23.8	23.8	24.2	23.7	34 diam straight pipe upstream
J	Water	0.51	62.3	1.57	1.57	1.59	1.52	
K	Water	0.51	62.3	1.79	1.80	1.84	1.79	
L	Blast-furnace gas	3.0	0.078	13.5	13.5	13.6	13.5	Straight length of gas main
M	Air	4.17	0.069	50.1	50.5	48.6	50.9	Immediately downstream from butterfly valve
N	Blast-furnace gas	2.0	0.077	57.1	57.4	57.1	58.5	3 diam after one bend and just above another
O	Air	1.54	0.059	54.1	54.0	54.2	54.0	Downstream from fan in straight section
P	Hot air	1.5	0.053	39.6	39.6	39.9	39.6	8 diam downstream and
Q	Air	1.94	0.075	39.0	39.2	39.6	39.0	2 diam upstream from bends
R	HCl air, 44% HCl	0.83	0.0829	27.5	27.6	27.6	27.6	5 diam downstream from bend
Average percentage error as compared with graphical integration..................					0.25	1.1	0.8	

Simpson's rule involves multiplication of the ordinates sampled by simpler constants than does Gauss's method for equal accuracy, but the number of samples is larger, and they must be chosen at equal intervals. For integration of algebraic functions, Simpson's rule is usually to be preferred, but Gauss's method offers attractive possibilities for engineering test work.

Various other methods have been developed for numerical integration. Several of these are described and compared by Scarborough,[4] whose book should be consulted for a more complete discussion of the subject.

BIBLIOGRAPHY

1. Lipka, J.: "Graphical and Mechanical Computation," John Wiley & Sons, Inc., New York, 1921.
2. Perry, J. H.: "Chemical Engineers Handbook," McGraw-Hill Book Company, Inc., New York, 1952.
3. Peirce, B. O., and R. M. Foster: "A Short Table of Integrals," 4th ed., Ginn & Company, Boston, 1956.
4. Scarborough, J. B.: "Numerical Mathematical Analysis," 2d ed., Johns Hopkins Press, Baltimore, 1950.
5. Avakian, A. S.: thesis in department of mathematics, Massachusetts Institute of Technology, 1933; supervised by R. D. Douglass.

PROBLEMS

1-1. A successful correlation of flooding data in packed towers employs a graph of

$$\frac{U_0^2 s \rho_G \mu^{0.2}}{g_c F^3 \rho_L} \quad \text{vs.} \quad \frac{L}{G}\sqrt{\frac{\rho_G}{\rho_L}}$$

where U_0 = superficial gas velocity, fps
s = packing surface, ft²/ft³
ρ_G = gas density, pcf
ρ_L = liquid density, pcf
μ = liquid viscosity, lb/(sec) (ft)
G = superficial gas rate, lb/(hr) (ft²)
L = liquor rate, lb/(hr) (ft²)
F = fraction free volume
g_c = dimensional constant, 32.174 mass lb-ft/sec² (force pounds)

If such a graph is used, trial and error are encountered when the allowable gas velocity is to be calculated, if the liquor rate, gas density, packing characteristics, and liquor characteristics are known. How might the relation be replotted in order to obtain a graph which may be used to calculate the allowable gas rate without trial and error?

1-2. Figure 1-21 represents a well-known correlation of data on heat transfer from a round pipe to oil flowing at low velocities. Discuss.

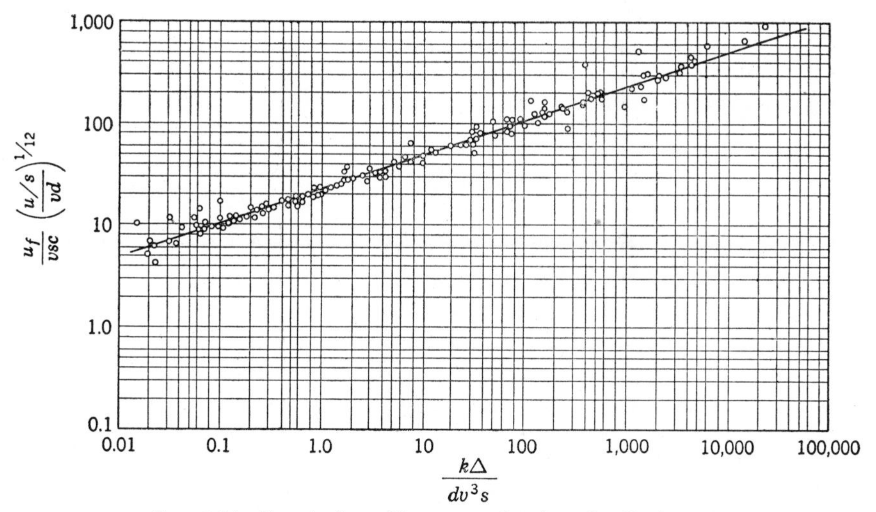

FIG. 1-21. Correlation of heat-transfer data for Prob. 1-2.

1-3. Determine methods of plotting the variables in the following equations so that data fitting the equations will fall on straight lines from which the values of the numerical constants may be computed:

(a) $$\ln y = (ax)e^x + bx$$
(b) $$V - \mu - a = 4\sqrt{a\mu}$$
(c) $$\ln p = A + \frac{B}{T} + \frac{C}{T^2}$$
(d) $$\frac{1}{\ln ay} = b + cx$$

1-4. From the data given below, calculate the value of the constants a and b in the following equation:

$$y^2 + a^2x^2 - 2abx - 2by - 2axy + b^2 = 0$$

Data:

x	2.24	4	23	56.1	90	144	210	272	324
y	376	360	256	182	132	81	42.2	25	9

1-5. The following data were reported for the vapor-liquid equilibrium for binary mixtures of benzene and toluene:

x	0.167	0.333	0.500	0.667	0.833
y	0.320	0.550	0.710	0.830	0.930

where x and y represent the mole fraction of benzene in the liquid and vapor, respectively. Indicate how these data might be plotted to give a straight line if the relative volatility were constant. Fit the best straight line to the points on such a graph by (*a*) visual inspection of the best straight line through the points and (*b*) the method of averages. Determine the average relative volatility of the mixture by the use of the straight lines obtained.

1-6. Given the equation

$$(1 - x^2)\frac{d^2y}{dx^2} - 2x\frac{dy}{dx} + n(n + 1)y = 0$$

replace x by the new independent variable θ, where $x = \cos\theta$; n is a constant.

1-7. In the differential equation

$$x^2\frac{d^2y}{dx^2} + x\frac{dy}{dx} + (x^2 - N^2)y = 0$$

replace y by the new variable $x^N Z$. N is a constant, and x is the independent variable.

1-8. Find d^2y when $y = \sin^2 x$ if (*a*) x is the independent variable and (*b*) $x = e^t$, and t is the independent variable.

1-9. If $y = e^x$, find d^3y if (*a*) x is the independent variable and (*b*) $x = \ln Z$, where Z is the independent variable.

1-10. Given the function $y = f(x)$, it is desired to utilize polar coordinates where

$$x = r\cos\theta$$
$$y = r\sin\theta$$

Find d^2y/dx^2 in terms of r and θ only, where θ is to be considered the new independent variable.

1-11. In the design of a certain piece of equipment, it is found that the two costs involved may each be expressed as a power function of a design variable x.

$$\text{Total costs} = A + B$$

where

$$A = ax^n \qquad B = \frac{b}{x^m}$$

Show that for the optimum design, i.e., the value of x corresponding to minimum total cost, the ratio of the individual costs A and B is m/n.

1-12. Find the limit approached by the functions

(*a*) $$\frac{x^3}{e^x - 1} \qquad \text{as } x \to 0$$

(*b*) $$(x)^x \qquad \text{as } x \to 0$$

1-13. In the design of counterflow heat exchangers in which the heat-transfer coefficient is substantially constant, no heat losses occur, and kinetic-energy effects are negligible, the use of a "log mean" temperature difference is recommended. The log mean temperature difference is defined by the following equation:

$$(\Delta t)_m = \frac{(\Delta t)_2 - (\Delta t)_1}{\ln (\Delta t)_2/(\Delta t)_1}$$

What is $(\Delta t)_m$ when $(\Delta t)_2 = (\Delta t)_1$?

1-14. In a plant producing nitric acid by the distillation of a mixture of sodium nitrate and sulfuric acid, the distillation proceeded as follows:

t	S, per cent	W, lb
0	0	0
¼	30.0	100
½	60.0	190
¾	75.5	290
1	83.0	380
1.5	93.0	570
2	97.0	760
3	95.5	1,150
4	94.2	1,530
5	92.5	1,910
6	91.0	2,290
7	89.0	2,680
8	86.7	3,060
9	84.2	3,440
10	81.3	3,820
11	77.6	4,210
12	73.0	4,590
13	67.7	4,970
14	60.0	5,350
15	42.0	5,740
15.167 (end)	35.0	5,800

where t = time measured from start of distillation, hr

S = strength of distillate condensing at time t, expressed in weight per cent of HNO_3

W = total weight of distillate from start of time t

(*a*) Prepare a plot showing the concentration of acid that would result from the continuous collection and mixing in one vessel of all the condensate as a function of the amount of condensate collected.

(*b*) What is the maximum weight of 90.0 per cent acid obtainable from a process which involves only segregation of the condensate as it is formed?

CHAPTER 2

INTERPRETATION OF ENGINEERING DATA

2-1. Introduction. The significance of conclusions based upon numerical results is necessarily determined by the reliability of the data and of the methods of calculation in which they are employed. The engineer is, therefore, always concerned with appraisals of reliability, and his appreciation of the principles of error and precision is one of the readiest gauges of his professional competence. Failure to realize the importance of such appraisals not only causes much misdirected effort and expense in both plant and laboratory but may also result in technical blunders and unsound decisions. These statements apply with no less force to the roughest sort of cost estimate than to the most precise determination of physical or chemical constants.

The fundamental principles determining the reliability of numerical results are two in number:

1. The degree of accuracy sought in any investigation should, in general, depend upon the projected use of the results, and the accuracy of the required data and calculations should be consistent with the desired accuracy in the result.

2. It is desirable to complete the investigation and obtain the required accuracy with a minimum of time and expense.

This second requirement is intimately related to the first, particularly in experimental investigations, because it is usually found that the costs of experiment mount rapidly as the desired degree of accuracy increases. It is most important that the result should be presented in a manner which indicates clearly and logically the limitations on its accuracy and general reliability.

The student or engineer encounters many questions in attempting for the first time to apply the principles that have just been stated in general form. For example, what types of errors are likely to enter into a given measurement or calculation? To what extent do the errors in the data influence the error in the final result? How accurately must the individual quantities entering a calculation be known in order to ensure a specified accuracy in the calculated result? What accuracy is it economically feasible to obtain in a given investigation? The present chapter will attempt to describe the practical techniques that have been developed to answer these questions. Although the general principles and theory to be followed in obtaining these answers are well known, they may, in some cases, particularly in certain engineering problems, be obtained most readily by an appeal to experience. The detailed procedures incident to an estimate of accuracy in any problem will necessarily vary. A discussion of the evaluation of the accuracy of certain physical or chemical data by the methods of least squares would be hopelessly out of place in the estimation of the accuracy of cost calculations on a commercial-size plant designed from pilot-plant data. In either case, however, if the results are to be of value, their accuracy must be investigated.

2-2. Explanation of Terms. The accuracy of a number representing the value of a quantity is the degree of concordance between this number and the number that represents the true value of the quantity; it may be expressed in either absolute or relative terms. This is a qualitative rather than a quantitative definition, and it must be recognized that the true value of many quantities can never be determined.

Classification of Errors. All measurements and calculations are subject to two broad classes of errors, viz., determinate and indeterminate errors.

Any error that is discovered and allowed for in magnitude and sign in the form of a correction allowing for its effect is a "determinate error." For example, comparison of an ordinary thermometer with a standard may reveal errors in the graduation of the former and result in a standard calibration. Every temperature measured with this thermometer would then be subject to an error of definite magnitude and sign that could be determined by reference to the standard calibration. Such errors are termed "biases." All errors that either cannot be or are not properly allowed for in magnitude and sign are known as "indeterminate errors." It is obvious that the corrections for determinate errors will themselves be subject to errors and constitute one class of indeterminate errors. In practical investigations, it is very common to encounter errors that might theoretically be determined as to both magnitude and sign by the adoption of a more refined method of measurement and calculation, were this economically justified. Such a situation often arises in preliminary investigations, and here it is essential to make the best possible estimate of the magnitude of the error; such estimated errors are classed as indeterminate. When estimates based upon past experience and best judgment are regarded as inadequate for the purpose at hand, it becomes necessary to devise practicable means for determining the accuracy. For instance, in much engineering work it often proves satisfactory to employ heat and material balances as a basis for indicating probable limitations on the accuracy of a result. Not infrequently, higher standards of accuracy will be necessitated as an investigation progresses, and with the adoption of more refined methods of measurement, errors that have previously been indeterminate may become determinate.

Accidental Errors of Measurement. A particularly important class of indeterminate errors is that of accidental errors. To illustrate the nature of these, consider the very simple and direct measurement of the weight of a crucible by means of an analytical balance. Suppose that several independent weighings are made and that the weight may be read to $\frac{1}{10}$ mg. When the results of the different weighings are compared, it will be found that even though they have been performed very carefully, they may differ from each other by several tenths of a milligram. Experience has shown that such deviations are inevitable in all measurements and that they result from small unavoidable errors of observation due to more or less fortuitous variation in the sensitivity of measuring instruments and the keenness of the senses of perception. Such errors are due to the combined effect of a large number of undetermined causes, and they are known as "accidental errors."

Precision and Constant Errors. The word "precision" is used to denote reliability, i.e., the extent to which a result is free from accidental errors. It is important to note that *a result may be extremely precise and at the same time highly inaccurate.* For instance, the weighings just mentioned might all agree to within 1 mg, but from this it would not be permissible to conclude that the weight is accurate to 1 mg until it can be shown definitely that the combined effects of uncorrected constant errors and the corrections for known errors are negligible compared with 1 mg. It is quite conceivable that the arms of the balance might be of unequal lengths and that the calibration of the weights might be grossly incorrect. Such errors as these are constantly present and can never be detected by repeated weighing on the same balance with the same set of metal weights. *Such constant errors can be detected only by performing the measurement with a number of different instruments and, if possible, by several independent methods and observers.* Comparison of the results of such a procedure will, in all probability, indicate such constant errors if they exist and make them to a large extent determinate. If such a procedure is impractical, the constant errors remain indeterminate and must be estimated. It is not uncommon to find that many plant measurements are highly precise and also highly inaccurate. For example, several samples withdrawn from one point in an apparatus and subjected to analysis might give agreement to a few tenths of 1 per cent, indicating high precision, and yet, if these samples should not be representative of the entire contents of the apparatus, the accuracy of their representation of the concentration of the contents of the apparatus might be of such a low order as to render them useless. Errors due to faulty sampling must be guarded against constantly in both plant and laboratory investigations.

Errors of Method. These arise as a result of approximations and assumptions made in the theoretical development of an equation used to calculate the desired result and must never be neglected in the estimation of over-all accuracy. For example, the over-all plate efficiency of a rectifying column performing a given separation is often calculated as the ratio of the number of theoretical plates to the number of actual plates. It is frequently assumed that the over-all efficiency is identical with the individual plate efficiency, but this is an approximation involving an error of method, the extent of which depends upon the location of operating line and equilibrium curve on a McCabe-Thiele diagram. In this case, the error of method might be determined by a more time-consuming calculation, should this be necessary. Unfortunately, many errors of method are indeterminate, and the estimates of their magnitudes can be improved only by lengthy research and investigation. This is usually true of calculations based upon a new theory or hypothesis.

It is also possible for errors to creep into a calculation through mistakes in the ordinary operations of arithmetic, but, fortunately, these can usually be made negligible by methods to be discussed in the next few paragraphs.

2-3. Significant Figures. A significant figure is any one of the digits 1, 2, 3, 4, 5, 6, 7, 8, 9, and 0 when it is not used merely to locate the position of the decimal point. For example, the numbers 13,002, 0.32016, 0.000021352, 20.301, and $10{,}905 \cdot 10^3$ each contain five significant figures. In the number 13,400, there is no way of telling whether or not the two ciphers are significant figures. In such cases, a direct statement of some sort is necessary.

A number is rounded to n significant figures by discarding all digits to the right of the nth place. To round 3.26589 to three, four, and five significant figures, respectively, write 3.27, 3.266, and 3.2659. In order to round a number with the least possible error, it is convenient to use the following rules:

1. Increase the digit in the nth place by 1 if the discarded number is greater than half a unit in the nth place.

2. Leave the digit in the nth place unaltered if the discarded number is less than half a unit in the nth place.

3. When the discarded number is exactly half a unit in the nth place, round off so as to leave the nth digit an even number. For example, when rounded to three significant figures, $3.645 = 3.64$, and $3.655 = 3.66$.

This last rule is arbitrary, but the errors due to rounding will tend to be neutralized when it is followed consistently.

Significant Figures and the Numerical Expression of Error. If x denotes the measured value of a quantity, it is customary to indicate the "estimated error," or uncertainty, in x by the terminology

$$x \pm \Delta x \tag{2-1}$$

The methods by which Δx is estimated and the interpretation of the result will be discussed later. Δx is referred to as the "absolute error" in x; the ratio $\pm(\Delta x/x)$ is termed the "relative error" in x.

Various rules have been proposed for determining the number of significant figures to be retained in the expression of an error. Any rule adopted in this connection is bound to be arbitrary, but in ordinary work it is fully sufficient to retain only two significant figures in the number expressing an error. The argument for a rule of this nature is illustrated by the following example: Consider the measured volume of liquid

$$7{,}500 \pm 15 \text{ gal}$$

The error shows that 7,500 is uncertain by 15 units in its fourth place. One-tenth of this error corresponds to a change of 1.5 units in the fourth place, which is already uncertain by 15 units, and a change of such magnitude may be considered negligible under these conditions. Consequently, a change of equal amount in the error is also negligible, but such a change will always show up as a change of at least one unit in the second significant figure of the error. If it is granted that a change of one unit in the second significant figure of the error is negligible, it must be granted that rejection of all significant figures beyond this place will produce a negligible effect in the number representing the error, since the process of rejection produces, at most, a change of this magnitude. A more tolerant rule would require carrying only one, rather than two, unreliable figures in a result. This latter is customary in engineering work, although the practice must necessarily vary with individual situations.

Once the error in a number is estimated and expressed to the proper number of significant figures, the number should be carried out to the place corresponding to the last significant figure of the expression for the error. More significant figures than this will produce a false appearance of accuracy, and fewer will result in a needless sacrifice of accuracy. According to this, write 15.04 ± 0.15, not 15.036 ± 0.15. The foregoing discussion applies not only to the over-all error of a result but also to the individual errors that go to make up the over-all error.

2-4. Significant Figures and the Operations of Arithmetic. From the principle that two or, in many cases, only one uncertain figure should be carried in a result, it is possible to formulate certain principles regarding the performance of the ordinary operations of arithmetic.

Addition and Subtraction. When a series of several quantities are added, the sum should contain only as many significant figures as the quantity having the greatest absolute error. The error in the sum may possibly be as great as the sum of the absolute errors in the numbers. The sum of the numbers 123, 32.3, 0.276, and 0.0324, each containing one uncertain figure, is 155.6084 if the numbers are added as they stand. Inspection shows, however, that this sum is uncertain in its third significant figure. When rounded to three significant figures, it should be written as 156. This does not have the deceiving appearance of high accuracy suggested by 155.6084.

The operation of subtraction is frequently the source of great loss of accuracy in cases where it is necessary to obtain the difference between two large numbers each of which is subject to error. For example, the enthalpy changes in two reactions may be 9,654 $\pm$ 40 cal and 9,435 $\pm$ 50 cal, respectively. If it is desired to calculate the enthalpy change in a reaction that is the difference between these two, the result is

$$(9{,}654 \pm 40) - (9{,}435 \pm 50) = 219 \pm 90 \text{ cal}$$

Irrespective of the fact that ± 90 represents the maximum error in the answer and of the argument that the two errors might offset rather than augment each other, one can never be certain that this has occurred, and it is clear that two numbers, each accurate to less than 0.5 per cent, have given rise to a number that is possibly uncertain by over 40 per cent.

All proposed calculations and measurements should be scrutinized carefully for situations of this nature, as their occurrence frequently necessitates a complete change of method if the desired accuracy is to be obtained.

Multiplication and Division. The relative error of a product or a quotient may possibly be as great as the sum of the relative errors of the numbers entering into the calculation. Therefore, the result should be given to as many significant figures as are contained in the least accurate number entering the calculation and no more. This is the most conservative practice. However, as will be shown later, there are many cases where the error is probably less than this maximum, and the treatment may be modified accordingly.

Engineering Calculations and Reports. Although rigid adherence to specific arbitrary rules for the use of significant figures is not essential to good engineering work, it is important to follow the *general* principles that have already been discussed, if misleading results are to be avoided. In most engineering work, the number of justifiable significant figures can be carried on a 10-in. slide rule, and the important exceptions to this procedure will be obvious to one who appreciates the principles of error.

Whereas in accurate and precise laboratory work (e.g., certain chemical analytical work) it may be customary to carry two uncertain figures in the result, only one uncertain figure is carried in most engineering work.

Many engineering problems involve economic factors, and the economic data that must enter the calculations are usually of a considerably lower order of accuracy than the technical data. An engineer must be prepared to accept the fact that in many fields where the technical theory and data have reached a point to justify carrying three or even four significant figures in the calculations, economic data such as estimates of operating and investment costs enter the calculations and by their inherent indeterminacy reduce the accuracy of the result to such an extent as to justify its expression by means of only two significant figures. Indeterminate factors of even greater consequence may sometimes be involved in the final decision as to whether a proposed expenditure should be authorized. The result of technical and economic calculations may be only one factor in a decision involving other factors as diverse as the results of patent litigation and political forecasts.

In the face of these truly indeterminate factors of such great importance, a decision must be made on the basis of business judgment, and under such circumstances it might appear questionable to maintain even ordinary engineering accuracy in the technical calculations. As a general rule, however, it is desirable to maintain as high a standard of accuracy in all calculations and measurements as is possible without undue effort and expense, since this ensures that no unnecessary inaccuracy from the technical results will creep into a decision. Furthermore, experience shows that many calculations and measurements become valuable for purposes entirely unforeseen at the time they were made.

Significant Figures and Financial Accounting Practice. It must be agreed that in the accounting for receipts and disbursements of money the accounts must be kept accurately, down to the smallest unit of money. This may necessitate the use of six, seven, eight, or more significant figures, implying a very high order of accuracy. Where balancing of accounts is not involved, however, accounting practice should follow the technical principles of precision of measurements. Practically all cases of estimates by accountants of future performance fall in this category.

2-5. Classification of Measurements. Several measurements on the same or different quantities are independent when (1) no mathematical relation necessarily exists among them and (2) the different measurements are entirely unbiased by each other or by other results.

Measurements that satisfy the second condition, but not the first, are known as "conditioned measurements." Thus, in the complete chemical analysis of a material, a mathematical relation must exist among the percentages of the various constituents, inasmuch as these must add up to 100.

Measurements satisfying the first, but not the second condition, are said to be "dependent." It is highly desirable, but often difficult, to avoid dependent measurements. For instance, when a series of readings are being made by resetting the indicator on a scale, if the previous results are remembered, one must contend with the temptation to make the new readings agree closely with the previous results rather than to exercise completely independent judgment. Such a practice renders it impossible to determine the true precision of the observations, since the various readings are not truly independent but are affected by preconceived notions.

All measurements may be classed as either direct or indirect. A direct measurement is made whenever the magnitude of the measured quantity is determined by direct observation from the measuring instrument. The measurement of length by a meter stick, time by a clock, and weight by a balance are examples of direct measurements.

In contrast to this direct procedure, the magnitude of a quantity is often measured by calculation from the magnitudes of other quantities directly measured, the calculation being made by means of some functional relationship existing among the quantities. The estimate of error in an indirect measurement is more difficult than the estimate of error in a direct measurement, since the errors in the direct measurements concerned may either augment or offset each other's effect on the error of the calculated result, depending upon their signs and the form of the functional relationship. Indirect measurements may be made for the purpose of computing a desired quantity from a group of directly measurable quantities by means of a known functional relationship containing known constants or for the purpose of determining the unknown constants in a functional relationship of known form.

2-6. Propagation of Errors. When the desired quantity M is related to the several directly measured quantities $M_1, M_2, \ldots, M_n$ by the equation

$$M = \gamma(M_1, M_2, \ldots, M_n) \tag{2-2}$$

M becomes an indirectly measured quantity. In general, the true value of M cannot be known because the true values of $M_1, M_2, \ldots, M_n$ are unknown, but the most probable value of M, denoted by Q, may be calculated by inserting the most probable values of $M_1, M_2, \ldots, M_n$, denoted by $q_1, q_2, \ldots, q_n$, into (2-2). Evidently, the errors in the directly measured quantities will result in an error in the calculated quantity, the value of which it is important to ascertain. If the original measurements are available, an obvious method of procedure would be to calculate a value of M corresponding to every set of measurements. The mean of all these calculated values could then be obtained and the characteristic errors of the mean calculated from the residuals.

Very often, however, the only data available are $q_1, q_2, \ldots, q_n$, together with their characteristic errors, from which it is necessary to estimate the characteristic errors in Q. It may be that $q_1, q_2, \ldots, q_n$ are the most probable values calculated from a set of observations, or they may be merely estimated values employed in the preliminary discussion of the proposed measurement. In these cases, it becomes necessary to devise a procedure for relating the errors in the measured quantities to the error in the calculated quantity.

Such a procedure makes possible the solution of the two fundamental problems of indirect measurements:

1. Given the errors of several directly measured quantities, to calculate the error of any function of these quantities
2. Given a prescribed error in the quantity to be indirectly measured, to specify the allowable errors in the directly measured quantities

The method is as follows: In terms of the most probable quantities, (2-2) may be written

$$Q = \gamma(q_1, q_2, \ldots, q_n) \tag{2-3}$$

The differential change in Q corresponding to a differential change in each of the q's is

$$dQ = \frac{\partial \gamma}{\partial q_1} dq_1 + \frac{\partial \gamma}{\partial q_2} dq_2 \cdots \frac{\partial \gamma}{\partial q_n} dq_n \tag{2-4}$$

where $\partial\gamma/\partial q_n$ denotes the *partial* derivative of γ with respect to q_n and is obtained by differentiating γ with respect to q_n, with all other q's regarded as constant.

If the differentials $dq_1, dq_2, \ldots, dq_n$ are replaced by small finite increments $\Delta q_1, \Delta q_2, \ldots, \Delta q_n$, there results as a good approximation†

† The limitations on this approximation by means of the first differential may become clearer from the following considerations: Errors of $\Delta q_1, \Delta q_2, \ldots, \Delta q_n$ in the quantities $q_1, q_2, \ldots, q_n$ will produce a corresponding error ΔQ in Q according to the equation

$$Q + \Delta Q = \gamma(q_1 + \Delta q_1, q_2 + \Delta q_2, \ldots, q_n + \Delta q_n)$$

Expansion of γ in the neighborhood of $q_1, q_2, \ldots, q_n$ by means of Taylor's theorem gives

$$Q + \Delta Q = \gamma(q_1, q_2, \ldots, q_n) + \frac{\partial \gamma}{\partial q_1}\Delta q_1 + \frac{\partial \gamma}{\partial q_2}\Delta q_2 + \cdots + \frac{\partial \gamma}{\partial q_n}\Delta q_n + \frac{\partial^2 \gamma}{\partial q_1^2}\frac{(\Delta q_1)^2}{1 \cdot 2} + \cdots$$

(Terms of higher order)

If the quantities Δq are small, the terms of higher order are negligible, and the expression reduces to (2-5).

for ΔQ the expression

$$\Delta Q = \frac{\partial \gamma}{\partial q_1} \Delta q_1 + \frac{\partial \gamma}{\partial q_2} \Delta q_2 + \cdots + \frac{\partial \gamma}{\partial q_n} \Delta q_n \tag{2-5}$$

The quantities Δq_1, Δq_2, . . . , Δq_n may be considered as errors in q_1, q_2, . . . , q_n, and (2-5) provides a means of computing the resulting error in the function. Equation (2-5) holds for any type of errors, provided only that they are small. On the other hand, (2-5) does not utilize all the information that may be available and consequently often *overestimates* the error in Q. The following example will illustrate the use of (2-5) and also point out its defects.

Example 2-1. During the course of an analysis of plant performance it becomes necessary to determine the average velocity of water flowing through a certain pipe. The most convenient method of measurement is an indirect one: measurement of the weight W of water issuing from the pipe during the time t, measurement of the pipe diameter D, and calculation of the average velocity from the density ρ of the water and the relation

$$V_{\text{av}} = \frac{W}{tA\rho} \cong \frac{4W}{\pi D^2 t\rho} \tag{2-6}$$

Before undertaking the measurement, the engineer decides to calculate the uncertainty in his result by means of the procedure based upon Eq. (2-5). Accordingly, he estimates the values of the variables and their uncertainty as follows:

1. Weight of water. Information concerning the weighing scales available sets 100 lb as a convenient figure for the weight of water to be collected. The particular scale to be used is not yet known. However, the engineer recognizes that some of the scales in the plant are in rather poor repair and, on the chance that one of these might be used, takes a "conservative" uncertainty of ± 5 lb.
2. Time of collection of sample. Previous information indicates that approximately 70 sec will be required to collect the 100-lb sample. An electric clock will be used to measure the elapsed time. The engineer feels that human and clock error combined will not exceed ± 1 sec.
3. The pipe area. The nominal diameter of the pipe is 1 in. Taking into account the deviations from roundness, caliper error, etc., the engineer mentally estimates an uncertainty in the pipe diameter not exceeding ± 0.03 in.
4. Water density. The water temperature will be about 60°F. The density at 60°F is 62.34 pcf. The estimated uncertainty in the temperature is ± 3°F corresponding to a density variation of less than 0.1 per cent. Since the uncertainty in the water density is an order of magnitude less than that of the other quantities, its effect may be neglected.

Approximate values of the data and the estimated maximum errors are then the following:

Variable	Approximate value	Measured to
W	100 lb	± 5 lb
t	70 sec	± 1.0 sec
D	1 in.	± 0.03 in.

The partial derivatives of (2-6) needed for use in an equation of the form of (2-5) are

$$\frac{\partial V_{\text{av}}}{\partial W} = \frac{4}{t\pi D^2\rho} \qquad \frac{\partial V_{\text{av}}}{\partial t} = -\frac{4W}{\pi D^2\rho t^2} \qquad \frac{\partial V_{\text{av}}}{\partial D} = -\frac{8W}{t\pi\rho D^3}$$

Consequently,

$$\begin{aligned}\Delta V &= \frac{4}{t\pi D^2\rho}\,\Delta W - \frac{4W}{\pi D^2\rho t^2}\,\Delta t - \frac{8W}{t\pi\rho D^3}\,\Delta D \\ &= \frac{4\cdot 144}{70\cdot 3.14\cdot 1\cdot 62.3}\,\Delta W - \frac{4\cdot 100\cdot 144}{3.14\cdot 1\cdot 62.3(70)^2}\,\Delta t - \frac{8\cdot 100\cdot 1{,}728}{70\cdot 3.14\cdot 62.3\cdot 1}\,\Delta D \\ &= 0.042\Delta W - 0.060\Delta t - 100\Delta D \end{aligned} \tag{2-7}$$

In order to obtain the maximum error, the sign of the ΔW will be taken as positive, and the signs of Δt and ΔD will be taken negative. Therefore,

$$\begin{aligned}\Delta V_{\max} &= 0.042\cdot 5 + 0.060\cdot 1.0 + 100\,\frac{0.03}{12} = 0.210 + 0.060 + 0.252 \\ &= 0.522 \text{ fps}\end{aligned}$$

Since the approximate value of the velocity is

$$V = \frac{4\cdot 100\cdot 144}{70\cdot 3.14\cdot 1\cdot 62.3} = 4.21 \text{ fps}$$

the maximum percentage error is $\pm(0.522/4.21)100 = \pm 12.4$ per cent.

Some important points are illustrated by the above example. It will be seen that the error in a calculated quantity that is a function of several directly measured quantities depends on (1) the nature of the function, (2) the magnitudes of the measured quantities, and (3) the magnitudes of the errors. Furthermore, variables such as ρ, whose values are known much more accurately than the rest of the variables, may be considered constants, since the error they introduce into the final result will be negligible.

It is evident that Eq. (2-5) quite probably overestimates the error involved in the measurement. It takes no account of the possibility of compensating errors. Even more serious is its failure to take into account the method used to obtain the original estimates of the uncertainties in the directly measured quantities. Presumably the engineer felt that errors exceeding those estimated were most improbable. Clearly, then, the *simultaneous* occurrence of three error extremes is distinctly less probable than the occurrence of more modest errors. For example, suppose that the probability of the assumed errors in W, t, and D in Example 2-1 are each 0.1, i.e., that 90 per cent of the time the actual errors are smaller than this. Then the probability of obtaining an error in V as large as the calculated 12.4 per cent is only one in a thousand. Modern statistical methods often permit the analyst to take such probability factors into account in the error analysis. When this is possible, the results are more realistic and valuable. Succeeding sections will discuss some of these methods.

Despite the fact that Eq. (2-5) usually overestimates the uncertainty in a dependent quantity, it is a valuable tool. In the case of formulas of the type of (2-6), consisting of the products of powers of the variables, it is possible to effect a simplification in the calculations by the use of fractional errors. To illustrate this, consider the general function

$$Q = q_a^a q_b^b \cdots q_n^n \tag{2-8}$$

Applying (2-5) gives

$$\Delta Q = (q_b^b \cdots q_n^n) a q_a^{a-1} \, \Delta q_a + (q_a^a q_c^c \cdots q_n^n) b q^{b-1} \, \Delta q_b \cdots$$

and

$$\frac{\Delta Q}{Q} = a \frac{\Delta q_a}{q_a} + b \frac{\Delta q_b}{q_b} + \cdots + n \frac{\Delta q_n}{q_n} \tag{2-9}$$

which states that the fractional error in the function Q is given by the sum of the fractional errors in the measured quantities, each multiplied by the respective power to which it appears in the function. When (2-9) is multiplied by 100, it is seen that the same rule applies to percentage errors. This rule provides a rapid solution of the preceding example.

Example 2-2. It is suggested that an attempt be made to measure V_{av} in (2-6) to within an error of ± 2.0 per cent. Under this condition, what are the allowable errors in the directly measured quantities W, t, and D?

Either (2-5) or (2-9) may be used, but it is apparent that there is no unique answer to the problem as stated. More conditions are necessary. For example, the value of two errors might be fixed, whereupon the value of the third is fixed. In this general type of problem, it must be recognized that the labor and expense involved in measuring the various quantities to a given degree of accuracy are different. Ideally, those quantities which are easiest to measure should be measured the most accurately, more tolerance being allowed in the more difficult measurements, so that the required accuracy in the final result will be obtained with a minimum of expense for labor and apparatus. Because there is no general relationship between the difficulty and the accuracy of measurements, this condition cannot be given exact mathematical expression. As a starting point, it is customary to impose the condition that the errors in each of the directly measured quantities contribute equally to the error in the function. This condition is known as the "principle of equal effects."

Applied to (2-5) with the understanding that the sign of each Δq will be taken such as to make all terms of the same sign and will result in the maximum allowable error in ΔQ, the principle of equal effects gives

$$\Delta Q = n \frac{\partial \gamma}{\partial q_1} \Delta q_1 = n \frac{\partial \gamma}{\partial q_2} \Delta q_2 = \cdots = n \frac{\partial \gamma}{\partial q_n} \Delta q_n \tag{2-10}$$

In the case of a function such as (2-8), it is convenient to work with fractional errors, and the principle of equal effects reduces (2-9) to

$$\frac{\Delta Q}{Q} = na\,\frac{\Delta q_a}{q_a} = nb\,\frac{\Delta q_b}{q_b} \qquad \text{etc.} \tag{2-11}$$

Employing (2-11) in the solution of Example 2-2, there results in the case of W

$$100\,\frac{\Delta V}{V} = 2.0 = 3 \cdot 1\,\frac{\Delta W}{W}\,100$$

from which $100(\Delta W/W) = 0.7$ or ± 0.7 lb/100 lb. Similar calculations for the allowable errors in t and D give ± 0.5 sec and ± 0.003 in., respectively. These results should not be considered inflexible but should serve as a basis for deciding the optimum errors to tolerate in each quantity to ensure an error of no more than 2 per cent in the calculated velocity, with minimum labor and apparatus.

Discussions of the kind illustrated by the two preceding examples often serve to reveal that some quantity is being measured with a higher degree of precision than necessary, in view of the magnitude of the errors inherent in the other quantities concerned, or that particular attention must be focused upon the accurate measurement of a certain quantity, because of its unusually large influence on the final calculated result. Unfortunately, in many important cases the functions are quite complex, and calculations involving data known only in the form of tables and curves are necessary. The estimation of errors in such cases is more difficult and not infrequently can be accomplished only by actual repetition of the calculations and comparison of the results obtained from different sets of values of the measured quantities.

Many important design calculations necessitate graphical integrations. For example, one may face the problem of designing a liquid-liquid heat exchanger wherein both individual film coefficients vary considerably with temperature. The heat-transfer area is calculated by graphical evaluation of the integral

$$A = \int_{\Delta_1}^{\Delta_2} \frac{dq}{U\Delta} \tag{2-12}$$

where q and Δ are related by a heat balance and U, the over-all coefficient at any point, is a complex function of the liquid's physical properties, which in turn depend upon the temperature. In this case, a reliable estimate of the error in A due to a given error in the terminal temperatures is best obtained by repetition of the entire calculation, if different values for the terminal temperatures are used. It is especially important to note that the smaller the temperature difference Δ, the more serious an error of given magnitude in the measured temperatures becomes.

Particularly serious in many calculations on mass-transfer processes are errors in equilibrium data. The common calculation on a McCabe-

Thiele diagram of the number of theoretical plates required to effect a given separation with a given reflux ratio provides an instructive example of the importance of accurate equilibrium data. Figure 2-1 represents a diagram for a binary mixture of low relative volatility. A small percentage error in the equilibrium data will cause a large percentage error in the vertical distance between operating line and equilibrium curve, with a resultant large percentage error in the number of

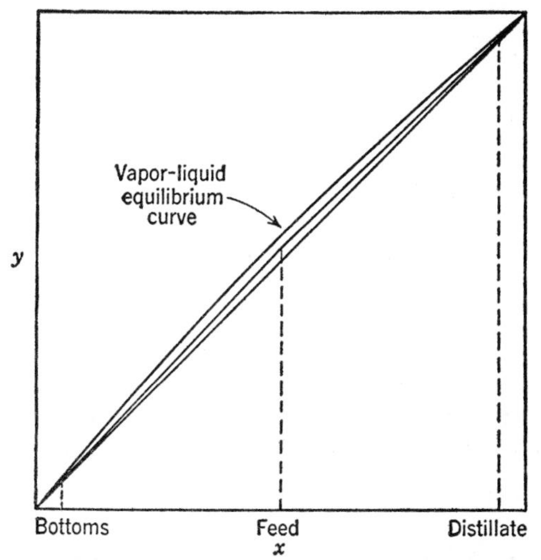

FIG. 2-1. McCabe-Thiele diagram for binary system separable with difficulty.

theoretical plates. The change in the number of theoretical plates due to error in the equilibrium data is best computed by drawing in the equilibrium curve in the position corresponding to the estimated error and repeating the construction for determination of the number of theoretical plates. A similar procedure should be employed in the estimation of the error produced by equilibrium data in the calculation of the number of transfer units in adsorption, extraction, or distillation.

2-7. Variance and Distribution of Random Errors. If an experimental measurement is repeated a number of times ("replicated"), the recorded values of the measured quantity almost invariably differ from one another. The data so obtained may be used for two important purposes: (1) to evaluate the precision of the measurement and (2) to obtain an estimate of the probability that the mean of the measurement differs from the true value of the measured quantity by some specified amount.

The "scatter" of the repeated measurements of the quantity x is commonly reported in terms of the "variance," or, alternatively, of the "standard deviation" of the set of measurements. These quantities are defined in such a way as to be immediately useful in the estimation of the probability of occurrence of random errors of specified magnitude in the measurements.

$$\text{Sample variance} = \bar{\sigma}^2 = \frac{\sum^n (x - \bar{x})^2}{n} = \frac{\sum^n x^2 - \left(\sum^n x\right)^2/n}{n} \tag{2-13}$$

$$\text{Sample standard deviation} = \sqrt{\bar{\sigma}^2} = \bar{\sigma} \tag{2-14}$$

$$\text{Sample mean} = \bar{x} = \frac{\sum^n x}{n} \tag{2-15}$$

The sample variance is simply the mean-square deviation of the n measured values of x from the sample mean $\bar{x}$. In this definition, positive and negative fluctuations about the mean do not cancel one another. In Eq. (2-13), which defines the sample variance, it will be found that the last form of the equation is the most convenient when actual calculations are involved.

The value of the variance becomes more reliable as more measurements are obtained, and the true precision of the measurement procedure is indicated by the value of the variance calculated from a very large amount of data. When the number of data points obtained becomes infinite, the infinite set is termed a "population" of values. For such a population, the population mean, $\bar{\bar{x}}$, is defined as

$$\text{Population mean} = \bar{\bar{x}} = \lim_{n \to \infty} \frac{\sum^n x}{n} \tag{2-16}$$

and the population variance, σ^2, is defined as

$$\text{Population variance} = \sigma^2 = \lim_{n \to \infty} \frac{\sum^n (x - \bar{\bar{x}})^2}{n} = \lim_{n \to \infty} \frac{\sum^n x^2 - \left(\sum^n x\right)^2 / n}{n} \tag{2-17}$$

The population standard deviation is the square root of the population variance.

The population mean $\bar{\bar{x}}$ is the "best," or most probable, value of x, provided that the variations in x are the result of small, random, independent, additive effects. If, however, errors of method and/or nonrandom errors are inherent in the measurements, $\bar{\bar{x}}$ may differ significantly from the true value of x. For example, suppose that each of 50 tests of the tensile strength of a rubber sheet is in error because of the faulty setting of the test machine. This introduces an error of method which will be included in $\bar{\bar{x}}$. Furthermore, flaws in the sheet may cause a large number of very low values to be recorded without any compensating high values. The distribution of the measured values of this set would not be "normal," and $\bar{\bar{x}}$ would not be the best value.

If in an infinite data set the variations in x are random, it was first shown by Gauss that the distribution of values of x about the population mean is given by

$$\mathbf{f} = \frac{\exp\{-\tfrac{1}{2}[(x - \bar{\bar{x}})/\sigma]^2\}}{\sigma\sqrt{2\pi}} \tag{2-18}$$

Here $\mathbf{f}$ is frequency, or probability of occurrence, of a value of magnitude x. Note that $\mathbf{f}$ is a maximum when $x = \bar{\bar{x}}$. For this reason the population mean ordinarily is regarded as the "best" value; it is the most probable value if the errors are random.

The distribution function may be simplified by expressing the variation in x in terms of the population standard deviation σ:

$$u = \frac{x - \bar{\bar{x}}}{\sigma} \tag{2-19}$$

whence

$$\mathbf{f} = \frac{\exp(-\tfrac{1}{2}u^2)}{\sigma\sqrt{2\pi}} \tag{2-20}$$

The *normal frequency distribution* of $\mathbf{f}$, Eq. (2-20), is plotted in Fig. 2-2. The curve is symmetrical about $u = 0 (x = \bar{\bar{x}})$. The probability that a value of $x = \bar{\bar{x}} + \epsilon$ will be observed is the same as the probability that a value of $x = \bar{\bar{x}} - \epsilon$ will be observed. The probability is small that a value of x will be observed which differs greatly from $\bar{\bar{x}}$. It is possible to calculate the probability that an observed value of x will differ from $\bar{\bar{x}}$ by *more than* $\pm c\sigma$.

The probability that a single measurement will give a value lying between $x - dx/2$ and $x + dx/2$ is

$$\mathbf{f}\,dx = \frac{\exp\{-\tfrac{1}{2}[(x - \bar{\bar{x}})/\sigma]^2\}}{\sigma\sqrt{2\pi}}\,dx \tag{2-21}$$

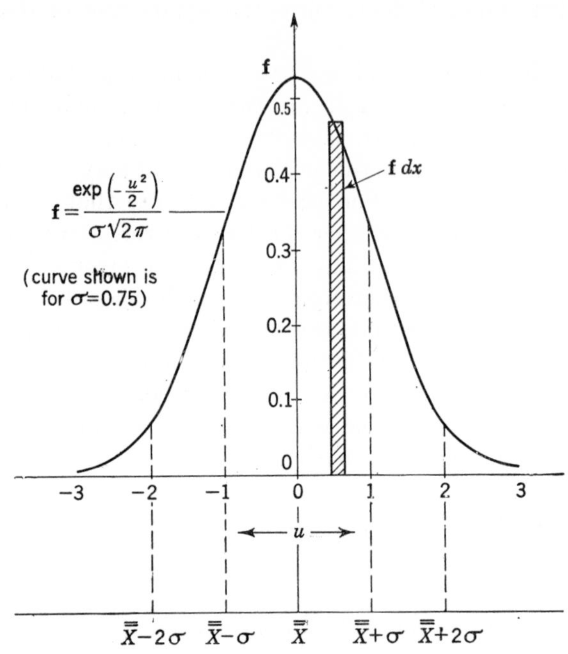

FIG. 2-2. Normal frequency distribution.

The probability that x lies *outside* $\bar{\bar{x}} \pm c\sigma$ is

$$\int_{-\infty}^{\bar{\bar{x}}-c\sigma} \mathrm{f}\,dx + \int_{\bar{\bar{x}}+c\sigma}^{\infty} \mathrm{f}\,dx = \int_{-\infty}^{-c} \frac{\exp(-\tfrac{1}{2}u^2)\,du}{\sqrt{2\pi}} + \int_{c}^{\infty} \frac{\exp(-\tfrac{1}{2}u^2)\,du}{\sqrt{2\pi}}$$

$$= \int_{-\infty}^{\infty} \frac{\exp(-\tfrac{1}{2}u^2)\,du}{\sqrt{2\pi}} - \int_{-c}^{c} \frac{\exp(-\tfrac{1}{2}u^2)\,du}{\sqrt{2\pi}}$$

$$= 1 - 2\int_0^c \frac{\exp(-\tfrac{1}{2}u^2)\,du}{\sqrt{2\pi}} \tag{2-22}$$

With the change of variable

$$u = z\sqrt{2} \tag{2-23}$$

the right-hand side of Eq. (2-22) becomes

$$1 - \frac{2}{\sqrt{\pi}} \int_0^{c/\sqrt{2}} \exp(-z^2)\,dz \tag{2-24}$$

The integral appearing in (2-24) is known as the "probability integral" and is tabulated in a number of references (for example, Peirce[19]). For a value of c equal to 2, (2-24) takes on the value $1 - 0.9545 = 0.0455$.

This means that the probability is less than 0.05 that a single measurement of x will differ from $\bar{\bar{x}}$ by more than twice the standard deviation, i.e., by more than $\pm 2\sigma$. Alternatively, 95 per cent of the measured values of x will differ from $\bar{\bar{x}}$ by less than $\pm 2\sigma$. The range $\pm 2\sigma$ is frequently called the "95 per cent confidence belt on x."

It is evident that, in a normal population, a knowledge of $\bar{\bar{x}}$ and σ provides a description of the precision of the method of measuring x. A large number of chemical analyses of a single sample might be used to obtain σ, which might then be used to estimate the precision of a single analysis of another material. This sort of information evidently has great practical value. The application of the method requires methods of estimating σ from a sample of finite size.

2-8. Properties of the Variance. The population variance defined by Eq. (2-17) is based on a hypothetical sample containing an infinite number of values of the replicated measurement, and $\bar{\bar{x}}$ is the mean of these values. For practical purposes, it is necessary to deal with a finite number of values of the quantity in question. The sample mean $\bar{x}$ is the best estimate of the population mean $\bar{\bar{x}}$. However, the sample variance $\bar{\sigma}^2$ is not the best estimate of the population variance σ^2. It may be shown that a better estimate is given by

$$s^2 = \frac{\sum^n (x - \bar{x})^2}{n - 1} = \frac{\sum^n x^2 - \left(\sum^n x\right)^2 / n}{n - 1} = \frac{n}{n - 1}\, \bar{\sigma}^2 \qquad (2\text{-}25)$$

where s^2 is termed the "sample estimate of the population variance." s^2 is the best estimate of the population variance that can be obtained from n values of x. The validity of these statements is a consequence of the properties of the normal-distribution function. A formal proof may be found in most of the statistics texts listed in the bibliography of this chapter.

In engineering work, a quantity Q is frequently calculated from measured values of several quantities q_i by means of a mathematical relation which can be represented formally by

$$Q = \gamma(q_1, q_2, q_3, \ldots, q_i) \qquad (2\text{-}26)$$

In a typical case, each measurement has been made n times under supposedly identical conditions. In fact, however, the action of random errors results in a series of values of q_1: $q_{11}, q_{12}, \ldots, q_{1k}$ which form the q_1 set and a corresponding series forming the q_2 set, etc. It is then desired to calculate the best value of Q and the variance of Q. If the errors in the q's are normally distributed, the best value of Q will be given by the expression

$$\bar{Q} = \gamma(\bar{q}_1, \bar{q}_2, \ldots, \bar{q}_i) \qquad (2\text{-}27)\dagger$$

The variance of Q is given by the expression

$$\sigma^2 Q = \sum \left(\frac{\partial Q}{\partial \bar{q}_i}\right)^2 \sigma^2(q_i) \qquad (2\text{-}29)$$

The partial derivative $\partial Q/\partial \bar{q}_i$ is to be evaluated numerically at the mean value of each set: $\bar{q}_1$, $\bar{q}_2$, etc.

The derivation of Eq. (2-29) is best illustrated by a simple case. Suppose that the relation between Q and the q's is

$$Q_k = q_{1k} + q_{2k} \qquad (2\text{-}30)$$

Then

$$\bar{Q} = \bar{q}_1 + \bar{q}_2 \qquad (2\text{-}31)$$

and

$$\Delta Q_k = Q_k - \bar{Q} = (q_{1k} - \bar{q}_1) + (q_{2k} - \bar{q}_2) = \Delta q_{1k} + \Delta q_{2k} \qquad (2\text{-}32)$$

$$(\Delta Q_k)^2 = (\Delta q_{1k})^2 + (\Delta q_{2k})^2 + 2(\Delta q_{1k})(\Delta q_{2k}) \qquad (2\text{-}33)$$

and

$$\sum^n (\Delta Q_k)^2 = \sum^n (\Delta q_{1k})^2 + \sum^n (\Delta q_{2k})^2 + 2\sum^n (\Delta q_{1k})(\Delta q_{2k}) \qquad (2\text{-}34)$$

If there is no correlation between Δq_{1k} and Δq_{2k}, the terms in the cross-product sum $\sum^n (\Delta q_{1k})(\Delta q_{2k})$ will vary in both magnitude and sign and will tend to cancel one another. Consequently, the cross-product sum will remain small even though n is large. On the other hand, the squared terms like $\sum^n (\Delta q_{1k})^2$ will increase with n, since no cancellation due to alteration of sign is possible. Hence, when n is large, the cross-product sum may be neglected in comparison with the squared sums, giving

$$\sum^n (\Delta Q_k)^2 = \sum^n (\Delta q_{1k})^2 + \sum^n (\Delta q_{2k})^2 \qquad (2\text{-}35)$$

The expressions for the variance of Q are easily obtained from (2-35). For example,

$$\begin{aligned} s^2(Q) &= \frac{\sum^n (\Delta Q_k)^2}{n-1} \\ &= \frac{\sum^n (\Delta q_{1k})^2}{n-1} + \frac{\sum^n (\Delta q_{2k})^2}{n-1} \\ &= s^2(q_1) + s^2(q_2) \end{aligned} \qquad (2\text{-}36)$$

† Actually, the best value is given by

$$\bar{Q} = \frac{\sum^n \gamma(q_{1k}, q_{2k}, \ldots, q_{ik})}{n} \qquad (2\text{-}28)$$

However, if the $q_{ik} - \bar{q}_i$ terms are small, (2-28) and (2-27) are essentially equivalent.

The derivation of the general variance equation (2-29) is obtained in an analogous manner if use is made of the relation (2-5) to calculate ΔQ_k. It is important to note that the validity of Eq. (2-29) is not dependent upon any assumption concerning a normal error frequency distribution. All that is required is that no correlation exist between the Δq_{ik} terms, an assumption frequently termed "the assumption of statistical independence." On the other hand, the usual methods of relating variance to probability do involve the assumption of a normal frequency distribution.

Variance of the Sample Mean. An important problem which can be answered by statistical techniques is that of assigning probability limits to deviations of the sample mean from the population mean. Phrased differently, the problem is "By how much can the mean of n measurements be expected to differ from the best value of $\bar{\bar{x}}$, the population mean?" This can be done if the "variance of the mean" can be estimated. The following discussion will clarify the problem.

Suppose that a given measurement were repeated 10 times. Call this set 1, and $n_1 = 10$. From this set of 10 values, the sample mean $\bar{x}_1$ and the sample estimate of the population variance s_1^2 can be calculated. These quantities are estimates of the population mean $\bar{\bar{x}}$ and population variance σ^2. If a second set of 10 measurements of the same quantity were made, the values of $\bar{x}_2$ and s_2^2, calculated from the second set of 10 values, would be expected to differ from the first set and from the population values. Suppose that a large number of sets, each of 10 measurements, were obtained. From this series of sets, a new set comprising the sample means $\bar{x}_1, \bar{x}_2, \ldots, \bar{x}_i$ could be generated. This set of means would exhibit some very important characteristics. If the grand mean $\bar{x}_m$ is calculated,

$$\bar{x}_m = \frac{\sum^{n_i} \bar{x}_i}{n_i} \tag{2-37}$$

(where n_i represents the number of sets), it will be found that $\bar{x}_m$ is a better estimate of the population mean $\bar{\bar{x}}$ than the individual $\bar{x}_i$'s. Further, *on the average*, the deviation of $\bar{x}_i$ from $\bar{x}_m$ will be less than the deviation of a single term x_{ik} from the mean of the ith set $\bar{x}_i$. The sample estimate of the variance of the set of *means*

$$s_m^{\,2} = \frac{\sum^{n_i} (\bar{x}_i - \bar{x}_m)^2}{n_i - 1} = \frac{\sum^{n_i} \bar{x}_i^{\,2} - \left(\sum^{n_i} \bar{x}_i\right)^2 \Big/ n_i}{n_i - 1} \tag{2-38}$$

will be *smaller* than the sample estimate of the population variance s^2. It is also found that the frequency distribution of the sample means

$\bar{x}_i$ about the population mean $\bar{\bar{x}}$ is essentially normal, even though the population frequency distribution is nonnormal! When these findings are treated analytically, it is found that the sample variance of the mean $s_m{}^2$ may be *estimated* from the variance calculated from a *single* set:

$$s_m{}^2 \cong \frac{s^2}{n} \tag{2-39}$$

In Eq. (2-39), s^2 is the sample estimate of the population variance calculated from a single set of measurements, n is the number of measurements in the single set, and $s_m{}^2$ is the estimated variance of the set of means. These quantities have extremely important applications.

2-9. Confidence Limits for Small Samples. Suppose that a single set of measurements containing a finite number of values $x_{i1}, x_{i2}, \ldots, x_{ik}$ has been made and the sample mean $\bar{x}_i$ calculated. It is now desired to associate the magnitude of deviations of $\bar{x}_i$ from the population mean $\bar{\bar{x}}$ with the probability of the occurrence of such deviations. In order to illustrate the concepts involved, first consider the *idealized* case in which the estimate of the variance of the mean $s_m{}^2$ given by Eq. (2-39) can be taken to be the true value of the variance of the *population* of *means*. Then, since the distribution function for the sample means tends to be normal, the distribution function for the sample means is formally the same as Eq. (2-18):

$$\begin{aligned} \mathbf{f}\, d\bar{x}_i &= \frac{\exp\{-\frac{1}{2}[(\bar{x}_i - \bar{\bar{x}})/\sigma_m]^2\}}{\sigma_m \sqrt{2\pi}}\, d\bar{x}_i \\ &\cong \frac{\exp\{-\frac{1}{2}[(\bar{x}_i - \bar{\bar{x}})/s_m]^2\}}{s_m \sqrt{2\pi}}\, d\bar{x}_i \\ &= \frac{\exp\{-\frac{1}{2}[(\bar{x}_i - \bar{\bar{x}})/(s/\sqrt{n})]^2\}}{(s/\sqrt{n})\sqrt{2\pi}}\, d\bar{x}_i \end{aligned} \tag{2-40}$$

Since the distribution function is known, the formal treatment is completely analogous to that given in Sec. 2-7. The calculations given there show that in 95 per cent of the cases, the measured mean $\bar{x}_i$ will differ from $\bar{\bar{x}}$ by less than $\pm 2s_m = \pm 2s/\sqrt{n}$. The range of $\pm 2s/\sqrt{n}$ may be called the "95 per cent confidence limit on $\bar{x}_i$."

It is evident that the above treatment is approximate, since the true variance of the population of means $\sigma_m{}^2$ was not known but estimated. It is known that if the sample set contains at least 20 entries ($n \geqq 20$), the error introduced by the above procedure is not serious. For smaller samples, however, s^2/n is not an adequate estimate of $\sigma_m{}^2$. The solution to this difficulty was first pointed out in *Biometrika*, vol. VI, 1908, by an author (W. S. Gossett) who signed his article "Student."

The dimensionless quantity of particular interest in a confidence-limit analysis is

$$t = \frac{\bar{x}_i - \bar{\bar{x}}}{s_m} \tag{2-41}$$

which is called "Student's t." In Eq. (2-41), $\bar{x}_i$ denotes the sample mean calculated from the ith set, $\bar{\bar{x}}$ is the population mean of the population sampled by the ith set, and s_m is the square root of the estimate of the variance of the mean calculated from the ith set. The dimensionless quantity t is analogous to the dimensionless quantity u discussed earlier. t is more useful since it involves estimates obtainable from a sample of finite size. "Student" (and later Fisher[9] in a more rigorous manner) derived the frequency distribution for t and found it to be of the form

$$\mathbf{f}_t\,dt = C_f \left\{ \left(1 + \frac{t^2}{f}\right)^{(f-1)/2} \right\} dt \tag{2-42}$$

C_f is a function of f only, and f is the "*degrees of freedom*," *defined as the number of values used to calculate the means on which t is based, less the number of means so calculated.* This concept of degrees of freedom will be made clearer in the examples to follow.

t is the difference between the measured sample mean and the true (but generally unknown) population mean divided by the sample estimate of the standard deviation of the population of means. If t is known, the sample data may be used to calculate the true population mean. Ordinarily t is not known. However, the distribution function for t is known, and this permits probability limits to be assigned to t intervals. Thus, suppose the probability that the true value of t lies between -2.35 and 2.35 were desired. Then

$$P = \int_{-2.35}^{2.35} \mathbf{f}_t\,dt$$

The resulting probability value is a function of the size of the sample or, more properly, of the degrees of freedom f. Thus, for the interval $-2.35 \leqq t \leqq 2.35$, for $f = 3$, $P \cong 0.9$; for $f = 7$, $P \cong 0.95$; for $f \to \infty$, $P \cong 0.98$. The interpretation of these results is as follows: For a sample of four entries from which one mean, $\bar{x}$, has been calculated, giving $f = 3$, in 9 chances out of 10, $-2.35s_m \leqq \bar{x} - \bar{\bar{x}} \leqq 2.35s_m$ or

$$\bar{x} - 2.35s_m \leqq \bar{\bar{x}} \leqq \bar{x} + 2.35s_m$$

On the other hand, for a sample of eight entries from which one mean has been calculated, giving $f = 7$, in 95 chances out of 100,

$$\bar{x} - 2.35s_m \leqq \bar{\bar{x}} \leqq \bar{x} + 2.35s_m$$

TABLE 2-1. DISTRIBUTION OF t†

$P = \int_{-t}^{t} \mathbf{f}_t \, dt =$ probability that the true value lies *inside* the limits $-t \rightarrow +t$. The probability that the true value lies *outside* the limits $-t \rightarrow +t$ is $1 - P$.

Degrees of freedom	P = 0.1	0.2	0.3	0.4	0.5	0.6	0.7	0.8	0.9	0.95	0.98	0.99
1	0.158	0.325	0.510	0.727	1.000	1.376	1.963	3.078	6.314	12.706	31.821	63.657
2	0.142	0.289	0.445	0.617	0.816	1.061	1.386	1.886	2.920	4.303	6.965	9.925
3	0.137	0.277	0.424	0.584	0.765	0.978	1.250	1.638	2.353	3.182	4.541	5.841
4	0.134	0.271	0.414	0.569	0.741	0.941	1.190	1.533	2.132	2.776	3.747	4.604
5	0.132	0.267	0.408	0.559	0.727	0.920	1.156	1.476	2.015	2.571	3.365	4.032
6	0.131	0.265	0.404	0.553	0.718	0.906	1.134	1.440	1.943	2.447	3.143	3.707
7	0.130	0.263	0.402	0.549	0.711	0.896	1.119	1.415	1.895	2.365	2.998	3.499
8	0.130	0.262	0.399	0.546	0.706	0.889	1.108	1.397	1.860	2.306	2.896	3.355
9	0.129	0.261	0.398	0.543	0.703	0.883	1.100	1.383	1.833	2.262	2.821	3.250
10	0.129	0.260	0.397	0.542	0.700	0.879	1.093	1.372	1.812	2.228	2.764	3.169
11	0.129	0.260	0.396	0.540	0.697	0.876	1.088	1.363	1.796	2.201	2.718	3.106
12	0.128	0.259	0.395	0.539	0.695	0.873	1.083	1.356	1.782	2.179	2.681	3.055
13	0.128	0.259	0.394	0.538	0.694	0.870	1.079	1.350	1.771	2.160	2.650	3.012
14	0.128	0.258	0.393	0.537	0.692	0.868	1.076	1.345	1.761	2.145	2.624	2.977
15	0.128	0.258	0.393	0.536	0.691	0.866	1.074	1.341	1.753	2.131	2.602	2.947
16	0.128	0.258	0.392	0.535	0.690	0.865	1.071	1.337	1.746	2.120	2.583	2.921
17	0.128	0.257	0.392	0.534	0.689	0.863	1.069	1.333	1.740	2.110	2.567	2.898
18	0.127	0.257	0.392	0.534	0.688	0.862	1.067	1.330	1.734	2.101	2.552	2.878
19	0.127	0.257	0.391	0.533	0.688	0.861	1.066	1.328	1.729	2.093	2.539	2.861
20	0.127	0.257	0.391	0.553	0.687	0.860	1.064	1.325	1.725	2.086	2.528	2.845
21	0.127	0.257	0.391	0.532	0.686	0.859	1.063	1.323	1.721	2.080	2.518	2.831
22	0.127	0.256	0.390	0.532	0.686	0.858	1.061	1.321	1.717	2.074	2.508	2.819
23	0.127	0.256	0.390	0.532	0.685	0.858	1.060	1.319	1.714	2.069	2.500	2.807
24	0.127	0.256	0.390	0.531	0.685	0.857	1.059	1.318	1.711	2.064	2.492	2.797
25	0.127	0.256	0.390	0.531	0.684	0.856	1.058	1.316	1.708	2.060	2.485	2.787
26	0.127	0.256	0.390	0.531	0.684	0.856	1.058	1.315	1.706	2.056	2.479	2.779
27	0.127	0.256	0.389	0.531	0.684	0.855	1.057	1.314	1.703	2.052	2.473	2.771
28	0.127	0.256	0.389	0.530	0.683	0.855	1.056	1.313	1.701	2.048	2.467	2.763
29	0.127	0.256	0.389	0.530	0.683	0.854	1.055	1.311	1.699	2.045	2.462	2.756
30	0.127	0.256	0.389	0.530	0.683	0.854	1.055	1.310	1.697	2.042	2.457	2.750
∞	0.12566	0.25335	0.38532	0.52440	0.67449	0.84162	1.03643	1.28155	1.64485	1.95996	2.32634	2.57582

† Reprinted from Table IV of R. A. Fisher, "Statistical Methods for Research Workers," 10th ed., Oliver & Boyd, Ltd., Edinburgh and London, 1946, by permission of the author and publishers.

Probability calculations of this kind have been carried out over a wide range of conditions, and the results are tabulated in Table 2-1. As an example of the use of Table 2.1, suppose that five measurements have been made. From these five measurements, $\bar{x}$ is found to be 56.6, and the sample estimate of the standard deviation of the population of means s_m is found to be 0.45. It is desired to determine the limits within which the true population mean $\bar{\bar{x}}$ will fall 95 times out of 100, termed the "95 per cent confidence limits on $\bar{\bar{x}}$." With a sample size of five and only one mean ($\bar{x}$) calculated, $f = 5 - 1 = 4$. In Table 2-1, the entry corresponding to 4 degrees of freedom and $P = 0.95$ is 2.776. This means that 95 per cent of the time, the true value of t lies between -2.776 and 2.776. Since t is defined to be $(\bar{x} - \bar{\bar{x}})/s_m$ and for this illustration $s_m = 0.45$, it follows that 95 per cent of the time $\bar{x} - \bar{\bar{x}}$ lies between -1.3 and 1.3. With $\bar{x} = 56.6$, 95 per cent of the time $\bar{\bar{x}}$ lies between 55.3 and 57.9, or $55.3 \leqq \bar{\bar{x}} \leqq 57.9$. An alternative interpretation of this result is that only 5 times in 100 will the analyst be in error if he assigns the limits $\bar{\bar{x}} = 56.6 \pm 1.3$ to his result. Clearly, the use of 95 per cent confidence limits here is arbitrary. In a specific case the analyst may wish to assign narrower limits of which he is less certain or broader limits to which he can assign a higher degree of probability.

In subsequent sections further important applications of the t distribution will be discussed.

Example 2-3. The analytical results presented in Table 2-2 represent "check" analyses for a particular component carried out on the same sample by two independent methods. Procedure 1 is a gravimetric method; procedure 2 is a volumetric method. It is desired to use these data to obtain the following information:

1. The confidence limits to be assigned to the results of procedures 1 and 2
2. The significance of the difference between the mean values of the results of procedures 1 and 2
3. The "best value" to be assigned to the sample analysis
4. The confidence limits of the best value

The investigational procedures are carried out below:

1. Confidence limits. First consider the results of procedure 1 only. The sample mean $\bar{x}_1$ is easily found to be 56.6. The confidence limits are specified by the "t table" (Table 2-1) after the probability limits are fixed, the degrees of freedom f ascertained, and the variance of the mean $s_m{}^2$ estimated. The *external* probability limit on t will be arbitrarily set at 0.05, corresponding to $P = 0.95$, or 95 per cent confidence limits. In calculating t, five measured values are used and one mean, $\bar{x}$, calculated. Hence, $f = 5 - 1 = 4$. From Table 2-1, for $f = 4$, values of t lying *outside* ± 2.8 only are 0.05 probable. Consequently, the 95 per cent confidence limits on t are $-2.8 \leqq t \leqq 2.8$. The calculation of the 95 per cent confidence limits of the sample mean $\bar{x}_1$ requires an estimate of $s^2_{m,1}$. By virtue of Eq. (2-39),

$$s^2_{m,1} = \frac{s^2(x_1)}{n_1} \tag{2-43}$$

By definition,

$$s^2(x_1) = \frac{\sum^{n_1} (x_1 - \bar{x}_1)^2}{n_1 - 1}$$

and

$$s_{m,1}^2 = \frac{\sum^{n_1} (x_1 - \bar{x}_1)^2}{n_1(n_1 - 1)} \tag{2-44}$$

Since

$$\sum^{n} (x - \bar{x})^2 = \sum^{n} (x^2) - \frac{\left(\sum^{n} x\right)^2}{n}$$

the "sum of squares" $\sum^{n_1} (x_1 - \bar{x}_1)^2$ is readily found to be 4.1 and

$$s_{m,1}^2 = \frac{4.1}{5 \cdot 4} = 0.20$$

and

$$s_{m,1} = \sqrt{0.20} = 0.45$$

The 95 per cent confidence limits on $\bar{x}_1$ are then $\bar{x}_1 - \bar{\bar{x}}_1 = \pm 2.8 \cdot 0.45 = \pm 1.3$, or $55.3 \leqq \bar{\bar{x}}_1 \leqq 57.9$. There is only 1 chance in 20 that the sample mean $\bar{x}_1$ differs from the true population mean $\bar{\bar{x}}_1$ by more than ± 1.3.

The confidence-limit calculations for procedure 2 are carried out in an analogous fashion. Table 2-2 contains the results of such calculations.

TABLE 2-2. ANALYTICAL RESULTS

	Procedure 1	Procedure 2
x_{i1}	55.3	52.6
x_{i2}	56.9	54.3
x_{i3}	55.8	58.0
x_{i4}	57.3	52.7
x_{i5}	57.7	60.0
$\bar{x}_i$	56.6	55.5
$\sum^{n_k} (x_{ik} - \bar{x}_i)^2$	4.1	43.9
$s_i^2(x_i)$	1.0	11
s_m^2	0.2	2.2
95% confidence limits	$55.3 \leqq \bar{\bar{x}}_1 \leqq 57.9$	$51.3 \leqq \bar{\bar{x}}_2 \leqq 59.9$

2. The significance of the difference between the mean values of the results of procedures 1 and 2 can also be analyzed by statistical procedures. The mean value obtained by procedure 1 is $\bar{x}_1 = 56.6$, with 95 per cent limits of $55.3 \leqq \bar{\bar{x}}_1 \leqq 57.9$. For procedure 2, $\bar{x}_2 = 55.5$, and the 95 per cent limits were found to be $51.3 \leqq \bar{\bar{x}}_2 \leqq 59.9$. The sample means are different, which might be taken to indicate a systematic difference or *bias* between the two methods of analysis. On the other hand, the 95 per cent confidence-limit analysis shows that the mean of sample 2, $\bar{x}_2$, is included within the confidence limits of sample 1, and vice versa. Consequently, it may be concluded that the difference between the two means has no statistical signifi-

cance, assuming, of course, that the 95 per cent confidence limit arbitrarily imposed is sufficiently reliable for the comparison at hand.

3. Since the difference between $\bar{x}_1$ and $\bar{x}_2$ is not statistically significant, the "best value" to be assigned to the sample analysis is a weighted combination of the two mean values. If, however, the difference between the means had been significant, it would have been concluded that one or both of the procedures were affected by non-random factors (errors of method or bias). In this case, without further information, the best value could not be estimated.

It can be shown that the best value obtainable from a series of sets of measurements *exhibiting statistically equivalent means* is given by

$$\bar{x}_{b.v.} = \frac{\sum^{n_i} W_i \bar{x}_i}{\sum^{n_i} W_i} \tag{2-45}$$

where

$$W_i = \frac{(n_k)_i}{\sigma_i^2} \tag{2-46}$$

and $(n_k)_i$ denotes the number of measurements in the ith set. The individual means are weighted in inverse proportion to their respective variance. However, the population variance is seldom known, and the following equation is used as an *approximation:*

$$W_i \cong \frac{(n_k)_i}{s_i^2} = \frac{1}{s_{m,1}^2} \tag{2-47}$$

Equation (2-47) applies only if the variances of the data sets are truly different. Since the sample variance is subject to sampling error, it is possible to have $s_{m,1}^2 \neq s_{m,2}^2$ when σ_1^2 is actually equal to σ_2^2. A test of this possibility, known as the "L_1 test," is given in Sec. 2-10. In the present example, the L_1 test shows that there is somewhat less than 1 chance in 20 that $\sigma_1^2 = \sigma_2^2$. This is a borderline case, but in this example it will be assumed that $\sigma_1^2 \neq \sigma_2^2$. Then,

$$\bar{x}_{b.v.} = \frac{56.6/0.2 + 55.5/2.2}{1/0.2 + 1/2.2} = 56.6$$

4. In order to determine the confidence limits of the best value, the variance of the best value is needed. This is given by the equation

$$\sigma_{m,b.v.}^2 = \frac{1}{\Sigma[(n_k)_i/\sigma_i^2]} \cong \frac{1}{\Sigma(1/s_{m,i}^2)} = s_{m,b.v.}^2 \tag{2-48}$$

Hence,

$$s_{m,b.v.}^2 = \frac{1}{1/0.2 + 1/2.2} = 0.18$$

The 95 per cent confidence limits are found from the t table (Table 2-1) with $f = 10 - 2 = 8$. The 8 degrees of freedom result from the 10 data points and two means ($\bar{x}_1$ and $\bar{x}_2$) used. The t limits are ± 2.3. Consequently, the 95 per cent confidence limits on the best value are $\pm 2.3\sqrt{0.18} = \pm 1.0$, or $55.6 \leqq \bar{\bar{x}} \leqq 57.6$.

2-10. Analysis of Variance. In Sec. 2-9 a statistical analysis of two sets of data was made. In the example given, it was easy to show that the two sample means were not statistically different. However, in other cases an examination of the respective confidence limits does not always lead to definite conclusions. It is possible to obtain more reliable estimates by the application of other techniques. Furthermore, additional information may be obtained and the procedures simultaneously applied to several sets of data. These methods are discussed by means of examples.

Example 2-4. Table 2-3 presents the results of three independent laboratory analyses of the same sample, carried out by methods A, B, and C. It is desired to find out whether or not the difference between the three mean values $\bar{x}_A$, $\bar{x}_B$, and $\bar{x}_C$ can be explained by random errors. The respective populations sampled by set A, set B, and set C fall into one of four categories:

1. They represent identical populations, the apparent differences resulting from sampling error.
2. They exhibit identical variances but different means.
3. They exhibit identical means but different variances.
4. They are totally different.

TABLE 2-3. ANALYTICAL RESULTS

	A	B	C
x_{i1}	55.3	53.6	50.5
x_{i2}	56.9	55.3	53.0
x_{i3}	55.8	57.0	53.8
x_{i4}	57.3	53.7	53.0
x_{i5}	57.7	58.0	55.3
$\bar{x}_i$	56.6	55.5	52.9
$\Sigma(x_i - \bar{x}_i)^2$	4.1	15.4	8.5
$s_i^2(x_i)$	1.0	3.9	2.1
$s^2_{m,i}$	0.2	0.8	0.4
95% confidence limits	$55.3 \lesseqgtr \bar{\bar{x}}_A \lesseqgtr 57.9$	$53.1 \lesseqgtr \bar{\bar{x}}_B \lesseqgtr 57.9$	$51.1 \lesseqgtr \bar{\bar{x}}_C \lesseqgtr 54.7$

The confidence-limit calculations shown in Table 2-3 fail to classify the populations adequately. The following procedures will often supply more definite information.

1. Each set of data is examined for deviations from a normal frequency distribution. Suitable tests (called the "a" and "b_1" tests) have been developed and are described in Freeman.[11] The chief difficulty with these tests is the relatively large sample size required. A sample containing only five entries would not give conclusive results. Fortunately, the remaining tests are not extremely sensitive to the form of the frequency-distribution curve, and a reasonable approximation to a normal distribution is satisfactory.

2. A calculation is carried out to determine the probability that the samples represent normal populations exhibiting the same population variance σ^2 but without regard to the population means. This is the L_1 test devised by Neyman and Pearson.[15] The quantity

$$L_1 = \frac{n_i(s_1{}^2 s_2{}^2 \cdots s_i{}^2)^{1/n_i}}{\Sigma s_i{}^2} \tag{2.49}$$

is calculated. n_i represents the number of sample sets. Each set must contain the same number of entries, $(n_k)_i$. Table 2-4 (prepared by Mahalanobis[13]) tabulates the 5 per cent and 1 per cent confidence values of L_1. In the present example,

$$L_1 = \frac{3(1 \cdot 3.9 \cdot 2.1)^{1/3}}{1 + 3.9 + 2.1} = 0.86$$

Referring to Table 2-4 for n_i, the number of sets, equal to 3 and for $(n_k)_i$, the number of measurements in one set, equal to 5, it is found that a value of L_1 of 0.5755 is associated with a 5 per cent probability. The calculated value of L_1, 0.86, shows that the probability that the population variances are the same is greater than 5 per cent. It is reasonable to conclude that the populations represented by the samples have the same variance.

3. If the L_1 test is affirmative, the probability that the population means are equal may be ascertained. This is the "F" test and is discussed shortly. If, however, the L_1 test had been negative, no further information could be obtained. Conclusions regarding the population means then must be based entirely upon the individual-sample confidence-limit calculations.

The F Test. The basis for the F test lies in the independent estimates of variance which the data provide. Consider a single sample, say set A. The variance s_A^2 of that set is a measure of the internal scatter of the measurements. A similar statement applies to s_B^2, etc. On the other hand, these variances are subject to sampling error and are not necessarily equal to the population variance. If the L_1 test has shown that the population variances are probably the same, a better estimate of the *internal* scatter of the measurements may be obtained by pooling the individual-sample variances. The better estimate of the within-sample or error variance is given by

$$s_e^2 = \frac{\sum^{n_1} (x_{1k} - \bar{x}_1)^2 + \sum^{n_2} (x_{2k} - \bar{x}_2)^2 + \cdots}{n_1 + n_2 + \cdots - n_i}$$

$$= \frac{\sum^{n_i} \sum^{n_k} (x_{ik} - \bar{x}_i)^2}{\sum^{n_i} (n_k)_i - n_i} \tag{2-50a}$$

where n_i denotes the number of *sets* and $(n_k)_i$ denotes the number of measurements in the ith set. The degrees of freedom associated with s_e^2 are

$$f_e = \sum^{n_i} (n_k)_i - n_i \tag{2-50b}$$

that is, the total number of measurements in all sets less the number of sets. The term $\sum^{n_i} \sum^{n_k} (x_{ik} - \bar{x}_i)^2$ is called the "within-groups sum of squares" and is most easily calculated by the identity

$$\sum^{n_i} \sum^{n_k} (x_{ik} - \bar{x}_i)^2 = \sum^{n_i} \sum^{n_k} x^2_{ik} - \sum^{n_i} \frac{\left(\sum^{n_k} x_{ik}\right)^2}{(n_k)_i} \tag{2-50c}$$

where $$\sum^{n_i} \frac{\left(\sum^{n_k} x_{ik}\right)^2}{(n_k)_i} = \frac{\left(\sum^{n_1} x_{1k}\right)^2}{n_1} + \frac{\left(\sum^{n_2} x_{2k}\right)^2}{n_2} + \cdots$$

TABLE 2-4. DISTRIBUTION OF L_1†

$(n_k)_i$ = number of measurements in one set

n_i, number of data sets	2	3	4	5	10	15	20	30	40	50
5 Per Cent Values										
2	0.0723	0.3107	0.4782	0.5842	0.7985	0.8673	0.9014	0.9349	0.9512	0.9612
3	0.0704	0.3040	0.4696	0.5755	0.7925	0.8632	0.8980	0.9325	0.9495	0.9598
4	0.0753	0.3152	0.4800	0.5849	0.7970	0.8662	0.9005	0.9341	0.9506	0.9608
5	0.0825	0.3278	0.4915	0.5950	0.8025	0.8699	0.9032	0.9358	0.9519	0.9618
10	0.1135	0.3738	0.5341	0.6318	0.8228	0.8813	0.9135	0.9427	0.9573	0.9659
20	0.1472	0.4191	0.5697	0.6658	0.8417	0.8961	0.9227	0.9498	0.9619	0.9694
25	0.1578	0.4320	0.5841	0.6747	0.8453	0.8989	0.9250	0.9508	0.9630	0.9697
50	0.1878	0.4723	0.6278	0.7137	0.8688	0.9150	0.9365	0.9584	0.9672	0.9730
1 Per Cent Values										
2	0.0126	0.1361	0.2818	0.3855	0.6782	0.7821	0.8359	0.8902	0.9171	0.9340
3	0.0169	0.1615	0.3138	0.4291	0.6992	0.7976	0.8476	0.8981	0.9234	0.9388
4	0.0233	0.1876	0.3414	0.4594	0.7194	0.8118	0.8586	0.9056	0.9291	0.9434
5	0.0285	0.2101	0.3703	0.4838	0.7350	0.8228	0.8645	0.9113	0.9334	0.9470
10	0.0624	0.2838	0.4483	0.5556	0.7791	0.8537	0.8908	0.9273	0.9458	0.9565
20	0.0997	0.3524	0.5138	0.6142	0.8131	0.8768	0.9082	0.9404	0.9546	0.9633
25	0.1129	0.3718	0.5304	0.6287	0.8201	0.8817	0.9121	0.9415	0.9565	0.9640
50	0.1517	0.4328	0.5915	0.6672	0.8576	0.9077	0.9307	0.9548	0.9643	0.9697

† Reproduced from *Sankhyā*, The Indian Journal of Statistics, vol. 1, pt. 1, June, 1933, by permission of the author, P. C. Mahalanobis and the publisher.

An independent estimate of the population variance may be obtained from the sample means. The variance of the population of *means* is

$$s^2_{m,p} = \frac{\sum^{n_i} (\bar{x}_i - \bar{x}_p)^2}{n_i - 1} \tag{2-51}$$

where

$$\bar{x}_p = \frac{\sum^{n_i} (n_k)_i \bar{x}_i}{\sum^{n_i} (n_k)_i} \tag{2-52}$$

However, the variance of the population of means is not in itself equal to the population variance. If all samples were of the same size, the second estimate of the population variance would be

$$s_p{}^2 = n_k s^2_{m,p} \tag{2-53}$$

If the samples are of different sizes [if $(n_k)_i \neq (n_k)_2$], a pooled result is to be used as an average sample size, n_0:

$$n_0 = \frac{1}{n_i - 1} \left\{ \sum^{n_i} (n_k)_i - \frac{\sum^{n_i} [(n_k)_i]^2}{\sum^{n_i} (n_k)_i} \right\} \tag{2-54}$$

and

$$s_p{}^2 = n_0 s^2_{m,p} \tag{2-55}$$

When $s_p{}^2$ is estimated by means of Eq. (2-53) or (2-55), $n_i - 1$ degrees of freedom are involved.

Now if the random factors which give rise to the within-sample or error variance $s_e{}^2$ are the only factors causing the differences between the sample means, the two independent variance estimates $s_e{}^2$ and $s_p{}^2$ should, except for sampling error, be equal. If, on the other hand, there are nonrandom factors which cause differences between the individual sample means, $s_p{}^2$ should be *greater* than $s_e{}^2$. The ratio of the two estimates

$$F = \frac{s_p{}^2}{s_e{}^2} \tag{2-56}$$

is then a measure of whether or not random experimental error can account for the observed differences between sample means. Values of F which are significantly greater than unity indicate that the differences between means are due to nonrandom factors.

Table 2-5 relates the F values to the probability that the observed value could arise from sampling error. When the F test is applied to the data of this example, the following results are obtained:

$$s_e{}^2 = \frac{4.1 + 15.4 + 8.5}{5 + 5 + 5 - 3} = \frac{28}{12} = 2.3$$

$$\bar{x}_p = 55.0$$

$$s^2_{m,p} = \frac{(56.6 - 55.0)^2 + (55.5 - 55.0)^2 + (52.9 - 55.0)^2}{3 - 1}$$

$$= \frac{7.2}{2} = 3.6$$

$$s_p{}^2 = 5 \cdot 3.6 = 18.0$$

$$F = \frac{s_p{}^2}{s_e{}^2} = \frac{18.0}{2.3} = 7.8$$

TABLE 2-5. DISTRIBUTION OF F†

5% (lightface) and 1% (**boldface**) values

Degrees of freedom for lesser mean square	Degrees of freedom for greater mean square																							
	1	2	3	4	5	6	7	8	9	10	11	12	14	16	20	24	30	40	50	75	100	200	500	∞
1	161	200	216	225	230	234	237	239	241	242	243	244	245	246	248	249	250	251	252	253	253	254	254	254
	4,052	**4,999**	**5,403**	**5,625**	**5,764**	**5,859**	**5,928**	**5,981**	**6,022**	**6,056**	**6,082**	**6,106**	**6,142**	**6,169**	**6,208**	**6,234**	**6,258**	**6,286**	**6,302**	**6,323**	**6,334**	**6,352**	**6,361**	**6,366**
2	18.51	19.00	19.16	19.25	19.30	19.33	19.36	19.37	19.38	19.39	19.40	19.41	19.42	19.43	19.44	19.45	19.46	19.47	19.47	19.48	19.49	19.49	19.50	19.50
	98.49	**99.00**	**99.17**	**99.25**	**99.30**	**99.33**	**99.34**	**99.36**	**99.38**	**99.40**	**99.41**	**99.42**	**99.43**	**99.44**	**99.45**	**99.46**	**99.47**	**99.48**	**99.48**	**99.49**	**99.49**	**99.49**	**99.50**	**99.50**
3	10.13	9.55	9.28	9.12	9.01	8.94	8.88	8.84	8.81	8.78	8.76	8.74	8.71	8.69	8.66	8.64	8.62	8.60	8.58	8.57	8.56	8.54	8.54	8.53
	34.12	**30.82**	**29.46**	**28.71**	**28.24**	**27.91**	**27.67**	**27.49**	**27.34**	**27.23**	**27.13**	**27.05**	**26.92**	**26.83**	**26.69**	**26.60**	**26.50**	**26.41**	**26.35**	**26.27**	**26.23**	**26.18**	**26.14**	**26.12**
4	7.71	6.94	6.59	6.39	6.26	6.16	6.09	6.04	6.00	5.96	5.93	5.91	5.87	5.84	5.80	5.77	5.74	5.71	5.70	5.68	5.66	5.65	5.64	5.63
	21.20	**18.00**	**16.69**	**15.98**	**15.52**	**15.21**	**14.98**	**14.80**	**14.66**	**14.54**	**14.45**	**14.37**	**14.24**	**14.15**	**14.02**	**13.93**	**13.83**	**13.74**	**13.69**	**13.61**	**13.57**	**13.52**	**13.48**	**13.46**
5	6.61	5.79	5.41	5.19	5.05	4.95	4.88	4.82	4.78	4.74	4.70	4.68	4.64	4.60	4.56	4.53	4.50	4.46	4.44	4.42	4.40	4.38	4.37	4.36
	16.26	**13.27**	**12.06**	**11.39**	**10.97**	**10.67**	**10.45**	**10.27**	**10.15**	**10.05**	**9.96**	**9.89**	**9.77**	**9.68**	**9.55**	**9.47**	**9.38**	**9.29**	**9.24**	**9.17**	**9.13**	**9.07**	**9.04**	**9.02**
6	5.99	5.14	4.76	4.53	4.39	4.28	4.21	4.15	4.10	4.06	4.03	4.00	3.96	3.92	3.87	3.84	3.81	3.77	3.75	3.72	3.71	3.69	3.68	3.67
	13.74	**10.92**	**9.78**	**9.15**	**8.75**	**8.47**	**8.26**	**8.10**	**7.98**	**7.87**	**7.79**	**7.72**	**7.60**	**7.52**	**7.39**	**7.31**	**7.23**	**7.14**	**7.09**	**7.02**	**6.99**	**6.94**	**6.90**	**6.88**
7	5.59	4.74	4.35	4.12	3.97	3.87	3.79	3.73	3.68	3.63	3.60	3.57	3.52	3.49	3.44	3.41	3.38	3.34	3.32	3.29	3.28	3.25	3.24	3.23
	12.25	**9.55**	**8.45**	**7.85**	**7.46**	**7.19**	**7.00**	**6.84**	**6.71**	**6.62**	**6.54**	**6.47**	**6.35**	**6.27**	**6.15**	**6.07**	**5.98**	**5.90**	**5.85**	**5.78**	**5.75**	**5.70**	**5.67**	**5.65**
8	5.32	4.46	4.07	3.84	3.69	3.58	3.50	3.44	3.39	3.34	3.31	3.28	3.23	3.20	3.15	3.12	3.08	3.05	3.03	3.00	2.98	2.96	2.94	2.93
	11.26	**8.65**	**7.59**	**7.01**	**6.63**	**6.37**	**6.19**	**6.03**	**5.91**	**5.82**	**5.74**	**5.67**	**5.56**	**5.48**	**5.36**	**5.28**	**5.20**	**5.11**	**5.06**	**5.00**	**4.96**	**4.91**	**4.88**	**4.86**
9	5.12	4.26	3.86	3.63	3.48	3.37	3.29	3.23	3.18	3.13	3.10	3.07	3.02	2.98	2.93	2.90	2.86	2.82	2.80	2.77	2.76	2.73	2.72	2.71
	10.56	**8.02**	**6.99**	**6.42**	**6.06**	**5.80**	**5.62**	**5.47**	**5.35**	**5.26**	**5.18**	**5.11**	**5.00**	**4.92**	**4.80**	**4.73**	**4.64**	**4.56**	**4.51**	**4.45**	**4.41**	**4.36**	**4.33**	**4.31**
10	4.96	4.10	3.71	3.48	3.33	3.22	3.14	3.07	3.02	2.97	2.94	2.91	2.86	2.82	2.77	2.74	2.70	2.67	2.64	2.61	2.59	2.56	2.55	2.54
	10.04	**7.56**	**6.55**	**5.99**	**5.64**	**5.39**	**5.21**	**5.06**	**4.95**	**4.85**	**4.78**	**4.71**	**4.60**	**4.52**	**4.41**	**4.33**	**4.25**	**4.17**	**4.12**	**4.05**	**4.01**	**3.96**	**3.93**	**3.91**
11	4.84	3.98	3.59	3.36	3.20	3.09	3.01	2.95	2.90	2.86	2.82	2.79	2.74	2.70	2.65	2.61	2.57	2.53	2.50	2.47	2.45	2.42	2.41	2.40
	9.65	**7.20**	**6.22**	**5.67**	**5.32**	**5.07**	**4.88**	**4.74**	**4.63**	**4.54**	**4.46**	**4.40**	**4.29**	**4.21**	**4.10**	**4.02**	**3.94**	**3.86**	**3.80**	**3.74**	**3.70**	**3.66**	**3.62**	**3.60**
12	4.75	3.88	3.49	3.26	3.11	3.00	2.92	2.85	2.80	2.76	2.72	2.69	2.64	2.60	2.54	2.50	2.46	2.42	2.40	2.36	2.35	2.32	2.31	2.30
	9.33	**6.93**	**5.95**	**5.41**	**5.06**	**4.82**	**4.65**	**4.50**	**4.39**	**4.30**	**4.22**	**4.16**	**4.05**	**3.98**	**3.86**	**3.78**	**3.70**	**3.61**	**3.56**	**3.49**	**3.46**	**3.41**	**3.38**	**3.36**

13	4.67 **9.07**	3.80 **6.70**	3.41 **5.74**	3.18 **5.20**	3.02 **4.86**	2.92 **4.62**	2.84 **4.44**	2.77 **4.30**	2.72 **4.19**	2.67 **4.10**	2.63 **4.02**	2.60 **3.96**	2.55 **3.85**	2.51 **3.78**	2.46 **3.67**	2.42 **3.59**	2.38 **3.51**	2.34 **3.42**	2.32 **3.37**	2.28 **3.30**	2.26 **3.27**	2.24 **3.21**	2.22 **3.18**	2.21 **3.16**
14	4.60 **8.86**	3.74 **6.51**	3.34 **5.56**	3.11 **5.03**	2.96 **4.69**	2.85 **4.46**	2.77 **4.28**	2.70 **4.14**	2.65 **4.03**	2.60 **3.94**	2.56 **3.86**	2.53 **3.80**	2.48 **3.70**	2.44 **3.62**	2.39 **3.51**	2.35 **3.43**	2.31 **3.34**	2.27 **3.26**	2.24 **3.21**	2.21 **3.14**	2.19 **3.11**	2.16 **3.06**	2.14 **3.02**	2.13 **3.00**
15	4.54 **8.68**	3.68 **6.36**	3.29 **5.42**	3.06 **4.89**	2.90 **4.56**	2.79 **4.32**	2.70 **4.14**	2.64 **4.00**	2.59 **3.89**	2.55 **3.80**	2.51 **3.73**	2.48 **3.67**	2.43 **3.56**	2.39 **3.48**	2.33 **3.36**	2.29 **3.29**	2.25 **3.20**	2.21 **3.12**	2.18 **3.07**	2.15 **3.00**	2.12 **2.97**	2.10 **2.92**	2.08 **2.89**	2.07 **2.87**
16	4.49 **8.53**	3.63 **6.23**	3.24 **5.29**	3.01 **4.77**	2.85 **4.44**	2.74 **4.20**	2.66 **4.03**	2.59 **3.89**	2.54 **3.78**	2.49 **3.69**	2.45 **3.61**	2.42 **3.55**	2.37 **3.45**	2.33 **3.37**	2.28 **3.25**	2.24 **3.18**	2.20 **3.10**	2.16 **3.01**	2.13 **2.96**	2.09 **2.89**	2.07 **2.86**	2.04 **2.80**	2.02 **2.77**	2.01 **2.75**
17	4.45 **8.40**	3.59 **6.11**	3.20 **5.18**	2.96 **4.67**	2.81 **4.34**	2.70 **4.10**	2.62 **3.93**	2.55 **3.79**	2.50 **3.68**	2.45 **3.59**	2.41 **3.52**	2.38 **3.45**	2.33 **3.35**	2.29 **3.27**	2.23 **3.16**	2.19 **3.08**	2.15 **3.00**	2.11 **2.92**	2.08 **2.86**	2.04 **2.79**	2.02 **2.76**	1.99 **2.70**	1.97 **2.67**	1.96 **2.65**
18	4.41 **8.28**	3.55 **6.01**	3.16 **5.09**	2.93 **4.58**	2.77 **4.25**	2.66 **4.01**	2.58 **3.85**	2.51 **3.71**	2.46 **3.60**	2.41 **3.51**	2.37 **3.44**	2.34 **3.37**	2.29 **3.27**	2.25 **3.19**	2.19 **3.07**	2.15 **3.00**	2.11 **2.91**	2.07 **2.83**	2.04 **2.78**	2.00 **2.71**	1.98 **2.68**	1.95 **2.62**	1.93 **2.59**	1.92 **2.57**
19	4.38 **8.18**	3.52 **5.93**	3.13 **5.01**	2.90 **4.50**	2.74 **4.17**	2.63 **3.94**	2.55 **3.77**	2.48 **3.63**	2.43 **3.52**	2.38 **3.43**	2.34 **3.36**	2.31 **3.30**	2.26 **3.19**	2.21 **3.12**	2.15 **3.00**	2.11 **2.92**	2.07 **2.84**	2.02 **2.76**	2.00 **2.70**	1.96 **2.63**	1.94 **2.60**	1.91 **2.54**	1.90 **2.51**	1.88 **2.49**
20	4.35 **8.10**	3.49 **5.85**	3.10 **4.94**	2.87 **4.43**	2.71 **4.10**	2.60 **3.87**	2.52 **3.71**	2.45 **3.56**	2.40 **3.45**	2.35 **3.37**	2.31 **3.30**	2.28 **3.23**	2.23 **3.13**	2.18 **3.05**	2.12 **2.94**	2.08 **2.86**	2.04 **2.77**	1.99 **2.69**	1.96 **2.63**	1.92 **2.56**	1.90 **2.53**	1.87 **2.47**	1.85 **2.44**	1.84 **2.42**
21	4.32 **8.02**	3.47 **5.78**	3.07 **4.87**	2.84 **4.37**	2.68 **4.04**	2.57 **3.81**	2.49 **3.65**	2.42 **3.51**	2.37 **3.40**	2.32 **3.31**	2.28 **3.24**	2.25 **3.17**	2.20 **3.07**	2.15 **2.99**	2.09 **2.88**	2.05 **2.80**	2.00 **2.72**	1.96 **2.63**	1.93 **2.58**	1.89 **2.51**	1.87 **2.47**	1.84 **2.42**	1.82 **2.38**	1.81 **2.36**
22	4.30 **7.94**	3.44 **5.72**	3.05 **4.82**	2.82 **4.31**	2.66 **3.99**	2.55 **3.76**	2.47 **3.59**	2.40 **3.45**	2.35 **3.35**	2.30 **3.26**	2.26 **3.18**	2.23 **3.12**	2.18 **3.02**	2.13 **2.94**	2.07 **2.83**	2.03 **2.75**	1.98 **2.67**	1.93 **2.58**	1.91 **2.53**	1.87 **2.46**	1.84 **2.42**	1.81 **2.37**	1.80 **2.33**	1.78 **2.31**
23	4.28 **7.88**	3.42 **5.66**	3.03 **4.76**	2.80 **4.26**	2.64 **3.94**	2.53 **3.71**	2.45 **3.54**	2.38 **3.41**	2.32 **3.30**	2.28 **3.21**	2.24 **3.14**	2.20 **3.07**	2.14 **2.97**	2.10 **2.89**	2.04 **2.78**	2.00 **2.70**	1.96 **2.62**	1.91 **2.53**	1.88 **2.48**	1.84 **2.41**	1.82 **2.37**	1.79 **2.32**	1.77 **2.28**	1.76 **2.26**
24	4.26 **7.82**	3.40 **5.61**	3.01 **4.72**	2.78 **4.22**	2.62 **3.90**	2.51 **3.67**	2.43 **3.50**	2.36 **3.36**	2.30 **3.25**	2.26 **3.17**	2.22 **3.09**	2.18 **3.03**	2.13 **2.93**	2.09 **2.85**	2.02 **2.74**	1.98 **2.66**	1.94 **2.58**	1.89 **2.49**	1.86 **2.44**	1.82 **2.36**	1.80 **2.33**	1.76 **2.27**	1.74 **2.23**	1.73 **2.21**
25	4.24 **7.77**	3.38 **5.57**	2.99 **4.68**	2.76 **4.18**	2.60 **3.86**	2.49 **3.63**	2.41 **3.46**	2.34 **3.32**	2.28 **3.21**	2.24 **3.13**	2.20 **3.05**	2.16 **2.99**	2.11 **2.89**	2.06 **2.81**	2.00 **2.70**	1.96 **2.62**	1.92 **2.54**	1.87 **2.45**	1.84 **2.40**	1.80 **2.32**	1.77 **2.29**	1.74 **2.23**	1.72 **2.19**	1.71 **2.17**
26	4.22 **7.72**	3.37 **5.53**	2.98 **4.64**	2.74 **4.14**	2.59 **3.82**	2.47 **3.59**	2.39 **3.42**	2.32 **3.29**	2.27 **3.17**	2.22 **3.09**	2.18 **3.02**	2.15 **2.96**	2.10 **2.86**	2.05 **2.77**	1.99 **2.66**	1.95 **2.58**	1.90 **2.50**	1.85 **2.41**	1.82 **2.36**	1.78 **2.28**	1.76 **2.25**	1.72 **2.19**	1.70 **2.15**	1.69 **2.13**
27	4.21 **7.68**	3.35 **5.49**	2.96 **4.60**	2.73 **4.11**	2.57 **3.79**	2.46 **3.56**	2.37 **3.39**	2.30 **3.26**	2.25 **3.14**	2.20 **3.06**	2.16 **2.98**	2.13 **2.93**	2.08 **2.83**	2.03 **2.74**	1.97 **2.63**	1.93 **2.55**	1.88 **2.47**	1.84 **2.38**	1.80 **2.33**	1.76 **2.25**	1.74 **2.21**	1.71 **2.16**	1.68 **2.12**	1.67 **2.10**

Table 2-5. Distribution of F† (*Continued*)
5% (lightface) and 1% (**boldface**) values

Degrees of freedom for lesser mean square	Degrees of freedom for greater mean square																							
	1	2	3	4	5	6	7	8	9	10	11	12	14	16	20	24	30	40	50	75	100	200	500	∞
28	4.20	3.34	2.95	2.71	2.56	2.44	2.36	2.29	2.24	2.19	2.15	2.12	2.06	2.02	1.96	1.91	1.87	1.81	1.78	1.75	1.72	1.69	1.67	1.65
	7.64	**5.45**	**4.57**	**4.07**	**3.76**	**3.53**	**3.36**	**3.23**	**3.11**	**3.03**	**2.95**	**2.90**	**2.80**	**2.71**	**2.60**	**2.52**	**2.44**	**2.35**	**2.30**	**2.22**	**2.18**	**2.13**	**2.09**	**2.06**
29	4.18	3.33	2.93	2.70	2.54	2.43	2.35	2.28	2.22	2.18	2.14	2.10	2.05	2.00	1.94	1.90	1.85	1.80	1.77	1.73	1.71	1.68	1.65	1.64
	7.60	**5.42**	**4.54**	**4.04**	**3.73**	**3.50**	**3.33**	**3.20**	**3.08**	**3.00**	**2.92**	**2.87**	**2.77**	**2.68**	**2.57**	**2.49**	**2.41**	**2.32**	**2.27**	**2.19**	**2.15**	**2.10**	**2.06**	**2.03**
30	4.17	3.32	2.92	2.69	2.53	2.42	2.34	2.27	2.21	2.16	2.12	2.09	2.04	1.99	1.93	1.89	1.84	1.79	1.76	1.72	1.69	1.66	1.64	1.62
	7.56	**5.39**	**4.51**	**4.02**	**3.70**	**3.47**	**3.30**	**3.17**	**3.06**	**2.98**	**2.90**	**2.84**	**2.74**	**2.66**	**2.55**	**2.47**	**2.38**	**2.29**	**2.24**	**2.16**	**2.13**	**2.07**	**2.03**	**2.01**
32	4.15	3.30	2.90	2.67	2.51	2.40	2.32	2.25	2.19	2.14	2.10	2.07	2.02	1.97	1.91	1.86	1.82	1.76	1.74	1.69	1.67	1.64	1.61	1.59
	7.50	**5.34**	**4.46**	**3.97**	**3.66**	**3.42**	**3.25**	**3.12**	**3.01**	**2.94**	**2.86**	**2.80**	**2.70**	**2.62**	**2.51**	**2.42**	**2.34**	**2.25**	**2.20**	**2.12**	**2.08**	**2.02**	**1.98**	**1.96**
34	4.13	3.28	2.88	2.65	2.49	2.38	2.30	2.23	2.17	2.12	2.08	2.05	2.00	1.95	1.89	1.84	1.80	1.74	1.71	1.67	1.64	1.61	1.59	1.57
	7.44	**5.29**	**4.42**	**3.93**	**3.61**	**3.38**	**3.21**	**3.08**	**2.97**	**2.89**	**2.82**	**2.76**	**2.66**	**2.58**	**2.47**	**2.38**	**2.30**	**2.21**	**2.15**	**2.08**	**2.04**	**1.98**	**1.94**	**1.91**
36	4.11	3.26	2.86	2.63	2.48	2.36	2.28	2.21	2.15	2.10	2.06	2.03	1.98	1.93	1.87	1.82	1.78	1.72	1.69	1.65	1.62	1.59	1.56	1.55
	7.39	**5.25**	**4.38**	**3.89**	**3.58**	**3.35**	**3.18**	**3.04**	**2.94**	**2.86**	**2.78**	**2.72**	**2.62**	**2.54**	**2.43**	**2.35**	**2.26**	**2.17**	**2.12**	**2.04**	**2.00**	**1.94**	**1.90**	**1.87**
38	4.10	3.25	2.85	2.62	2.46	2.35	2.26	2.19	2.14	2.09	2.05	2.02	1.96	1.92	1.85	1.80	1.76	1.71	1.67	1.63	1.60	1.57	1.54	1.53
	7.35	**5.21**	**4.34**	**3.86**	**3.54**	**3.32**	**3.15**	**3.02**	**2.91**	**2.82**	**2.75**	**2.69**	**2.59**	**2.51**	**2.40**	**2.32**	**2.22**	**2.14**	**2.08**	**2.00**	**1.97**	**1.90**	**1.86**	**1.84**
40	4.08	3.23	2.84	2.61	2.45	2.34	2.25	2.18	2.12	2.07	2.04	2.00	1.95	1.90	1.84	1.79	1.74	1.69	1.66	1.61	1.59	1.55	1.53	1.51
	7.31	**5.18**	**4.31**	**3.83**	**3.51**	**3.29**	**3.12**	**2.99**	**2.88**	**2.80**	**2.73**	**2.66**	**2.56**	**2.49**	**2.37**	**2.29**	**2.20**	**2.11**	**2.05**	**1.97**	**1.94**	**1.88**	**1.84**	**1.81**
42	4.07	3.22	2.83	2.59	2.44	2.32	2.24	2.17	2.11	2.06	2.02	1.99	1.94	1.89	1.82	1.78	1.73	1.68	1.64	1.60	1.57	1.54	1.51	1.49
	7.27	**5.15**	**4.29**	**3.80**	**3.49**	**3,26**	**3.10**	**2.96**	**2.86**	**2.77**	**2.70**	**2.64**	**2.54**	**2.46**	**2.35**	**2.26**	**2.17**	**2.08**	**2.02**	**1.94**	**1.91**	**1.85**	**1.80**	**1.78**
44	4.06	3.21	2.82	2.58	2.43	2.31	2.23	2.16	2.10	2.05	2.01	1.98	1.92	1.88	1.81	1.76	1.72	1.66	1.63	1.58	1.56	1.52	1.50	1.48
	7.24	**5.12**	**4.26**	**3.78**	**3.46**	**3.24**	**3.07**	**2.94**	**2.84**	**2.75**	**2.68**	**2.62**	**2.52**	**2.44**	**2.32**	**2.24**	**2.15**	**2.06**	**2.00**	**1.92**	**1.88**	**1.82**	**1.78**	**1.75**
46	4.05	3.20	2.81	2.57	2.42	2.30	2.22	2.14	2.09	2.04	2.00	1.97	1.91	1.87	1.80	1.75	1.71	1.65	1.62	1.57	1.54	1.51	1.48	1.46
	7.21	**5.10**	**4.24**	**3.76**	**3.44**	**3.22**	**3.05**	**2.92**	**2.82**	**2.73**	**2.66**	**2.60**	**2.50**	**2.42**	**2.30**	**2.22**	**2.13**	**2.04**	**1.98**	**1.90**	**1.86**	**1.80**	**1.76**	**1.72**

48	4.04	3.19	2.80	2.56	2.41	2.30	2.21	2.14	2.08	2.03	1.99	1.96	1.90	1.86	1.79	1.74	1.70	1.64	1.61	1.56	1.53	1.50	1.47	1.45
	7.19	**5.08**	**4.22**	**3.74**	**3.42**	**3.20**	**3.04**	**2.90**	**2.80**	**2.71**	**2.64**	**2.58**	**2.48**	**2.40**	**2.28**	**2.20**	**2.11**	**2.02**	**1.96**	**1.88**	**1.84**	**1.78**	**1.73**	**1.70**
50	4.03	3.18	2.79	2.56	2.40	2.29	2.20	2.13	2.07	2.02	1.98	1.95	1.90	1.85	1.78	1.74	1.69	1.63	1.60	1.55	1.52	1.48	1.46	1.44
	7.17	**5.06**	**4.20**	**3.72**	**3.41**	**3.18**	**3.02**	**2.88**	**2.78**	**2.70**	**2.62**	**2.56**	**2.46**	**2.39**	**2.26**	**2.18**	**2.10**	**2.00**	**1.94**	**1.86**	**1.82**	**1.76**	**1.71**	**1.68**
55	4.02	3.17	2.78	2.54	2.38	2.27	2.18	2.11	2.05	2.00	1.97	1.93	1.88	1.83	1.76	1.72	1.67	1.61	1.58	1.52	1.50	1.46	1.43	1.41
	7.12	**5.01**	**4.16**	**3.68**	**3.37**	**3.15**	**2.98**	**2.85**	**2.75**	**2.66**	**2.59**	**2.53**	**2.43**	**2.35**	**2.23**	**2.15**	**2.06**	**1.96**	**1.90**	**1.82**	**1.78**	**1.71**	**1.66**	**1.64**
60	4.00	3.15	2.76	2.52	2.37	2.25	2.17	2.10	2.04	1.99	1.95	1.92	1.86	1.81	1.75	1.70	1.65	1.59	1.56	1.50	1.48	1.44	1.41	1.39
	7.08	**4.98**	**4.13**	**3.65**	**3.34**	**3.12**	**2.95**	**2.82**	**2.72**	**2.63**	**2.56**	**2.50**	**2.40**	**2.32**	**2.20**	**2.12**	**2.03**	**1.93**	**1.87**	**1.79**	**1.74**	**1.68**	**1.63**	**1.60**
65	3.99	3.14	2.75	2.51	2.36	2.24	2.15	2.08	2.02	1.98	1.94	1.90	1.85	1.80	1.73	1.68	1.63	1.57	1.54	1.49	1.46	1.42	1.39	1.37
	7.04	**4.95**	**4.10**	**3.62**	**3.31**	**3.09**	**2.93**	**2.79**	**2.70**	**2.61**	**2.54**	**2.47**	**2.37**	**2.30**	**2.18**	**2.09**	**2.00**	**1.90**	**1.84**	**1.76**	**1.71**	**1.64**	**1.60**	**1.56**
70	3.98	3.13	2.74	2.50	2.35	2.23	2.14	2.07	2.01	1.97	1.93	1.89	1.84	1.79	1.72	1.67	1.62	1.56	1.53	1.47	1.45	1.40	1.37	1.35
	7.01	**4.92**	**4.08**	**3.60**	**3.29**	**3.07**	**2.91**	**2.77**	**2.67**	**2.59**	**2.51**	**2.45**	**2.35**	**2.28**	**2.15**	**2.07**	**1.98**	**1.88**	**1.82**	**1.74**	**1.69**	**1.62**	**1.56**	**1.53**
80	3.96	3.11	2.72	2.48	2.33	2.21	2.12	2.05	1.99	1.95	1.91	1.88	1.82	1.77	1.70	1.65	1.60	1.54	1.51	1.45	1.42	1.38	1.35	1.32
	6.96	**4.88**	**4.04**	**3.56**	**3.25**	**3.04**	**2.87**	**2.74**	**2.64**	**2.55**	**2.48**	**2.41**	**2.32**	**2.24**	**2.11**	**2.03**	**1.94**	**1.84**	**1.78**	**1.70**	**1.65**	**1.57**	**1.52**	**1.49**
100	3.94	3.09	2.70	2.46	2.30	2.19	2.10	2.03	1.97	1.92	1.88	1.85	1.79	1.75	1.68	1.63	1.57	1.51	1.48	1.42	1.39	1.34	1.30	1.28
	6.90	**4.82**	**3.98**	**3.51**	**3.20**	**2.99**	**2.82**	**2.69**	**2.59**	**2.51**	**2.43**	**2.36**	**2.26**	**2.19**	**2.06**	**1.98**	**1.89**	**1.79**	**1.73**	**1.64**	**1.59**	**1.51**	**1.46**	**1.43**
125	3.92	3.07	2.68	2.44	2.29	2.17	2.08	2.01	1.95	1.90	1.86	1.83	1.77	1.72	1.65	1.60	1.55	1.49	1.45	1.39	1.36	1.31	1.27	1.25
	6.84	**4.78**	**3.94**	**3.47**	**3.17**	**2.95**	**2.79**	**2.65**	**2.56**	**2.47**	**2.40**	**2.33**	**2.23**	**2.15**	**2.03**	**1.94**	**1.85**	**1.75**	**1.68**	**1.59**	**1.54**	**1.46**	**1.40**	**1.37**
150	3.91	3.06	2.67	2.43	2.27	2.16	2.07	2.00	1.94	1.89	1.85	1.82	1.76	1.71	1.64	1.59	1.54	1.47	1.44	1.37	1.34	1.29	1.25	1.22
	6.81	**4.75**	**3.91**	**3.44**	**3.14**	**2.92**	**2.76**	**2.62**	**2.53**	**2.44**	**2.37**	**2.30**	**2.20**	**2.12**	**2.00**	**1.91**	**1.83**	**1.72**	**1.66**	**1.56**	**1.51**	**1.43**	**1.37**	**1.33**
200	3.89	3.04	2.65	2.41	2.26	2.14	2.05	1.98	1.92	1.87	1.83	1.80	1.74	1.69	1.62	1.57	1.52	1.45	1.42	1.35	1.32	1.26	1.22	1.19
	6.76	**4.71**	**3.88**	**3.41**	**3.11**	**2.90**	**2.73**	**2.60**	**2.50**	**2.41**	**2.34**	**2.28**	**2.17**	**2.09**	**1.97**	**1.88**	**1.79**	**1.69**	**1.62**	**1.53**	**1.48**	**1.39**	**1.33**	**1.28**
400	3.86	3.02	2.62	2.39	2.23	2.12	2.03	1.96	1.90	1.85	1.81	1.78	1.72	1.67	1.60	1.54	1.49	1.42	1.38	1.32	1.28	1.22	1.16	1.13
	6.70	**4.66**	**3.83**	**3.36**	**3.06**	**2.85**	**2.69**	**2.55**	**2.46**	**2.37**	**2.29**	**2.23**	**2.12**	**2.04**	**1.92**	**1.84**	**1.74**	**1.64**	**1.57**	**1.47**	**1.42**	**1.32**	**1.24**	**1.19**
1000	3.85	3.00	2.61	2.38	2.22	2.10	2.02	1.95	1.89	1.84	1.80	1.76	1.70	1.65	1.58	1.53	1.47	1.41	1.36	1.30	1.26	1.19	1.13	1.08
	6.66	**4.62**	**3.80**	**3.34**	**3.04**	**2.82**	**2.66**	**2.53**	**2.43**	**2.34**	**2.26**	**2.20**	**2.09**	**2.01**	**1.89**	**1.81**	**1.71**	**1.61**	**1.54**	**1.44**	**1.38**	**1.28**	**1.19**	**1.11**
∞	3.84	2.99	2.60	2.37	2.21	2.09	2.01	1.94	1.88	1.83	1.79	1.75	1.69	1.64	1.57	1.52	1.46	1.40	1.35	1.28	1.24	1.17	1.11	1.00
	6.64	**4.60**	**3.78**	**3.32**	**3.02**	**2.80**	**2.64**	**2.51**	**2.41**	**2.32**	**2.24**	**2.18**	**2.07**	**1.99**	**1.87**	**1.79**	**1.69**	**1.59**	**1.52**	**1.41**	**1.36**	**1.25**	**1.15**	**1.00**

† Reprinted from Table 10-5-3 of G. W. Snedcor, "Statistical Methods Applied to Experiments in Agriculture and Biology," 5th ed., Collegiate Press, Inc., of Iowa State College, Ames, Iowa, 1956, with the permission of the author and the publisher.

Since s_e^2 is associated with $15 - 3 = 12$ degrees of freedom and s_p^2 with $3 - 1 = 2$ degrees, Table 2-5 gives

$$F = 3.88 \quad \text{at the 5 per cent level}$$
$$F = 6.93 \quad \text{at the 1 per cent level}$$

Since the calculated value of F is greater than that corresponding to the 1 per cent level, in less than 1 per cent of the cases could the observed difference in sample means be explained on the basis of the scatter of the observed data. It is reasonable to conclude that the sample means are actually different. Further information could now be obtained by applying the F test to the *most closely corresponding* pair of means to see if they differ significantly.

The F test is based upon the following considerations: In a single set of data, two types of error are possible—random errors and errors of method or bias. The magnitude of the random errors is estimated by the within-sample or error variance. However, from the single set of data, no estimate of the error due to method is possible. Suppose, however, that other sets of data are available which are likewise subject to both random and method errors. For each data set, the within-sample or error variance may be calculated. These individual error variances may then be pooled to give the best estimate of the error variance of all the data points. In this pooled estimate, the effect of errors of method has not been included. This pooled error-variance estimate s_e^2 is defined by Eq. (2-50*a*) and the degrees of freedom associated with it, f_e, by Eq. (2-50*b*).

Now consider the derived set of data which can be formed from the individual means, $\bar{x}_i$, of each data set. These means will differ from each other because of random error and method error *if the method error varies from set to set.* The method error would be expected to vary from set to set if different experimental techniques were applied to each set. The sample estimate of the population variance obtained from the derived set of means then includes both random error and error due to different experimental methods if such occurs. Consequently, s_p^2 [see Eq. (2-55)] obtained from the derived set of means would be expected to be bigger than s_e^2 if errors of method vary from one set of data to the next. The statistic F is the ratio of the variance which contains both random and method error to the variance which includes random errors only. The magnitude of F formed in this way is a measure of the importance of errors of method which *differ* from one data set to the next. If F is near unity, the method error is nearly the same for each set of data. If F is large, then significant differences in errors of method are highly probable. Because of sampling error, F ordinarily differs from unity even when errors of method are not significant. The probability that F can take on values different from unity as a result of random errors can be computed. It is these probability values that are tabulated in Table 2-5. Note that the probability associated with a given value of F is a function of the degrees of freedom associated with each variance.

2-11. Design of Experiments. Experiments which are designed to make use of statistical-analysis methods in their interpretation are more efficient than those in which a haphazard approach is employed. More information is obtained for a given expenditure of effort, and the reliability of the results can be estimated with greater confidence. This section will discuss the principles of what is called "the factorial design of experiments," a technique which makes effective use of statistical methods. The subject is introduced by means of an example.

Example 2-5. Tests were run in a pilot-plant installation in order to obtain an early estimate of the effect of concentration C and temperature R on the yield x of a chemical reaction. The results of the tests are shown in Table 2-6.

TABLE 2-6. VALUES OF YIELD IN PER CENT

	C_1	C_2
R_1	12	15
	10	16
R_2	18	23
	20	20

The tests were made at two levels of temperature, R_1 and R_2, and two levels of concentration, C_1 and C_2. All possible combinations were covered. In a two-level experiment with two independent variables, $2^2 = 4$ tests were needed, forming a "block" of experiments. In order to estimate the precision of the tests, the entire block was repeated, a process called "replication." However, the tests themselves were made at random with respect to time. The purpose of this was to minimize the probability of false interpretations due to other factors. It was suspected, for example, that the pressure level, throughput rate, etc., affected the yield. Although these factors were to be held "constant," it was realized that both slow and abrupt changes in operating level could occur. A random test order minimized the possibility of introducing spurious effects due to such changes, since it was unlikely that both runs at the same level (say R_1C_2) would be made when the system was out of control. On the other hand, the effect of such control variations would appear in the error estimate and in this way would provide valuable information regarding the reproducibility of the process.

Casual inspection of the data would indicate the conclusions shown in Table 2-7. Table 2-7 indicates an *interaction* between temperature and concentration, i.e., that the effect on yield of a change in one variable is dependent upon the level of the other variable. Thus, a change from C_1 to C_2 at R_1 produces an indicated yield change of 4.5, whereas at R_2 the change from C_1 to C_2 produces an indicated yield change of only 2.5. The results shown in Table 2-7 would be very useful if they could be trusted. However, the replication runs disclose considerable scatter, and this scatter may be

TABLE 2-7. CONCLUSIONS BASED UPON CASUAL INSPECTION
Data of Table 2-6, Example 2-5

Level (variable held constant)	Average yield	Average yield *change* resulting from change in other variable	Range of variable changed
R_1	13.25	4.5	$C_1 \rightarrow C_2$
R_2	20.25	2.5	$C_1 \rightarrow C_2$
C_1	15.0	8.0	$R_1 \rightarrow R_2$
C_2	18.5	6 0	$R_1 \rightarrow R_2$

responsible for some of the effects shown in Table 2-7. Clearly, a more thorough analysis of the data is indicated. The method of analysis is that of the analysis of variance.

In a given run, the observed yield may be considered to be the sum of four effects:

$$x_{crk} = U_c + Y_r + Z_{cr} + e_{crk} \tag{2-57}$$

U_c denotes a pure concentration effect,

$$U_c = \phi(C_c) \tag{2-58}$$

Y_r denotes a pure temperature effect,

$$Y_r = \alpha(R_r) \tag{2-59}$$

Z_{cr} denotes the interaction effect due to the combined action of the concentration and temperature,

$$Z_{rc} = \gamma(C_c,R_r) \tag{2-60}$$

e_{crk} denotes the experimental error and is *assumed* to be normally distributed. The concentration level is denoted by the subscript c, the temperature level by the subscript r, and the replication number by the subscript k. Four variances may be defined:

$$s_e^2 = \frac{\sum^{n_c}\sum^{n_r}\sum^{n_k} (e_{crk} - e_{cr\bar{k}})^2}{n_c n_r (n_k - 1)} \tag{2-61}$$

where $e_{cr\bar{k}}$ denotes the mean error averaged over n_k at fixed c and r.

The interaction variance

$$\sigma_I^2 = \frac{\sum^{n_c}\sum^{n_r} (Z_{cr} - \bar{Z})^2}{n_c n_r} \tag{2-62}$$

where $\bar{Z}$ denotes the grand mean.

The temperature variance

$$\sigma_r^2 = \frac{\sum^{n_r} (Y_r - \bar{Y})^2}{n_r} \tag{2-63}$$

The concentration variance

$$\sigma_c^2 = \frac{\sum^{n_c} (U_i - \bar{U})^2}{n_c} \tag{2-64}$$

Now each of these variance terms is a measure of the importance of the effect under study.

If each of these variances can be calculated from the experimental data, the F test may be used to determine the significance of the observed effect. It is not possible completely to isolate each variance, but satisfactory estimates may be obtained as follows:

The error variance can be determined easily. Simple algebra gives as a result of (2-57)

$$e_{crk} - e_{cr\bar{k}} = x_{crk} - x_{cr\bar{k}} \tag{2-65}$$

Consequently,

$$s_e^2 = \frac{\sum^{n_c}\sum^{n_r}\left[\sum^{n_k}(x_{crk} - x_{cr\bar{k}})^2\right]}{n_c n_r(n_k - 1)}$$
$$= \frac{\sum^{n_c}\sum^{n_r}\sum^{n_k} x_{crk}^2 - \sum^{n_c}\sum^{n_r}\left[\left(\sum^{n_k} x_{crk}\right)^2 / n_k\right]}{n_c n_r(n_k - 1)} \tag{2-66}$$

The numerator of (2-66) is called the "error sum of squares" or "within-group sum of squares," with the symbol S_e. The denominator is the degrees of freedom, f_e, associated with the error variance.

In the present example, the error sum of squares is

$$S_e = 12^2 + 10^2 + 18^2 + 20^2 + 15^2 + 16^2 + 23^2 + 20^2 - \left(\frac{22^2}{2} + \frac{38^2}{2} + \frac{31^2}{2} + \frac{43^2}{2}\right)$$
$$= 9.0$$

The degrees of freedom associated with the error variance is

$$f_e = n_c n_r(n_k - 1) = 2 \cdot 2(2 - 1) = 4$$

Then

$$s_e^2 = 9/4 = 2.25$$

The remainder of the analysis is best followed in terms of the block diagram of Fig. 2-3*a* and *b*. The numbers in the center of each square denote the mean value at the given level obtained from the original data:

$$x_{11\bar{k}} = \frac{\sum^{n_k} x_{11k}}{n_k} \qquad \text{etc.}$$

In this example, at the C_1R_1 level, $x_{11\bar{k}} = 11$. The numbers in the arc-enclosed corners of the rectangles denote the number of replications used to calculate the mean.

Now consider the interaction. *If there were no interaction*, it is easily shown by means of (2-57) that the mean of the yields at the C_1R_1 and C_2R_2 levels would, except for sampling error, be the same as the mean of the yields at the C_1R_2 and C_2R_1 levels. These two means may be considered to form a sample of a population of means and may be used to form a second estimate of the variance of the population of measurements. Shortly, the ratio of the interaction estimate of the population variance to the error estimate of the population variance will be used in an F test to determine the importance of interaction. The basis for this results from the following considerations: If no interaction occurs, then the interaction estimate of the population variance should result from random errors only. In this case, the two separate population-

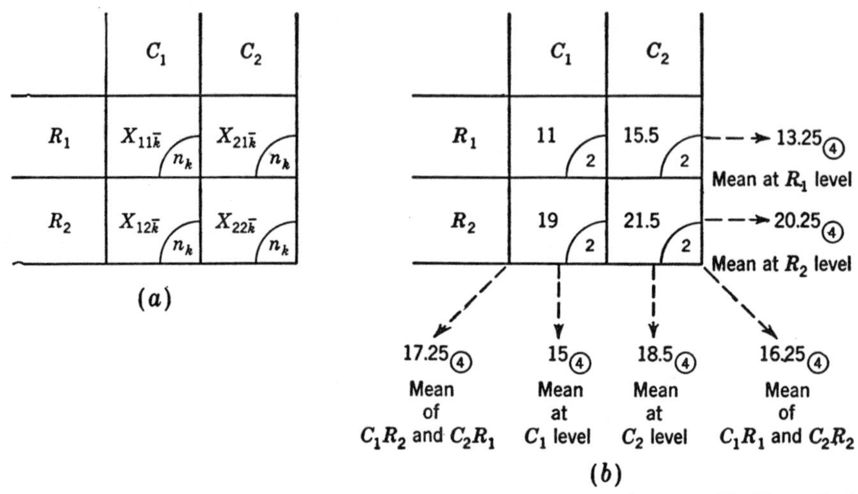

FIG. 2-3. (a) Schematic block for two-level factorial experiment. (b) Numerical block for two-level factorial experiment.

variance estimates obtained from the error variance and interaction variance, respectively, should be statistically equivalent. On the other hand, the presence of significant interaction will cause the interaction estimate of the population variance to be larger than the error estimates. Consequently, a comparison of the two variance estimates can be used to determine the significance of the interaction.

The interaction estimate of the population variance is given by

$$(s_p^2)_I = \frac{(n_0)_I \sum^{n_i} [\bar{x}_i - (\bar{x}_m)_I]^2}{n_i - 1} \tag{2-67}$$

In Eq. (2-67), $\bar{x}_i$ represents an interaction mean, $(\bar{x}_m)_I$ the mean of the interaction means, n_i the number of interaction means, and $(n_0)_I$ the average number of original data points (x_{crk}) used to form an interaction mean term [see Eq. (2-54)]. Equation (2-67) is analogous to Eq. (2-55).

For the present example, the interaction means are

$$\frac{x_{11\bar{k}} + x_{22\bar{k}}}{2} = 16.25 \qquad \text{and} \qquad \frac{x_{12\bar{k}} + x_{21\bar{k}}}{2} = 17.25$$

$$(\bar{x}_m)_I = \frac{16.25 + 17.25}{2} = 16.75 \qquad (n_0)_I = 4$$

The interaction sum of squares is

$$(n_0)_I \sum^{n_i} [\bar{x}_i - (\bar{x}_m)_I]^2 = 4[(17.25 - 16.75)^2 + (16.25 - 16.75)^2] = 20$$

The interaction degrees of freedom are $f_I = n_i - 1 = 2 - 1 = 1$. The interaction estimate of the population variance is then

$$(s_p^2)_I = \frac{2.0}{1} = 2.0$$

The interpretation and utility of this estimate of the population variance (frequently called the "interaction mean square") will be made clearer by the following: In a two-factor, two-level experiment there are two interaction means:

$$\frac{x_{11\bar{k}} + x_{22\bar{k}}}{2} = \tfrac{1}{2}(U_1 + U_2 + Y_1 + Y_2 + Z_{11} + Z_{22} + e_{11\bar{k}} + e_{22\bar{k}}) \qquad (2\text{-}68)$$

and

$$\frac{x_{12\bar{k}} + x_{21\bar{k}}}{2} = \tfrac{1}{2}(U_1 + U_2 + Y_1 + Y_2 + Z_{12} + Z_{21} + e_{12\bar{k}} + e_{21\bar{k}}) \qquad (2\text{-}69)$$

The mean of the interaction means is

$$\begin{aligned}(x_m)_I &= \frac{x_{11\bar{k}} + x_{22\bar{k}} + x_{12\bar{k}} + x_{21\bar{k}}}{4} \\ &= \tfrac{1}{2}(U_1 + U_2 + Y_1 + Y_2 + 2\bar{Z} + 2\bar{e})\end{aligned} \qquad (2\text{-}70)$$

The interaction estimate of the population variance is

$$\begin{aligned}(s_p^2)_I &= \frac{(n_0)_I \sum^{n_i} [\bar{x}_i - (\bar{x}_m)_I]^2}{n_i - 1} \\ &= 4\left[\left(\frac{Z_{11} - \bar{Z}}{2} + \frac{Z_{22} - \bar{Z}}{2} + \frac{e_{11\bar{k}} - \bar{e}}{2} + \frac{e_{22\bar{k}} - \bar{e}}{2}\right)^2 \right. \\ &\quad \left. + \left(\frac{Z_{12} - \bar{Z}}{2} + \frac{Z_{21} - \bar{Z}}{2} + \frac{e_{12\bar{k}} - \bar{e}}{2} + \frac{e_{21\bar{k}} - \bar{e}}{2}\right)^2\right]\end{aligned} \qquad (2\text{-}71)$$

If the right-hand side of (2-71) is expanded and the errors are assumed statistically independent, it is found that

$$(s_p^2)_I \cong \sigma_e^2 + n_k \sigma_I^2 \qquad (2\text{-}72)$$

The sample estimate of the population variance obtained from the interaction terms and calculated by means of Eq. (2-67) *is then (approximately) the sum of the error variance and the product of the number of times each observation is repeated and the interaction variance.*

The interaction estimate of the population variance may then be used in an F test to determine the significance of the interaction. For this example,

$$F_I = \frac{(s_p^2)_I}{s_e^2} = \frac{2}{2.25} = 0.9$$

and is based upon 1 and 4 degrees of freedom. It is not necessary to use the table of F values to interpret this result. Reference to Eq. (2-72) shows that $\sigma_I^2 \cong 0$, and the interaction is not significant.

It follows then that some of the indications of Table 2.7 are spurious. The effect of a change in concentration is most probably independent of temperature level and vice versa.

Since it has been shown that the interaction variance σ_I^2 is not significant, it follows from Eq. (2-72) that $(s_p^2)_I$ provides an independent estimate of s_e^2. The two estimates may then be pooled to form a better estimate in accordance with Eq. (2-50*a*):

$$(s_e^2)_{b.v.} = \frac{f_e s_e^2 + f_I (s_p^2)_I}{f_e + f_I} \tag{2-73}$$

Then

$$(s_e^2)_{b.v.} = \frac{4 \cdot 2.25 + 1 \cdot 2.0}{4 + 1} = 2.2$$

This pooled estimate of the error variance is valid only if the interaction is not significant.

Now consider the pure concentration effect. It will be treated in a manner which takes into account the possibility that the interaction term is significant. The average yield at the C_1 level is the mean of the yields at the C_1R_1 and C_1R_2 levels. In the block diagram of Fig. 2-3, the average yield at the C_1 level is the mean of the column corresponding to a constant C_1 level. The average yield at the C_2 level is the mean of the yields at the C_2R_1 and C_2R_2 levels. In general, these two average yields will be different and may be used to estimate the population variance:

$$(s_p^2)_c = \frac{(n_0)_c \sum^{n_c} (x_{c\bar{r}\bar{k}} - \bar{x}_{\bar{c}\bar{r}\bar{k}})^2}{n_c - 1} \tag{2-74}$$

In Eq. (2-74), $(n_0)_c$ denotes the average number of data points (x_{crk}) used to form a concentration mean term, $x_{c\bar{r}\bar{k}}$ denotes the average yield at a given concentration level, $\bar{x}_{\bar{c}\bar{r}\bar{k}}$ the grand mean yield, and n_c the number of concentration means.

Expansion of (2-74) in a manner analogous to the expansion of (2-67) in the interaction case gives, approximately,

$$(s_p^2)_c \cong \sigma_e^2 + n_k \sigma_I^2 + n_k n_r \sigma_c^2 \tag{2-75}$$

Consequently, an F test may be used to determine the significance of the concentration effect. It is evident, however, that if the interaction is significant, the test should be

$$F_c = \frac{(s_p^2)_c}{(s_p^2)_I} \tag{2-76}$$

in order to determine whether or not the concentration has an effect independent of the concentration-temperature interaction. If the interaction is not significant, then the F test should be based upon the pooled estimate of the error variance:

$$F_c = \frac{(s_p^2)_c}{(s_e^2)_{b.v.}} \tag{2-77}$$

Equation (2-77) should be used to test the data of this example. The concentration (or column) sum of squares is

$$(n_0)_c \sum^{n_c} (x_{c\bar{r}\bar{k}} - x_{\bar{c}\bar{r}\bar{k}})^2 = S_c = 4[(15-16.75)^2 + (18.5 - 16.75)^2] = 24.5.$$

The concentration degrees of freedom is

$$f_c = n_c - 1 = 2 - 1 = 1$$

Hence,

$$(s_p^2)_c = \frac{24.5}{1} = 24.5$$

Using the pooled error-variance estimate,

$$F_c = \frac{(s_p^2)_c}{(s_e^2)_{b.v.}} = \frac{24.5}{2.2} = 11.1$$

with 5 and 1 degrees of freedom. This F value lies between the 5 and 1 per cent limits. The observed pure concentration effect is probably significant.

The pure temperature effect is treated in the same fashion. The estimate of the population variance is

$$(s_p^2)_r = \frac{(n_0)_r \sum^{n_r} (x_{\bar{c}r\bar{k}} - \bar{x}_{\bar{c}\bar{r}\bar{k}})^2}{n_r - 1} \tag{2-78}$$

and

$$(s_p^2)_r \cong \sigma_e^2 + n_k\sigma_I^2 + n_k n_c \sigma_r^2 \tag{2-79}$$

In the case of this example, the temperature (or row) sum of squares is

$$(n_0)_r \sum^{n_r} (x_{\bar{c}r\bar{k}} - \bar{x}_{\bar{c}\bar{r}\bar{k}})^2 = S_r = 4[(13.25 - 16.75)^2 + (20.25 - 16.75)^2] = 98$$

The temperature (or row) degrees of freedom is $f_r = n_r - 1 = 2 - 1 = 1$. Hence,

$$(s_p^2)_r = 98$$

$$F_r = \frac{(s_p^2)_r}{(s_e^2)_{b.v.}} = \frac{98}{2.2} = 44.5$$

For 5 and 1 degrees of freedom, this value of F falls far below the 1 per cent level, and the temperature effect is highly significant.

There is additional information to be obtained from the data of this example. The confidence limits of the average yield may be determined by a t test. For example, at the C_1R_1 level, the average yield $x_{11\bar{k}} = 11$. The pooled estimate of the error variance of the *population* has been found to be 2.2 with 5 degrees of freedom. From Table 2-1, for 5 degrees of freedom, values of t lying outside ± 2.6 are only 0.05 probable. In order to apply the t limits, the error variance of the *population of means* is needed. Since two values ($n_k = 2$) were used to calculate $x_{11\bar{k}}$, the best estimate is obtained from Eq. (2-43) as

$$s_{m,e}^2 = \frac{2.2}{2} = 1.1$$

Then

$$x_{11\bar{k}} - \bar{\bar{x}}_{11} = \pm 2.6\sqrt{1.1} = \pm 2.7$$

or

$$8.3 \lesseqgtr \bar{\bar{x}}_{11} \lesseqgtr 13.7$$

with 95 per cent confidence. The reader may readily verify the greater precision obtained by use of the pooled error variance in place of an estimate based upon the replicated values at the C_1R_1 level only.

The t test may also be used to determine the confidence limits of the observed effect of a change in the level of R or of C. For example, consider the change from R_1 to R_2. Since the interaction has been shown to be negligible, the best measure of the yield change resulting from a temperature change is the difference between the mean yield at the R_2 level, irrespective of the value of C, and the corresponding mean at the R_1 level. This difference is $x_{\bar{c}2\bar{k}} - x_{\bar{c}1\bar{k}} = 20.25 - 13.25 = 7.0$. Since four values were used to calculate one of the means (say $x_{\bar{c}2\bar{k}}$), the error variance of the population of *means* is obtained by dividing the population variance by four: $s_{m,e}^2 = 2.2/4 = 0.55$.

The error variance for the *difference* of two means is obtained by use of Eq. (2-30) or (2-36) and is the *sum* of the two variances: 0.55 + 0.55 = 1.1. Again 5 degrees of freedom are associated with this variance. The 95 per cent confidence limits on the temperature effect are

$$(\bar{x}_{\bar{c}2\bar{k}} - \bar{x}_{\bar{c}2\bar{k}}) - (\bar{\bar{x}}_{\bar{c}2\bar{k}} - \bar{\bar{x}}_{\bar{c}1\bar{k}}) = \pm 2.6\sqrt{1.1} = \pm 2.7$$

or

$$4.3 \leqq \bar{\bar{x}}_{\bar{c}2\bar{k}} - \bar{\bar{x}}_{\bar{c}1\bar{k}} \leqq 9.7$$

with 95 per cent confidence. Note that the prior proof of the absence of interaction permits the 95 per cent confidence limits to be narrowed. If the interaction had been significant, the temperature effect would have been a function of the concentration level. Consequently, two different changes would need to be reported, one at the C_1 level and one at the C_2 level. At a fixed C and R level, only two points are available to form the mean; hence the estimate of the error variance of the population of such means would necessarily be greater than the estimate when the interaction is negligible.

The above variance analysis of a factorially designed experiment may be summarized as follows:

1. The precision of the experiment is estimated from the internal scatter of the data. Each set of measurements at a given level is considered to be a sample of the population of measurements. It is assumed that the population error distribution is normal; unless many replications are used, this assumption cannot be verified. An estimate of the population error variance is calculated for each set of replicated measurements, and these variances are then combined by means of Eq. (2-50) to give a pooled estimate of the population error variance s_e^2.

2. The interaction means are computed. Equations (2-51) to (2-53) or (2-54) and (2-55) are then used in conjunction with the interaction means to calculate an independent estimate of the population variance $(s_p^2)_I$. $(s_p^2)_I$ and the previously calculated error variance s_e^2 are then used in an F test to determine the significance of the interaction. If the interaction is insignificant, $(s_p^2)_I$ is pooled with s_e^2 to give a better estimate of the population error variance. In a two-factor, two-level experiment there are only two interaction terms. In a higher-order experimental block there are more. For example, in the three-factor, two-level block of Fig. 2-4 there are the following types of interaction:

Order	Interaction	Level
First..........	PC	R_1
First..........	PC	R_2
First..........	RC	P_1
First..........	RC	P_2
First..........	PR	C_1
First..........	PR	C_2
Second........	PCR	Average

Consider the first-order interaction PC terms at the R_1 level. They are $P_1C_1R_1$, $P_1C_2R_1$, $P_2C_1R_1$, $P_2C_2R_1$. They are to be combined to form two means, one at the extremes of the PC levels, $(P_1C_1R_1 + P_2C_2R_1)/2$, and the other at the intermediate level, $(P_1C_2R_1 + P_2C_1R_1)/2$. These means are used to provide an independent estimate of the variance of the entire population of data. This variance estimate is used with the error variance in an F test to determine the significance of the interaction

	R_1		R_2	
	P_1	P_2	P_1	P_2
C_1	X_{111}	X_{121}	X_{112}	X_{122}
C_2	X_{211}	X_{221}	X_{212}	X_{222}

FIG. 2-4. Block for three-factor, two-level experiment.

under study. The terms to be combined to form first-order interaction averages are easily identified. Higher-order interactions can be confusing. Mistakes will be avoided if a plus sign is associated with a high level of a variable and a minus sign with a low level. Each term is then assigned a sign by algebraic convention, and terms of like sign are averaged. For example, in the second-order interaction of a three-factor, two-level block, a typical term is $P_1C_2R_1 = (-)(+)(-) = +$. The complete set of terms is

$$
\begin{array}{lll}
P_1C_1R_1 = (-) & P_1C_2R_1 = (+) & P_2C_1R_1 = (+) \\
P_2C_2R_1 = (-) & P_1C_1R_2 = (+) & P_1C_2R_2 = (-) \\
P_2C_1R_2 = (-) & P_2C_2R_2 = (+) &
\end{array}
$$

The two interaction averages are then

$$\frac{P_1C_2R_1 + P_2C_1R_1 + P_1C_1R_2 + P_2C_2R_2}{4}$$

and

$$\frac{P_1C_1R_1 + P_2C_2R_1 + P_1C_2R_2 + P_2C_1R_2}{4}$$

3. The data corresponding to a fixed level of one variable are then averaged to give a mean value of the yield at that fixed level. The mean values corresponding to the remaining fixed levels of the one variable are calculated. The result is a sample of a population of means. This

sample is then used to calculate an independent estimate of the variance of the entire population of data using Eqs. (2-51) to (2-53) or (2-54) and (2-55). This variance is used in an F test to determine the significance of the change in variable level.

4. Significant effects are subjected to a confidence-limit analysis.

The foregoing discussion of the application of the analysis of variance to the examination of experimental data has been developed in a form designed to emphasize the principles involved. The calculating techniques used, however, have been unnecessarily cumbersome. In the following, procedures will be presented which are adapted to machine computation, and the terminology currently in use will be employed.

	C_1	C_2	C_3
R_1	X_{111} X_{112} X_{113}	X_{211} X_{212} X_{213}	X_{311} X_{312} X_{313}
R_2	X_{121} X_{122} X_{123}	X_{221} X_{222} X_{223}	X_{321} X_{322} X_{323}

(a)

	C_1	C_2	C_3
R_1	T_{11}	T_{21}	T_{31}
R_2	T_{12}	T_{22}	T_{32}

(b)

FIG. 2-5. (a) Experimental block. (b) Block formed from totals of (a).

Consider an experimental block such as shown in Fig. 2-5. C and R (with suitable subscripts) denote values of the independent variables. The symbol C_c denotes the level of the column treatment; the symbol R_r denotes the level of the row treatment. X_{crk} denotes the value of the dependent variable corresponding to the C_c column level, R_r row level, and kth replication at these levels. n_c denotes the number of column levels (three in Fig. 2-5), n_r the number of row levels (two in Fig. 2-5), and n_k the number of replications for each treatment combination (three in Fig. 2-5a).

Define the sum of squares for the total as

$$S_T = \sum^{n_c} \sum^{n_r} \sum^{n_k} X_{crk}^2 - \frac{\left(\sum^{n_c} \sum^{n_r} \sum^{n_k} X_{crk} \right)^2}{n_c n_r n_k} \tag{2-80a}$$

and the degrees of freedom for S_T as

$$f_T = n_c n_r n_k - 1 \tag{2-80b}$$

Define the total T_{cr} as

$$T_{cr} = \sum^{n_k} X_{crk} = X_{cr1} + X_{cr2} + \cdots \tag{2-81}$$

Using these totals, a new block is prepared, as shown in Fig. 2-5*b*. Define the sum of squares for subtotals as

$$S_s = \frac{\sum^{n_c}\sum^{n_r} (T_{cr})^2}{n_k} - \frac{\left(\sum^{n_c}\sum^{n_r}\sum^{n_k} X_{crk}\right)^2}{n_c n_r n_k} \tag{2-82a}$$

and the degrees of freedom for S_s as

$$f_s = n_c n_r - 1 \tag{2-82b}$$

Algebraic manipulation now gives

$$S_e = S_T - S_s \tag{2-83a}$$
$$f_e = f_T - f_s = n_c n_r (n_k - 1) \tag{2-83b}$$
$$s_e^2 = \frac{S_e}{f_e} \tag{2-61}$$

where S_e *is the error sum of squares,* f_e *is the error degrees of freedom, and* s_e^2 *is the sample estimate of the population error variance, also called the "error mean square,"* and is identical with the quantity defined earlier by Eq. (2-61).

The sum of squares for columns is

$$S_c = \frac{\sum^{n_c}\left(\sum^{n_r} T_{cr}\right)^2}{n_r n_k} - \frac{\left(\sum^{n_c}\sum^{n_r}\sum^{n_k} X_{crk}\right)^2}{n_c n_r n_k} \tag{2-84}$$

and the degrees of freedom for S_c are

$$f_c = n_c - 1$$

The column estimate of the population variance or column mean square is

$$(s_p^2)_c = \frac{S_c}{f_c} \tag{2-74}$$

The sum of squares for rows is

$$S_r = \frac{\sum^{n_r}\left(\sum^{n_c} T_{cr}\right)^2}{n_c n_k} - \frac{\left(\sum^{n_c}\sum^{n_r}\sum^{n_k} X_{crk}\right)^2}{n_c n_r n_k} \tag{2-85a}$$

and the degrees of freedom for S_r are

$$f_r = n_r - 1 \tag{2-85b}$$

TABLE 2-8. ANALYSIS OF VARIANCE

	Sum of squares	Degrees of freedom	Mean square	Estimate of
Column means (n_c = no. of columns)............	S_c [Eq. (2-84a)]	$f_c = n_c - 1$	$(s_p^2)_c = \frac{S_c}{f_c}$	$\sigma_e^2 + n_k\sigma_I^2 + n_kn_r\sigma_c^2$
Row means (n_r = no. of rows)..................	S_r [Eq. (2-85a)]	$f_r = n_r - 1$	$(s_p^2)_r = \frac{S_r}{f_r}$	$\sigma_e^2 + n_k\sigma_I^2 + n_kn_c\sigma_r^2$
Interaction..................................	$S_I = S_s - S_c - S_r$	$f_I = (n_c - 1)(n_r - 1)$	$(s_p^2)_I = \frac{S_I}{f_I}$	$\sigma_e^2 + n_k\sigma_I^2$
Subtotal.....................................	S_s [Eq. (2-82a)]	$f_s = n_cn_r - 1$		
Within or error..............................	$S_e = S_T - S_s$	$f_e = n_cn_r(n_k - 1)$	$s_e^2 = \frac{S_e}{f_e}$	σ_e^2
Total (n_k = no. of replications per cell)..........	S_T [Eq. (2-80a)]	$f_T = n_cn_rn_k - 1$		

The row estimate of the population variance or row mean square is

$$(s_p^2)_r = \frac{S_r}{f_r} \tag{2-78}$$

Algebraic manipulation now gives

$$S_I = S_s - S_c - S_r \tag{2-86a}$$
$$f_I = f_s - f_c - f_r = (n_c - 1)(n_r - 1) \tag{2-86b}$$
$$(s_p^2)_I = \frac{S_I}{f_I} \tag{2-71}$$

where S_I = interaction sum of squares
f_I = interaction degrees of freedom
$(s_p^2)_I$ = interaction estimate of population variance or interaction mean square

The above relations are summarized in Table 2-8. Note how neatly the analysis fits together when handled in this fashion. The algebraic manipulations required to show that this computation scheme is legitimate are carried out in references 1, 4, 7, 10, and 16.

Recalculation of Example 2-5. The use of the computation method for analysis of variance just presented will be illustrated by the recomputation of Example 2-5. Table 2-6 presents the basic data. $n_c = 2$, $n_r = 2$, $n_k = 2$.

$$S_T = \sum^{n_c}\sum^{n_r}\sum^{n_k} X_{crk}^2 - \frac{\left(\sum^{n_c}\sum^{n_r}\sum^{n_k} X_{crk}\right)^2}{n_c n_r n_k} \tag{2-80a}$$

$$\sum^{n_c}\sum^{n_r}\sum^{n_k} X_{crk}^2 = 12^2 + 10^2 + 18^2 + 20^2 + 15^2 + 16^2 + 23^2 + 20^2 = 2{,}378$$

$$\left(\sum^{n_c}\sum^{n_r}\sum^{n_k} X_{crk}\right)^2 = (12 + 10 + 18 + 20 + 15 + 16 + 23 + 20)^2 = (134)^2 = 17{,}956$$

$$S_T = 2{,}378 - \frac{17{,}956}{2 \cdot 2 \cdot 2} = 2{,}378 - 2{,}244.5 = 133.5$$

Now form the cell totals.

$$T_{11} = 12 + 10 = 22$$
$$T_{12} = 18 + 20 = 38 \qquad \text{etc.}$$

The block formed from totals is shown in Table 2-9.

TABLE 2-9

	C_1	C_2
R_1	22	31
R_2	38	43

$$S_s = \frac{\sum^{n_c}\sum^{n_r} (T_{cr})^2}{n_k} - \frac{\left(\sum^{n_c}\sum^{n_r}\sum^{n_k} X_{crk}\right)^2}{n_c n_r n_k} \tag{2-82a}$$

$$\sum^{n_c}\sum^{n_r} (T_{cr})^2 = (22)^2 + (38)^2 + (31)^2 + (43)^2 = 4{,}738$$

$$S_s = \frac{4{,}738}{2} - 2{,}244.5 = 124.5$$

$$S_e = S_T - S_s = 133.5 - 124.5 = 9.0$$

$$f_e = 2 \cdot 2(2 - 1) = 4$$

$$s_e^2 = \frac{S_e}{f_e} = \frac{9.0}{4} = 2.25$$

$$S_c = \frac{\sum^{n_c}\left(\sum^{n_r} T_{cr}\right)^2}{n_c n_k} - \frac{\left(\sum^{n_c}\sum^{n_r}\sum^{n_k} X_{crk}\right)^2}{n_c n_r n_k} \tag{2-84}$$

$$\sum^{n_c}\left(\sum^{n_r} T_{cr}\right)^2 = (22 + 38)^2 + (31 + 43)^2 = 9{,}076$$

$$S_c = \frac{9{,}076}{2 \cdot 2} - 2{,}244.5 = 24.5$$

$$f_c = n_c - 1 = 2 - 1 = 1$$

$$(s_p^2)_c = \frac{S_c}{f_c} = 24.5$$

$$S_r = \frac{\left(\sum^{n_r}\sum^{n_c} T_{cr}\right)^2}{n_c n_k} - \frac{\left(\sum^{n_c}\sum^{n_r}\sum^{n_k} X_{crk}\right)^2}{n_c n_r n_k} \tag{2-85a}$$

$$\sum^{n_r}\left(\sum^{n_c} T_{cr}\right)^2 = (22 + 31)^2 + (38 + 43)^2 = 9{,}370$$

$$S_r = \frac{9{,}370}{2 \cdot 2} - 2{,}244.5 = 98.0$$

$$f_r = n_r - 1 = 2 - 1 = 1$$

$$(s_p^2)_r = \frac{S_r}{f_r} = 98.0$$

$$S_I = S_s - S_c - S_r \tag{2-86a}$$

$$S_I = 124.5 - 24.5 - 98.0 = 2.0$$

$$f_I = (n_c - 1)(n_r - 1) = 1$$

$$(s_p^2)_I = \frac{S_I}{f_I} = 2.0$$

These values all agree with the ones calculated previously.

The application of the analysis of variance to factorial experiments involving several factors and levels follows the principles just outlined. In higher-order experiments it is customary to dispense with replication and use the population variance estimated from the highest-order interaction term as the error variance. Further details will be found in Anderson and Bancroft,[1] Brownlee,[2] Cochran and Cox,[4] Dixon and Massey,[7] Fisher,[10] and Freeman.[11]

2-12. Least Squares. Procedures for fitting analytical expressions to experimental data were discussed in Chap. 1. Further information concerning the precision of the fitting process may be obtained by the application of statistical methods. In essence, the statistical method proceeds by evaluating the constants in the analytical expression chosen to represent the data in a manner which makes the variance a minimum. This gives the "best" value of the constants in the analytical expression. The variance is then used to determine the precision of the fitting process.

Fitting a Straight Line. Suppose that several measurements have been made, giving values of the dependent variable $y_1, y_2, \ldots, y_i$ as a function of the independent variable $x_1, x_2, \ldots, x_i$. Suppose further that the uncertainty in x is insignificant in comparison with the uncertainty in y. It is desired to fit a straight line to these results:

$$Y = a + b(x - \bar{x}) \tag{2-87}$$

where upper-case Y denotes the value *predicted* by the equation. The best-fitting straight line through the data is the line which makes the sum of squares of deviations of measured y values from predicted Y values a minimum. This sum of squares is

$$\sum^{n_i} (Y_i - y_i)^2 = \sum^{n_i} [a + b(x_i - \bar{x}) - y_i]^2 \tag{2-88}$$

The sum of squares is a function of the constants a and b. It will be a minimum when

$$\frac{\partial}{\partial a}\left[\sum^{n_i} (Y_i - y_i)^2\right] = 0$$

$$\frac{\partial}{\partial b}\left[\sum^{n_i} (Y_i - y_i)^2\right] = 0$$

When the indicated differentiations are carried out and the resulting expressions solved for a and b, there results

$$a = \frac{\sum^{n_i} y_i}{n_i} = \bar{y} \tag{2-89}$$

$$b = \frac{\sum^{n_i} (x_i - \bar{x})(y_i - \bar{y})}{\sum^{n_i} (x_i - \bar{x})^2}$$

$$= \frac{\sum^{n_i} y_i(x_i - \bar{x})}{\sum^{n_i} (x_i - \bar{x})^2}$$

$$= \frac{\sum^{n_i} x_i y_i - \left(\sum^{n_i} x_i\right)\left(\sum^{n_i} y_i\right) / n_i}{\sum^{n_i} x_i^2 - \left(\sum^{n_i} x_i\right)^2 / n_i} \tag{2-90}$$

Implicit in the above analysis is the assumption that the error in y is independent of the magnitude of x. The best measure of the precision with which the points fit the line is the *variance of estimate*

$$s_e^2(y_i) = \frac{\sum^{n_i} (Y_i - y_i)^2}{n_i - 2} \tag{2-91}$$

The $n_i - 2$ degrees of freedom result from the use of two quantities, a and b, which are calculated from the data. An estimate of the population variance is

$$s_p^2(y_i) = \frac{\sum^{n_i} (y_i - \bar{y})^2}{n_i - 1} \tag{2-92}$$

An estimate of the validity of the linear behavior of the data may be obtained by means of an F test:

$$F = \frac{s_p^2(y_i)}{s_e^2(y_i)} \tag{2-93}$$

If the F test is satisfactory, $s_e^2(y_i)$ may be taken as the best estimate of the precision of the data.

Now both a and b as calculated are subject to error, and this error is a direct consequence of the error in the measured values of y. Estimates of the error variance of a and b are readily evaluated:

$$s_e^2(a) = s_e^2(\bar{y}) = \frac{s_e^2(y_i)}{n_i} \tag{2-94}$$

$$s_e^2(b) = s_e^2 \frac{\sum^{n_i} x_i y_i - (\Sigma x_i \Sigma y_i / n_i)}{\sum^{n_i} (x_i - \bar{x})^2}$$

$$= \frac{\sum^{n_i} x_i^2 - n_i \bar{x}^2}{\left[\sum^{n_i} (x_i - \bar{x})^2\right]^2} s_e^2(y_i)$$

$$= \frac{s_e^2(y_i)}{\sum^{n_i} (x_i - \bar{x})^2} \tag{2-95}$$

The expansions leading to (2-95) are a consequence of Eq. (2-29). As a result of the error in a and b, the predicted value of Y may be in error. The estimate of the error variance of Y_i is

$$\begin{aligned} s_e^2(Y_i) &= s_e^2[a + b(x_i - \bar{x})] \\ &= s_e^2(a) + (x_i - \bar{x})^2 s_e^2(b) \\ &= s_e^2(y_i)\left[\frac{1}{n_i} + \frac{(x_i - \bar{x})^2}{\sum^{n_i} (x_i - \bar{x})^2}\right] \end{aligned} \tag{2-96}$$

The number of degrees of freedom associated with $s_e^2(Y_i)$ is $n_i - 2$. The t test may be used to determine the confidence limits of Y_i:

$$\text{Confidence limits on } Y_i = (\pm t)\sqrt{s_e^2(Y_i)} \tag{2-97}$$

The confidence limits are functions of x_i. If the limits are plotted on the same graph as the least-squares straight line, they will form a hyperbolic envelope enclosing the straight line. If 95 per cent confidence limits were chosen, the true line will lie inside the confidence envelope 95 per cent of the time. However, the confidence envelope will not include 95 per cent of the data points.

Weighted Least Squares. The previous treatment requires modification if the absolute precision of y_i is a function of x_i (or y_i). This is often the case and may arise in two ways. In some types of measurements, the absolute precision varies with the magnitude of y_i. For example, many instrument scales may, because of a nonlinear graduation, be read with greater absolute precision at small values of y than at large values. Under such conditions, the variance of the measurement will be a function of y or x. In the second case, the variance may be a function of the variable level as a result of the method of treating the data. For example, suppose that the quantity z_i is measured as a function of x_i and the variance $s_e^2(z_i)$ is independent of the magnitude of z. However, the relation between z and x is of the form

$$Z = e^{a+bx} \tag{2-98}$$

where Z is the predicted value of z. It is desired to determine the best values of the constants a and b. The most convenient method to treat the data is to convert Eq. (2-98) to a linear form by taking logs:

$$\ln Z = Y = a + bx \tag{2-99}$$

The new dependent variable is then

$$y_i = \ln z_i \tag{2-100}$$

In view of the variance expression (2-29),

$$\sigma^2 y_i = \left(\frac{1}{z_i}\right)^2 \sigma^2 z_i = e^{-2y_i}\sigma^2 z_i \tag{2-101}$$

Consequently, the variance of y_i is a function of y_i.

The necessary modification of the least-squares treatment in such cases is to weight the measurements by the reciprocal of their variances:

$$W_i = \frac{1}{s^2 y_i} \qquad \text{(2-102)}\dagger$$

The equation for the constants in the equation

$$Y_i = a + b(x_i - \bar{x}_w) \tag{2-103}$$

are

$$a = \bar{y}_w \tag{2-104}$$

$$b = \frac{\sum^{n_i} W_i(x_i - \bar{x}_w)(y_i - \bar{y}_w)}{\Sigma W_i(x_i - \bar{x}_w)^2}$$
$$= \frac{\left(\sum^{n_i} W_i\right)\left(\sum^{n_i} W_i x_i y_i\right) - \left(\sum^{n_i} W_i x_i\right)\left(\sum^{n_i} W_i y_i\right)}{\left(\sum^{n_i} W_i\right)\left(\sum^{n_i} W_i x_i^2\right) - \left(\sum^{n_i} W_i x_i\right)^2} \tag{2-105}$$

where

$$\bar{x}_w = \frac{\sum^{n_i} W_i x_i}{\sum^{n_i} W_i} \tag{2-106}$$

$$\bar{y}_w = \frac{\sum^{n_i} W_i y_i}{\sum^{n_i} W_i} \tag{2-107}$$

Then

$$s_e^2 a = \frac{1}{\sum^{n_i} W_i} \tag{2-108}$$

$$s_e^2 b = \frac{1}{\sum^{n_i} W_i(x - \bar{x}_w)^2} \tag{2-109}$$

$$s_e^2 Y_i = \frac{1}{\sum^{n_i} W_i} + \frac{(x_i - \bar{x}_w)^2}{\sum^{n_i} W_i(x_i - \bar{x}_w)^2} \tag{2-110}$$

† Strictly, the population variance $\sigma^2 y_i$ should be used. The sample estimate is used as an approximation.

Generalization of Least-squares Treatment. The least-squares treatment may be applied to linear relations in which both x and y are subject to error (Wald[18]). The equation to be fitted need not be linear, although in such cases the calculations may become very tedious. General methods for fitting a polynomial of the form

$$Y = a + bx + cx^2 + \cdots \tag{2-111}$$

to give the best values and the precision of the constants have been developed and are given in Snedecor,[16] Freeman,[11] Deming,[6] and Fisher.[9]

2-13. Other Applications. Statistical methods may be applied to other engineering problems: statistical quality control, minimization of risk in buying or selling according to quality specifications, etc. Space does not permit a discussion of the method of handling such situations. The references at the end of this chapter should be consulted.

BIBLIOGRAPHY

1. Anderson, R. L., and T. A. Bancroft: "Statistical Theory in Research," 1st ed., McGraw-Hill Book Company, Inc., New York, 1952.
2. Brownlee, K. A.: "Industrial Experimentation," Chemical Publishing Company, Inc., New York, 1947.
3. Burr, Irving W.: "Engineering Statistics and Quality Control," McGraw-Hill Book Company, Inc., New York, 1953.
4. Cochran, W. G., and G. M. Cox: "Experimental Designs," John Wiley & Sons, Inc., New York, 1950.
5. Cramer, H.: "Mathematical Methods of Statistics," Princeton University Press, Princeton, N.J., 1946.
6. Deming, W. E.: "Statistical Adjustment of Data," John Wiley & Sons, Inc., New York, 1946.
7. Dixon, Wilfrid, J., and Frank J. Massey, Jr.: "Introduction to Statistical Analysis," 1st ed., McGraw-Hill Book Company, Inc., New York, 1951.
8. Ezekiel, Mordecai: "Methods of Correlation Analysis," 2d ed., John Wiley & Sons, Inc., New York, 1941.
9. Fisher, R. A.: "Statistical Methods for Research Workers," 10th ed., Oliver & Boyd, Ltd., Edinburgh and London, 1946; Hafner Publishing Company, New York, 1946.
10. Fisher, R. A.: "Designs of Experiments," 4th ed., Oliver & Boyd, Ltd., Edinburgh and London, 1947; Hafner Publishing Company, New York, 1947.
11. Freeman, H. A.: "Industrial Statistics," John Wiley & Sons, Inc., New York, 1946.
12. Grant, E. L.: "Statistical Quality Control," McGraw-Hill Book Company, Inc., New York, 1946.
13. Mahalanobis, P. C.: Tables for L Tests, *Sankhyā* (The Indian Journal of Statistics), vol. 1, pt. 1, June, 1933.
14. Mood, Alexander McFarlane: "Introduction to the Theory of Statistics," 1st ed., McGraw-Hill Book Company, Inc., New York, 1950.
15. Neyman, J., and E. S. Pearson: On the Problem of k Samples, *Bull. acad. polonaise sci. lettres, Ser. A*, 1931.

16. Snedecor, G. W.: "Statistical Methods Applied to Experiments in Agriculture and Biology," Collegiate Press, Inc., of Iowa State College, Ames, Iowa, 1946.
17. Taylor, Charles R.: Graphical Correlation, *Iron Age*, vol. 160, no. 14, p. 78, Nov. 6, 1947.
18. Wald, A.: The Fitting of a Straight Line if Both Variables Are Subject to Error, *Ann. Math. Stat.*, September, 1940, p. 240.
19. Peirce, B. O.: "A Short Table of Integrals," Ginn & Company, Boston, 1929.

PROBLEMS

2-1. Each of the following quantities is in error to the extent indicated:

$$
\begin{aligned}
a &= 5.20 \pm 0.02 \\
b &= 0.00175 \pm 0.00003 \\
c &= 75 \pm 2 \\
x &= 15{,}400 \pm 200 \\
y &= (5.33 \pm 0.03)10^4 \\
z &= 4.98 \pm 0.03
\end{aligned}
$$

Determine the maximum values of the absolute and relative errors in the quantities resulting from performance of the following operations:

(*a*) $a + b + x + y$. (*b*) $a + b - z$.
(*c*) $y^{3.2}$. (*d*) $\log (a - z)$.
(*e*) antilog $(a - z)$. (*f*) $\dfrac{x}{y}$.
(*g*) abc. (*h*) $\displaystyle\int_{w=z}^{w=c} \frac{dw}{a - w}$.
(*i*) $\displaystyle\int_{w=z}^{w=c} \frac{dw}{a - \ln (w)}$. (*j*) $a + z = p$, $a - z = q$. Find error in solution for p and q.
(*k*) $p^2 + ap + z = 0$. Find error in solution for p.

2-2. The volume of stack gas leaving an oil-fired boiler is to be calculated from the Orsat analysis of the gas, the analysis of the oil fired, and the rate of firing. Approximate values of these quantities are as follows:

Orsat analysis:	
CO_2	11.1%
O_2	5.3%
CO	0.2%
N_2	83.4%
Oil rate	105 gal/hr
Specific gravity of oil	0.90
Temperature of flue gas	430°F
Temperature of air	80°F
Humidity of air	Unknown
Oil, % C	88.0
Percentage O	Negligible
Barometer	758 mm Hg
Pressure of stack gas	5 in. H_2O

How accurately must each of the quantities be measured if the volume of stack gas is desired accurate to 5.0 per cent. On the assumption that the partial pressure of water vapor in the entering air is 3.0 ± 0.2 mm Hg and if the tolerances just calculated are used, what is the accuracy in the calculated value for partial pressure of water in the stack gases?

2-3. A binary mixture of two completely miscible components obeying Raoult's law has a relative volatility of approximately 2.50. The original mixture containing

60 mole % A (low boiler) is to be rectified to an overhead product of 98 mole % A and a bottoms of 98 mole % B. The feed will enter at its boiling point, and the ratio of reflux to overhead product will be 2.30. How accurately must the relative volatility be known to ensure that the number of calculated perfect plates required be accurate to 10 per cent?

2-4. A sharp-edged rectangular weir was used to measure the flow of water from a trombone cooler under conditions such that the Francis formula,

$$q = C(L - 0.2H)^{1.5}$$

may be applied. For q in cubic feet per second and L and H in feet, $C = 3.33 \pm 3$ per cent. The value of L was 8.0 in. with an estimated precision of ± 0.1 in. H was measured to be 4.0 in., but because of turbulence the precision of this measurement was estimated to be only ± 0.4 in.

(*a*) What is the "maximum possible systematic error" in the calculated value of q?

(*b*) What is the probable precision of the calculated value of q?

(*c*) Express the calculated value of q in the proper and accepted form.

(*d*) If the weir is to be calibrated at these conditions by collecting and weighing the water which flows over the weir in a period of approximately ½ min, with what precision should the individual measurements be made if the desired precision of q is ± 0.5 per cent?

2-5. In a control test a sample of basic solution containing approximately 2.00 milliequivalents of base is titrated with standard acid. The normality of the standard acid can be measured to ± 0.001 and the volume of acid to ± 0.05 cm³. What acid normality and volume should be used under these circumstances to assure minimum error in the determination of the number of basic milliequivalents, and what will this error be?

(*a*) Assume applicability of the differential formula.

(*b*) Assume applicability of the probability formula.

2-6. A sample is analyzed with the following results: 4.69 per cent, 4.58 per cent, 4.41 per cent, 4.62 per cent, 4.65 per cent. What is the most probable value of the analysis? What confidence limits would you assign? Would your answers be changed if the above results had been obtained by five different analytical procedures?

2-7. In viscose production, four cakes were purified by a test procedure, and four cakes were purified in the regular manner. The following average residual shrinkages were obtained:

Cake	Test	Control
1	1.577	1.817
2	1.703	1.716
3	1.553	1.785
4	1.540	1.735
Average....	1.593	1.763

Is the difference in averages significant?

2-8. The following values have been obtained by counting the number of defects in a given area of sheet film:

Sample number	*Number of defects*
1	18
2	19
3	11
4	14
5	15
6	12
7	7
8	12
9	17
10	19

Is it reasonable to suppose that these samples came from a population having $\bar{X} = 17.5$?

2-9. What are 99 per cent confidence limits for the mean of the following subgroup. Could this subgroup have been drawn from a population that averaged zero?

0.983	−1.381
2.585	0.935
1.349	0.259
−3.370	0.184
1.925	0.494

2-10. In a routine analytical test, it has been determined that the standard deviation of the tests is $\sigma = 3.0$. On a single sample, four separate analyses are performed. What are the 95 per cent confidence limits to be applied to the mean of the quadruplicate tests?

2-11. If batch-to-batch variations in viscosity for 40-RV flake is $\sigma = 1.0$ RV, on the average how many out-of-limit batches will there be if the specification limits are 40 ± 2.0 RV?

2-12. The following two sets of data have been obtained:

Test A: 134, 146, 104, 119, 124, 161, 107, 83, 113, 129, 97, 123
Test B: 70, 118, 101, 85, 107, 132, 94

Is there a significant difference?

2-13. The standard deviation for the carboxyl analysis is 0.7 unit. A technical man wants a certain sample analyzed so that he is sure of the carboxyl concentration within ± 0.5 unit, with only 1 chance in 100 of being wrong. How many analyses would be necessary for such accuracy?

2-14. A jelly manufacturer knows that his filling machine can place jelly in a jar with a standard deviation of 0.2 oz. How much jelly, on the average, must be placed in jelly jars to be able to state on the jar "net 10 oz" if the Bureau of Food and Drugs inspector states that only one jar in a hundred can go below the stated 10 oz?

2-15. When dropped from 6 ft, tennis balls made by a certain manufacturer have been found to bounce 4 ft (48 in.), with a standard deviation $\sigma = 2$ in. If, in a certain contract the minimum bounce from a 6-ft drop is set at 45 in., what per cent of the balls would be rejected?

2-16. Fifty machines from manufacturer A require 12 overhauls per month, while 30 machines from manufacturer B require four overhauls per month. Is there a significant difference between machines from A and B?

2-17. How many pairs of analyses will be required to confirm that polymer blend A has an analysis of 40 amine ends and B has 44 ends, if $\sigma = 0.4$?

2-18. Several investigators independently measured the same quantity. When the several sets of data were analyzed, the within-group or error variance s_e^2 was found to be 12, with 16 degrees of freedom. The second estimate of the population variance, s_p^2, obtained from the mean values reported by each investigator was found to be 10, with 4 degrees of freedom. Interpret these results.

2-19. Analyze the following data obtained in a factorially designed experiment:

	C_1	C_2	C_3
R_1	14 17 15	12 13 12	15 16 14
R_2	19 18 18	18 17 15	20 18 17

2-20. Analyze the following data obtained in a factorially designed experiment:

	C_1	C_2	C_3	C_4
R_1	41	54	57	48
R_2	36	49	52	43
R_3	46	59	62	53

2-21. It was desired to meter the flow of cooling water in a heat exchanger, and rather than shut down the equipment to install an orifice or other standard flowmeter, pressure taps were installed across a length of pipe through which the water entered the heat exchanger. This makeshift flowmeter was then calibrated by collecting the effluent water in a large vessel for various readings of the water manometer connected across the test section of pipe. From the calibration data it is desired to formulate an empirical expression relating the weight rate of flow to the manometer reading. Using statistical analysis, the following are to be obtained:

(*a*) The values of the constants in the empirical equation expressing the flowmeter calibration

(*b*) The limits of these constants within which it is 95 per cent probable that the true values of the constants will lie

(*c*) A graphical or analytical representation of the weight rate of flow, within which it is 95 per cent probable that the true value of the flow rate at that manometer reading must lie, as a function of the manometer reading

The following data were taken:

Manometer reading, in. H_2O	*Flow rate, lb/sec*
1.0	14.1
2.0	16.5
3.0	22.9
5.0	31.3
10.0	46.0
12.0	44.8
20.0	63.2
30.0	81.5
40.0	86.6

CHAPTER 3

MATHEMATICAL FORMULATION OF THE PHYSICAL PROBLEM

3-1. Introduction. The mathematical treatment of engineering processes involves three basic steps: the expression of the problem in mathematical language, the appropriate mathematical operations, and the interpretation of the results. The first frequently involves setting up a differential equation, and, although the procedure is not inherently complicated, it is with this first step that engineers often experience the greatest difficulty.

The practical value of a differential equation lies in the fact that it affords a connection between a simple basic physical or chemical law and a frequently complex relation of several variables of engineering importance in a practical problem. The application of even the simplest physical law to a process taking place under variable conditions may result in a relation of some complexity.

3-2. Formulation of the Differential Equation. In those processes which involve continuous changes in the values of the variables, a differential equation represents the correct relation between the local values of the variables and their rate of change. Although the formulation of the differential equation representing the physical situation cannot be reduced to a stereotyped procedure, certain broad techniques are generally applicable. Basically, the formulation procedure derives from the fact that in a process undergoing continuous changes the relation between the dependent and independent variables is continuous. Consequently, the conditions required for the existence of the derivative of the function relating the dependent and independent variable, $y = f(x)$, are fulfilled. The derivative $dy/dx = f'(x)$ represents the rate of change of y with x. Furthermore, *if the increment dx is infinitesimal,* the incremental change in y caused by an infinitesimal change in x is

$$dy = f'(x)\,dx$$

The use of these relations is the key step in the formulation process.

The law of the conservation of mass, the law of conservation of energy, and many other basic physical laws may be stated in the form

Input of conserved quantity into system
− output of conserved quantity from system
= accumulation of conserved quantity in system

or, alternatively,

Rate of input of conserved quantity into system
− rate of output of conserved quantity from system
= rate of accumulation of conserved quantity in system

Henceforth, these statements will be shortened to

$$\text{Input} - \text{output} = \text{accumulation}$$

When the basic physical laws are expressible in this form, the differential equation is easily formulated.

3-3. Application of the Law of Conservation of Mass

Example 3-1. Mixing in a Flow Process. Consider a flow process in which a precipitation is being carried out by mixing two streams A and B to form a third stream C in which the precipitate is carried away. Figure 3-1 represents the process diagrammatically. The reaction is extremely rapid, and agitation will be assumed to be so efficient that the material in the tank has substantially the same composition at all points. Streams A and B enter at the rate of a and b cfm, respectively. Change in volume due to reaction may be neglected, so that stream C leaves at the rate of $a + b$ cfm.

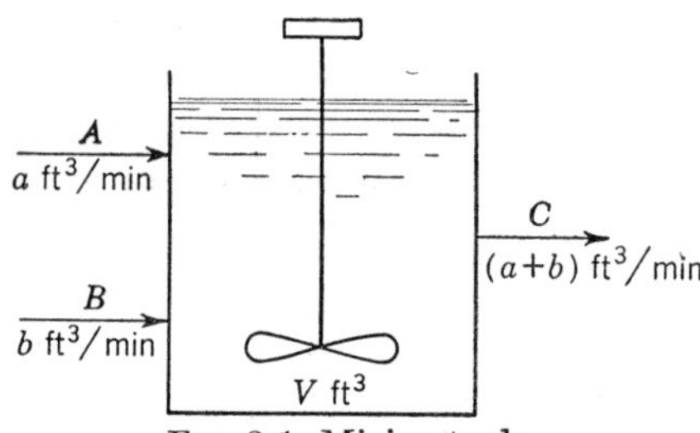

Fig. 3-1. Mixing tank.

To ensure proper quality of the precipitate, it is necessary to maintain the acidity of the tank contents at n_0 pcf acid, with an allowable variation of $\pm p$ pcf. The acidity of the bath is maintained by a negligibly small volume of acid carried by stream A. On the assumption that the acid supply fails suddenly at a time when the concentration is n_0, it is desired to develop the law relating acid concentration and time in order to estimate what time would elapse before this concentration falls below the allowable level. During this time, all other flows will be assumed constant.

Under the conditions granted, the law of importance in this situation is the law of conservation of matter. This is expressed in the form of a so-called "material balance"

$$\text{Input} - \text{output} = \text{accumulation}\dagger$$

In the case of the problem outlined, a material balance on the acid is significant, with input zero immediately upon failure of the acid supply. Picture what is happening over a very short interval of time during the process. The concentration of acid in

† When output is greater than input, accumulation is negative and is often called "depletion"; accumulation = −depletion.

the tank at any instant is n pcf, and the quantity of acid in the tank is Vn. Acid is flowing out of the tank at the rate of $n(a + b)$ lb/min. Over a short period of time Δt, the acid concentration changes by Δn to become $n + \Delta n$, and the accumulation of acid in the tank is the difference between the amount present at t and the amount present at $t + \Delta t$.

$$\text{Accumulation} = V(n + \Delta n) - Vn = V\,\Delta n$$

The output of acid over time interval Δt is given by the product of average concentration over this time n_{av} and the volume of effluent, $(a + b)\,\Delta t$. By the material balance,

$$\begin{aligned} -\text{Output} &= \text{accumulation} \\ -n_{av}(a + b)\,\Delta t &= V\,\Delta n \end{aligned} \tag{3-1}$$

Before an equation of this type becomes useful, it is necessary to determine the value of n_{av}. What type of average should be employed? Granting, for purposes of illustration, the applicability of an arithmetic average value of n at the beginning and end of the interval Δt, an obvious stepwise method of solution suggests itself. Starting with $n = n_0$, assume a small change in n, Δn, and calculate n_{av} as $(2n + \Delta n)/2$. Substitute this value of n_{av} into (3-1), and solve for Δt, obtaining the time t at which $n = n + \Delta n = n_1$. Repeat this process, assigning an increment to n_1, Δn_1, and, upon substitution of a new value of

$$n_{av} = \frac{2n_1 + \Delta n_1}{2}$$

obtain a corresponding time increment. Repeat the process until enough increments of n have been taken to cause it to reach its critical value $n_0 - p$, at which time the sum of all the time increments will approximate t_c. Obviously, the greater the accuracy desired in t_c, the smaller must be the increments Δn, so that the arithmetic mean value of n over an interval Δt will more nearly equal the true value of n_{av}. The great disadvantage of this method of solution is the excessive labor required to obtain an answer that is at best only approximate.

Although there are many cases in engineering where the relationship between the variables is so complex that this stepwise method becomes the only practical procedure, it is fortunate that in many cases an exact solution may be obtained by application of the calculus. Equation (3-1) may be rearranged to read

$$\frac{\Delta n}{\Delta t} = \frac{-n_{av}}{V}(a + b) \tag{3-2}$$

n is a function of t, and as $\Delta t \to 0$, $\Delta n \to 0$.

It will be recalled that, by definition of the derivative,

$$\lim_{\Delta t \to 0} \frac{\Delta n}{\Delta t} = \frac{dn}{dt}$$

Furthermore, as $\Delta t \to 0$, n_{av} approaches the value of n at time t, and (3-2) may be written

$$\frac{dn}{dt} = -\frac{n}{V}(a + b) \tag{3-3}$$

Multiplying by dt/n gives

$$\frac{dn}{n} = -\frac{a + b}{V}\,dt \tag{3-4}$$

Each side of (3-4) contains only one variable and may be integrated. The limits of the integrations are obtained if it is recalled that when $n = n_0$, $t = 0$, and when $n = n_0 - p$, $t = t_c$.

$$-\int_{n=n_0}^{n=n_0-p} \frac{dn}{n} = \int_{t=0}^{t=t_c} \frac{a+b}{V}\,dt \tag{3-5}$$

Integration and rearrangement give

$$t_c = \frac{V}{a+b} \ln \frac{n_0}{n_0 - p} \tag{3-6}$$

Equations of the type of (3-1), involving variables and finite increments of the variables, are known as "difference equations," and equations of the type of (3-3), involving variables and their infinitesimal increments or differentials, are known as "differential equations."

Example 3-2. Starting an Equilibrium Still. Consider the case of starting the equilibrium still shown schematically in Fig. 3-2. The still is purifying benzene and toluene from a small amount of essentially nonvolatile impurity and is initially charged with 20 lb moles of feed stock of composition x_F = 0.32 mole fraction benzene. Feed is supplied at the rate of 10 lb moles/hr, and the heat input is adjusted so that the total moles of liquid in the still remain constant at 20. It is desired to estimate the time required for the composition of overhead product y_D to fall to 0.40 mole fraction benzene. No liquid stock is removed from the still during this period.

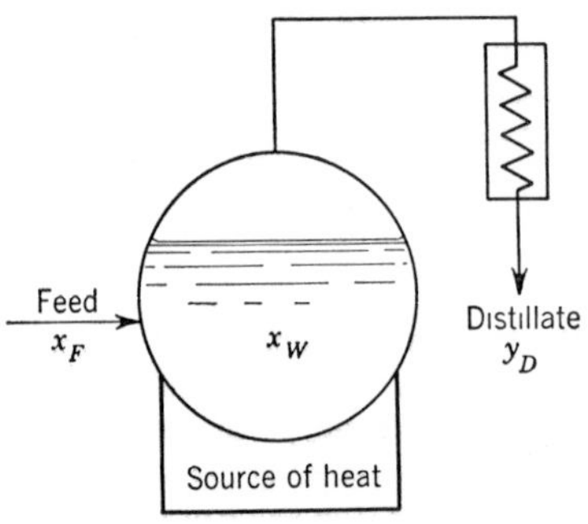

FIG. 3-2. Batch distillation.

Since no chemical changes occur, the mass and moles of benzene and toluene are individually conserved. Consequently, two material balances may be written, with the still taken to be the system:

$$\text{Rate of input of all components} = 10 \text{ moles/hr}$$
$$\text{Rate of output of all components} = D \text{ moles/hr}$$
$$\text{Rate of accumulation of all components} = \frac{dM}{dt} \text{ moles/hr} = 0$$

Hence
$$10 - D = \frac{dM}{dt} = 0 \tag{3-7}$$

and
$$D = 10 \tag{3-8}$$

$$\text{Rate of input of benzene} = 10 \cdot 0.32 = 3.2 \text{ moles/hr}$$
$$\text{Rate of output of benzene} = Dy_D = 10y_D \text{ moles/hr}$$
$$\text{Rate of accumulation of benzene} = \frac{d}{dt} MX_w$$
$$= M\frac{dX_w}{dt} + X_w\frac{dM}{dt}$$
$$= M\frac{dX_w}{dt} = 20\frac{dX_w}{dt} \text{ moles/hr}$$

Hence
$$3.2 - 10y_D = 20\frac{dX_w}{dt} \tag{3-9}$$

$$dt = \frac{20\,dX_w}{3.2 - 10y_D} \tag{3-10}$$

The left-hand side of this equation may be integrated directly, but before the right-hand side may be integrated it must be reduced to a function of one variable, that is, y_D must be known as a function of x_w. In the present case, the result of any refluxing in the neck of the still will be neglected and y_D assumed equal to the composition of equilibrium vapor off the still. The relation between x_w and y_D is given by vapor-liquid equilibrium data. If the relation is known, the right-hand side of (3-10) may be evaluated. Benzene and toluene may be assumed to follow Raoult's law, and the relative volatility α† may be taken as constant at an average value of 2.48.

Under these circumstances,

$$y_D = \frac{2.48x_w}{1 + 1.48x_w} \tag{3-11}$$

Substituting for y_D in (3-10) and simplifying give

$$dt = \frac{20dx_w}{3.2 - 20.1x_w} + \frac{29.6x_w\,dx_w}{3.2 - 20.1x_w} \tag{3-12}$$

The limits of integration are inserted by recalling that when $t = 0$, $x_w = x_F = 0.32$, and when $t = t$, x_w is obtained as 0.21 by substituting $y_D = 0.40$ in (3-11):

$$t = 20\int_{0.32}^{0.21} \frac{dx_w}{3.2 - 20.1x_w} + 29.6\int_{0.32}^{0.21} \frac{x_w\,dx_w}{3.2 - 20.1x_w} \tag{3-13}$$

Integration gives

$$t = \frac{-20}{20.1}[\ln(3.2 - 20.1x_w)]_{0.32}^{0.21} - 29.6\left[\frac{x_w}{20.1} + \frac{3.2}{(20.1)^2}\ln(3.2 - 20.1x_w)\right]_{0.32}^{0.21}$$

Substitution of limits gives

$$t = \frac{-20}{20.1}\,2.3\log\frac{3.2 - 20.1\cdot 0.21}{3.2 - 20.1\cdot 0.32} - 29.6\left[\frac{0.21 - 0.32}{20.1} + \frac{3.2}{(20.1)^2}\,2.3\log\frac{3.2 - 20.1\cdot 0.21}{3.2 - 20.1\cdot 0.32}\right]$$

$$= 1.58 \text{ hr}$$

It is important to emphasize the general procedure to be followed in setting up a differential material balance such as (3-9). No difficulty will be experienced with regard to signs of differentials if the general rule is followed of substituting the differential input, output, and accumulation directly into the material balance,

$$\text{Input} - \text{output} = \text{accumulation}$$

† If y_a and y_b are the vapor compositions of the two components a and b and if x_a and x_b are the corresponding equilibrium liquid compositions, the relative volatility α is defined by the equation

$$\frac{y_a}{y_b} = \alpha\frac{x_a}{x_b}$$

which, in the case of a binary, reduces to $y_a/(1 - y_a) = \alpha[x_a/(1 - x_a)]$. When Raoult's law is obeyed, α is the ratio of the vapor pressures of the two components.

if no attention is paid to the inherent sign of the differentials themselves, and if they are always considered positive. For example, x_w in the preceding problem decreases as t increases, so that dx_w is *inherently* negative when dt is positive. In (3-9), however, it must be noted that the signs in front of all the differentials are taken as positive, the one negative sign always being present before the output term in the material balance when output is written on the left-hand side of the equation.

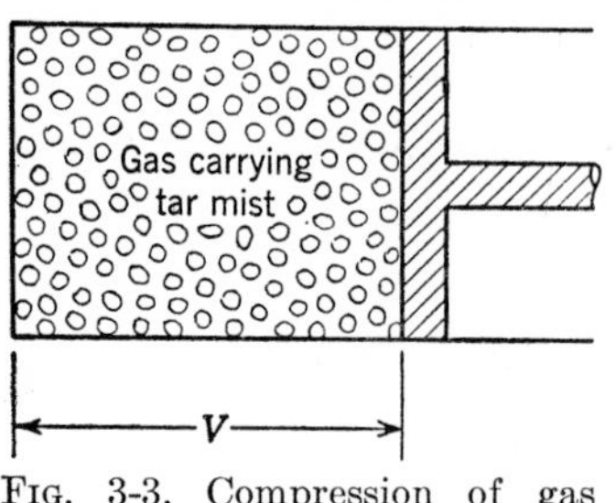

FIG. 3-3. Compression of gas carrying tar mist.

3-4. Application of the Law of Conservation of Energy. Many problems require the use of the law of conservation of energy. It will be seen from the following illustration that the principles involved in setting up differential energy balances and differential material balances are identical.

Example 3-3. Gas Compression. A gas containing an entrained mist of nonvolatile tar is located inside the cylinder of a reciprocating compressor. (See Fig. 3-3.) It is desired to determine the work required to compress the gas adiabatically and reversibly from its present pressure of 0.33 atm to a pressure of 1.0 atm. The following information is available:

1. The gas.
   a. Molecular weight is 24.
   b. Specific heat at constant volume is constant at 6.2 Btu/(lb mole)(°F).
   c. Obeys the perfect-gas equation of state.
   d. Initial temperature is 170°F or 630°R.
   e. Initial pressure is 0.33 atm.
   f. Final pressure is to be 1.0 atm.
2. The tar.
   a. The tar is always present as a mist in the ratio 0.2 lb of tar per pound of tar-free gas.
   b. The volume of the tar may be neglected in comparison with the volume of the associated tar-free gas.
   c. The temperature of the tar is always the same as the temperature of the associated tar-free gas.
   d. The specific heat of the tar is 0.5 Btu/(lb)(°F).
3. The compression cylinder.
   a. The initial cylinder volume is 0.4 ft.^{3}
   b. The initial cylinder pressure is 0.33 atm.
   c. The final cylinder pressure is to be 1.0 atm.
   d. The compression process is reversible and adiabatic in the sense that friction and heat transfer to or from the cylinder, piston, and associated machinery may be neglected.

Solution. Let the thermodynamic system comprise the constant mass m of the gas and associated tar originally inside the cylinder. When applied to this system, the first law of thermodynamics may be written in the form

$$dE = dq - dW \tag{3-14}$$

But $dq = 0$, and integration of (3-14) gives

$$-W = \text{work done by surroundings on system} = E_2 - E_1 \tag{3-15}$$

Now

$$dE = mc_V\,dT \tag{3-16}$$

where c_V is the specific heat per unit mass of *tar and gas* and m is the mass of tar and gas.

$$dW = p\,dV \tag{3-17}$$

where V is the volume of the system at the pressure p. Since the tar volume may be neglected, V may be taken to be the volume of the tar-free gas. Hence

$$V = \frac{R}{24}\frac{m}{1.2}\frac{T}{p} \tag{3-18}$$

and

$$dW = \frac{R}{24}\frac{m}{1.2}\left(dT - T\frac{dp}{p}\right) \tag{3-19}$$

when consistent units are employed. Combination of (3-14) with $dq = 0$, (3-16), (3-17), and (3-19) gives

$$mc_V\,dT = \frac{R}{24}\frac{m}{1.2}\left(T\frac{dp}{p} - dT\right)$$

or

$$\left(\frac{28.8c_V}{R} + 1\right)\frac{dT}{T} = \frac{dp}{p} \tag{3-20}$$

Integration of (3-20) gives

$$\frac{28.8c_V + R}{R}\ln\frac{T_2}{T_1} = ln\frac{p_2}{p_1} \tag{3-21}$$

Now

$$c_V = \frac{(6.2/24) + 0.2 \cdot 0.5}{1.2} = 0.30 \text{ Btu/(lb)(°R)}$$

and $R = 1.99$ Btu/(lb mole)(°R). Hence (3-21) becomes

$$T_2 = 630(3)^{1/5.4} = 772°\text{R}$$

The mass m of the system may be found from (3-18):

$$m = \frac{28.8}{R}\frac{V_1 p_1}{T_1} \tag{3-22}$$

In order to introduce consistent units, take

$$R = 1.99 \cdot 778 = 1{,}544 \frac{\text{ft-lb force}}{\text{(lb mole)(°R)}}$$

$$V_1 = 0.4 \text{ ft}^3$$

$$p_1 = 0.33 \cdot 14.696 \cdot 144 = 696 \frac{\text{lb force}}{\text{ft}^2}$$

$$T_1 = 170 + 460 = 630°\text{R}$$

Then

$$m = \frac{28.8 \cdot 0.4 \cdot 696}{1{,}455 \cdot 630} = 0.00825 \text{ lb mass}$$

The work of compression is given by combining (3-15) and the integral of (3-16)

$$-W = E_2 - E_1 = mc_V(T_2 - T_1) \tag{3-23}$$

Hence, the work done by the surroundings on the system during the compression is

$$-W = 0.00825 \cdot 0.30(722 - 630) = 0.36 \text{ Btu} \quad \text{or} \quad 277 \text{ ft-lb force}$$

3-5. Summary of Steps. The preceding illustrations demonstrate the use of five well-defined steps in the solution of a problem involving differential equations:

1. Recall the basic physical laws that govern the situation, and decide which quantities must be regarded as unknown variables. If every quantity can be expressed in terms of a single independent variable, an ordinary differential equation may be expected.
2. Formulate the differential equation according to the controlling physical laws.
3. Eliminate all but two variables from the equation formulated in 2. A thorough understanding of the physical situation should ensure no difficulty in these first three steps.
4. Obtain the general solution by suitable integration of the differential equation in two variables resulting from 3. This is primarily a mathematical problem and is covered more fully in Chap. 4.
5. Substitute the boundary conditions in the general solution, and determine the value of the constants for a particular solution applicable to the case at hand.

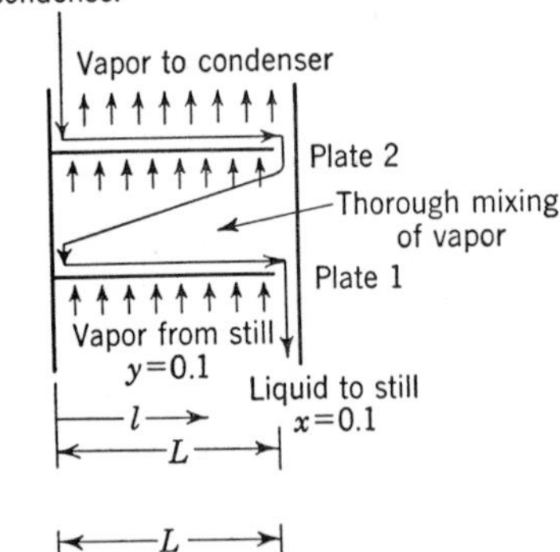

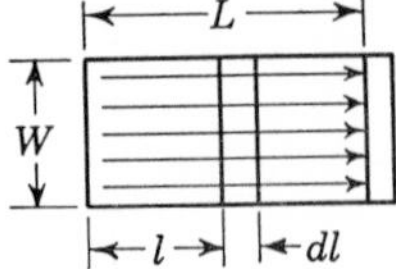

FIG. 3-4. Rectifying column with cross down pipe.

Example 3-4. Concentration Gradient across a Bubble Plate. The usual calculations in the design of rectifying columns assume perfect mixing on each plate and that the composition of the liquid leaving the plate is the same as that of the liquid at every point on the plate. Actually, conditions as pictured by this simple concept do not exist on the plates in any real column. To study the effects of actual concentration gradients on the rectification of a binary mixture, it has been proposed to employ a column (Fig. 3-4) consisting of two horizontal rectangular perforated plates. Liquid reflux at its boiling point is to return from the condenser and flow from left to right across the second (upper) plate. The reflux leaving the right side of the second plate returns to the first (lower) plate, again flowing from left to right.

The two components A and B are miscible in all proportions, and their mixtures will be assumed to obey Raoult's law. Over the range of temperatures involved, the vapor pressure of A is just twice that of B so that the relative volatility of A to B (α) is 2.00.

The column is to be operated at total reflux; i.e., all the vapor passing up through the plates is condensed and returned as reflux. The still will be charged with liquid of such composition that when the steady state is reached the vapor entering the first plate will be of uniform composition (10 mole % A). For the puposes of the present calculation, assume the plate efficiency to be 100 per cent at all points on the plates; i.e., the vapor leaving the liquid at any point on the plates is in equilibrium with the liquid at that point. The vapors leaving the lower plate will be considered to be thoroughly mixed before entering the top plate. It is required to calculate the average composition (mole % A) of the vapors leaving the top plate.

This composition may be readily calculated by suitable integration of differential material balances. The usual assumptions as to constancy of O and V will be made. The following nomenclature will be employed (see Fig. 3-4):

x_n = liquid composition at any point on nth plate (mole fraction)
y_n = composition of vapor evolved from any point on nth plate (mole fraction)
y_n^* = composition of vapor in equilibrium with liquid of composition x_n
$\bar{y}_n$ = composition of completely mixed vapor leaving nth plate.
l = distance from left to right on the plate
L = length of plate
W = width of plate
V = moles of vapor flowing up column per unit width of plate
O = moles of liquid flowing down column per unit width of plate

Concentration gradients in the direction W will be neglected so that the actual value of W is immaterial and a plate may be considered one unit in width.

Now apply the law of conservation of mass to component A, with the system taken to be a differential volume cutting out a differential length dl on the nth plate.

Input of Component A

$$\text{Carried by vapor entering system from plate below} = \frac{dl}{L} V y_{n-1}^*$$

$$\text{Carried by liquid entering section} = O x_n$$

Output of Component A

$$\text{Carried by vapor leaving system} = \frac{dl}{L} V y_n^*$$

$$\begin{aligned}\text{Carried by liquid leaving section} &= O x_n + \frac{d}{dl}(O x_n)\, dl \\ &= O(x_n + dx_n)\end{aligned}$$

The reasoning behind this last relation is as follows: At the position $l = l$, the rate of flow of A in the liquid is Ox_n. Since this rate of flow is a continuously changing process, the derivative $d(Ox_n)/dl$ is the rate at which the rate of flow is changing with distance along the plate. The increment in the rate of flow corresponding to an infinitesimal change dl in position along the plate is then $[d(Ox_n)/dl]\, dl$, and the rate of flow at the position $l + dl$ is $Ox_n + [d(Ox_n)/dl]\, dl$.

Accumulation of Component A. The accumulation in the section is zero because the column is operating in the steady state.

Substitution into the material balance gives

$$\text{Input} - \text{output} = \text{accumulation}$$

$$\frac{dl}{L} V y_{n-1}^* + O x_n - O(x_n + dx_n) - \frac{dl}{L} V y_n^* = 0 \tag{3-24}$$

When (3-24) is rearranged,

$$\frac{V\,dl}{L}(y_n^* - y_{n-1}^*) = -O dx_n \tag{3-25}$$

On application of (3-25) to plate 1, y_{n-1}^*, the composition of vapor from the still, becomes constant at 0.1, and y_n^* is related to x_n by the equilibrium equation

$$y_n^* = \frac{\alpha x_n}{1 + (\alpha - 1)x_n} = \frac{2x_n}{1 + x_n} \tag{3-26}$$

Substitution in (3-25) and separation of the variables result in

$$-\frac{(1 + x_1)\,dx_1}{1.9x_1 - 0.1} = \frac{V}{O}\,d\,\frac{l}{L} \tag{3-27}$$

When l varies from O to L, the variable l/L varies from 0 to 1. V/O is equal to unity because the column is operating at total reflux. Again, by a material balance, the composition of liquid leaving the first plate must be equal to 0.1, the composition of vapor rising from the still, and the composition of liquid entering the first plate must equal $\bar{y}_1$, the composition of mixed vapor leaving plate 1. Integrating between the limits of $x_1 = \bar{y}_1$, $l/L = 0$, $x_1 = 0.10$, $l/L = 1$, there results

$$\frac{1}{1.90}[2.30 \log (1.90x_1 - 0.10)]_{0.1}^{\bar{y}_1} + \left[\frac{x_1}{1.9}\right]_{0.1}^{\bar{y}_1} + \frac{0.1}{(1.90)^2}[2.30 \log (1.90x_1 - 0.10)]_{0.1}^{\bar{y}_1} = 1$$

Inserting the limits and collecting terms give

$$\frac{2.00 \cdot 2.30}{(1.90)^2} \log \frac{1.90\bar{y}_1 - 0.1}{0.090} + \frac{\bar{y}_1 - 0.10}{1.90} = 1 \tag{3-28}$$

Solving (3-28) for $\bar{y}_1$ gives $\bar{y}_1 = 0.29$. Note that $\bar{y}_1$ is independent of the value of L. The next step is to apply the differential material balance (3-25) to plate 2.

$$\frac{V\,dl}{L}(y_2^* - \bar{y}_1) = -O dx_2 \tag{3-29}$$

Substituting for y_2^* in terms of x_2 from (3-26) and separating the variables result in an equation exactly analogous to (3-27):

$$\int_{x_2=0.294}^{x_2=\bar{y}_2} \frac{(1 + x_2)\,dx_2}{1.71x_2 - 0.29} = \frac{V}{O}\int_{l=0}^{l=L} \frac{dl}{L} \tag{3-30}$$

Integration, substitution of limits, and solution for $\bar{y}_2$ in the preceding manner give $\bar{y}_2 = 0.58 = x_{\text{condenser}}$.

3-6. Rate Equations. The fundamental laws governing each of the unit operations in chemical engineering are most readily expressed in the form of simple rate equations. In heat transmission by conduction, for example, the basic law states that the rate of heat transmission with respect to time, dQ/dt, is directly proportional to the area normal to the direction of flow and to dT/dx, the rate of change of temperature with distance in the direction of flow. The basic law in the theory of viscous flow in pipes states that in a given pipe the rate of change of pressure with length of pipe is proportional to the velocity of flow. Diffusion theory proceeds from the basic law that the rate of mass transferred by diffusion with respect to time is directly proportional to the area normal to the direction of flow and to dc/dx, the rate of change of concentration with distance in the direction of flow. The rate at which a chemical reaction proceeds is a function of the concentrations of the reacting substances.

In view of the fact that the mathematical statements of all the preceding laws and many others not mentioned are essentially parallel, it is not surprising to find that the quantitative treatment of many unit operations is quite similar. Particular cases, however, demand the combination of these simple rate equations with differential material and energy balances in such an endless variety of combinations that the various integrated equations often bear but superficial, if any, semblance to each other. The following examples have been selected for the purpose of illustrating the methods of combining these simple relationships in the solution of more complex problems.

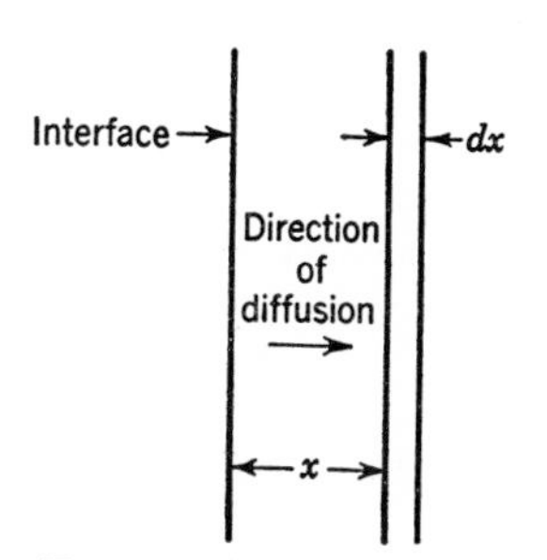

FIG. 3-5. Diffusion with chemical reaction.

Example 3-5. Diffusion with Chemical Reaction. A gas is absorbed by a solution with which it reacts chemically. The rate of diffusion in the liquid will be assumed proportional to the concentration gradient, and the diffusing gas is eliminated as it diffuses by a chemical reaction of the first order in which the rate of reaction is proportional to the concentration of the solute gas in the liquid. It is required to obtain an expression for the concentration in the liquid as a function of the distance from the interface.

Although no new principles are involved in formulating the differential equation arising from this problem, it will be convenient to illustrate the use of a slightly different technique which has general applicability in cases of this nature.

Figure 3-5 represents an enlarged section of the liquid film adjacent to the gas-liquid interface. Conditions at any point in a plane a distance x from the interface and normal to the direction of diffusion are the same, and a material balance will be set up describing the diffusion in and out of the differential element of thickness dx. Considering the diffusion across one unit of area into this differential layer, the rate of input of material is

$$\text{Rate of input} = -D\frac{dc}{dx}$$

where D is the diffusion coefficient and c the concentration of gas in the liquid at the point x. Since the concentration gradient is inherently negative in the direction of flow, it is necessary to place a negative sign before the gradient in order to substitute with no change of signs into the material balance as ordinarily written. The output of solute from the volume element of width dx is due to two effects: the diffusional transport of matter from the element and the consumption of the solute by the chemical reaction. Since the parameters vary in a continuous manner, the rate at which the solute diffuses out of the system across unit area of a plane through the point $x + dx$ is

$$-\left[D\frac{dc}{dx} + \frac{d}{dx}\left(D\frac{dc}{dx}\right)dx\right]$$

While diffusion is taking place across the layer dx, the diffusing material disappears by chemical reaction at a rate proportional to the amount of material present. The volume of the element under consideration is dx, since unit area is being considered, and the quantity of diffusing material in the element at any time is given by the product of this volume and the concentration c. The output rate due to chemical reaction is $-d(c\,dx)/dt$ and

$$-\frac{d(c\,dx)}{dt} = kc\,dx$$

where k is the reaction-rate constant.

Since the process is taking place under steady-state conditions, the accumulation is zero. The material balance is then

$$\frac{d}{dx}\left(D\frac{dc}{dx}\right) = kc \tag{3-31}$$

If the diffusion constant D is independent of x, Eq. (3-31) becomes

$$\frac{d^2c}{dx^2} = \frac{k}{D}c \tag{3-32}$$

This is a linear differential equation of the second order and has a well-known solution, which will be discussed in Chap. 4.

Example 3-6. Flow of Heat from a Fin. A copper fin L ft long is triangular in cross section. It is w ft thick at the base and tapers off to a line (see Fig. 3-6). The base of this wedge-shaped piece of metal is maintained at a constant temperature T_B, and the fin loses heat by convection to the surrounding air which is at a temperature T_a. The surface coefficient of heat transfer is h Btu/(hr)(ft²)(°F). What is the relation between the temperature of the fin metal, T, and the distance from the base, $L - x$?

In this example, heat is the only form of energy involved and consequently may be regarded as a conserved quantity. The rate at which heat is conducted in the x direction is given by the Fourier relation

$$q = -kA\frac{dT}{dx}$$

where A is the area normal to x.†

† A rigorous analysis would require allowance for the temperature gradient in the direction normal to x. In the present case this is ignored and the temperature assumed to be a function of x only. Most fins are constructed so that this simplification is justified.

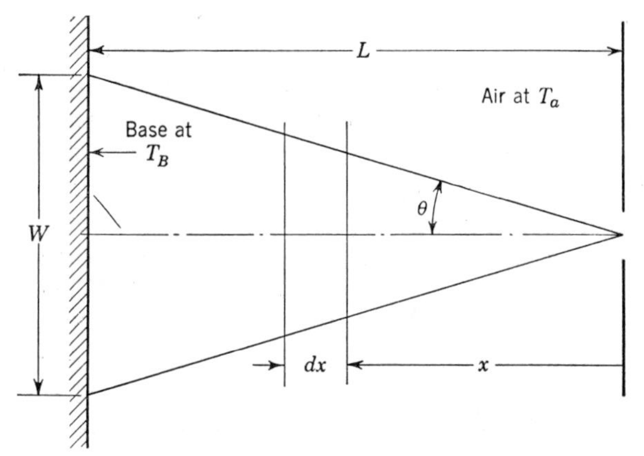

FIG. 3-6. Wedge-shaped fin.

Consider an element of volume of length dx and unit width at the point x. The rate at which heat is conducted into this element is

$$\text{Input rate} = -kA\frac{dT}{dx} = -k\frac{w}{L}x\frac{dT}{dx}$$

At $x + dx$, the rate of heat output by conduction is

$$-\left[k\frac{w}{L}x\frac{dT}{dx} + \frac{d}{dx}\left(k\frac{w}{L}x\frac{dT}{dx}\right)dx\right]$$

Note particularly that it is the increment in the complete heat-conduction rate that is involved and not just the change in the temperature gradient.

The rate at which heat passes from the fin surface to the surrounding air is

$$h\,ds(T - T_a) = h(2dx \sec\theta)(T - T_a)$$

The temperature-difference term $T - T_a$ is taken so that the expression will be positive if T is greater than T_a and hence physically would result in an output of heat from the element. Since unit width is involved, the surface area in contact with the air, ds, is equal to the perimeter which is $2dx \sec\theta$.

The process occurs at steady state, and hence the accumulation is zero. Then, rate of input $-$ rate of output $= 0$, and

$$\frac{d}{dx}\left(k\frac{w}{L}x\frac{dT}{dx}\right) - h\,2\sec\theta(T - T_a) = 0 \tag{3-33}$$

If the thermal conductivity is independent of x and the substitution $y \equiv T - T_a$ is introduced, Eq. (3-33) becomes

$$x\frac{d^2y}{dx^2} + \frac{dy}{dx} - \frac{2hL\sec\theta)}{kw}y = 0 \tag{3-34}$$

This is one form of Bessel's equation, and its solution is discussed in Chap. 5.

3-7. Rate Equations for Homogeneous Chemical Reactions. Consider the general irreversible reaction

$$aA + bB + cC \rightarrow dD + eE + fF \tag{3-35}$$

where a denotes the moles of species A reacting with b moles of B and c moles of C and where the mechanism of reaction is exactly as indicated by the stoichiometrical formulation. In deriving an expression for the rate of reaction, it is immaterial whether the rate of *disappearance* of any molecular species on the left or the rate of *appearance* of any molecular species on the right is followed. For purposes of illustration, the rate of disappearance of species A will be followed. At any time t, let N_A represent the moles of A present so that dN_A/dt will represent the moles of A disappearing per unit time. As t increases, N_A decreases, so that if dN_A/dt is to be equated to quantities inherently positive it must be preceded by a negative sign. $-dN_A/dt$ is first of all proportional to N_A, the moles of A present at any time, because clearly, if N_A is doubled, twice as many molecules will be available for reaction, resulting in double the number of moles disappearing per unit time; insertion of N_A is, in reality, insertion of a factor to allow for the scale of the operation. Reaction (3-35) may be written

$$(1)A + (a - 1)A + bB + cC \rightarrow dD + eE + fF \tag{3-36}$$

Focusing attention upon one molecule of A (3-36) makes it clear that, for reaction to occur, this molecule must collide simultaneously with $a - 1$ molecules of A, b molecules of B, and c molecules of C. The rate of disappearance of A is proportional to the number of simultaneous collisions of the type just mentioned, and the number of these collisions is proportional to the product of the concentrations of the reacting molecules, the concentration of each species being raised to a power corresponding to the number of molecules of that species which must collide with the molecule of A under consideration. The differential equation expressing the rate of disappearance of A at constant temperature is then

$$-\frac{dN_A}{dt} = kN_A C_A^{a-1} C_B{}^b C_C{}^c \tag{3-37}$$

where C represents concentration as moles per unit volume.

Equation (3-37) is the *general* equation applying to reaction rate in homogeneous systems.

The initial quantities of all species concerned in the reaction being known, it is possible by use of material balances, together with the stoichiometrical relations involved, to express all concentrations as functions of N_A; the two variables N_A and t, together with their differentials, may then be collected to give a solution of (3-37) by integration.

In the case of an adiabatic reaction, the temperature of the reacting materials will vary, and k, being a function of temperature, will vary. It is possible by application of the first law of thermodynamics, knowing the enthalpy of reaction and the heat capacities of the substances concerned, to express the temperature of reaction as a function of N_A. Under these circumstances, procedure by analytical methods is usually impractical, and it is best to write (3-37) in the form

$$\int_0^t dt = \int_{N_{A_0}}^{N_A} \frac{-dN_A}{kN_A C_A^{a-1} C_B{}^b C_C{}^c} \tag{3-38}$$

and evaluate graphically. In a reaction occurring at constant pressure under flow conditions and involving a change in volume, the time of contact t may be a function of N_A. An example of this kind is given subsequently.

In the special case of reaction at constant volume V, since $C_A = N_A/V$ and $dN_A = V\,dC_A$, Eq. (3-37) reduces to

$$-\frac{dC_A}{dt} = kC_A{}^a C_B{}^b C_C{}^c \tag{3-39}$$

Equation (3-39) is the form of the reaction-rate law usually seen in textbooks on physical chemistry. Obviously it is not generally applicable to the industrially important case of reaction at constant pressure unless there is no change in total moles, a condition implying constant volume.

By application of the stoichiometrical relations between the reactants and products given by (3-35), (3-37) is readily transformed to apply to any one of the molecular species present. For example, the rate of appearance of D is found by substituting in (3-37) the relation

$$-dN_A = \frac{a}{d}\,dN_D \tag{3-40}$$

whereupon (3-37) becomes

$$\frac{dN_D}{dt} = k\frac{d}{a} N_A C_A^{a-1} C_B{}^b C_C{}^c \tag{3-41}$$

In the application of equations of the type of (3-39) and (3-41), complication may arise from several sources. The general reaction represented by (3-35) may be reversible, in which case species A is being consumed by the forward reaction and generated by the reverse reaction. To follow the net change in quantity of any species with time, (3-37) must be applied to both forward and reverse reactions.

A frequent complication arises when the mechanism of the reaction may not be the same as indicated by the stoichiometric formulation. There may be intermediate products formed and subsequently broken down to the final products. In this case, rate equations are written for the controlling reactions, and substitution of terms is based on the over-all stoichiometry.

Finally, it is possible for the reactants as well as the intermediate products to be consumed by many side reactions in addition to the main reaction. If the quantity of material affected by these side reactions is appreciable, unless their nature is known, the development of a quantitative description of rate of conversion of any of the reactants must proceed on an empirical basis.

Example 3-7. Consecutive Reversible Reactions at Constant Volume. Consider the set of reversible reactions

$$A \rightleftharpoons B \tag{a}$$
$$B \rightleftharpoons C \tag{b}$$

Assume 1 mole of A present at the start, and let N_A, N_B, and N_C denote the moles of A, B, and C, respectively, present at time t. Then, since the reaction is assumed to occur at constant volume, N_A, N_B, and N_C are proportional to concentrations. Let k_1 and k_2 be the velocity constants of the forward and reverse reactions, respectively, of (a); likewise, let k_3 and k_4 apply to (b). The net rate of disappearance of A is given by application of (3-39) to the forward and reverse reactions (a):

$$\frac{dN_A}{dt} = -k_1N_A + k_2N_B \tag{3-42}$$

and the net rate of disappearance of B is given by

$$\frac{dN_B}{dt} = -(k_2 + k_3)N_B + k_1N_A + k_4N_C \tag{3-43}$$

At all times, the material balance and the stoichiometrical relations dictate the relation

$$N_A + N_B + N_C = 1$$

Differentiate (3-42) with respect to t:

$$\frac{d^2N_A}{dt^2} = -k_1\frac{dN_A}{dt} + k_2\frac{dN_B}{dt} \tag{3-44}$$

Substitute dN_B/dt from (3-43):

$$\frac{d^2N_A}{dt^2} = -k_1\frac{dN_A}{dt} - k_2(k_2 + k_3)N_B + k_2k_1N_A + k_2k_4N_C \tag{3-45}$$

Replacing N_C by its equivalent $(1 - N_A - N_B)$, substituting for N_B from (3-42), and collecting terms give

$$\frac{d^2N_A}{dt^2} + (k_1 + k_2 + k_3 + k_4)\frac{dN_A}{dt} + (k_1k_3 + k_2k_4 + k_1k_4)N_A - k_2k_4 = 0 \tag{3-46}$$

This is a linear equation with constant coefficients which may be solved by the methods of Chap. 4.

3-8. Flow Systems. The analysis of a flow system is based upon the principles developed earlier; the approach differs only in the details of the technique. The analysis of a flow system may proceed from either of two different points of view. In the first approach, frequently termed the "Eulerian method," the analyst takes a position fixed in space. Attention is focused on a small volume element likewise fixed in space, and the law of conservation of mass, first law of thermodynamics, reaction-rate expressions, etc., are applied to this stationary system. In a steady-state situation, the objective of the analysis is to determine the properties of the fluid as a function of position. This method of approach is illustrated in the following example.

Example 3-8. Flow Process from the Eulerian Point of View. Consider a conduit through which a fluid is flowing at a steady rate. Let x denote the distance from the conduit entrance to an arbitrary position measured along the center line of the conduit

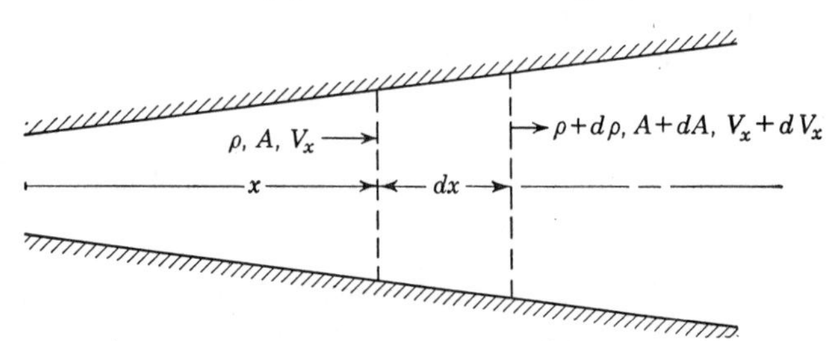

FIG. 3-7. Control volume fixed in space.

in the direction of flow. Let V_x denote the velocity of the fluid in the x direction at the point x, A denote the area normal to the x direction, and ρ denote the fluid density at point x. Apply the law of conservation of mass to an infinitesimal element of volume *fixed in space* and of length dx. (See Fig. 3-7.) If V_x and ρ are essentially constant across the area A, the rate of input of mass is

$$\text{Rate of mass input} = \rho A V_x = w$$

The rate of mass output is $\rho A V_x + (d/dx)(\rho A V_x)\,dx = w + (dw/dx)\,dx$. At steady state the accumulation is zero. Hence, the differential equation resulting from the application of the law of the conservation of mass to the volume element fixed in space is

$$d(\rho A V_x) = dw = 0 \tag{3-47}$$

Alternative forms of (3-47) are

$$\rho A\,dV_x + \rho V_x\,dA + A V_x\,d\rho = 0 \tag{3-48}$$

and

$$\frac{d\rho}{\rho} + \frac{dA}{A} + \frac{dV_x}{V_x} = 0 \tag{3-49}$$

The integrated form of (3-47) is

$$\rho A V_x = w = \text{const} \tag{3-50}$$

The equations (3-47) to (3-50) are commonly termed the "continuity equations." Since A is related to x through the geometry of the conduit, (3-50) can be used to relate ρV_x to x. Further information is needed to determine the separate values of ρ and V_x.

The alternative point of view is associated with Lagrange, who used it extensively. Here, the analyst takes a position astride a small volume element which *moves with the fluid.* In a steady-state situation, the objective of the analysis is to determine the properties of the fluid comprising the moving volume element as a function of the time which has elapsed since the volume element first entered the system. This elapsed time is frequently termed the "contact time." It is important to note that the point of view taken introduces the concept of time into the analysis, even though the flow itself may be a steady-state process. If the system is at steady state, then the properties of the fluid comprising a volume element are determined *solely* by the *elapsed* time which is the *difference* between the absolute time at which the element is examined and the absolute time at which the element entered the system. In the unsteady-state case, *both* the elapsed time and the absolute time would have an effect on the properties of the fluid comprising the element.

In the following example, the Lagrangian approach is applied to the situation considered in Example 3-8.

Example 3-9. Flow Process from the Lagrangian Point of View. Consider the flow system examined in Example 3-8. Focus attention on an infinitesimal mass element which moves with the fluid through the flow system. The dimensions (volume) of this element may vary as it moves with the fluid, but by definition the mass contained inside the volume remains constant. Let τ denote the elapsed time:

$$\tau = t - t_0 \tag{3-51}$$

where t is the absolute time at which the element is observed and t_0 is the absolute time at which the element entered the system. At elapsed time τ, the volume of the element is $A\ \delta a$ (where δa is the length of the element parallel to x), the density is ρ,

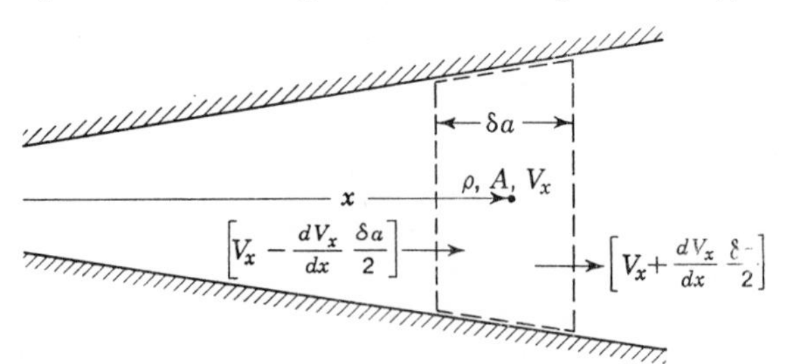

FIG. 3-8. Control volume moving with fluid.

and the velocity of the element relative to the stationary conduit walls is V_x. The properties A, ρ, V_x refer to conditions at the center of the element. (See Fig. 3-8.) When applied to the volume element, the law of conservation of mass becomes

$$\frac{d}{dt}(\rho A\ \delta a) = 0 \tag{3-52}$$

The integral of (3-52) is

$$\rho A\ \delta a = \delta m = \text{const} \tag{3-53}$$

In the above situation, the area of the element A is also the cross section of the conduit. A is generally specified as a function of the distance inside the conduit, x. This distance may be related to the elapsed time τ. The velocity V_x of the element is the time rate of change of the position of the element with respect to the conduit:

$$V_x = \frac{dx}{dt} \tag{3-54}$$

If $t = t_0$ when $x = 0$,

$$t - t_0 = \tau = \int_0^x \frac{dx}{V_x} \tag{3-55}$$

The time rate of change of the length of the element, $(d\,\delta a)/dt$, is related to the spatial rate of change of the velocity of the element in the following way: At time t, the velocity of the forward face of the element relative to the velocity of the center of the element is

$$\left(V_x + \frac{dV_x}{dx}\frac{\delta a}{n}\right) - V_x$$

Similarly, the relative velocity of the trailing face of the element is

$$\left(V_x - \frac{n-1}{n}\frac{\delta a}{2}\frac{dV_x}{dx}\right) - V_x$$

The difference between these two relative velocities is the time rate of change of the length of the element:

$$\frac{dV_x}{dx}\,\delta a = \frac{d\,\delta a}{dt} \tag{3-56}$$

Combining (3-56) and (3-54) gives

$$\frac{dV_x}{V_x} = \frac{d\,\delta a}{\delta a} \tag{3-57}$$

Differentiation of (3-53) gives

$$\frac{d\rho}{\rho} + \frac{dA}{A} + \frac{d\,\delta a}{\delta a} = 0 \tag{3-58}$$

Combining (3-58) and (3-57) generates

$$\frac{d\rho}{\rho} + \frac{dA}{A} + \frac{dV_x}{V_x} = 0 \tag{3-49}$$

which is one form of the continuity equation obtained earlier by application of the Euler approach.

The relations used above may always be used to transform the one-dimensional steady-state Lagrangian equations into the one-dimensional steady-state Eulerian equations. Similar, but more complicated, expressions apply to the multidimensional and/or unsteady-state transformations.

Example 3-10. Chemical Reaction in a Flow System. For the kinetically second-order irreversible decomposition of acetaldehyde

$$2CH_3CHO \rightarrow 2CH_4 + 2CO \tag{3-59}$$

Hinshelwood and Hutchison† report

$$k = 0.331/(\text{sec})(\text{g moles/liter}) \text{ at } 518^\circ\text{C}$$

Assume this reaction to be carried out at 1 atm by boiling the aldehyde in a flask and introducing the vapors into one end of a silica reaction tube while withdrawing products from the other end. Assume that piston flow occurs. All the contents of the reaction tube will be maintained at 518°C. If the reaction tube is 3.3 cm internal diameter and 80 cm long and if aldehyde is introduced at the rate of 50 g/hr, what fraction of the aldehyde distilled will be decomposed. What is the contact time τ?

Adopt the Eulerian point of view. Consider an infinitesimal volume element of area $A = \pi d^2/4$ and length dx, and let it be fixed in space at a point x distance from the entrance to the tube. Let

M_0 = molal flow rate of aldehyde at $x = 0$
M_a = molal flow rate of aldehyde at $x = x$
M_c = molal flow rate of CO at $x = x$

The stoichiometric equation (3-59) then demands that the molal flow rate of methane equal the molal flow rate of CO, M_c. The law of conservation of mass requires

$$M_a(MW)_a + M_c(MW)_{CO} + M_c(MW)_{CH_4} = M_0(MW)_a \tag{3-60}$$

where MW refers to the molecular weight of a chemical component. However, the stoichiometry demands that

$$(MW)_a = (MW)_{CO} + (MW)_{CH_4} \tag{3-61}$$

Hence

$$M_c = M_0 - M_a \tag{3-62}$$

At any point x, the total molal flow rate is then

$$M_T = 2M_0 - M_a \tag{3-63}$$

and the mole fraction of aldehyde is

$$y_a = \frac{M_a}{2M_0 - M_a} \tag{3-64}$$

With reference to the stationary volume element, a material balance applied to the aldehyde gives

$$\text{Input mole rate of aldehyde} = M_a$$

$$\text{Output mole rate of aldehyde} = M_a + \frac{d}{dx} M_a\, dx + \text{rate of aldehyde disappearance as a result of chemical reaction}$$

In the infinitesimal volume $A\,dx$, the total moles of gas are

$$dN = \frac{pA\,dx}{RT} \tag{3-65}$$

The moles of aldehyde are

$$N_a = dN y_a = \frac{(pA\,dx)M_a}{RT(2M_0 - M_a)} \tag{3-66}$$

The rate of disappearance of aldehyde in the volume element is

$$-\frac{dN_a}{dt} = kN_aC_a = k\left(\frac{p}{RT}\right)^2\left(\frac{M_a}{2M_0 - M_a}\right)^2 A\,dx \tag{3-67}$$

† *Proc. Roy Soc. (London)*, ser. A, vol. 111, p. 380, 1926.

The aldehyde material balance then becomes

$$\frac{dM_a}{dx}\,dx + k\left(\frac{p}{RT}\right)^2\left(\frac{M_a}{2M_0 - M_a}\right)^2 A\,dx = 0 \tag{3-68}$$

With the substitution

$$f = \frac{M_a}{M_0} = \text{fraction aldehyde undecomposed} \tag{3-69}$$

Eq. (3-68) becomes

$$\left(\frac{2-f}{f}\right)^2 df = -\frac{k}{M_0}\left(\frac{p}{RT}\right)^2 A\,dx \tag{3-70}$$

Integrating (3-70) between the limits $f = 1$ at $x = 0$ and $f = f_a$ at $x = l$ gives

$$4 - \frac{4}{f_a} - 4\ln f_a + (f_a - 1) = -\frac{k}{M_0}\left(\frac{p}{RT}\right)^2 Al \tag{3-71a}$$

In computing a value of f_a from (3-71a), consistent units must be used throughout. Thus, if the cgs system is used,

$$\begin{aligned}
k &= 0.331/(\text{sec})(\text{g moles/liter})\\
M_0 &= \frac{50}{3{,}600 \cdot 44} = 3.16 \cdot 10^{-4}\ \text{g mole/sec}\\
p &= 1\ \text{atm}\\
T &= 518 + 273 = 791^\circ\text{K}\\
R &= 0.08206\ \text{liter-atm/(g mole)}(^\circ\text{K})\\
Al &= \frac{\pi(3.3)^2 80}{4 \cdot 1{,}000} = 0.683\ \text{liter}\\
-\frac{k}{M_0}\left(\frac{p}{RT}\right)^2 Al &= \frac{-0.331}{3.16 \cdot 10^{-4}}\left(\frac{1}{0.08206 \cdot 791}\right)^2 0.683 = -0.17
\end{aligned}$$

$$3 - \frac{4}{f_a} - 9.2\log f_a + f_a = -0.17 \tag{3-71b}$$

Solving by trial gives $f_a = 0.87$; 13 per cent of the aldehyde will be decomposed upon passage through the tube. The contact time is

$$\tau = \int_0^l \frac{dx}{V_x} \tag{3-72}$$

$$\begin{aligned}
V_x &= \frac{Q}{A} = \frac{M_T RT}{Ap} = \frac{(2M_0 - M_a)RT}{Ap}\\
&= \frac{M_0(2-f)RT}{Ap}
\end{aligned} \tag{3-73}$$

The manipulations are easier if dx is eliminated from (3-72) by using (3-70) and the limits are changed. Then the introduction of (3-73) gives

$$\begin{aligned}
\tau &= -\frac{1}{k}\frac{RT}{p}\int_1^{f_a}\frac{2-f}{f^2}\,df\\
&= -\frac{1}{k}\frac{RT}{p}\left[2\left(1 - \frac{1}{f_a}\right) - \ln f_a\right]
\end{aligned} \tag{3-74}$$

At $f_a = 0.87$, $\tau = 31.2$ sec.

Example 3-11. Chemical Reaction in a Packed Tower. A System of Simultaneous Differential Equations. A countercurrent packed absorption tower is to be used for carrying out the liquid-phase reaction

$$A + B \rightarrow C$$

This reaction is irreversible, and the reaction rate may be expressed as follows:

$$-\frac{dN_A}{dt} = kN_AX_B \tag{3-75}$$

Components B and C are nonvolatile and never appear in the gas phase. Substance B is introduced into the tower dissolved in a nonvolatile solvent. Compound A is volatile and is introduced into the tower as a vapor carried by an insoluble, inert gas. The rate of transfer of A from the gas phase to the liquid phase is controlled by the gas-phase resistance and is proportional to $K_ga(Y_A - Y_A^*)$. Y_A^* is the equilibrium gas-phase concentration corresponding to the liquid-phase concentration, X_A. Y_A^* is related to X_A by the equation

$$Y_A^* = mX_A \tag{3-76}$$

Assuming isothermal conditions, develop the differential equation for the decrease in the concentration of A in the gas phase as a function of the tower height z and the following known quantities:

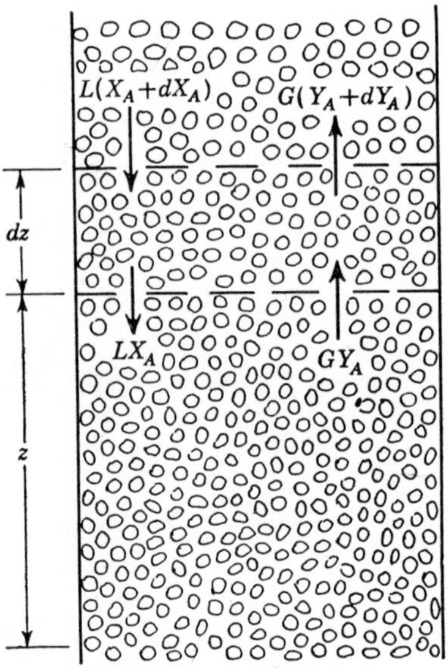

a = area for mass transfer per unit of tower volume, ft²/ft³
G = inert-gas rate, lb moles/hr
H = moles of inert solvent held up by packing per unit of tower volume
k = reaction-rate constant, lb moles/-(lb mole/hr)
K_G = mass-transfer coefficient, lb moles/hr (unit $Y_A - Y_A^*$)(ft²)
L = inert-solvent rate, lb moles/hr
m = proportionality constant
S = tower cross section, ft²

FIG. 3-9. Chemical reaction in a packed tower.

Nomenclature for Variables

t = time, hr
z = distance from bottom of tower, ft
N = moles of a component; N_A refers to moles of substance A, etc.
Y = gas-phase concentration, moles per mole of inert gas; Y_A refers to substance A
X = liquid-phase concentration, moles per mole of inert solvent; X_A refers to substance A, etc.

Consider a stationary volume element of cross section S and length dz located a distance z from the bottom of the tower. (See Fig. 3-9.) Write a material balance for component A:

$$\text{Input rate} = L\left[X_A + \frac{d}{dz}(X_A)\,dz\right] + GY_A$$

$$\text{Output rate} = LX_A + G\left[Y_A + \frac{d}{dz}(Y_A)\,dz\right] + kX_AS\,dz\,HX_B$$

$$\text{Accumulation} = 0$$

In the above formulation, note particularly the use of the convention that all variables are assumed to increase as the independent variable increases. This procedure keeps the algebraic signs of the terms consistent. The over-all material balance on component A is then

$$L\frac{dX_A}{dz} - G\frac{dY_A}{dz} - kSHX_AX_B = 0 \tag{3-77}$$

A similar material balance written for component B is

$$L\frac{dX_B}{dz} - kSHX_AX_B = 0 \tag{3-78}$$

Now consider the gas phase only. Write a material balance on component A:

$$\begin{aligned} \text{Input rate} &= GY_A \\ \text{Output rate} &= G\left[Y_A + \frac{d}{dz}(Y_A)\,dz\right] + K_gaS\,dz(Y_A - Y_A^*) \\ \text{Accumulation} &= 0 \end{aligned}$$

Then, the gas-phase material balance is

$$G\frac{dY_A}{dz} + K_gaS(Y_A - Y_A^*) = 0 \tag{3-79}$$

Combining (3-79) and (3-76) gives

$$G\frac{dY_A}{dz} + K_gaS(Y_A - mX_A) = 0 \tag{3-80}$$

The system of Eqs. (3-77), (3-78), and (3-80) must now be used to eliminate two of the three dependent variables, Y_A, X_A, and X_B, in order to obtain an equation which contains only one dependent variable. Y_A is taken as the retained dependent variable and X_A and X_B as the variables to be eliminated. If Eqs. (3-77), (3-78), and (3-80) are examined, it will be seen that in an algebraical sense the variables to be eliminated are X_A, X_B, *and* their derivatives dX_A/dz and dX_B/dz. At the moment, then, it is clear that not enough relations are available to accomplish the elimination. This difficulty is overcome in the following manner:

Rewrite Eq. (3-80):

$$X_A = \frac{1}{m}\left(Y_A + \frac{G}{K_gaS}\frac{dY_A}{dz}\right) \tag{3-80}$$

and differentiate:

$$\frac{dX_A}{dz} = \frac{1}{m}\left(\frac{dY_A}{dz} + \frac{G}{K_gaS}\frac{d^2Y_A}{dz^2}\right) \tag{3-81}$$

Now differentiate (3-81):

$$\frac{d^2X_A}{dz^2} = \frac{1}{m}\left(\frac{d^2Y_A}{dz^2} + \frac{G}{K_gaS}\frac{d^3Y_A}{dz^3}\right) \tag{3-82}$$

Rewrite (3-77):

$$\frac{L}{kHS}\frac{dX_A}{dz} - \frac{G}{kHS}\frac{dY_A}{dz} = X_AX_B \tag{3-77}$$

and differentiate:

$$\frac{L}{kHS}\frac{d^2X_A}{dz^2} - \frac{G}{kHS}\frac{d^2Y_A}{dz^2} = X_B\frac{dX_A}{dz} + X_A\frac{dX_B}{dz} \tag{3-83}$$

Rewrite (3-78):

$$\frac{dX_B}{dz} = \frac{kSH}{L} X_A X_B \tag{3-78}$$

The differentiation process coupled with the original relations has generated six equations, (3-77), (3-78), and (3-80) to (3-83), which involve five "unknowns," X_A, dX_A/dz, d^2X_A/dz^2, X_B, and dX_B/dz. Consequently, algebraic elimination is possible.

Combine (3-83) and (3-78) and solve explicitly for X_B:

$$X_B = \frac{(L/kHS)(d^2X_A/dz^2) - (G/kHS)(d^2Y_A/dz^2)}{(kHS/L)X_A^2 + dX_A/dz} \tag{3-84}$$

Eliminate X_B from (3-77) by using (3-84), and then eliminate X_A, dX_A/dz, and d^2X_A/dz^2 from the result by using (3-80) to (3-82). With the notation

$$P = kHS \tag{3-85}$$

$$R = K_g aS \tag{3-86}$$

the end result is

$$\frac{L}{m}\left(\frac{dY_A}{dz} + \frac{G}{R}\frac{d^2Y_A}{dz^2}\right) - G\frac{dY_A}{dz} = \frac{[Y_A + (G/R)(dY_A/dz)][(L/m)(d^2Y_A/dz^2) + (LG/mR)(d^3Y_A/dz^3) - G(d^2Y_A/dz^2)]}{(P/mL)[Y_A + (G/R)(dY_A/dz)]^2 + (dY_A/dz) + (G/R)(d^2Y_A/dz^2)} \tag{3-87}$$

Equation (3-87) is the desired relation since it involves only Y_A, z, their derivatives, and known constants. The relation is so complex, however, that elementary integration procedures are not applicable.

The differentiation procedure used above to reduce a system of differential equations to a single equation involving a single dependent variable is generally applicable. If the reduction is feasible, the number of additional relations generated by the differentiation of the system of equations will exceed the number of additional "unknowns" (higher derivatives) introduced until finally the solution is possible by algebraic techniques. In the case of linear equations, the elimination processes may be formalized by means of operational procedures. These are described in subsequent chapters.

3-9. The General Problem. A general problem that arises constantly in chemical engineering and applied chemistry is the following: Two or more phases containing several components are to be contacted with each other in a suitable apparatus. Each component may be considered transferable to the other phases with attendant heat effects, and chemical reactions may occur with liberation or absorption of heat. Given the initial conditions, it is required to determine either the state of the system at a given point in the system (after a given time of contact) or the size of equipment required (time of contact required) for a given state to be reached. This general problem might include any operation from multicomponent rectification to the catalytic oxidation of ammonia with subsequent oxidation and absorption of the nitrogen oxides. To describe this system, a differential section of the apparatus in which the contact is occurring is considered, and one or more of the following four types of relationships are applied:

1. *Material Balances.* There will be a differential material balance applying to each component. If chemical reactions are occurring, corresponding stoichiometric relations will be available for use.

2. *First Law of Thermodynamics.* A differential form of the first law may be applied to the section as a whole. If no heat is flowing to or from the surroundings and no work is being done by the system, the first law may reduce to an enthalpy balance which, however, may be complicated by heat effects associated with the various changes of state occurring in the section.

3. *Rate Equations.* Rate equations apply to the transfer of heat across all phase interfaces, to the mass transfer of each individual component across phase interfaces, and to each of the individual chemical reactions occurring in the various phases.

4. *Equilibriums.* All available evidence indicates that equilibrium substantially prevails in the region of the interfacial boundary between phases. These equilibriums are constantly shifting until the phases as a whole reach final equilibrium.

A very large number of simultaneous differential equations may be required to supply an exact description of a process—so many, in fact, that their manipulation becomes impractical. Consequently, mathematical simplification must be introduced in the form of reasonable simplifying assumptions, and the practical success of such a procedure will depend entirely upon the validity of the simplifying assumptions as revealed by preliminary knowledge or calculation.

PROBLEMS

3-1. A tank contains 100 ft^3 of fresh water; 2 ft^3 of brine, having a concentration of 1 pcf of salt, is run into the tank per minute, and the mixture, kept uniform by mixing, runs out at the rate of 1 ft^3/min. What will be the exit brine concentration when the tank contains 150 ft^3 of brine?

3-2. N_0 grams of a solid material was placed in W g of water at time t_0. The liquid was continuously stirred and maintained at a constant temperature. At the end of t_1 sec, N_1 g of solid remained undissolved. At the end of a very long period of time, N_2 g of solid remained undissolved. Set up the differential equations required to determine the rate of solution of the solid in terms of N_0, N_1, N_2, and t_0, t_1. Do not integrate your expression.

DATA AND NOTES: It is assumed that the rate of solution is proportional to (1) the surface area of the material and (2) the concentration driving force, where the concentration is expressed as grams of solid per gram of water. The original solid consisted of S spheres, each of initial diameter D_0 ft.

3-3. The compositions of coexisting liquid and vapor phases of a binary mixture are to be determined experimentally by means of an Othmer-type still. This apparatus (shown in Fig. 3-10) consists of a boiler connected to a condenser by means of a thermally insulated vapor tube. The condensate collects in a liquid sample trap, the overflow from this trap being returned to the boiler. After boiling has commenced, liquid is condensed in the trap until the trap is filled. At this point, the condensate from the trap runs back into the still. For the following conditions, obtain

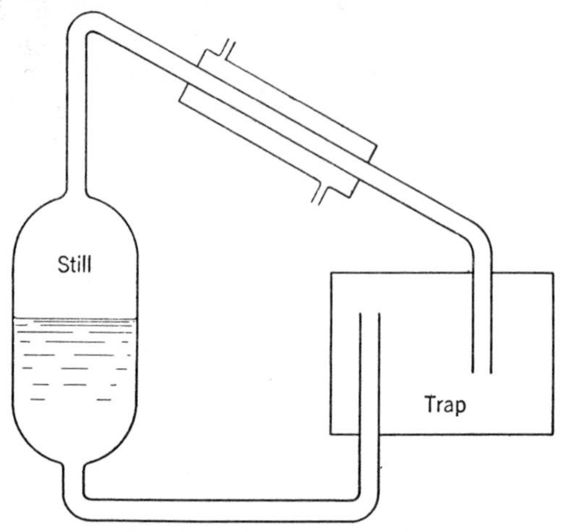

FIG. 3-10. Othmer-type still.

an equation relating the *difference* in the composition between the vapor leaving the still and the liquid in the trap *to* the total amount of liquid condensed (consider both the periods before and after the trap is filled):

S and T moles of mixture containing x_0 mole fraction of the more volatile component are initially charged to the still. The contents of both the still and trap are completely mixed at all times. When full, the trap holds T moles of liquid. The equilibrium relation is

$$y = mx + b$$

Let D represent the total moles of liquid that have been condensed since the start of boiling; Z = mole fraction of the more volatile component in the trap. Holdup in the vapor and liquid lines may be neglected.

3-4. In an experimental study of the saponification of methyl acetate by sodium hydroxide, it is found that 25 per cent of the ester is converted to alcohol in 12 min when the initial concentrations of both ester and caustic are 0.01 m. What conversion of ester would be obtained in 1 hr if the initial ester concentration were 0.025 m and the initial caustic concentration 0.015 m?

3-5. Bodenstein† reports the following data on the dissociation of HI at constant volume at the boiling point of sulfur:

Minutes	10	20	30	40	50	60
Fraction HI dissociated	0.084	0.0917	0.1315	0.1571	0.1771	0.1878

Minutes	70	80	Equilibrium
Fraction HI dissociated	0.1963	0.2043	0.2143

† *Ber. deut. chem. Ges.*, vol. 26, p. 2609, 1893.

Test the applicability of the second-order rate equation by plotting the data, using coordinates such that a straight line should result if the reaction is second-order.

Calculate the average reaction-rate constant in the units of minutes and mole fractions.

3-6. Daniels and Johnston† report the following data on the thermal decomposition of N_2O_5 at constant volume at 55°C:

Time, min	*Total pressure, mm Hg*	*Time, min*	*Total pressure, mm Hg*
0	331.2	14	589.4
3	424.5	16	604.0
4	449.0	18	616.3
6	491.8	22	634.0
8	524.8	26	646.0
10	551.3	Infinite	673.7

The reaction is

$$2N_2O_5 \rightarrow 2N_2O_4 + O_2$$

but the tetroxide dissociates to form NO_2, with which it is in constant equilibrium. The main reaction has been shown to be not reversible, and the N_2O_5 has been shown to exist as such, and not as a polymer.

Making due allowance for the equilibrium between N_2O_4 and NO_2, determine the order of the reaction (test two possibilities) by plotting the data in such a way as to obtain a straight line for the correct mechanism.

3-7. (*a*) A container is maintained at a constant temperature of 800°F and is fed with a pure gas A at a *steady rate* of 1 lb mole/min; the product gas stream is withdrawn from the container at the rate necessary to keep the total pressure constant at a value of 3 atm. The container contents are vigorously agitated, and the gas mixture is always well mixed. The following *irreversible* second-order gas-phase reaction occurs in the container:

$$2A \rightarrow B$$

At a temperature of 800°F, the reaction-rate constant for the reaction has the numerical value of 1,000 ft³/lb mole/min. Both A and B are perfect gases. Because of their low temperature, no reaction occurs in the lines leading to and from the vessel.

If, under steady-state conditions, the product stream is to contain $33\frac{1}{3}$ mole % B, how large (in cubic feet) should be the volume of the reaction container?

(*b*) After the steady state of (*a*) has been attained, the valve on the exit pipe of the isothermal vessel is abruptly closed. The *feed rate* is controlled so that the total tank pressure is maintained at 3 atm. If the mixing is still perfect, how many minutes will it take (after the instant of closing the valve) for the tank contents to be 90 mole % B?

3-8. The dried gas from an ammonia oxidation catalyst chamber contains 9 per cent NO, 9 per cent oxygen, and 82 per cent nitrogen (by volume). This gas is passed at 25°C and 1 atm into a vertical wetted-wall absorption column, the walls of which are wet with a dilute solution of sodium hydroxide. The tower is 200 cm tall, 5 cm ID, and the inlet gas velocity is 97.1 cm/sec. What is the percentage recovery of nitrogen oxides in the column?

For purposes of calculation, the following assumptions may be made:

1. The concentrations at any point will be based on the total moles taken as con-

† *J. Am. Chem. Soc.*, vol. 53, p. 62, 1929.

stant and equal to the arithmetic mean of initial value and the final value corresponding to complete absorption.

2. The NO_2 concentrations will be so low that N_2O_4 formation may be neglected.
3. The NO is not absorbed.
4. The NO_2 is absorbed as such, with gas film controlling and no partial pressure of NO_2 over the solution.

The absorption coefficient k_G is $3.1 \cdot 10^{-7}$ g moles/(sec) (cm²)(mm Hg).† Bodenstein gives the reaction-rate constant for the oxidation of NO at 25°C as $k = 1.77 \cdot 10^{12}$ cm⁶ (g mole)²/min and shows that the reaction is homogeneous and third-order. The rate equation is

$$\frac{d[NO_2]}{dt} = k[NO]^2[O_2]$$

3-9. A perfect gas consisting of 100 per cent pure A enters a tube of diameter D at a rate of M_0 lb moles/(ft²hr) at a pressure p and temperature T^0(°R). The *irreversible* reaction

$$A \rightarrow B + \tfrac{1}{2}C$$

takes place *only* at the walls of the tube and is *controlled entirely* by the rate at which A is transferred to the tube wall. The transfer coefficient K_G, which is independent of composition and temperature, is in units of feet per reciprocal hour when the driving force is moles of A per unit volume of the bulk stream. At the temperature T^0, the enthalpy change of the reaction is

$$\Delta H^0 = H_B{}^0 + \frac{H_c{}^0}{2} - H_A{}^0$$

The molal specific heats of A, B, and C are independent of temperature and obey the relation

$$MC_{pA} = MC_{pB} + \frac{MC_{pC}}{2}$$

The outer wall of the tube is insulated, and no heat is lost from the system. No heat is conducted axially along the tube. The pressure drop in the tube may be neglected. What is the relation between the moles of component A flowing per hour past any point in the tube and the position L downstream in the tube?

3-10. A simple adiabatic converter for the oxidation of SO_2 to SO_3 is to operate upon raw gas entering at 400°C and 1.70 fps and containing 0.60 per cent SO_3, 10.1 per cent SO_2, 10.0 per cent O_2, and 79.3 per cent N_2. Using the following data, estimate the thickness of catalyst mass necessary to convert 57.0 per cent of the entering SO_2 to SO_3. Pressure = 1 atm.

Under these conditions for the catalyst mass being used, the net rate of oxidation of SO_2 may be calculated from the equation‡

$$-\frac{dx}{dt} = k\frac{x^2}{(y)^{0.2}}\ln\frac{r_e}{r}$$

y = moles SO_3 at time t per 100 moles entering gas
x = moles SO_2 at time t per 100 moles entering gas
t = time of contact, sec (based on catalyst bulk volume and superficial gas velocity)

† See Chambers and Sherwood, *Ind. Eng. Chem.*, vol. 29, p. 1415, 1937.
‡ Chang and Chang, *Eng. Quart., Univ. Chekiang*, vol. 3, no. 4, p. 315, 1936.

T = temperature of gas after time of contact t
r_e = molal ratio of SO_3 to SO_2 at equilibrium at temperature T
r = molal ratio of SO_3 to SO_2 at time t
k = reaction-rate constant at temperature T
K = equilibrium constant = $p_{SO_3}/p_{SO_2}\sqrt{p_{O_2}}$ (p in atm)

T(°C)..........	400	425	450	475	500
k..............	6.6	13.6	5.2	2.7	2.0

$$RT \ln K = 22{,}600 - 21.36T \qquad T \text{ in °K}$$

$$SO_2 + \tfrac{1}{2}O_2 \rightarrow SO_3 \qquad \Delta H = 22{,}500 \text{ g-cal/g mole}$$

$$MC_p \text{ for } N_2 = 6.8 \text{ g-cal/(g mole)(°C)}$$

$$MC_p \text{ for } O_2 = 6.8 \text{ g-cal/(g mole)(°C)}$$

$$MC_p \text{ for } SO_2 = 11.0 \text{ g-cal/(g. mole)(°C)}$$

$$MC_p \text{ for } SO_3 = 14.4 \text{ g-cal/(g mole)(°C)}$$

CHAPTER 4

SOLUTION OF ORDINARY DIFFERENTIAL EQUATIONS

4-1. Functional Relationships. Whenever a relationship exists between two variables x and y such that for every value of x there exists at least one value of y, y is said to be a function of x, and the functional relation may be expressed symbolically as

$$f(x,y) = 0 \tag{4-1}$$

Equation (4-1) may be solved for either x or y, and it will be supposed that an arbitrary constant c enters into the solution. Consider, for example, a parabola with vertex at the origin

$$f(x,y) = cx^2 - y = 0 \tag{4-2}$$

$$y = cx^2 \tag{4-3}$$

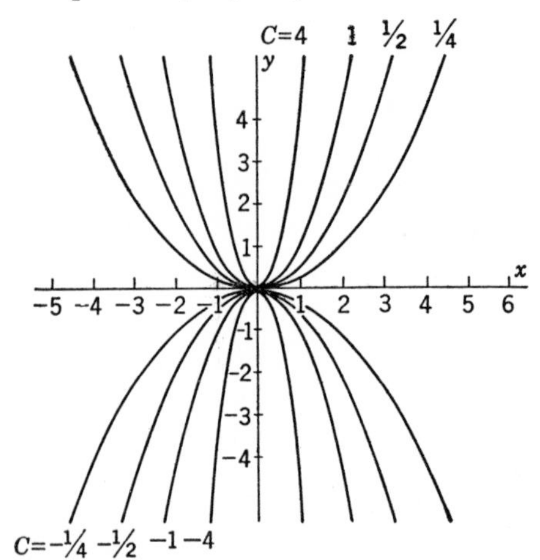

Fig. 4-1. Family of parabolas with vertices at the origin.

Equation (4-3) represents a family of parabolas with vertices at the origin, as shown in Fig. 4-1, and in order to specify any particular parabola, it is necessary to specify the value of c. This fact may be indicated in the general equation (4-1) by the inclusion of c as follows:

$$f(x,y,c) = 0 \tag{4-4}$$

4-2. Mathematical Origin of Differential Equations. It is possible to represent the family of curves (4-3) by differentiation and elimination of the constant c.

$$\frac{dy}{dx} = 2cx \tag{4-5}$$

$$\frac{dy}{dx} = 2\frac{y}{x} \tag{4-6}$$

Equation (4-6) is said to be the differential equation of the family of curves represented by (4-3), and (4-3) is known as the "primitive" of the differential equation (4-6). If the primitive contains only one arbitrary constant, the differential equation will contain the variables and a first derivative. If the primitive contains two arbitrary constants, it is necessary to differentiate twice in order to eliminate the two constants, and the differential equation will involve a second derivative.

For example, the differential equation of the family of parabolas with vertices on the x axis and axes parallel to the y axis is obtained as follows:

$$y = c(x - a)^2 \tag{4-7}$$

$$\frac{dy}{dx} = 2c(x - a) \tag{4-8}$$

$$\frac{d^2y}{dx^2} = 2c \tag{4-9}$$

Eliminating c and a gives

$$2y \frac{d^2y}{dx^2} = \left(\frac{dy}{dx}\right)^2 \tag{4-10}$$

4-3. Definition of Terms. The *order* of a differential equation is the order of the highest derivative occurring in the equation.

The *degree* of a differential equation is the power to which the highest derivative is raised when the equation has been rationalized and cleared of fractions.

The differential equation

$$\frac{d^2y}{dx^2} = f(x) \left[1 + \left(\frac{dy}{dx}\right)^2\right]^{3/2} \tag{4-11}$$

represents the curvature of a liquid interface. It is of the second order and the second degree, as may be seen by applying the preceding rules after squaring both sides.

The solution of a differential equation involving the most general possible relationship among the variables is known as the "general solution" or the "primitive."

When the constants in the general solution are allowed to take on values corresponding to the conditions in a specific problem, the resulting equation is known as a "particular solution" of the differential equation, because it is now applicable to but one particular case. $y = cx^2$ is the general solution of (4-6), and $y = \frac{1}{9}x^2$ is a particular solution.

The limitations introduced in obtaining a particular solution from a general solution are known collectively as the "boundary conditions" of the problem. Thus, c was determined as $\frac{1}{9}$ from the fact that out of the whole family represented by Eq. (4-3) the particular parabola of interest passed through the point (3,1).

Differential equations are divided into two general classes. Equations containing ordinary total derivatives or differentials are known as "ordinary differential equations," and equations containing partial derivatives or differentials are known as "partial differential equations."

4-4. Ordinary Differential Equations. Ordinary differential equations arise from those systems in which all the quantities may be taken as a function of a single independent variable. It is easy to test for the presence of a single independent variable from knowledge of the physical situation at hand.

For example, in a single-pass countercurrent heat exchanger operating under fixed end conditions it is necessary only to specify the position in the exchanger by means of one variable, such as the distance x from the cold end, and all the other variables at this point are fixed by the physical laws controlling the system. In a continuously operated absorption tower, liquid and gas compositions vary throughout the tower, but if the end conditions are fixed, it suffices (if channeling effects are neglected and concentrations are assumed equal at all points across any cross section) to specify position in order to calculate the properties of both the liquid and gas streams.

4-5. Partial Differential Equations. Suppose now that the end conditions in the two preceding cases are not steady or constant with time. The rate of flow of the cold stream to the heat exchanger may vary with time, or the rate of flow of the solvent liquid to the tower may vary with time, according to some known relation. In order to specify conditions completely at any point in either apparatus, the values of two independent variables—position and either time or rate of flow of the variable stream—must be specified. In this case, the instantaneous behavior of the system will be expressed by an equation involving the variables and their partial derivatives. The solution of partial differential equations is discussed in later chapters.

4-6. Classification of Ordinary Differential Equations. Ordinary differential equations arise generally from a description of situations involving but one independent variable. The most important cases in practice involve, in addition to the independent variable, one dependent variable, and the most general differential equation in two variables may be written

$$f\left(x,y,\frac{dy}{dx},\frac{d^2y}{dx^2},\frac{d^3y}{dx^3},\ \ldots\ ,\frac{d^ny}{dx^n}\right) = 0 \tag{4-12}$$

Any value of y which, when substituted into (4-12), reduces the left-hand side to zero is called a "solution" of the differential equation. A differential equation containing a derivative of the nth order can be derived from an equation between x and y involving n independent arbitrary constants by n successive differentiations of the equation. The resulting n equations together with the original equation constitute a set of $n + 1$ equations from which the n constants can be eliminated to obtain (4-12). This process of obtaining a differential equation from its primitive is comparatively easy to perform.

Conversely, it is shown in treatises on differential equations that any equation of the type of (4-12) possesses a solution and that the most general form of this solution is a relationship between x, y, and n arbitrary constants, i.e.,

$$F(x,y,c_1,c_2,c_3, \ldots ,c_n) = 0 \tag{4-13}$$

If (4-13) is solved for y and substituted in (4-12), it must reduce the left-hand side of the latter to zero. From a practical standpoint, it may not help a great deal to know that (4-12) has a solution, for experience has shown that the problem of obtaining this solution is frequently difficult, solutions having been obtained for only comparatively few of the many possible types of equations. Furthermore, the fact that a solution exists does not imply that this solution is expressible in the form of elementary functions.

In the present chapter, the most important equations of common occurrence will be listed and their solutions indicated, and an attempt will be made to indicate the types of equations that can and that cannot be solved by elementary methods.

The equations that have proved most amenable to solution are those involving the dependent variable and its derivatives to the first power only. These are known as "linear differential equations."

Although very few nonlinear equations have been solved, there are several standard forms that occur repeatedly and admit of ready solution. The majority of these are equations of the first order and first degree, and their nonlinear character arises from the dependent variable being present to other powers than the first.

EQUATIONS OF THE FIRST ORDER

Any differential equation of the first order and first degree in dy/dx may be written in the form

$$M\,dx + N\,dy = 0 \tag{4-14}$$

where M and N are functions of x and y. Several classes of equations of this form are easily solved.

4-7. Separable Equations. If (4-14) can be reduced to the form

$$f_1(x)\,dx + f_2(y)\,dy = 0 \tag{4-15}$$

the variables are said to be "separated," and the solution is

$$\int f_1(x)\,dx + \int f_2(y)\,dy = C \tag{4-16}$$

The two constants accompanying the two indefinite integrals on the left are not independent but may be combined as the single constant C.

Very frequently the possibility of separation of the variables may not be immediately evident but may be feasible with sufficient ingenuity. Consider the equation

$$x(1 + y^2)^{1/2} + y(1 + x^2)^{1/2}\frac{dy}{dx} = 0 \tag{4-17}$$

After being written in the form

$$x(1 + y^2)^{1/2}\,dx = -y(1 + x^2)^{1/2}\,dy$$

the equation is seen to be separable and is easily transformed to the standard form

$$\frac{x\,dx}{(1 + x^2)^{1/2}} + \frac{y\,dy}{(1 + y^2)^{1/2}} = 0$$

the solution of which is

$$(1 + x^2)^{1/2} + (1 + y^2)^{1/2} = C$$

4-8. Equations Made Separable by Change of Variable. In some cases, separation of the variables may be effected by a suitable change of variable. For example, the equation

$$(xy^2 + y)\,dx - x\,dy = 0 \tag{4-18}$$

may be made separable by placing $xy = v$; then $dv = x\,dy + y\,dx$, and substitution gives

$$\left(\frac{v^2}{x} + \frac{v}{x}\right)dx + \frac{v\,dx}{x} - dv = 0$$

$$\frac{dx}{x} - \frac{dv}{v(v + 2)} = 0$$

$$\ln x + \frac{1}{2}\ln\frac{v + 2}{v} = C_1$$

$$x\left(\frac{v + 2}{v}\right)^{1/2} = e^{c_1}$$

$$x^2\left(1 + \frac{2}{xy}\right) = e^{2c_1} = C$$

This example illustrates the point that a solution may be written in several different equivalent forms.

No general rule for reduction of an equation to simpler form by substitution and change of variable may be given. Often some of the quantities that occur in the statement of the problem may suggest themselves as good variables, and often some expression in the differential equation will indicate a simplifying substitution. For example, if y enters the equation only as y^2 and $y(dy/dx)$, take $y^2 = u$, whereupon

$$y \frac{dy}{dx} = \frac{1}{2} \frac{du}{dx}$$

and a simplification has been effected.

4-9. Homogeneous Equations. Any function $f(x,y)$ is said to be a "homogeneous function" of the nth degree if, when x and y are multiplied by t, the function is multiplied by t^n. Thus if the function f is homogeneous in x and y,

$$f(tx,ty) = t^n f(x,y) \tag{4-19}$$

If the coefficients M and N in (4-14) are homogeneous and of the same degree in x and y, it is easily shown that the substitution $y = ux$ will make the equation separable. Equation (4-14) may be written

$$f_1(x,y)\, dx + f_2(x,y)\, dy = 0 \tag{4-20}$$

If ux is substituted for y, the result is to multiply each term of $f_1(1,u)$ and $f_2(1,u)$ by x; since f_1 and f_2 are assumed to be homogeneous and of the same degree, (4-20) may be written

$$x^n f_1(1,u)\, dx + x^n f_2(1,u)(u\, dx + x\, du) = 0$$

After separation of the variables, the equation becomes

$$\frac{dx}{x} + \frac{f_2(1,u)\, du}{f_1(1,u) + uf_2(1,u)} = 0$$

the solution of which is

$$\ln x + \int \frac{f_2(1,u)\, du}{f_1(1,u) + uf_2(1,u)} = C \tag{4-21}$$

4-10. Equations of First Order and First Degree with Linear Coefficients. In the equation

$$(ax + by + c)\, dx + (gx + hy + k)\, dy = 0 \tag{4-22}$$

the coefficients of dx and dy are linear in x and y. Such an equation can usually be made homogeneous by the substitutions

$$\begin{aligned} x &= w + m \qquad & dx &= dw \\ y &= v + n \qquad & dy &= dv \end{aligned} \tag{4-23}$$

the constants m and n being evaluated to satisfy the equations

$$\begin{aligned} am + bn + c &= 0 \\ gm + hn + k &= 0 \end{aligned} \tag{4-24}$$

If $a/g = b/h$, Eq. (4-24) will be inconsistent. In this case, eliminating y in (4-22) by the substitution $w = ax + by$ will provide a separable equation in x and w.

4-11. Exact Equations. Consider the function

$$3x^2y^2 + 2y^3x + y^2 + C = 0 \tag{4-25}$$

Differentiating gives

$$6y^2x\,dx + 6x^2y\,dy + 2y^3\,dx + 6y^2x\,dy + 2y\,dy = 0 \tag{4-26}$$

The quantity on the left of (4-26) is called an "exact differential" of the function (4-25), because it is the result of differentiation only, no other algebraic operations having been performed. Consequently, by one integration of (4-26), it is possible to obtain (4-25).

When (4-26) is written in the standard form (4-14),

$$(6y^2x + 2y^3)\,dx + (6x^2y + 6y^2x + 2y)\,dy = 0 \tag{4-27}$$

Considering x constant, differentiate M, the coefficient of dx, with respect to y; that is, obtain the partial derivative of M with respect to y:

$$\frac{\partial M}{\partial y} = 12xy + 6y^2$$

Obtain next the partial derivative of N, the coefficient of dy, with respect to x:

$$\frac{\partial N}{\partial x} = 12xy + 6y^2$$

$\partial M/\partial y = \partial N/\partial x$, and *this is the necessary and sufficient condition that any function* $(M\,dx + N\,dy)$ *be an exact differential.*

Returning now to (4-27), divide through by $2y$, obtaining

$$(3yx + y^2)\,dx + (3x^2 + 3yx + 1)\,dy = 0 \tag{4-28}$$

or

$$M_1\,dx + N_1\,dy = 0$$

After this division,

$$\frac{\partial M_1}{\partial y} = 3x + 2y \qquad \frac{\partial N_1}{\partial x} = 6x + 3y$$

and since the two partial derivatives are no longer equal, (4-28) is no longer an exact differential and cannot be integrated directly to give its primitive (4-25). However, the differential equation (4-28) could be made exact and consequently directly integrated if it were multiplied by the factor $2y$ previously canceled. Such a factor is called an "integrating factor," and it may be shown that for any equation of the form of (4-14) an infinite number of integrating factors exist. Although no general method is known for finding integrating factors, they are known in a few cases. The most important of these is the linear differential equation of the first order.

4-12. Linear Equation of the First Order. The most general form of this equation may be written

$$\frac{dy}{dx} + Py = Q \tag{4-29}$$

where P and Q are constants or functions of x only. If (4-29) is multiplied by the integrating factor $e^{\int P\,dx}$, it becomes

$$e^{\int P\,dx}\frac{dy}{dx} + e^{\int P\,dx}Py = e^{\int P\,dx}Q \tag{4-30}$$

The left-hand side is clearly the derivative of the quantity $ye^{\int P\,dx}$, so that the equation is exact, and its solution is

$$ye^{\int P\,dx} = \int e^{\int P\,dx}Q\,dx + C \tag{4-31}$$

4-13. Bernoulli's Equation. Bernoulli's equation may be written

$$\frac{dy}{dx} + Py = Qy^n \tag{4-32}$$

where P and Q are again constants or functions of x. This equation may be made linear upon division by y^n and substitution of $y^{1-n} = z$. Integration may then be accomplished by the method of Sec. 4-12.

4-14. Other Integrating Factors. The following special cases are often mentioned with reference to (4-14):

If $[(\partial M/\partial y) - (\partial N/\partial x)]/N = f(x)$, then $e^{\int f(x)\,dx}$ is an integrating factor.

If $[(\partial M/\partial y) - (\partial N/\partial x)]/M = f(y)$, then $e^{\int -f(y)\,dy}$ is an integrating factor.

If $M = yf_1(xy)$, and $N = xf_2(xy)$, then $1/(xM - yN)$ is an integrating factor.

In practice, unless the equation falls under one of the standard forms, the determination of an integrating factor is difficult, and no attempt should be made to find one until other methods of integration have failed.

4-15. Integration of Exact Equations. When the integral of an exact equation is not obvious upon inspection, the following procedure may be used: With the notation of the standard form

$$M\,dx + N\,dy = 0 \tag{4-14}$$

perform the following operations:

1. Evaluate $\int M\,dx$, holding y constant.
2. Evaluate R, where

$$R \equiv N - \frac{\partial}{\partial y} \int M\,dx$$

that is, R is the difference between the function of x and y, N, and the function obtained by differentiating with respect to y, holding x constant, the quantity obtained in 1.

3. Evaluate $\int R\,dy$, holding x constant.
4. The desired solution is then

$$(\int M\,dx)_y + (\int R\,dy)_x = C \tag{4-33}$$

In some cases, the following steps are easier to carry out:

1*a*. Evaluate $\int N\,dy$, holding x constant.

2*a*. Evaluate S:

$$S \equiv M - \frac{\partial}{dx} \int N\,dy$$

3*a*. Evaluate $\int S\,dx$, holding y constant.

4*a*. The desired solution is then

$$(\int N\,dy)_x + (\int S\,dx)_y = C \tag{4-34}$$

4-16. Equations of the First Order and Higher Degree. The general equation of this type is

$$f\left(x,y,\frac{dy}{dx}\right) = 0 \tag{4-35}$$

Case I. Equations Solvable for dy/dx. After solving for dy/dx, treat each of the solutions as it occurs under some case mentioned previously. A typical equation of this class is

$$\left(\frac{dy}{dx}\right)^2 + \frac{dy}{dx} - 6 = 0 \tag{4-36}$$

This is equivalent to

$$\left(\frac{dy}{dx} + 3\right)\left(\frac{dy}{dx} - 2\right) = 0 \tag{4-37}$$

which has two solutions

$$\frac{dy}{dx} = -3, \text{ from which } y = -3x + c_1$$

and

$$\frac{dy}{dx} = 2, \text{ from which } y = 2x + c_2$$

These solutions may be used separately or, if desired, they may be combined by multiplication, which in this case would give

$$(y + 3x - c_1)(y - 2x - c_2) = 0 \tag{4-38}$$

There is an important difference between first-order equations of the first and higher degrees in dy/dx. Whereas an equation of first order and first degree determines one slope at any point x, y, an equation of first order and second degree, having two values of dy/dx, determines two slopes at any point. In general, a first-order equation of degree n in dy/dx will determine n slopes at every point, although these may not all apply to real curves.

The constants c_1 and c_2 arising from the two solutions of (4-36) are not arbitrary in the sense that the solution can be made to fit two independent initial conditions. Since (4-36) determines two slopes at every point, the general solution must represent two curves through every point and, hence, the two values of c for every point.

Case II. Equations Solvable for y. Solution of the general equation for y gives one or more equations of the form

$$y = f_1\left(x, \frac{dy}{dx}\right) \tag{4-39}$$

Differentiating with respect to x and substituting $dy/dx = p$ result in

$$p = f_2\left(x,p, \frac{dp}{dx}\right) \tag{4-40}$$

If (4-40) may be integrated to give

$$f_3(x,p,c) = 0 \tag{4-41}$$

then p may be eliminated between (4-39) and (4-41) to give a relation between x, y, and c, which in general is the solution of (4-39). Inasmuch as the process of elimination may introduce extraneous factors, the solution should always be checked by differentiation. If desired, the two equations (4-39) and (4-41) may be retained as a solution, p being a parameter the value of which determines x and y.

Do not integrate (4-41) to eliminate p.

Case III. Equations Solvable for x. Solution of the general equation for x gives one or more equations of the form

$$x = f_4\left(y, \frac{dy}{dx}\right) \tag{4-42}$$

Differentiate with respect to y, and substitute $dx/dy = 1/p$, to obtain

$$\frac{1}{p} = f_5\left(y,p,\frac{dp}{dy}\right) \tag{4-43}$$

If the integral relation between p and y can be obtained from (4-43), the elimination of p between this integral and (4-42) will give a solution in terms of x and y. Otherwise, a parametric solution is indicated, as in case II.

An example of an equation that can be solved for either dy/dx, y, or x is

$$x - k\left(\frac{dy}{dx}\right)^2 = y\,\frac{dy}{dx} \tag{4-44}$$

In a case like this, there is no guide but experience to indicate which procedure is most satisfactory. As a first trial, solve for dy/dx by the quadratic formula

$$\frac{dy}{dx} = \frac{-y \pm \sqrt{y^2 + 4kx}}{2k} \tag{4-45}$$

Equation (4-45) does not appear particularly promising of solution by any of the previous methods, and so try solving (4-44) for y, placing $dy/dx = p$. This gives

$$y = \frac{x - kp^2}{p} \tag{4-46}$$

Differentiating (4-44) with respect to x in order to eliminate y gives

$$\frac{dy}{dx} = p = \frac{p[1 - 2kp(dp/dx)] - [(x - kp^2)(dp/dx)]}{p^2} \tag{4-47}$$

Multiplying each side by p^2 and factoring out dp/dx result in the much simpler form

$$p - p^3 = (x + kp^2)\,\frac{dp}{dx} \tag{4-48}$$

If (4-48) is written in the form

$$\frac{dx}{dp} = \frac{x + kp^2}{p - p^3} \tag{4-49}$$

it is apparent that the result is a linear equation of the first order in x. This may be integrated by the method of Sec. 4-12 to give a relation between x and p, which with (4-46) constitutes a solution in the form of a pair of parametric equations in which any value of p determines a value for x and for y.

If the original equation (4-44) had been solved for x, the solution would turn out to be very similar.

4-17. Clairaut's Equation. This equation, usually written

$$y = x\frac{dy}{dx} + f\left(\frac{dy}{dx}\right) \tag{4-50}$$

is an important case of an equation solved for y. Placing $dy/dx = p$ and differentiating with respect to x give

$$p = p + x\frac{dp}{dx} + f'(p)\frac{dp}{dx} \tag{4-51}$$

Factoring out dp/dx produces

$$[x + f'(p)]\frac{dp}{dx} = 0 \tag{4-52}$$

In (4-52), either $dp/dx = 0$ or $x + f'(p) = 0$. If $dp/dx = 0$, $p = c$, and by substitution in (4-50) the *general solution*

$$y = cx + f(c) \tag{4-53}$$

is obtained.

Another solution may be obtained by eliminating p from (4-50) and $x + f'(p) = 0$. Although this solution will satisfy the original equation, it obviously contains no arbitrary constant and, hence, cannot be the general solution. This leads to the consideration of a third type of solution of a differential equation, the singular solution.

4-18. Singular Solutions. It is instructive to consider the geometric interpretation of a singular solution, and for this purpose it will be convenient to take the special case of Clairaut's equation

$$y = px + 2p \tag{4-54}$$

Differentiating with respect to x results in

$$\frac{dp}{dx}(x + 2) = 0 \tag{4-55}$$

The general solution, containing one arbitrary constant, is

$$y = cx + 2c \tag{4-56}$$

The other factor gives $x = -2$, which when substituted into (4-54) gives $y = 0$. As may be verified by substituting in (4-54), $y = 0$, $x = -2$ is a solution, and since this solution can in no way be obtained from the general solution (4-56), regardless of the value assigned to the arbitrary constant, it is called a "singular solution." Equation (4-56) represents a family of straight lines having y intercepts always twice the corresponding slopes (Fig. 4-2). The singular solution is the locus of the intersection of this family of lines, which in this case is a single point.

A more common type of singular solution arises from the solution of the equation

$$y = px + p^2 \tag{4-57}$$

the general solution of which is

$$y = cx + c^2 \tag{4-58}$$

and the singular solution is

$$y = -\frac{x^2}{4} \tag{4-59}$$

Equation (4-58) represents a family of straight lines, the slopes of which are the square roots of their respective y intercepts. As shown in Fig. 4-3, these lines are tangent to the parabola $y = -(x^2/4)$, and the parabola is known as the "envelope locus" of the family of lines. Although the envelope locus cannot be obtained from the general solution, every point on the envelope locus will also satisfy a particular solution obtained from the general solution by assigning the proper value to c.

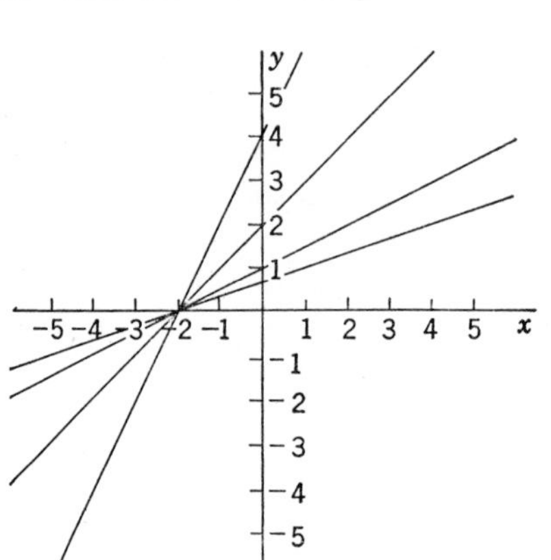

FIG. 4-2. Illustration of a singular solution.

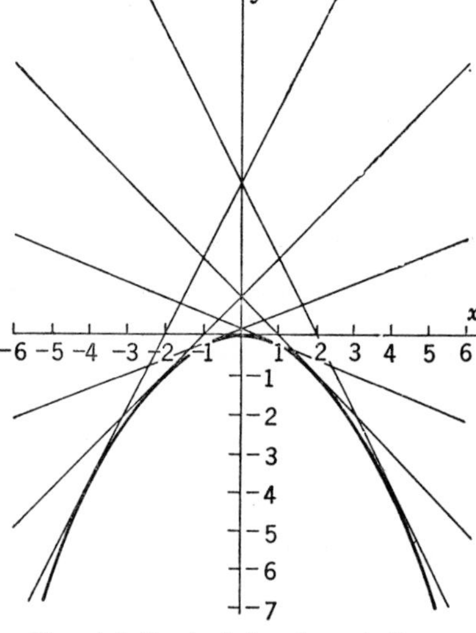

FIG. 4-3. Loci of singular solutions.

Any suspected singular solution should always be checked by insertion in the original equation, for the operations of elimination and substitution may give rise to types of loci that are not solutions.

SPECIAL TYPES OF SECOND-ORDER EQUATIONS

4-19. Equations with Missing Terms. It is not uncommon to encounter second-order equations that are lacking in either y, x, or both y and x.

By proper substitutions, the solution of such an equation may be made to depend upon a first-order equation.

Case I. Equations Not Containing y. If the equation does not contain dy/dx, it may be written in the form

$$\frac{d^2y}{dx^2} = f(x) \tag{4-60}$$

and solved directly by two integrations.

If dy/dx is present, the equation can be written in the form

$$\frac{d^2y}{dx^2} = f\left(x, \frac{dy}{dx}\right) \tag{4-61}$$

The substitution $dy/dx = p$ will reduce (4-61) to an equation of the first order in p and x. To illustrate, consider

$$\frac{d^2y}{dx^2} + x\frac{dy}{dx} = x \tag{4-62}$$

Let $dy/dx = p$, so that $d^2y/dx^2 = dp/dx$. After substitution, the separable equation

$$\frac{dp}{1-p} = x\,dx \tag{4-63}$$

is obtained.

The integral of (4-63) is

$$-\ln(1-p) = \frac{x^2}{2} + C_1' \tag{4-64}$$

which may be written in the exponential form

$$p = \frac{dy}{dx} = 1 - C_1e^{-(x^2/2)} \tag{4-65}$$

Integration of (4-65) gives the final solution

$$y = x - C_1\int e^{(x^2/2)}\,dx + C_2 \tag{4-66}$$

The integral in (4-66) cannot be expressed in terms of a finite number of elementary functions.

Case II. Equations Not Containing x. The substitution of $p = dy/dx$ will reduce the equation to one of the first order. Instead of replacing d^2y/dx^2 by dp/dx, as in case I, note that

$$\frac{dp}{dx} = \frac{dp}{dy}\frac{dy}{dx} = p\frac{dp}{dy} \tag{4-67}$$

As an example of this technique, consider the solution of the equation

$$y \frac{d^2y}{dx^2} = 1 + \left(\frac{dy}{dx}\right)^2 \tag{4-68}$$

The preceding substitutions reduce this to

$$yp \frac{dp}{dy} = 1 + p^2 \tag{4-69}$$

in which the variables can be separated to give

$$\frac{p\,dp}{1 + p^2} = \frac{dy}{y} \tag{4-70}$$

The integral is

$$\tfrac{1}{2} \ln (1 + p^2) = \ln y + C_1'$$

Solving for p produces

$$p = \pm \sqrt{c_1 y^2 - 1} = \frac{dy}{dx} \tag{4-71}$$

A second integration results in the general solution

$$\pm x = \int \frac{dy}{\sqrt{c_1 y^2 - 1}} + c_2 = \frac{1}{c_1} \ln (y \sqrt{c_1} + \sqrt{c_1 y^2 - 1}) + c_2 \tag{4-72}$$

LINEAR DIFFERENTIAL EQUATIONS

4-20. General Properties of Linear Equations. The general linear equation of the nth order may be written

$$\frac{d^ny}{dx^n} + X \frac{d^{n-1}y}{dx^{n-1}} + \cdots + X_{n-1} \frac{dy}{dx} + X_n y = Q \tag{4-73}$$

$X \cdots X_{n-1}\, X_n$ and Q are functions of x or constants.

If I is any solution of this equation, the general solution may be represented by

$$y = I + Z \tag{4-74}$$

where Z is some function of x.

If the value of y given by (4-74) is substituted in the general equation, the result is

$$\left(\frac{d^nI}{dx^n} + X \frac{d^{n-1}I}{dx^{n-1}} + \cdots + X_{n-1} \frac{dI}{dx} + X_n I\right) + \left(\frac{d^nZ}{dx^n} + X \frac{d^{n-1}Z}{dx^n} + \cdots + X_{n-1} \frac{d^{n-1}Z}{dx^n} + X_n Z\right) = Q \tag{4-75}$$

But since I is assumed to be a solution of (4-73), the first member of (4-75) is equal to Q, and the second member must equal zero. The equation

$$\frac{d^n y}{dx^n} + X\frac{d^{n-1}y}{dx^n} + \cdots + X_{n-1}\frac{d^{n-1}y}{dx^n} + X_n y = 0 \qquad (4\text{-}76)$$

is called the "reduced equation," and Z is its general solution.

Property I. The general solution of (4-73) is the sum of any solution I, called the "particular integral," and Z, called the "complementary function." It is important to note that the equation giving rise to the complementary function is the same as the left-hand side of the original equation placed equal to zero. If, in the original equation, $Q = 0$, it is obvious that the complementary function will constitute the general solution.

Property II. If Y_1 is any particular solution of the reduced equation (4-76), then c_1Y_1 is also a solution, and the general solution of the reduced equation is

$$y = c_1Y_1 + c_2Y_2 + \cdots + c_nY_n \qquad (4\text{-}77)$$

provided that the particular solutions $Y_1, Y_2, \ldots, Y_n$ are linearly independent.† That (4-77) is a solution of (4-76) may be verified by direct substitution, and it is the general solution since it contains n arbitrary constants.

Property III. If a single solution of the reduced equation is known, the order of this equation may be lowered by unity; and if m solutions of the reduced equation are known, its order may be reduced by m. This simplifying process will be demonstrated on an equation of the second degree.

Assume that Y_1 is known to be a particular solution of the equation

$$\frac{d^2y}{dx^2} + f_1(x)\frac{dy}{dx} + f_2(x)y = 0 \qquad (4\text{-}78)$$

Place $y = Y_1z$; then

$$\frac{dy}{dx} = Y_1\frac{dz}{dx} + z\frac{dY_1}{dx}$$

and

$$\frac{d^2y}{dx^2} = Y_1\frac{d^2z}{dx^2} + \frac{dz}{dx}\frac{dY_1}{dx} + z\frac{d^2Y_1}{dx^2} + \frac{dY_1}{dx}\frac{dz}{dx}$$

† The condition of linear independence means that it is impossible to find a set of constants $c_1, c_2, \ldots, c_n$, other than zero, such that for all values of x,

$$c_1Y_1 + c_2Y_2 + \cdots + c_nY_n = 0$$

Substituting in (4-78) and collecting terms produce

$$Y_1 \frac{d^2z}{dx^2} + f_1(x) Y_1 \frac{dz}{dx} + 2 \frac{dz}{dx} \frac{dY_1}{dx} + z \left[\frac{d^2Y_1}{dx^2} + f_1(x) \frac{dY_1}{dx} + f_2(x) Y_1 \right] = 0 \quad (4\text{-}79)$$

Since Y_1 is a solution of (4-78), the coefficient of z is zero; if dz/dx is placed equal to u, (4-79) becomes

$$Y_1 \frac{du}{dx} + u \left[Y_1 f_1(x) + 2 \frac{dY_1}{dx} \right] = 0 \quad (4\text{-}80)$$

As Y_1, being a solution of (4-78), is a known function of x, (4-80) is a linear equation of the first order in u. Similar methods of operation are applicable to equations of higher order.

4-21. Linear Equations with Constant Coefficients. The general linear equation with constant coefficients may be written

$$\frac{d^ny}{dx^n} + a_1 \frac{d^{n-1}y}{dx^{n-1}} + \cdots + a_{n-1} \frac{dy}{dx} + a_n y = Q \quad (4\text{-}81)$$

where the a's are constants and Q may be constant or a function of x.

To solve this equation, it is convenient to introduce the notation of differential operators in which the operation of differentiating y with respect to x is written Dy. The symbol D thus stands for d/dx, and d^ny/dx^n is written D^ny. In this notation, (4-81) becomes

$$D^ny + a_1 D^{n-1}y + \cdots + a_{n-1}Dy + a_n y = Q \quad (4\text{-}82)$$

An equivalent expression is

$$(D^n + a_1 D^{n-1} + \cdots + a_{n-1}D + a_n)y = Q \quad (4\text{-}83)$$

The expression in parentheses is known as a "linear differential operator of order n," and it may be shown that this operator, which looks like an algebraic polynomial in D, follows the laws of operation associated with polynomials. For example,

$$D(y^2 + y^3) = Dy^2 + Dy^3 \qquad D\,Dy = D^2y$$
$$(D^2 - 1)y = (D - 1)(D + 1)y = (D + 1)(D - 1)y$$

The validity of these last transformations is indicated by the operations

$$(D^2 - 1)y = \frac{d^2y}{dx^2} - y$$

$$(D - 1)(D + 1)y = (D - 1)\left(\frac{dy}{dx} + y\right) = D\left(\frac{dy}{dx} + y\right) - \frac{dy}{dx} - y$$
$$= \frac{d^2y}{dx^2} + \frac{dy}{dx} - \frac{dy}{dx} - y = \frac{d^2y}{dx^2} - y$$

The result is the same, irrespective of which factor is taken first.

With reference to the general operator in (4-83), it is clear that it may be considered as an algebraic polynomial of degree n. It is shown in algebra that such a polynomial has n roots $r_1, r_2, \ldots, r_n$, so that

$$(D^n + a_1 D^{n-1} + \cdots + a_{n-1}D + a_n)y \equiv (D - r_1)(D - r_2) \cdots (D - r_n)y \quad (4\text{-}84)$$

Furthermore, the positions with respect to y of the factors on the right may be interchanged with no effect on the result.

The transformation from Eq. (4-82) to (4-83) is a result of the *distributive* property of linear operators. The identity (4-84) is a result of the *commutative* property exhibited by D when it operates on a *linear equation with constant coefficients*. This is not a general property of linear operators. If D operates on an equation with variable coefficients, the commutative property is ordinarily not observed.

Although the solution of (4-81) or its equivalent (4-83) may be obtained by an iterative integration method, it is usually more convenient to make use of the properties outlined in Sec. 4-20. The complementary function is determined and then the particular integral. The general solution is the sum of these two functions.

4-22. Determination of the Complementary Function. The reduced equation associated with (4-83) may be written

$$(D - r_1)(D - r_2) \cdots (D - r_n)y = 0 \quad (4\text{-}85)$$

Obviously this equation may be satisfied if any one of the factors operating on y becomes zero. Hence,

$$(D - r_1)y = 0 \quad (4\text{-}86)$$

from which $y = C_1 e^{r_1 x}$, C_1 being an arbitrary constant. Similarly, $(D - r_n)y = 0$, and another solution is $y = C_n e^{r_n x}$. As indicated by property II, Sec. 4-20, the sum of all these solutions

$$y = C_1 e^{r_1 x} + C_2 e^{r_2 x} + \cdots + C_n e^{r_n x} \quad (4\text{-}87)$$

is a solution, and it is the general solution of (4-85), because it contains n independent arbitrary constants. Since it is the general solution of the reduced equation, it is the complementary function of (4-83).

A special case arises when two or more of the roots r are equal. If t roots are equal to r_1, the part of the solution due to these roots is

$$(C_1 + C_2 x + \cdots + C_t x^{t-1})e^{r_1 x} \quad (4\text{-}88)$$

Another special case arises when some of the roots of (4-85) are imaginary. It is established in algebra that if imaginary roots occur, they occur in conjugate pairs, so that if there is one root $\alpha + i\beta$, there will be another $\alpha - i\beta$. If these two roots of the operator are designated as r_1 and r_2, the part of the solution due to them may be written

$$\begin{aligned} c_1 e^{r_1 x} + c_2 e^{r_2 x} &= c_1 e^{(\alpha + i\beta)x} + c_2 e^{(\alpha - i\beta)x} \\ &= e^{\alpha x}(A \cos \beta x + B \sin \beta x) \end{aligned} \tag{4-89}$$

The use of trigonometric functions in place of imaginary exponentials is a consequence of the properties of the exponential functions tabulated below.

4-23. Exponential Functions. Power-series development may be used to obtain the following relations between exponential, trigonometric, and hyperbolic functions. These properties often enable the formal solution of a differential equation to be put in a more usable form.

$$e^{i\beta x} = \cos \beta x + i \sin \beta x \tag{4-90}$$

$$e^{-i\beta x} = \cos \beta x - i \sin \beta x \tag{4-91}$$

$$\sin \beta x = \frac{e^{i\beta x} - e^{-i\beta x}}{2i} \tag{4-92}$$

$$\cos \beta x = \frac{e^{i\beta x} + e^{-i\beta x}}{2} \tag{4-93}$$

$$e^{\alpha x} = \cosh \alpha x + \sinh \alpha x \tag{4-94}$$

$$e^{-\alpha x} = \cosh \alpha x - \sinh \alpha x \tag{4-95}$$

$$\sinh \alpha x = \frac{e^{\alpha x} - e^{-\alpha x}}{2} \tag{4-96}$$

$$\cosh \alpha x = \frac{e^{\alpha x} + e^{-\alpha x}}{2} \tag{4-97}$$

$$\sin x = -i \sinh ix \tag{4-98}$$

$$\sinh ix = i \sin x \tag{4-99}$$

$$\cos x = \cosh ix \tag{4-100}$$

$$\cosh^2 x = 1 + \sinh^2 x \tag{4-101}$$

$$\sinh (x \pm y) = \sinh x \cosh y \pm \cosh x \sinh y \tag{4-102}$$

$$\cosh (x \pm y) = \cosh x \cosh y \pm \sinh x \sinh y \tag{4-103}$$

$$\sinh 2x = 2 \sinh x \cosh x \tag{4-104}$$

$$\cosh 2x = \cosh^2 x + \sinh^2 x \tag{4-105}$$

$$2 \sinh^2 \frac{x}{2} = \cosh x - 1 \tag{4-106}$$

$$2 \cosh^2 \frac{x}{2} = \cosh x + 1 \tag{4-107}$$

$$\sinh^{-1} x = \ln (x + \sqrt{x^2 + 1}) \tag{4-108}$$

$$\cosh^{-1} x = \ln (x + \sqrt{x^2 - 1}) \tag{4-109}$$

$$d(\sinh x) = \cosh x \, dx \tag{4-110}$$

$$d(\cosh x) = \sinh x \, dx \tag{4-111}$$

4-24. Determination of the Particular Integral. In this section two methods of finding the particular integral of a linear differential equation will be presented. The first procedure, known as the "method of undetermined coefficients," is easy to apply. However, it is limited to equations exhibiting both constant coefficients and a particular form of the function $Q(x)$. The second procedure, the "method of variation of parameters," applies to any linear equation.

Method of Undetermined Coefficients. Consider the general linear equation with constant coefficients

$$\frac{d^n y}{dx^n} + a_1 \frac{d^{n-1} y}{dx^{n-1}} + \cdots + a_{n-1} \frac{dy}{dx} + a_n y = Q \qquad (4\text{-}81)$$

In most cases of practical interest, Q will be a sum of one or more terms of the following types:

Constant
x^n (n a positive integer)
e^{rx}
$\cos kx$
$\sin kx$

or, less frequently, products of factors of these types. When this is the case, the particular integral must be of the form shown in Table 4-1, subject to the following qualifications:

1. When Q consists of the sum of several terms, the appropriate form of the particular integral is the sum of the particular integrals corresponding to these terms individually.
2. Whenever a term in any of the trial integrals listed is already a part of the complementary function of the equation, the indicated form of the particular integral should be multiplied by x.

TABLE 4-1. FORM OF PARTICULAR INTEGRAL

Q	*Form of particular integral*
Constant a	Constant A
ax^n	$A_n x^n + A_{n-1} x^{n-1} + \cdots A_1 x + A_0$
be^{rx}	Be^{rx}
$c \cos kx$, $d \sin kx$	$C \cos kx + D \sin kx$
$gx^n e^{rx} \cos kx$, $hx^n e^{rx} \sin kx$	$(G_n x^n + G_{n-1} x^{n-1} + \cdots + G_0) e^{rx} \cos kx + (H_n x^n + H_{n-1} x^{n-1} + \cdots + H_0) e^{rx} \sin kx$

Since the form of the particular integral is known, the constants may be evaluated by substitution in the differential equation. The following example will illustrate the procedure.

Consider the equation

$$\frac{d^3 y}{dx^3} - 3 \frac{d^2 y}{dx^2} + 3 \frac{dy}{dx} = x \qquad (4\text{-}112)$$

The reduced equation is

$$(D^3 - 3D^2 + 3D)y = D(D^2 - 3D + 3)y = 0 \qquad (4\text{-}113)$$

The roots are $r_1 = 0$, $r_2 = (3 - i\sqrt{3})/2$, $r_3 = (3 + i\sqrt{3})/2$. The complementary function is then

$$Z = C + e^{3/2 x}\left(A \sin \frac{\sqrt{3}}{2} x + B \cos \frac{\sqrt{3}}{2} x\right) \qquad (4\text{-}114)$$

Table 4-1 indicates that the form of the particular integral is

$$I = A_1 x + A_0$$

However, the complementary function contains a constant term. Consequently, qualification 2 requires the form of the particular integral to be

$$I = x(A_1 x + A_0) = A_1 x^2 + A_0 x \qquad (4\text{-}115)$$

Substitution of (4-115) in (4-112) gives

$$-6A_1 + 6A_1 x + 3A_0 = x \qquad (4\text{-}116)$$

Therefore, by equating the coefficients of like powers of x, $6A_1 = 1$, $-6A_1 + 3A_0 = 0$, and $A_1 = 1/6$, $A_0 = 1/3$,

$$I = 1/6 x^2 + 1/3 x \qquad (4\text{-}117)$$

The complete solution of (4-112) is then

$$y = Z + I = C + e^{3/2x}\left(A \sin \frac{\sqrt{3}}{2} x + B \cos \frac{\sqrt{3}}{2} x\right) + \frac{1}{6} x^2 + \frac{1}{3} x \qquad (4\text{-}118)$$

Variation of Parameters. The determination of the particular integral by the method of variation of parameters will be developed for a linear second-order equation. The extension to higher-order linear equations is straightforward. Consider the general linear second-order equation

$$\frac{d^2y}{dx^2} + a_1(x)\frac{dy}{dx} + a_2(x)y = Q(x) \qquad (4\text{-}119)$$

The coefficients a_1 and a_2 may be functions of x. It is assumed that the two functions which form the complementary solution are known:

$$Z = c_1 y_1 + c_2 y_2 \qquad (4\text{-}120)$$

It is now assumed that a particular integral of (4-119) is of the form

$$I = u y_1 + V y_2 \qquad (4\text{-}121)$$

where u and V are functions of x. Two equations are needed to determine u and V. One equation results from the condition that (4-121) must satisfy (4-119). The second equation is arbitrary in the sense that it can be set by the analyst. However, the success of the method rests upon the procedure used to obtain the second equation. This is done as follows: Differentiate (4-121) in order to obtain the conditions needed to satisfy (4-119):

$$\frac{dI}{dx} = (uy_1' + Vy_2') + (u'y_1 + V'y_2) \tag{4-122}$$

It is evident that a second differentiation will introduce the second derivatives of the unknown functions. In order to avoid this complication, the equation at the disposal of the analyst is taken to be

$$u'y_1 + V'y_2 = 0 \tag{4-123}$$

Equation (4-123) is then the second condition imposed on u and V. Further differentiation yields

$$\frac{d^2I}{dx^2} = (uy_1'' + Vy_2'') + (u'y_1' + V'y_2') \tag{4-124}$$

When (4-121) to (4-124) are introduced into the original differential equation (4-119), there results

$$u[y_1'' + a_1(x)y_1' + a_2(x)y_1] + V[y_2'' + a_1(x)y_2' + a_2(x)y_2] + u'y_1' + V'y_2' = Q(x) \tag{4-125}$$

The bracketed expressions vanish because y_1 and y_2 are solutions of the homogeneous equation. Hence, the condition on u and V is

$$u'y_1' + V'y_2' = Q(x) \tag{4-126}$$

When Eqs. (4-123) and (4-126) are solved for u' and V', there results

$$\begin{aligned} u' &= \frac{du}{dx} = -\frac{y_2}{y_1y_2' - y_2y_1'}\,Q(x) \\ V' &= \frac{dV}{dx} = \frac{y_1}{y_1y_2' - y_2y_1'}\,Q(x) \end{aligned} \tag{4-127}$$

The functions y_1, y_2, and $Q(x)$ are known. Hence, u and V are found by direct integration, and the particular integral

$$I = uy_1 + Vy_2 \tag{4-121}$$

is determined.

4-25. The Euler Equation. The solution of the linear equation

$$x^n\frac{d^ny}{dx^n} + a_1x^{n-1}\frac{d^{n-1}y}{dx^{n-1}} + \cdots + a_{n-1}x\frac{dy}{dx} + a_ny = Q(x) \tag{4-128}$$

can be reduced to the solution of an equation with constant coefficients by the substitution $x = e^z$. The method may be illustrated by solution of the equation

$$x^3 \frac{d^3y}{dx^3} + x \frac{dy}{dx} = \ln x \tag{4-129}$$

$$\frac{dy}{dx} = \frac{dy}{dz}\frac{dz}{dx} = \frac{dy}{dz} e^{-z} \tag{4-130a}$$

$$\frac{d^2y}{dx^2} = \frac{dz}{dx}\frac{d}{dz}\left(e^{-z}\frac{dy}{dz}\right) = e^{-2z}\frac{d^2y}{dz^2} - e^{-2z}\frac{dy}{dz} \tag{4-130b}$$

$$\begin{aligned}\frac{d^3y}{dx^3} &= \frac{dz}{dx}\frac{d}{dz}\left(e^{-2z}\frac{d^2y}{dz^2} - e^{-2z}\frac{dy}{dz}\right)\\ &= e^{-3z}\frac{d^3y}{dz^3} - 2e^{-3z}\frac{d^2y}{dz^2} - e^{-3z}\frac{d^2y}{dz^2} + 2e^{-3z}\frac{dy}{dz}\\ &= e^{-3z}\frac{d^3y}{dz^3} - 3e^{-3z}\frac{d^2y}{dz^2} + 2e^{-3z}\frac{dy}{dz}\end{aligned} \tag{4-130c}$$

Substituting these values in (4-129) gives

$$\frac{d^3y}{dz^3} - 3\frac{d^2y}{dz^2} + 3\frac{dy}{dz} = z \tag{4-131}$$

Equation (4-131) was solved in Sec. 4-24.

4-26. Simultaneous Linear Differential Equations. The mathematical formulation of a physical problem frequently results in the generation of a system of differential equations. If the problem is determinate, the system will involve n independent equations and n dependent variables. Such a situation arose in Example 3-11. The method of attack employed there was of general applicability, but when the equations are *linear with constant coefficients* a method based upon the use of differential operators is more convenient. Consider the system of equations

$$3\frac{d^2y}{dx^2} + 3\frac{dy}{dx} + 2y + \frac{d^2z}{dx^2} + 2\frac{dz}{dx} + 3z = 0 \tag{4-132}$$

$$2\frac{d^2y}{dx^2} - \frac{dy}{dx} - 2y + \frac{d^2z}{dx^2} + \frac{dz}{dx} + z = 8 \tag{4-133}$$

Rewrite them in operational form:

$$(3D^2 + 3D + 2)y + (D^2 + 2D + 3)z = 0 \tag{4-134}$$

$$(2D^2 - D - 2)y + (D^2 + D + 1)z = 8 \tag{4-135}$$

Multiply Eq. (4-134) by the operational coefficient of z, $(D^2 + D + 1)$ in Eq. (4-135), and multiply Eq. (4-135) by the operational coefficient of z, $(D^2 + 2D + 3)$ in Eq. (4-134). The operational coefficients of z in the two transformed equations are now the same. If the equations are subtracted, z is eliminated, and the result is

$$[(D^2 + D + 1)(3D^2 + 3D + 2) - (D^2 + 2D + 3)(2D^2 - D - 2)]y = -(D^2 + D + 3)8$$

or

$$(D^4 + 3D^3 + 6D^2 + 12D + 8)y = -24 \tag{4-136}$$

The general solution of (4-136) is readily found to be

$$y = Ae^{-x} + Be^{-2x} + C \cos 2x + J \sin 2x - 3 \tag{4-137}$$

The solution for z might now be found by substituting (4-137) in either (4-132) or (4-133) followed by integration of the resulting differential equation. It is easier, however, to start anew and to eliminate y from (4-134) and (4-135) by a procedure analogous to that used to eliminate z. The advantage of this procedure is the fact that the characteristic equation of the z differential equation is always the same as the characteristic equation found for the y differential equation. Thus, if y is eliminated between (4-134) and (4-135), the result is

$$[(2D^2 - D - 2)(D^2 + 2D + 3) - (3D^2 + 3D + 2)(D^2 + D + 1)]z = -(3D^2 + 3D + 2)8$$

or

$$(D^4 + 3D^3 + 6D^2 + 12D + 8)z = 16 \tag{4-138}$$

Comparison of (4-138) and (4-136) demonstrates the equality of the respective characteristic equations. The general solution of (4-138) is

$$z = Ee^{-x} + Fe^{-2x} + G \cos 2x + H \sin 2x + 2 \tag{4-139}$$

It appears that (4-137) and (4-139) are the complete solutions of the original system of differential equations, and eight constants, A, B, C, J, E, F, G, and H, are to be evaluated by means of appropriate boundary conditions. Actually, only four arbitrary constants are involved, a situation which might be surmised (but not proved) by examination of the original differential equations in operator form.

The relation between the constants is found by substituting the integrated solutions into *all but one* of the equations forming the original system. In the present case, choose (4-133). Then, after substitution, there results

$$(D^2 + D + 1)(Ee^{-x} + Fe^{-2x} + G \cos 2x + H \sin 2x + 2) + (2D^2 - D - 2)(Ae^{-x} + Be^{-2x} + C \cos 2x + J \sin 2x - 3) = 8$$

or

$$(A + E)e^{-x} + (8B + 3F)e^{-2x} + (-10C - 2J - 3G + 2H) \cos 2x + (2C - 10J - 2G - 3H) \sin 2x + 8 = 8 \tag{4-140}$$

Equation (4-140) must be identically true. This is the case only if

$$\begin{aligned} A + E &= 0 \\ 8B + 3F &= 0 \\ -10C - 2J - 3G + 2H &= 0 \\ 2C - 10J - 2G - 3H &= 0 \end{aligned} \tag{4-141}$$

Consequently, four relations exist among the eight constants; they may be solved to give

$$\begin{aligned} E &= -A \\ F &= -\tfrac{8}{3}B \\ G &= -2(C + J) \\ H &= 2(C - J) \end{aligned} \tag{4-142}$$

The integrated equations (4-137) and (4-139) now become

$$y = Ae^{-x} + Be^{-2x} + C \cos 2x + J \sin 2x - 3 \tag{4-137}$$

$$z = -Ae^{-x} - \tfrac{8}{3}Be^{-2x} - 2(C + J) \cos 2x + 2(C - J) \sin 2x + 2 \tag{4-143}$$

Straightforward substitution of (4-137) and (4-143) into the remaining unused differential equation (4-132) shows that it is identically satisfied without further restrictions on the constants. Equations (4-137) and (4-143) are then the complete general solutions of the original system of differential equations (4-132) and (4-133).

The method of solution of a system of simultaneous linear differential equations with constant coefficients may be summarized as follows:

1. Write the equations in operator form.
2. Eliminate all but one dependent variable by algebraic manipulation.
3. Solve the resulting differential equation.
4. Repeat steps 2 and 3 for each dependent variable. The characteristic equation of each differential equation will be found to be the same.
5. The number of independent arbitrary constants is equal to the highest order of the operator D found in the characteristic equation of step 2. The relations between the remaining apparently arbitrary constants appearing in the formal solution and the truly independent constants are found by substitution of the integrated relations into all but one of the differential equations forming the *original* system and demanding that the resulting relations be identically satisfied.

BIBLIOGRAPHY

1. Bateman, H.: "Differential Equations," Longmans, Green & Co., Inc., New York, 1926.
2. Cohen, A.: "An Elementary Treatise on Differential Equations," D. C. Heath and Company, Boston, 1933.
3. Forsyth, A. R.: "A Treatise on Differential Equations," Macmillan & Co., Ltd., London, 1929.
4. Ince, E. L.: "Ordinary Differential Equations," Longmans, Green & Co., Inc., New York, 1927.
5. Poole, E. G. C.: "Introduction to the Theory of Linear Differential Equations," Clarendon Press, Oxford, 1936.
6. Wylie, C. R., Jr.: "Advanced Engineering Mathematics," McGraw-Hill Book Company, Inc., New York, 1951.

PROBLEMS

4-1. Solve the following equations:

(a) $$(2x + y^{-1})\,dx + (y^{-1} - xy^{-2})\,dy = 0$$

(b) $$y\,dx - x\,dy = x^3(x^2 - y^2)^{1/2}\,dx$$

(c) $$\frac{dy}{dx} - \frac{2}{x}y = y^4$$

(d) $$y = 2x\frac{dy}{dx} + y^2\left(\frac{dy}{dx}\right)^3$$

(e) $$\frac{d^3y}{dx^3} + \frac{2d^2y}{dx^2} - \frac{3dy}{dx} = 0$$

(f) $$\frac{d^5y}{dx^5} - \frac{2d^4y}{dx^4} + \frac{d^3y}{dx^3} = 0$$

(g) $$\frac{d^2y}{dx^2} + 4y = 2e^{2x} + 20x$$

(h) $$(1 + x^2)\frac{d^2y}{dx^2} + x\frac{dy}{dx} + ax = 0$$

(i) Solve simultaneously

(1) $$\frac{dx}{dt} + \frac{dy}{dt} + y - x = e^{2t}$$

(2) $$\frac{d^2x}{dt^2} + \frac{dy}{dt} = 3e^{2t}$$

4-2. Solve the following equations:

(a) $$\rho\frac{d\theta}{d\rho} - \frac{2}{\rho}\frac{d\rho}{d\theta} = 0$$

(b) $$(x^2y + x)\,dy + (xy^2 - y)\,dx = 0$$

(c) $$\frac{dy}{dx} + y\cot x = \csc x$$

(d) $$\left(\frac{dy}{dx}\right)^2 - 4x\frac{dy}{dx} + 6y = 0$$

(e) $$\frac{d^3y}{dx^3} - \frac{3d^2y}{dx^2} + \frac{9dy}{dx} + 13y = 0$$

(f) $$\frac{d^2y}{dx^2} + y = \tan x$$

(g) $$y\frac{d^2y}{dx^2} + \left(\frac{dy}{dx}\right)^2 = \frac{dy}{dx}$$

4-3. Three tanks of 10,000-gal capacity are each arranged so that when water is fed into the first an equal quantity of solution overflows from the first to the second tank, likewise from the second to the third, and from the third to some point out of the system. Agitators keep the contents of each tank uniform in concentration. To start, let each of the tanks be full of a solution of concentration C_0 lb/gal. Run water into the first tank at 50 gpm, and let the overflows function as described above. Calculate the time required to reduce the concentration in the first tank to $C_0/10$. Calculate the concentrations in the other two tanks at this time.

4-4. A gas-turbine stator blade is to be kept cool by passing air up through a conduit provided in the blade core (see Fig. 4-4). Air, at a flow rate of W lb/hr, with a specific heat of C_p Btu/(lb)(°F) enters the root of the blade at a temperature of T_{g1}. It flows from the root to the tip of the blade where it joins the working fluid. The heat-transfer coefficient between the air and the inside of the blade passage is h_i Btu/(hr)(ft²)(°F). The air is heated in flowing through the blade. The outside of the blade is bathed in the working fluid which, because of its large mass flow, remains at the temperature T_0. The heat-transfer coefficient between the working fluid and the outside of the blade is h_0 Btu/(hr)(ft²)(°F). The blade itself is L ft long, is t ft thick, and has an outer perimeter of P ft. The blade is thin, and hence the inner perimeter may be considered substantially equal to the outer perimeter. The thermal conductivity of the blade metal is k Btu/(hr)(ft²)(°F/ft), and its specific heat is C Btu/(lb)(°F). It is desired to determine the steady-state temperature of the blade as a function of the distance Z. At $Z = 0$, the blade temperature is T_1. It may be assumed that at any section the blade temperature is uniform, but heat flows by conduction from the tip to the root of the blade. Assume that dT/dZ is zero at the blade tip ($Z = L$). Derive, integrate, and evaluate the constants in your expression.

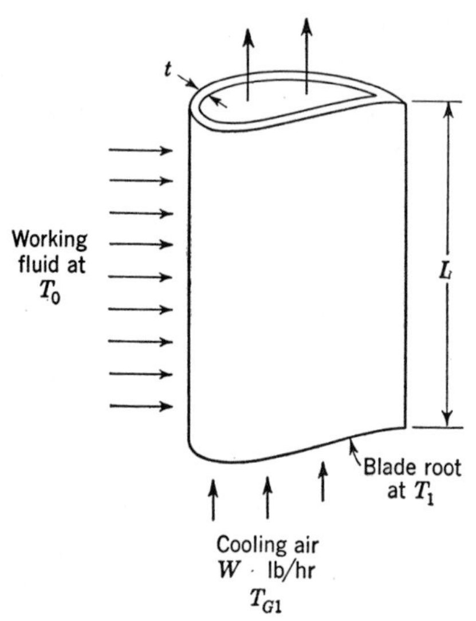

FIG. 4-4. Air-cooled stator blade of Prob. 4.4.

4-5. In a shell-and-tube heat exchanger the cold liquid enters at one end, makes one pass, reverses in direction, and leaves from the same end at which it enters. The hot liquid enters at the opposite end and flows through in one pass. Considering the temperature of the hot liquid uniform over any cross section and neglecting variations in the over-all coefficient of heat transfer from hot to cold liquid, derive an expression for the true average temperature difference in such an exchanger in terms of the terminal temperatures of the flowing fluids.

NOTE: Consider a differential cross section of the exchanger, set up the proper heat balances and rate equations, and integrate the resulting set of simultaneous differential equations.

4-6. It has been proposed that in the chlorination of methane the chlorine attacks the methane and its substituted chlorine derivatives in a statistical fashion. Every molecule is assumed to have an equal chance of undergoing collision with a chlorine molecule, and the probability that such a collision results in reaction is proportional to the number of hydrogen atoms remaining on the carbon atom.

Assume that the above postulated mechanism is correct. If chlorine is added to a system originally containing pure methane, determine the fraction of methane converted to each of the four possible chlorination products (CH_3Cl, CH_2Cl_2, $CHCl_3$, CCl_4) as a function of the total fraction of methane converted.

4-7. A fluidized-particle bed is to be used to carry out a chemical reaction. The proposed conditions are as follows:

1. The bed temperature will be constant, independent of bed height and of time.
2. The density and velocity of the gas flowing through the bed and the fluidized-particle concentration C_c will be substantially independent of bed height and time.
3. Conditions within the bed are not a function of the radius of the tower or of time.
4. The presence of the particles causes mixing of the gas. The mixing is superimposed on the bulk flow and corresponds to the type of mixing which occurs in tur-

bulent motion. In a fluidized bed, the gas mixing due to particle movement is termed "eddy mixing." At a given cross section of the bed, the rate at which a given component crosses unit area at a particular height is

$$\text{Bulk flow of component } i = w_i$$

where w_i is the eddy diffusion rate and is given by

$$w_i = -E\rho \frac{dY_i}{dx} = \frac{\text{moles of } i}{(\text{ft}^2)(\text{sec})}$$

Outside the bed the eddy mixing rate is negligible.

5. The reaction is *irreversible*. It takes place only in the presence of the catalytic fluidized particles. The reaction is

$$A \rightarrow B$$

and the reaction-rate expression is

$$-\frac{dN_A}{dt} = kN_A C_c = \frac{\text{moles}}{\text{sec}}$$

where C_c is the catalyst (fluidized-bed) concentration.

6. For $x < 0$,

$$Y_A = Y_{A0}$$
$$Y_B = Y_{B0}$$
$$C_c = 0$$
$$E = 0$$

7. For $x > L$ (L = tower height),

$$C_c = 0$$
$$E = 0$$

Derive the *integrated* relation between Y_B and x. Evaluate *all constants* of integration.

Symbols

C_c = volume concentration of particle, ft³/ft³
E = eddy diffusivity, ft²/sec
k = reaction-rate constant
L = height of fluidized bed, ft
N_A = moles of A
t = time, sec
v = gas velocity, ft/sec
w_i = eddy diffusion rate, moles/(ft²)(sec)
x = distance from bottom of fluidized bed, ft
Y = mole fraction
ρ = density of gas mixture, moles/ft³

4-8. A gas probe consists of an inner tube of perimeter a ft inside a double annulus. The middle wall is of perimeter b ft, and the outer jacket is of perimeter S ft. The over-all length is L ft. (See Fig. 4-5.)

The probe is placed in a large reservoir of hot gas at θ°F and of constant specific heat c_p Btu/(lb)(°F), such that G lb/hr flow into the tube. Cooling fluid of constant specific heat n Btu/(lb)(°F) enters the outer annulus at a temperature T_0 °F and at a flow rate of w lb/hr. At the end of the probe, it reverses itself and flows in the inner annulus.

Conditions are such that the over-all heat-transfer coefficients at the three bounding surfaces are constant at the following values:

Bounding surface	Coefficient, Btu/(hr)(ft²)(°F)	Perimeter, ft
Inner tube wall.	V	a
Middle wall.	μ	b
Outer jacket.	H	S

It is required to determine the temperature of the cooling fluid as it leaves the probe at $l = 0$ ft.

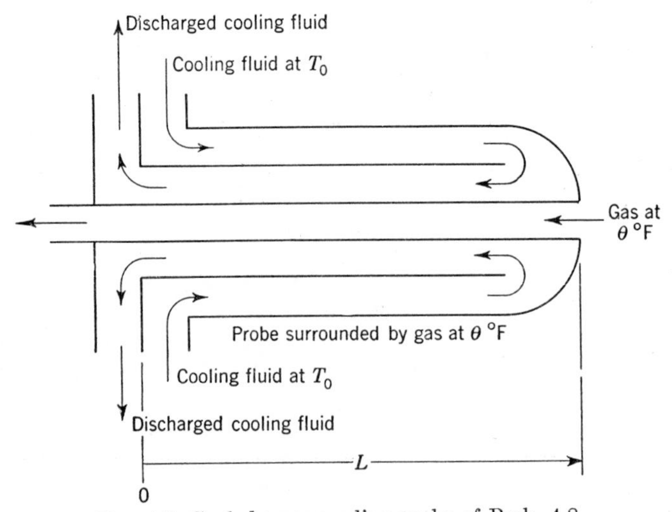

FIG. 4-5. Cooled gas sampling probe of Prob. 4.8.

4-9. Two similar vertical cylindrical tanks 6 ft in diameter and 10 ft high are placed side by side with their bottoms at the same level. They are connected at the bottom by a tube 2 ft long and 0.4 in. ID. Tank A is full of oil, and tank B is empty. Tank A has an outlet at the bottom, consisting of a short tube 1 ft long and 0.4 in. in diameter. Both this outlet tube and the connecting tube between the tanks are horizontal. Both tubes are opened simultaneously. What is the maximum oil level reached in tank B?

Assume streamline or "viscous" flow to occur in the small connecting and outlet tubes.

CHAPTER 5

SERIES AND NUMERICAL SOLUTIONS OF ORDINARY DIFFERENTIAL EQUATIONS

5-1. Introduction. This chapter deals with the two most powerful methods of integrating ordinary differential equations—expansion in series and numerical integration. In the majority of the more difficult equations found in practice an analytical solution is obtained only by recourse to series expansion. When this technique fails, a numerical solution usually can be obtained by one of the procedures to be discussed.

5-2. Series Solutions. A significant portion of the types of differential equations whose solutions can be expressed in closed form has been discussed in preceding sections. Even in such cases, many of the "closed" solutions are obtained in terms of functions which actually represent infinite series. Logarithmic, trigonometric, and hyperbolic functions are cases in point. It is not too surprising, then, to find that the solutions of a large class of differential equations are obtained in the form of an infinite series of terms. Certain equations of this class appear so frequently that the particular forms of infinite series which represent their solutions have been given specific names and symbols and the numerical values of the series tabulated as a function of x. Bessel functions, Legendre polynomials, etc., are typical examples. The next few sections of this chapter will be devoted to the properties and uses of infinite series, particularly as applied to the solution of differential equations.

5-3. Power Series. An expression of the type

$$A_0 + A_1(x - x_0) + \cdots + A_n(x - x_0)^n + \cdots = \sum_{n\to 0}^{\infty} A_n(x - x_0)^n \quad (5\text{-}1)$$

is termed a "power series." Such a series is said to "converge" if it approaches a finite value as n approaches infinity. The simplest test for convergence is the *ratio test;* if the *absolute* value of the ratio of the $(n + 1)$st term to the nth term in any infinite series approaches a limit j as $n \to \infty$, then the series converges for $j < 1$, diverges for $j > 1$, and the test fails for $j = 1$. As applied to (5-1), convergence requires

$$j = \lim_{n\to\infty} \left| \frac{A_{n+1}}{A_n} \right| |x - x_0| = L|x - x_0| < 1 \tag{5-2}$$

The quantity

$$\frac{1}{L} = \lim_{n\to\infty} \left| \frac{A_n}{A_{n+1}} \right|$$

is frequently called the "radius of convergence," since it represents the maximum $(x - x_0)$ value that can be used and convergence obtained. The ratio test is not the only method of testing for convergence; however, the alternative methods will not be discussed here.

Within the interval of convergence, a power series may be treated as a continuous function with continuous derivatives of all orders. Particularly useful properties are the following:

1. Within the interval of convergence of the *original* power series, series formed by termwise differentiation or integration of the original series are convergent.
2. The product of two power series converges inside the *common* interval of convergence of the original series.
3. The ratio of two power series converges inside the *common* interval of convergence of the original series, *provided* that the divisor does not vanish in the common interval.

Operations with series are more easily accomplished if the notation is compressed. Let y denote a function of x, $f(x)$, which is represented in the interval of convergence by the power series

$$y = f(x) = A_0 + A_1(x - x_0) + \cdots + A_n(x - x_0)^n + \cdots$$
$$= \sum_{n=0}^{\infty} A_n(x - x_0)^n \tag{5-3}$$

Then

$$\frac{dy}{dx} = A_1 + 2A_2(x - x_0) + \cdots + nA_n(x - x_0)^{n-1} + \cdots$$
$$= \sum_{n=0}^{\infty} nA_n(x - x_0)^{n-1} \tag{5-4}$$

$$\frac{d^2y}{dx^2} = 2A_2 + 6A_3(x - x_0) + \cdots + n(n-1)A_n(x - x_0)^{n-2} + \cdots$$
$$= \sum_{n=0}^{\infty} n(n-1)A_n(x - x_0)^{n-2} \tag{5-5}$$

$$\frac{d^ky}{dx^k} = \sum_{n=0}^{\infty} n(n-1) \cdots (n - k + 1)A_n(x - x_0)^{n-k} \tag{5-6}$$

5-4. Taylor Series. The power series (5-3) may be put into the useful form known as "Taylor series" as follows: Differentiate (5-3) n times and set $x = x_0$. Each member of the resulting set of equations then determines one constant:

$$A_0 = y_0 = f(x_0) \tag{5-7}$$

$$A_1 = \left(\frac{dy}{dx}\right)_{x_0} = f'(x_0) \tag{5-8}$$

$$A_n = \frac{f^n(x_0)}{n!} \tag{5-9}$$

Consequently, (5-3) becomes

$$y = f(x) = \sum_{n=0}^{\infty} \frac{f^n(x_0)}{n!} (x - x_0)^n \tag{5-10}$$

Equation (5-10) is the Taylor-series expansion of $f(x)$ near $x = x_0$. Since the original power series must converge if the expansion is to be valid, many functions cannot be represented by a Taylor series. For (5-10) to be valid, *all derivatives* of $f(x)$ must exist at $x = x_0$. A function which can be represented by a Taylor series or the completely equivalent power series about x_0 is said to be "regular" at $x = x_0$.

Taylor series have a number of important applications which will prove useful later.

5-5. Linear Second-order Equations with Variable Coefficients. The solution of the general linear homogeneous second-order differential equation is often best obtained by use of power-series methods. When written in the *standard form*

$$\frac{d^2y}{dx^2} + a_1(x)\frac{dy}{dx} + a_2(x)y = 0 \tag{5-11}$$

the properties of the coefficients $a_1(x)$ and $a_2(x)$ have an important bearing upon the behavior of an attempted power-series solution. Unless the functions $a_1(x)$ and $a_2(x)$ can be represented by a convergent power series in the interval of interest, the analyst can anticipate difficulties. The behavior of the series solutions in the neighborhood of a point x_0 can be predicted from the behavior of the functions $a_1(x)$ and $a_2(x)$ near x_0. It is customary to classify the point x_0 as follows:

1. x_0 is termed an "ordinary" point of the differential equation if both $a_1(x)$ and $a_2(x)$ can be represented by convergent power series which include $x = x_0$ in the interval of convergence, i.e., if $a_1(x)$ and $a_2(x)$ are *regular* at $x = x_0$.

2. x_0 is termed a "singular" point of the differential equation if either $a_1(x)$ or $a_2(x)$ fails to prove regular at $x = x_0$.

3. x_0 is termed a "regular singular" point of the differential equation if 2 holds but the products $(x - x_0)a_1(x)$ and $(x - x_0)^2a_2(x)$ *both* prove to be regular at $x = x_0$.

4. x_0 is termed an "irregular singular" point of the differential equation if 2 holds but 3 fails.

Singular points are most easily recognized by examination of the function $a(x)$ *and its derivatives* at $x = x_0$. Thus, $a(x) = x$ exhibits only ordinary points. $a(x) = 1 + 1/x$ becomes infinite at $x = 0$, which is the only singular point. However, $x(1 + 1/x)$ is regular at $x = 0$. $a(x) = 1/x(1 - x)$ exhibits singular points at $x = 0$ and $x = 1$.

The results of attempts to find power-series solutions of Eq. (5-11) can be predicted as follows:

1. If x_0 is an *ordinary* point of (5-11), *two linearly independent* power-series solutions which are regular at $x = x_0$ will be obtained. Each solution will be of the *form*

$$y = \sum_{n=0}^{\infty} A_n(x - x_0)^n \tag{5-12}$$

2. If x_0 is a *regular singular* point of (5-11), a power-series solution which is regular at $x = x_0$ cannot be guaranteed. However, the method of the next section will always generate at least *one* solution of the form

$$y = (x - x_0)^s \sum_{n=0}^{n=\infty} A_n(x - x_0)^n \tag{5-13}$$

in which s is a number whose value is determined in the course of the analysis. If s turns out to be zero or a positive integer, the series solution will turn out to be regular at $x = x_0$.

3. If x_0 is an *irregular singular* point of (5-11), a power-series solution may or may not exist.

5-6. The Method of Frobenius. The method of Frobenius often is the most convenient method of obtaining the power-series solution of the linear homogeneous second-order differential equation (5-11). The method proceeds to find solutions which are valid in the region of the point $x = 0$. Solutions valid in the region of a point $x = x_0$ may be obtained by transformation of the differential equation by use of the new variable $z = x - x_0$. In the remainder of this section it is assumed that such a transformation has already been accomplished. Now put Eq. (5-11) in the form

$$Ly \equiv R(x)\frac{d^2y}{dx^2} + \frac{1}{x}P(x)\frac{dy}{dx} + \frac{1}{x^2}V(x)y = 0 \tag{5-14}$$

where it is assumed that:

1. $R(x) \neq 0$ in the interval about $x = 0$ of interest.
2. The equation has been divided by a suitable constant which makes $R(0) = 1$.
3. $R(x)$, $P(x)$, and $V(x)$ are *regular* at $x = 0$.

Then $xa_1(x) \equiv P(x)/R(x)$ and $x^2a_2(x) \equiv V(x)/R(x)$ are regular at $x = 0$, and the point $x = 0$ is, at worst, a regular singular point.

Now represent the known functions $R(x)$, $P(x)$, $V(x)$ by the power series

$$R(x) = \sum_{k=0}^{\infty} R_k x^k \tag{5-15}$$

$$P(x) = \sum_{k=0}^{\infty} P_k x^k \tag{5-16}$$

$$V(x) = \sum_{k=0}^{\infty} V_k x^k \tag{5-17}$$

The numerical values of the coefficients R_k, P_k, V_k can be found in any practical problem.

The solution of (5-14) is assumed to be of the form

$$y = x^s \sum_{n=0}^{\infty} A_n x^n \tag{5-18}$$

where by definition A_0 is the first term of the series and *must not vanish.* Equation (5-18) is differentiated to determine the series representing dy/dx and d^2y/dx^2, and the results, together with the series expressions for $R(x)$, $P(x)$, and $V(x)$, are substituted in (5-14). The result is

$$\begin{aligned} L(y) &= \left(\sum_{k=0}^{\infty} R_k x^k\right)\left[\sum_{n=0}^{\infty} (n+s)(n+s-1)A_n x^{n+s-2}\right] \\ &\quad + \left(\sum_{k=0}^{\infty} P_k x^k\right)\left[\sum_{n=0}^{\infty} (n+s)A_n x^{n+s-2}\right] + \left(\sum_{k=0}^{\infty} V_k x^k\right)\left(\sum_{n=0}^{\infty} A_n x^{n+s-2}\right) \\ &= \sum_{k=0}^{\infty}\sum_{n=0}^{\infty} [(n+s)(n+s-1)R_k + (n+s)P_k + V_k]A_n x^{k+n+s-2} \\ &= 0 \end{aligned} \tag{5-19}$$

Equation (5-19) will be satisfied identically if the coefficients of each distinct power of x are identically zero. In its present form, the coefficients of a given power of x in Eq. (5-19) are obscure. This is remedied by the following device: Let

$$k + n = l \tag{5-20}$$

Then the coefficients of x^{l+s-2} are desired when l takes on a fixed value. Consider the term

$$\sum_{k=0}^{\infty} \sum_{n=0}^{\infty} V_k A_n x^{k+n+s-2} \tag{5-21}$$

For $l = 0$, $n = 0$, $k = 0$ is the only pair satisfying (5-20). Hence, the coefficient of x^{s-2} is $V_0 A_0$. For $l = 1$, only the pairs $n = 0$, $k = 1$; $n = 1$, $k = 0$ satisfy (5-20). The coefficient of x^{s-1} is $V_1 A_0 + V_0 A_1$. A continuation of this process will show that the coefficients are obtained from the expression

$$\sum_{l=0}^{\infty} \sum_{k=0}^{k=l} V_k A_{l-k} x^{l+s-2} \tag{5-22}$$

The conditions which satisfy (5-19) are then

$$\sum_{k=0}^{k=l} [(l + s - k)(l + s - k - 1)R_k + (l + s - k)P_k + V_k]A_{l-k} \equiv 0 \tag{5-23}$$

for *each* fixed value of l between 0 and ∞. Since $A_{l-k} = A_n$, Eq. (5-23) determines the coefficients (except for arbitrary constants) in the power-series solution (5-18) of the differential equation (5-14). The relation arising from $l = 0$ determines s. Thus, for $l = 0$, (5-23) becomes

$$[s(s - 1)R_0 + sP_0 + V_0]A_0 = 0$$

However, by definition A_0 cannot equal zero, and $R_0 = 1$. Hence, the *indicial* equation which fixes s is

$$s^2 + (P_0 - 1)s + V_0 = 0 \tag{5-24}$$

In general, Eq. (5-24) will determine two numerical values of s, s_1, and s_2, and hence *two distinct* series solutions of the differential equation (5-14). The exceptions will be discussed shortly. The leading term A_0 of either series is arbitrary and cannot be evaluated unless suitable boundary conditions are prescribed for the differential equation. They correspond to the two arbitrary constants which arise from the integration of a second-order differential equation. The remainder of the coefficients $A_1, A_2, \ldots, A_n, \ldots$ are determined in terms of A_0 for a particular value of s as a result of Eq. (5-23). Thus, for $l = 1$, (5-23) gives

$$A_1 = -\frac{s(s-1)R_1 + sP_1 + V_1}{(s+1)s + (s+1)P_0 + V_0} A_0$$

The condition for $l = 2$ determines A_2 in terms of A_1, and so on. With the notation

$$f(s) = s^2 + (P_0 - 1)s + V_0 \tag{5-25}$$
$$g_k(s) = R_k(s-k)^2 + (P_k - R_k)(s-k) + V_k \tag{5-26}$$

the recurrence formula relating A_n to preceding A's and hence to A_0 is readily obtained from (5-23) as

$$A_n = \frac{-\sum_{k=1}^{n} g_k(s+n)A_{n-k}}{f(s+n)} \tag{5-27}$$

where it is understood that $n \geqq 1$. Clearly, (5-27) breaks down if $f(s+n)$ turns out to be zero. This exceptional case is discussed in Sec. 5-7.

5-7. Exceptional Cases: $s_1 - s_2 = 0$ or Integer. If the indicial equation (5-24) results in $s_1 = s_2$, the method of Frobenius will involve only one arbitrary constant and hence will not represent the complete solution of the second-order equation. A similar situation may arise if the difference between the two indices, $s_1 - s_2$, is an integer. In this case, the term $f(s+n)$ in the denominator of the recurrence formula (5-27) will vanish for a particular value of n, say $n = N$, and A_N cannot be determined.

Analysis of these situations discloses that when the method of Frobenius is used to determine a series solution of a linear homogeneous second-order differential equation the following alternatives arise:

1. If s_1 and s_2 do not differ by zero or a real integer, two independent solutions of the form of (5-18) are obtained.
2. If $s_1 = s_2$, only one solution of the form of (5-18) is obtained.
3. If $s_1 - s_2 = N$, where N is a real integer, use of the larger value of $s(s_1)$ will always give one solution of the form of (5-18). If the smaller value of $s(s_2)$ is used, either *no* solution of the form of (5-18) is obtained or *two independent* solutions, one of which is identical to that found from s_1, of the form of (5-18) will be obtained. The latter occurs when $x = 0$ is an ordinary point.
4. In all cases in which only one solution

$$y_1 = \sum_{n=0}^{\infty} A_n x^{n+s_1} \equiv A_0 u_1(x) \tag{5-28}$$

can be found, the second *independent* solution is of the form

$$y_2 = Cu_1(x) \ln x + \sum_{n=0}^{\infty} B_n x^{n+s_2} \tag{5-29}$$

Differentiation of (5-29) followed by substitution in the original differential equation will determine the coefficients B_n in terms of the arbitrary constant C.

5-8. Bessel's Equation. The linear second-order equation

$$x^2 \frac{d^2y}{dx^2} + x \frac{dy}{dx} + (x^2 - p^2)y = 0 \tag{5-30}$$

occurs with surprising frequency in practical problems. As a result, the numerical values of the series solutions of (5-30) have been calculated and tabulated as a function of x and p. The equation is known as "Bessel's equation," and the solutions are termed "Bessel functions."

A solution of (5-30) of the form

$$y = x^s \sum_{n=0}^{n=\infty} A_n x^n \tag{5-18}$$

may be obtained by the method of Frobenius. It is first put in the form

$$\frac{d^2y}{dx^2} + \frac{1}{x}\frac{dy}{dx} + \frac{1}{x^2}(x^2 - p^2)y = 0 \tag{5-31}$$

Comparison with (5-14) shows that

$$\begin{aligned} R(x) &= 1 \\ P(x) &= 1 \\ V(x) &= x^2 - p^2 \end{aligned} \tag{5-32}$$

In the series expansions of (5-32) corresponding to (5-15) to (5-17), the coefficients are readily found to be

$$\begin{array}{llll} R_0 = 1 & R_1 = R_2 = \cdots = R_n = 0 & & \\ P_0 = 1 & P_1 = P_2 = \cdots = P_n = 0 & & \\ V_0 = -p^2 & V_1 = 0 & V_2 = 1 & V_3 = V_4 = \cdots = V_n = 0 \end{array} \tag{5-33}$$

The indicial equation (5-24) gives

$$\begin{aligned} s^2 &= p^2 \\ s_1 = p \qquad & s_2 = -p \end{aligned} \tag{5-34}$$

The solutions are developed by means of the recurrence formula (5-27) and are found to be

$$y_1(x) = A_0 x^p \left\{1 + \sum_{k=1}^{\infty} \frac{(-1)^k x^{2k}}{[(1+p)(2+p) \cdots (k+p)]2^{2k}k!}\right\} \tag{5-35}$$

corresponding to $s = p$ and

$$y_2(x) = B_0 x^{-p} \left\{1 + \sum_{k=1}^{\infty} \frac{(-1)^k x^{2k}}{[(1-p)(2-p) \cdots (k-p)]2^{2k}k!}\right\} \tag{5-36}$$

corresponding to $s = -p$. Equations (5-35) and (5-36) may be put into a more frequently used form by introducing the gamma function.

For positive values of p, the integral

$$\Gamma(p) = \int_0^\infty e^{-x} x^{p-1}\, dx \qquad p > 0 \tag{5-37}$$

defines the "gamma function." Numerical values of $\Gamma(p)$ are given in many tables, including Peirce.[10] Important properties of this function are

$$\Gamma(p+1) = p\Gamma(p) \qquad p > 0 \tag{5-38}$$

If N is a positive integer,

$$\Gamma(p+N) = (p+N-1)(p+N-2) \cdots (p+1)(p)\Gamma(p) \qquad p > 0 \tag{5-39}$$

$$\Gamma(p-1) = \frac{1}{p-1}\Gamma(p) \qquad p > 1 \tag{5-40}$$

If p is a positive integer n, then

$$\Gamma(n+1) = n! \tag{5-41}$$

$$\Gamma(1) = 0! = 1 \tag{5-42}$$

It is customary to extend (5-41) to noninteger values of p and to define the factorial of a positive quantity by the relation

$$\Gamma(p+1) \equiv p! \tag{5-43}$$

For negative values of p, $\Gamma(p)$ is not defined by (5-37), since the integral does not exist. It is customary to extend the definition of the gamma function to negative values of p *by defining* $\Gamma(p)$ for negative values of p by the relation

$$\Gamma(p) \equiv \frac{\Gamma(p+N)}{(p+N-1)(p+N-2) \cdots (p+1)(p)} \tag{5-44}$$

where N is a positive integer and $1 < p + N < 2$. Note, however, that the denominator of (5-44) vanishes if p equals zero or a negative integer, and hence $\Gamma(p)$ is not defined if p is zero or a negative integer.

When the gamma function is introduced, Eq. (5-35) becomes

$$y_1(x) = 2^p\Gamma(1+p)A_0 \sum_{k=0}^{\infty} \frac{(-1)^k(x/2)^{2k+p}}{k!(k+p)!} \tag{5-45}$$

or, with the notation

$$J_p(x) = \sum_{k=0}^{\infty} \frac{(-1)^k(x/2)^{2k+p}}{k!(k+p)!} \tag{5-46}$$

$$y_1(x) = C_1J_p(x) \tag{5-47}$$

The function denoted by $J_p(x)$ is called the "Bessel function of the first kind of order p."

If p is *neither* zero nor a positive integer, the second solution may be obtained from (5-36) as

$$y_2(x) = C_2J_{-p}(x) \tag{5-48}$$

$$J_{-p}(x) = \sum_{k=0}^{\infty} \frac{(-1)^k(x/2)^{2k-p}}{k!(k-p)!} \tag{5-49}$$

Consequently, if p is neither zero nor a positive integer, the complete solution of Bessel's equation, (5-30), *is*

$$y = C_1J_p(x) + C_2J_{-p}(x) \tag{5-50}$$

If p is zero or a positive integer n, the two solutions are not independent; it is found that

$$J_{-n}(x) = (-1)^nJ_n(x) \tag{5-51}$$

In this circumstance, the method of Frobenius does not provide the complete solution. However, the method outlined in Sec. 5-7, alternative 4, may be used to determine the second solution. It is found to be

$$y_2(x) = C_2Y_n(x) \tag{5-52}$$

where $Y_n(x)$, known as "Bessel's function of the second kind of order n," or "Weber's form," is defined by

$$Y_n(x) = \frac{2}{\pi}\left\{\left(\ln\frac{x}{2}+\gamma\right)J_n(x) - \frac{1}{2}\sum_{k=0}^{n-1}\frac{(n-k-1)!(x/2)^{2k-n}}{k!} + \frac{1}{2}\sum_{k=0}^{\infty}(-1)^{k+1}[\phi(k)+\phi(k+n)]\frac{(x/2)^{2k+n}}{k!(n+k)!}\right\} \tag{5-53}$$

where γ is Euler's constant

$$\gamma = 0.5772157 \cdots \tag{5-54}$$

and

$$\phi(k) = \sum_{m=1}^{k} \frac{1}{m} = 1 + \frac{1}{2} + \cdots + \frac{1}{k} \qquad k \geqq 1 \tag{5-55}$$

$$\phi(0) = 0 \tag{5-56}$$

Consequently, if p is zero or an integer n, the complete solution of Bessel's equation, (5-30), *is*

$$y = c_1 J_n(x) + c_2 Y_n(x) \tag{5-57}$$

The linear second-order equation

$$x^2 \frac{d^2y}{dx^2} + x \frac{dy}{dx} - (x^2 + p^2)y = 0 \tag{5-58}$$

is transformed into Bessel's equation (5-30) by the substitution $ix = z$. Consequently, the solution of (5-58) is

$$y = c_1 J_p(ix) + c_2 J_{-p}(ix) \tag{5-59}$$

if p is neither zero nor a positive integer, and

$$y = c_1 J_n(ix) + c_2 Y_n(ix) \tag{5-60}$$

if p is zero or a positive integer n. It is customary, however, to replace (5-59) and (5-60) by more convenient forms. Thus when p is neither zero nor a positive integer, the solution of (5-58) is written as

$$y = c_1 I_p(x) + c_2 I_{-p}(x) \tag{5-61}$$

and when p is zero or a positive integer n, the solution of (5-58) is taken as

$$y = c_1 I_n(x) + c_2 K_n(x) \tag{5-62}$$

$I_p(x)$, known as the "modified Bessel function of the first kind of order p," is defined by the expression

$$I_p(x) = i^{-p} J_p(ix) = \sum_{k=0}^{\infty} \frac{(x/2)^{2k+p}}{k!(k+p)!} \tag{5-63}$$

$K_n(x)$, known as the "modified Bessel function of the second kind of order n," is defined by the expression

$$K_n(x) = \frac{\pi}{2} i^{n+1} [J_n(ix) + i Y_n(ix)] \tag{5-64}$$

5-9. Generalized Form of Bessel's Equation. The generalized differential equation

$$x^2 \frac{d^2y}{dx^2} + x(a + 2bx^r) \frac{dy}{dx} + [c + dx^{2s} - b(1 - a - r)x^r + b^2x^{2r}]y = 0 \quad (5\text{-}65)$$

may be reduced to the form of Bessel's equation (5-30) by proper transformation of variables. The solution of (5-65) may then be obtained in terms of Bessel functions. The generalized solution of (5-65) is then

$$y = x^{(1-a)/2}e^{-(bx^r/r)} \left[c_1Z_p \left(\frac{\sqrt{|d|}}{s} x^s \right) + c_2Z_{-p} \left(\frac{\sqrt{|d|}}{s} x^s \right) \right] \quad (5\text{-}66)$$

where

$$p = \frac{1}{s} \sqrt{\left(\frac{1-a}{2} \right)^2 - c} \quad (5\text{-}67)$$

Z_p denotes one of the Bessel functions. If $\sqrt{d}/s$ is *real* and p is *not* zero or an integer, Z_p denotes J_p, Z_{-p} denotes J_{-p}; if p is zero or an integer, Z_p denotes J_n, Z_{-p} denotes Y_n. If $\sqrt{d}/s$ is *imaginary* and p is *not* zero or an integer, Z_p denotes I_p, Z_{-p} denotes I_{-p}; if p is zero or an integer n, Z_p denotes I_n, and Z_{-p} denotes K_n.

Example 5-1. As an illustration, consider the equation

$$x^2 \frac{d^2y}{dx^2} + x(1 - 2\beta) \frac{dy}{dx} + \beta^2x^{2\beta}y = 0 \quad (5\text{-}68)$$

When (5-68) is compared with (5-65), the following will be observed:

1. $1 - 2\beta$ will equal $a + 2bx^r$ if $b = 0$ and $a = 1 - 2\beta$.
2. $\beta^2x^{2\beta}$ will equal $c + dx^{2z} - b(1 - a - r)x^r + b^2x^{2r}$ if 1 above is granted and $c = 0$, $d = \beta^2$, and $s = \beta$.

Consequently, $a = 1 - 2\beta$, $b = 0$, $c = 0$, $d = \beta^2$, $s = \beta$. Then $p = (1/\beta)\sqrt{\beta^2} = 1$, $\sqrt{d}/s = 1$. The solution of (5-68) is then

$$y = x^\beta[c_1J_1(x^\beta) + c_2Y_1(x^\beta)] \quad (5\text{-}69)$$

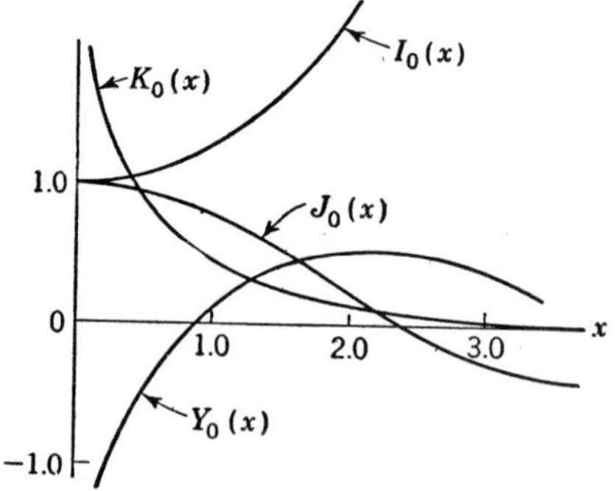

FIG. 5-1. Bessel functions of zero order.

5-10. Properties of Bessel Functions. Bessel functions are particularly useful because the numerical values of the series representing the functions have been calculated and reported as a function of the independent variable. Bessel-function tables are given in Watson,[13] Jahnke and Emde,[5] McLachlan,[8] and Gray, Mathews, and MacRobert.[4] A small list of values of Bessel functions is given in Table 5-1. The behavior of the zero-order Bessel functions is shown in Fig. 5-1. Unfortunately, the symbols used to denote the various functions are not well standardized and vary with the reference source. For convenience, Table 5-2 compares the Bessel-function notation used by various authors.

The limits approached by the various Bessel functions as $x \to 0$ or $x \to \infty$ are of considerable importance in the solution of practical problems.

For *small values of* x, the following approximations are useful:

$$J_p(x) \cong \frac{1}{2^p p!} x^p \qquad J_{-p}(x) \cong \frac{2^p}{(-p)!} x^{-p} \tag{5-70}$$

TABLE 5-1. NUMERICAL VALUES OF THE ZERO-ORDER BESSEL FUNCTIONS OF THE FIRST AND SECOND KINDS

x	$J_0(x)$	$Y_0(x)$	$I_0(x)$	$K_0(x)$
0.0	1.0000	$-\infty$	1.0000	∞
0.5	0.9385	−0.4445	1.0635	0.9244
1.0	0.7652	+0.0883	1.2661	0.4210
1.5	0.5118	0.3824	1.6467	0.2138
2.0	0.2239	0.5104	2.280	0.1139
2.5	−0.0484	0.4981	3.290	0.06235
3.0	−0.2601	0.3769	4.881	0.03474
3.5	−0.3801	0.1890	7.378	0.01960
4.0	−0.3971	−0.0169	11.302	0.01116
4.5	−0.3205	−0.1947	17.48	0.00640
5.0	−0.1776	−0.3085	27.24	0.00369
5.5	−0.0068	−0.3395	42.69	
6.0	+0.1506	−0.2882	67.23	
6.5	0.2601	−0.1732	106.29	
7.0	0.3001	−0.0259	168.6	
7.5	0.2663	+0.1173	268.2	
8.0	0.1717	0.2235	427.6	
8.5	0.0419	0.2702	683.2	
9.0	−0.0903	0.2499	1093.6	
9.5	−0.1939	0.1712	1753	
10.0	−0.2459	0.0557		

$$\begin{aligned} Y_n(x) &\cong -\frac{2^n(n-1)!}{\pi} x^{-n} \qquad n \neq 0 \\ Y_0(x) &\cong \frac{2}{\pi} \ln x \end{aligned} \tag{5-71}$$

$$\begin{aligned} I_p(x) &\cong \frac{1}{2^p p!} x^p \\ I_{-p}(x) &\cong \frac{2^p}{(-p)!} x^{-p} \end{aligned} \tag{5-72}$$

$$K_n(x) \cong 2^{n-1}(n-1)!x^{-n} \qquad n \neq 0$$
$$K_0(x) \cong -\ln x \tag{5-73}$$

Examination of the above relations shows that *only* $J_p(x)$ *and* $I_p(x)$ *are finite at* $x = 0$. However, the *power series* included in *all* Bessel functions converge for *all* finite values of x; the divergence of certain of the functions themselves at $x = 0$ results from the fact that in such cases the power series are multiplied by a negative power of x or by a term involving the logarithm of x.

For large values of x ($x \to \infty$), the following approximations are useful:

$$J_p(x) \cong \sqrt{\frac{2}{\pi x}} \cos\left(x - \frac{\pi}{4} - \frac{p\pi}{2}\right) \tag{5-74}$$

$$Y_n(x) \cong \sqrt{\frac{2}{\pi x}} \sin\left(x - \frac{\pi}{4} - \frac{n\pi}{2}\right) \tag{5-75}$$

$$I_p(x) \cong \frac{e^x}{\sqrt{2\pi x}} \tag{5-76}$$

$$K_n(x) \cong \sqrt{\frac{\pi}{2x}}\, e^{-x} \tag{5-77}$$

Both J_p and Y_n oscillate like damped sinusoidal functions and approach zero as $x \to \infty$. The amplitude of the oscillations about zero decreases as x increases, and the distance between successive zeros of both functions decreases toward a limit of π as x increases. The zeros of $J_{p+1}(x)$ separate the zeros of $J_p(x)$; that is, between any two values of x that make $J_{p+1}(x)$ equal zero there is one and only one value of x that makes $J_p(x)$ zero. This same statement applies to $Y_{n+1}(x)$ and $Y_n(x)$. Tables 5-3 and 5-4 give the values of x that make $J_0(x)$ and $J_1(x)$ equal zero.

In contrast to the behavior of $J_p(x)$ and $Y_n(x)$, I_p increases continuously with x, and K_n decreases continuously. Figure 5-1 indicates the behavior of the Bessel functions of zero order.

Bessel functions of order equal to half an odd integer can be represented in terms of elementary functions:

$$J_{1/2}(x) = \sqrt{\frac{2}{\pi x}} \sin x$$
$$J_{-1/2}(x) = \sqrt{\frac{2}{\pi x}} \cos x \tag{5-78}$$

$$I_{1/2}(x) = \sqrt{\frac{2}{\pi x}} \sinh x$$
$$I_{-1/2}(x) = \sqrt{\frac{2}{\pi x}} \cosh x \tag{5-79}$$

TABLE 5-2. BESSEL-FUNCTION NOTATION

Description	This book	Jahnke and Emde	McLachlan	Watson	Whittaker and Watson	Gray, Mathews, and MacRobert
Function of the first kind	$J_p(x)$	$J_p(x)$	$J_p(x)$	$J_p(x)$	$J_p(x)$	$J_p(x)$
Function of the second kind (Weber)	$Y_n(x)$	$N_n(x)$	$Y_n(x)$	$Y_n(x)$	$Y_n(x)$	$\frac{2}{\pi}[Y_n - (\ln 2 - \gamma)J_p]$
Function of the second kind (Neumann)	$\frac{\pi}{2} Y_n + (\ln 2 - \gamma)J_p$	....	$\mathbf{Y}n(x)$	$Y^{(n)}(x)$	$Y^{(n)}(x)$	$Y_n(x)$
Modified function of first kind	$I_p(x)$	$i^{-p}J_p(ix)$	$I_p(x)$	$I_p(x)$	$I_p(x)$	$I_p(x)$
Modified function of second kind	$K_n(x)$	$\frac{1}{2}\pi i^{n+1}H_n{}^1(ix)$	$K_n(x)$	$K_n(x)$	$\frac{K_n(x)}{\cos n\pi}$	$K_n(x)$

Equations (5-78) and (5-79) together with the recurrence formulas

$$J_{n+\frac{1}{2}}(x) = \frac{2n-1}{x} J_{n-\frac{1}{2}}(x) - J_{n-\frac{3}{2}}(x) \tag{5-80}$$

TABLE 5-3. VALUES OF x FOR WHICH $J_0(x) = 0$ AND CORRESPONDING VALUES OF $J_1(x)$

Value of x for which $J_0(x) = 0$	Differences in x values	Corresponding value of $J_1(x)$
2.4048		+0.5191
	3.1153	
5.5201		−0.3403
	3.1336	
8.6537		+0.2715
	3.1378	
11.7915		−0.2325
	3.1394	
14.9309		+0.2065

TABLE 5-4. VALUES OF x FOR WHICH $J_1(x) = 0$ AND CORRESPONDING VALUES OF $J_0(x)$

Value of x for which $J_1(x) = 0$	Differences in x values	Corresponding value of $J_0(x)$
3.8317		−0.4028
	3.1839	
7.0156		+0.3001
	3.1579	
10.1735		−0.2497
	3.1502	
13.3237		+0.2184
	3.1469	
16.4706		−0.1965

$$I_{n+\frac{1}{2}}(x) = -\frac{2n-1}{x} I_{n-\frac{1}{2}}(x) + I_{n-\frac{3}{2}}(x) \tag{5-81}$$

enable the complete set of Bessel functions of order equal to half an odd integer to be evaluated.

The following relations may be proved by recourse to the defining equations and are of considerable utility.

$$\frac{d}{dx}[x^p Z_p(\alpha x)] = \begin{cases} \alpha x^p Z_{p-1}(\alpha x) & Z = J, Y, I \\ -\alpha x^p Z_{p-1}(\alpha x) & Z = K \end{cases} \tag{5-82}$$

$$\frac{d}{dx}[x^{-p} Z_p(\alpha x)] = \begin{cases} -\alpha x^{-p} Z_{p+1}(\alpha x) & Z = J, Y, K \\ \alpha x^{-p} Z_{p+1}(\alpha x) & Z = I \end{cases} \tag{5-83}$$

$$\frac{d}{dx}[Z_p(\alpha x)] = \begin{cases} \alpha Z_{p-1}(\alpha x) - \frac{p}{x} Z_p(\alpha x) & Z = J, Y, I \\ -\alpha Z_{p-1}(\alpha x) - \frac{p}{x} Z_p(\alpha x) & Z = K \end{cases} \tag{5-84}$$

$$\frac{d}{dx}[Z_p(\alpha x)] = \begin{cases} -\alpha Z_{p+1}(\alpha x) + \frac{p}{x} Z_p(\alpha x) & Z = J, Y, K \\ \alpha Z_{p+1}(\alpha x) + \frac{p}{x} Z_p(\alpha x) & Z = I \end{cases} \tag{5-85}$$

$$2\frac{d}{dx} I_p(\alpha x) = \alpha[I_{p-1}(\alpha x) + I_{p+1}(\alpha x)] \tag{5-86}$$

$$2\frac{d}{dx} K_n(\alpha x) = -\alpha[K_{n-1}(\alpha x) + K_{n+1}(\alpha x)] \tag{5-87}$$

$$Z_p(\alpha x) = \frac{\alpha x}{2p}[Z_{p+1}(\alpha x) + Z_{p-1}(\alpha x)] \qquad Z = J, Y \tag{5-88}$$

$$I_p(\alpha x) = \frac{-\alpha x}{2p}[I_{p+1}(\alpha x) - I_{p-1}(\alpha x)] \tag{5-89}$$

$$K_n(\alpha x) = \frac{\alpha x}{2p}[K_{n+1}(\alpha x) - K_{n-1}(\alpha x)] \tag{5-90}$$

$$\left.\begin{aligned} J_{-n}(\alpha x) &= (-1)^n J_n(\alpha x) \\ I_{-n}(\alpha x) &= I_n(\alpha x) \\ K_{-n}(\alpha x) &= K_n(\alpha x) \end{aligned}\right\} \quad \text{when } n \text{ is zero or integer} \tag{5-91}$$

5-11. Examples of Bessel-function Solutions

Example 5-2. The Wedge-shaped Fin. In Example 3-6 the differential equation resulting from the analysis of heat flow through and from a wedge-shaped fin was derived. The resulting equation was

$$x\frac{d^2y}{dx^2} + \frac{dy}{dx} - \frac{2h \sec \theta L}{kw} y = 0 \tag{3-34}$$

where x = distance from tip of fin, ft
$y = T - T_a$
T = local fin temperature at x
T_a = temperature of surrounding air
h = heat-transfer coefficient from outside surface of fin to surrounding air, Btu/(hr)(ft²)(°F)
k = thermal conductivity of fin material
L = total length of fin, ft
w = thickness of fin at base, ft
θ = half wedge angle of fin

In order to integrate (3-34), put it in the form of (5-65) by multiplying by x and use the notation

$$\alpha = \frac{2h \sec \theta L}{kw}$$

Then (3-34) becomes

$$x^2\frac{d^2y}{dx^2} + x\frac{dy}{dx} - \alpha xy = 0 \tag{5-92}$$

When (5-92) is compared with (5-65), it is found that $a = 1, b = 0, c = 0, d = -\alpha$, $s = \frac{1}{2}, p = 0, r = 0$. The solution of (5-92) is then

$$y = T - T_a = c_1 I_0(2\sqrt{\alpha x}) + c_2 K_0(2\sqrt{\alpha x}) \tag{5-93}$$

However, the fin temperature must remain finite at the tip where $x = 0$. Since $K_0(0) \to \infty$, whereas $I_0(0) = 1$, c_2 must equal zero, and

$$T - T_a = c_1 I_0(2\sqrt{\alpha x}) \tag{5-94}$$

In order to determine c_1, an additional boundary condition must be specified. As an illustration, take

$T_a = 100°F$
$T_L = 200°F$, temperature at $x = L$,
$L = 1$ ft
$h = 2$ Btu/(hr)(ft²)(°F)
$k = 220$ Btu/(hr)(ft)(°F)
$\sec\theta = 1$
$w = \frac{1}{12}$ ft

Then $$\alpha = \frac{2 \cdot 2 \cdot 1 \cdot 12}{220 \cdot 1 \cdot 1} = 0.218$$

and $$T - T_a = c_1 I_0 2\sqrt{0.218}\, x \tag{5-95}$$

At $x = L = 1$,

$$200 - 100 = c_1 I_0(2\sqrt{0.218}) = 1.230c_1$$

and $c_1 = 81.2$. The value of 1.230 for $I_0(2\sqrt{0.218})$ is obtained from tables of the function. With the value of c_1 determined, Eq. (5-95) becomes

$$T = 81.2 I_0(0.934\sqrt{x}) + 100 \tag{5-96}$$

The fin temperature is obtained as a function of x from (5-96) by use of tabulated values of I_0. The results are shown in Table 5-5.

TABLE 5-5

x	$0.934\sqrt{x}$	$I_0(0.934\sqrt{x})$	T(°F)
0	0	1.000	181.2
0.2	0.417	1.043	184.8
0.4	0.590	1.089	188.5
0.5	0.660	1.112	190.4
0.6	0.722	1.134	192.2
0.8	0.832	1.180	195.9
1.0	0.934	1.230	200.0

The heat transferred from the fin to the surrounding air may be determined by either of two procedures. A direct approach is made by integration of the expression for the local rate of heat transfer

$$q = \int_0^A h(T - T_a)\, dA = \int_0^L h(T - T_a)\, 2 \sec\theta\, dx \tag{5-97}$$

When (5-94) is introduced, (5-97) becomes

$$q = 2c_1 h \sec \theta \int_0^L [I_0(2\sqrt{\alpha x})]\, dx \tag{5-98}$$

The integral in (5-98) is evaluated by use of Eq. (5-82) and the transformation

$$z = 2\sqrt{\alpha x} \tag{5-99}$$

Then

$$\int_0^L I_0(2\sqrt{\alpha x})\, dx = \frac{1}{2\alpha} \int_0^{2\sqrt{\alpha L}} z[I_0(z)]\, dz = \frac{1}{2\alpha} \Big[zI_1(z) \Big]_0^{2\sqrt{\alpha L}} = \frac{L}{\sqrt{\alpha L}} I_1(2\sqrt{\alpha L}) \tag{5-100}$$

and the heat-transfer rate per foot of fin width is

$$q = \frac{2c_1 h(\sec \theta)L}{\sqrt{\alpha L}} I_1(2\sqrt{\alpha L}) \tag{5-101}$$

An alternative approach is to make use of the fact that the heat transferred to the fin by conduction at $x = L$ is equal (at steady state) to the heat transferred by the fin to the air. Then

$$q = kw \left(\frac{dT}{dx}\right)_{x=L}$$

Differentiation of (5-94) gives

$$\frac{dT}{dx} = c_1 \frac{d}{dx} [I_0(2\sqrt{\alpha x})] = \frac{c_1 2\alpha}{z} \frac{d}{dz} [I_0(z)]$$
$$= \frac{c_1 2\alpha}{z} I_1(z)$$

when (5-86) and (5-99) are introduced. Then

$$q = \frac{kwc_1\alpha}{\sqrt{\alpha L}} I_1(2\sqrt{\alpha L}) = \frac{2c_1 h(\sec \theta)L}{\sqrt{\alpha L}} I_1(2\sqrt{\alpha L}) \tag{5-101}$$

The effectiveness η of a fin is defined as the ratio of the actual heat-transfer rate to the rate which would occur if the entire fin were at the base temperature. For the present example,

$$\eta = \frac{q}{\int_0^L 2h(\sec \theta)(T_L - T_a)\, dx} = \frac{c_1[I_1(2\sqrt{\alpha L})}{\sqrt{\alpha L}(T_L - T_a)}$$
$$= \frac{I_1(2\sqrt{\alpha L})}{\sqrt{\alpha L}[I_0(2\sqrt{\alpha L})]} \tag{5-102}$$

Example 5-3. Heat Loss from Oven Wall. As another example of the use of Bessel functions, consider the problem of heat loss from the surface of an oven wall due to "through metal," which conducts heat from the inside, the heat being dissipated to the air from the sheet-metal protective covering of the insulated housing. The metal covering is of steel 0.005 ft thick, having a thermal conductivity of 25 Btu/(hr)(ft²)(°F/ft). The surface coefficient of heat transfer is 2.5 Btu/(hr)(ft²)(°F), and the head of the bolt is ⅝ in. in diameter. The room air is at 70°F, and the temperature of the head of the bolt is constant at 150°F. Neglecting heat loss except by conduction along the bolt, determine the temperature of the outer metal wall at several points up to 1 ft from the bolt.

In setting up the necessary differential equation, it is important to choose the proper variables. On first thought, it would appear that three variables are involved: temperature and two variables to define position. Since the temperature function is symmetrical about the bolt head, the single variable r, representing the radial distance from the bolt center, may be used to define position. When the differential heat balance on an annular surface element is set up, the rate of heat input q at radius r is given by

$$q = -ka\, 2\pi r \frac{dT}{dr}$$

At $r + dr$, the rate of heat output is

$$q + dq = -ka\, 2\pi r \frac{dT}{dr} + \frac{d}{dr}\left(-ka\, 2\pi r \frac{dT}{dr}\right) dr + h(T - T_a) 2\pi r\, dr$$

At steady state the accumulation is zero. Then, when the thermal conductivity k and the metal thickness a are constant, there results

$$\frac{d^2T}{dr^2} + \frac{1}{r}\frac{dT}{dr} - \frac{h(T - T_a)}{ka} = 0 \tag{5-103}$$

Let $y = T - T_a$ and $\beta = h/ka$; then (5-103) becomes

$$r^2 \frac{d^2y}{dr^2} + r\frac{dy}{dr} - \beta r^2 y = 0 \tag{5-104}$$

The solution of (5-104) is found to be

$$y = C_1 I_0(r\sqrt{\beta}) + c_2 K_0(r\sqrt{\beta}) \tag{5-105}$$

This time $I_0(r\sqrt{\beta})$ is eliminated, for it increases with increasing positive values of the variable and could not represent the temperature at large distances from the bolt, which obviously decreases to approach 70°F as an asymptote. Consequently, the result is

$$T - 70 = C_2 K_0(r\sqrt{\beta}) \tag{5-106}$$

It is also possible to determine C_1 as zero and hence eliminate $I_0(r\sqrt{\beta})$ by consideration of the boundary condition $r = \infty$, $y = T - T_a = 0$. At $r = \infty$, then

$$C_1 = \frac{-C_2 K_0(r\sqrt{\beta})}{I_0(r\sqrt{\beta})}$$

Since at $r = \infty$, $K_0(r\sqrt{\beta}) = 0$ and $I_0(r\sqrt{\beta}) = \infty$, $C_1 = 0$.

At this point in the solution, it is well to check the validity of the result through the use of Eqs. (5-82) to (5-91).

$$\frac{dy}{dr} = -\sqrt{\beta}\, K_1(r\sqrt{\beta}) \qquad r \neq 0$$

$$\frac{d^2y}{dr^2} = -\frac{\sqrt{\beta}}{r} K_1(r\sqrt{\beta}) + \beta K_2(r\sqrt{\beta}) \qquad r \neq 0$$

On substitution in (5-104), the left-hand side becomes

$$-r\sqrt{\beta}K_1(r\sqrt{\beta}) + r^2\beta K_2(r\sqrt{\beta}) - r\sqrt{\beta}K_1(r\sqrt{\beta}) - r^2\beta K_0(r\sqrt{\beta})$$

From (5-90) it is seen that the sum of the first and third terms is equal to the sum of the second and fourth terms, and the expression reduces to zero. The use of $K_0(r\sqrt{\beta})$ as a solution of (5-104) is substantiated in this way.

Proceeding with the numerical solution,

$$\beta = \frac{h}{ka} = \frac{2.5}{25 \cdot 0.005} = 20$$

and r_0, the radius of the bolt head, is $5/(2 \cdot 8 \cdot 12) = 0.026$ ft. Hence

$$150 - 70 = C_2K_0(0.026\sqrt{20}) = C_2K_0(0.116) = 2.3C_2$$

from which $C_2 = 34.8$. The final form is

$$T - 70 = 34.8K_0(r\sqrt{20})$$

The following table gives the temperatures calculated for various values of r:

TABLE 5-6

r, ft	$r\sqrt{20}$	$K_0(r\sqrt{20})$	T(°F)
0.05	0.223	1.65	128
0.1	0.446	1.04	106
0.2	0.892	0.493	87
0.4	1.785	0.155	75.5
0.6	2.68	0.051	71.8
0.8	3.57	0.018	70.6
1.0	4.47	0.006	70.2

In the case of an actual oven wall, there is simultaneous conduction through the wall itself. This does not complicate the problem appreciably if it is assumed that such heat conduction is wholly normal to the surface.

5-12. Important Second-order Equations. In addition to Bessel's equation, there are a number of other frequently occurring second-order differential equations whose series solutions have been studied and their numerical values tabulated. The more important of these equations are listed here for reference, although their solutions will not be examined in detail.

Legendre Functions. The following four equations are alternative forms of equations whose solution is in the form of Legendre polynomials. In these equations p is real and nonnegative. If $p = -n$, the solution is the same as for $p = n + 1$, and so it is possible to handle equations in which p is negative.

$$\begin{aligned}
(1 - x^2)\frac{d^2y}{dx^2} - 2x\frac{dy}{dx} + p(p+1)y &= 0 \\
\frac{d}{dx}\left[(1 - x^2)\frac{dy}{dx}\right] + p(p+1)y &= 0 \\
\frac{1}{\sin\theta}\frac{d}{d\theta}\left(\sin\theta\frac{dy}{d\theta}\right) + p(p+1)y &= 0 \\
\frac{d^2y}{d\theta^2} + \frac{dy}{d\theta}\cot\theta + p(p+1)y &= 0
\end{aligned} \tag{5-107}$$

These equations often arise in problems in which some potential (such as voltage or temperature) has spherical boundaries. The method of Frobenius leads to a solution of the form

$$y = C_1u_p(x) + C_2v_p(x) \tag{5-108}$$

where

$$u_p(x) = 1 - \frac{p(p+1)}{2!}x^2 + \frac{p(p-2)(p+1)(p+3)}{4!}x^4 - \frac{p(p-2)(p-4)(p+1)(p+3)(p+5)}{6!}x^6 \cdots$$

$$v_p(x) = x - \frac{(p-1)(p+2)}{3!}x^3 + \frac{(p-1)(p-3)(p+2)(p+4)}{5!}x^5 \cdots$$

Note that if p is an even integer or zero, $u_p(x)$ is a polynomial with a finite number of terms. If p is an odd integer, then $v_p(x)$ has a finite number of terms. Thus, one or the other of the solutions is always an infinite series, and if p is not an integer, they are both infinite series.

From the original equation, it can be seen that u_p and v_p will converge if $-1 < x < 1$. Otherwise, they will diverge (unless they terminate), and in particular they will diverge at $x = \pm 1$.

The above series result if a straightforward approach to the solution is made. However, usually it turns out that p is an integer, and when this is the case, it is convenient to use a somewhat different notation:

For integral values of p, let $p = n$.

If n is even or zero, $\qquad P_n(x) \equiv \dfrac{u_n(x)}{u_n(1)}$

If n is odd, $\qquad P_n(x) \equiv \dfrac{v_n(x)}{v_n(1)} \qquad$ (5-109)

If n is even, $\qquad u_0(1) = 1$

$$u_n(1) = (-1)^{n/2}\frac{2 \cdot 4 \cdot 6 \cdots n}{1 \cdot 3 \cdot 5 \cdots (n-1)}$$

If n is odd, $v_1(1) = 1$ (5-110)

$$v_n(1) = (-1)^{(n-1)/2} \frac{2 \cdot 4 \cdot 6 \cdots (n-1)}{1 \cdot 3 \cdot 5 \cdots n}$$

Hence

$$\begin{aligned} P_0(x) &= 1 \\ P_1(x) &= x \\ P_2(x) &= \tfrac{1}{2}(3x^2 - 1) \\ P_3(x) &= \tfrac{1}{2}(5x^3 - 3x) \\ P_4(x) &= \tfrac{1}{8}(35x^4 - 30x^2 + 3) \end{aligned}$$

The function $P_n(x)$ defines one of the solutions of Legendre's equation for integer n. The second solution, called "Legendre functions of the second kind," is denoted by $Q_n(x)$, where

$$Q_n(x) = \begin{cases} [-v_n(1)]u_n(x) & n \text{ odd} \\ u_n(1)v_n(x) & n \text{ even} \end{cases} \qquad (5\text{-}111)$$

but is defined only for $-1 < x < 1$, since $u_n(x)$ is an infinite series when n is odd and $v_n(x)$ is an infinite series when n is even, and neither series converges outside the interval $-1 < x < 1$. Although the $Q_n(x)$ are infinite series, they may be expressed in the form

$$\begin{aligned} Q_0(x) &= \frac{1}{2} \ln \frac{1+x}{1-x} = \tanh^{-1} x \\ Q_1(x) &= xQ_0(x) - 1 \\ Q_2(x) &= P_2(x)Q_0(x) - \tfrac{3}{2}x \\ Q_3(x) &= P_3(x)Q_0(x) - \tfrac{5}{2}x^2 + \tfrac{2}{3} \end{aligned}$$

etc.

In general, *both* $P_n(x)$ and $Q_n(x)$ satisfy the recurrence formula

$$nS_n(x) = (2n-1)xS_{n-1}(x) - (n-1)S_{n-2}(x) \qquad (5\text{-}112)$$

so that $S_n(x)$ may be found from $S_{n-1}(x)$ and $S_{n-2}(x)$. The formal solution of Legendre's equation for integer n is then

$$y = AP_n(x) + BQ_n(x) \qquad (5\text{-}113)$$

where only $P_n(x)$ is finite outside the interval $-1 < x < 1$.

Hypergeometric Functions. The solution of Gauss's equation

$$x(1-x)\frac{d^2y}{dx^2} + [\nu - (\alpha + \beta + 1)x]\frac{dy}{dx} - \alpha\beta y = 0 \qquad (5\text{-}114)$$

is expressed in the form

$$y = A_0F(\alpha,\beta;\nu;x) + B_0x^{1-\nu}F(\alpha - \nu + 1, \beta - \nu + 1; 2 - \nu; x) \qquad (5\text{-}115)$$

$F(\alpha,\beta;\nu;x)$ denotes the "hypergeometric" series

$$F(\alpha,\beta;\nu;x) = 1 + \frac{\alpha\beta}{1\nu} x + \frac{\alpha(\alpha+1)\beta(\beta+1)}{1 \cdot 2\nu(\nu+1)} x^2 + \cdots + \frac{[\alpha(\alpha+1) \cdots (\alpha+k-1)][\beta(\beta+1) \cdots (\beta+k-1)]}{[1 \cdot 2 \cdot \cdots \cdot k][\nu(\nu+1) \cdots (\nu+k-1)]} x^k + \cdots \tag{5-116}$$

The series multiplied by A_0 in (5-115) does not exist (in general) when ν is zero or a negative integer, and the series multiplied by B_0 does not exist when ν is a positive integer greater than unity.

Laguerre Polynomial. The equation

$$x \frac{d^2y}{dx^2} + (c - x) \frac{dy}{dx} - ay = 0 \tag{5-117}$$

is satisfied by the "confluent hypergeometric function" of Kummer, $M(a,c;x)$, if c is nonintegral:

$$y = AM(a,c;x) + \beta x^{1-c} M(1 + a - c, 2 - c; x) \tag{5-118}$$

If $c = 1$ and $a = -n$, where n is a positive integer or zero, one solution is the nth Laguerre polynomial

$$y = L_n(x) \tag{5-119}$$

If $c = k + 1$ and $a = k - n$, where k and n are integral, one solution is the associated Laguerre polynomial

$$y = AL_n^k(x) = \frac{Ad^k}{dx^k} L_n(x) \qquad \text{if } k \leqq n \tag{5-120}$$

Hermite Polynomial. The equation

$$\frac{d^2y}{dx^2} - 2x \frac{dy}{dx} + 2ny = 0 \tag{5-121}$$

is satisfied by the Hermite polynomial of degree n,

$$y = AH_n(x) \tag{5-122}$$

if n is a positive integer or zero.

Tschebyscheff Polynomial. The equation

$$(1 - x^2) \frac{d^2y}{dx^2} - x \frac{dy}{dx} + n^2y = 0 \tag{5-123}$$

is satisfied by the nth Tschebyscheff polynomial

$$y = AT_n(x) \tag{5-124}$$

if n is a positive integer or zero.

Jacobi Polynomial. The equation

$$x(1-x)\frac{d^2y}{dx^2} + [a - (1+b)x]\frac{dy}{dx} + n(a+n)y = 0 \qquad (5\text{-}125)$$

is satisfied by the nth Jacobi polynomial

$$y = AJ_n(a,b,x) \qquad (5\text{-}126)$$

5-13. Numerical-solution Methods. Because of the complexity of the functions encountered in many chemical-engineering processes, particularly those functions representing thermodynamic and kinetic data, the differential equations encountered are often incapable of solution by the methods presented in the preceding sections. In such cases, it is necessary to resort to one of the many so-called "approximate methods" of solution. Although these do not lead to analytical solutions, it is possible to calculate numerical values for particular solutions to any desired number of significant figures.

If an equation of the first order is expressed in the form

$$\frac{dy}{dx} = f(x,y) \qquad (5\text{-}127)$$

its solution may be expressed as

$$y = \int f(x,y)\,dx + c \qquad (5\text{-}128)$$

Let it be specified that a particular solution must pass through the point $x = x_0$, $y = y_0$; then (5-128) becomes

$$y = y_0 + \int_{x_0}^{x} f(x,y)\,dx \qquad (5\text{-}129)$$

an equivalent form of which is

$$y = y_0 + \int_{x_0}^{x} \frac{dy}{dx}\,dx \qquad (5\text{-}130)$$

Equations (5-129) and (5-130) are known as "integral equations," because both variables appear under the integral sign and their direct, exact solution by means of ordinary quadrature formulas, which contain only one variable, is impossible. It will be shown, however, that by several alternative methods of successive approximation the solution may be obtained to any desired degree of accuracy, provided that several simple conditions are fulfilled.

5-14. Modified Euler Method. Starting at the point x_0, y_0, calculate $(dy/dx)_0$. If the increment to be calculated is small, the corresponding portion of the y, x curve to be calculated will approximate a straight line. This means that the value of dy/dx over this interval remains nearly constant at its value $(dy/dx)_0$ calculated at the beginning of the interval. Granting these conditions, (5-129) and (5-130) may be integrated to give

$$y_1^{(1)} = y_0 + \left(\frac{dy}{dx}\right)_0 (x_1 - x_0) = y_0 + f(x_0,y_0)(x_1 - x_0) \quad (5\text{-}131)$$

$y_1^{(1)}$ denotes the first approximation to the true value of y at x_1. When $y_1^{(1)}$ is calculated, a new interval may be taken and (5-131) applied to give

$$y_2^{(1)} = y_1^{(1)} + \left(\frac{dy}{dx}\right)_1 (x_2 - x_1) = y_1 + f(x_1,y_1^{(1)})(x_2 - x_1) \quad (5\text{-}132)$$

This process, repeated a sufficient number of times, will yield y as a function of x over the desired interval. A great disadvantage of the method lies in the fact that, if dy/dx is changing rapidly over an interval, its value at the beginning of the interval may be a poor approximation for its average value over the interval, and $y_1^{(1)}$ may be much in error compared with the true value. Errors of this nature accumulate with succeeding intervals until the value of y becomes so much in error as to be completely useless. Figure 5-2 shows a plot of the exponential function $y = 0.1e^x + 1$, which is a particular solution of the equation

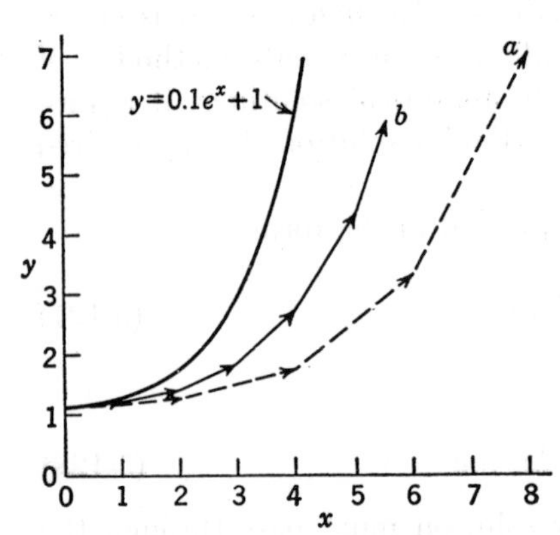

FIG. 5-2. Solution of a differential equation by method of finite increments.

$$\frac{dy}{dx} = y - 1 \quad (5\text{-}133)$$

passing through the point (0,1.1). The dotted line connecting a series of arrows shows the result of Euler's method when the increment in x is taken as 2 and the value of dy/dx taken as its value at the beginning of each interval. If the increment is halved, the approximation as indicated by curve b is better but still inadmissible.

An improved result may be effected by either or a combination of two methods:

1. Decreasing the size of the increment Δx
2. Using a better value for the average value of $dy/dx = f(x,y)$ over each interval

Each of the following methods to be discussed has as its basis for an improvement in accuracy the use of a better average value for the slope dy/dx over each interval. Regardless of the method used, it is usually possible to secure an improvement in the accuracy of this stepwise process by decreasing the size of the interval $x_n - x_{n-1}$, although this improvement in accuracy is secured at the expense of considerable additional labor in calculation.

TABLE 5-7

Interval	1	2	3	4
y_a	1.10	1.29	1.85	3.49
x_a	0.00	1.00	2.00	3.00
Δx	1.00	1.00	1.00	1.00
x_b	1.00	2.00	3.00	4.00
$\left(\frac{dy}{dx}\right)_a = y_a - 1$	0.10	0.29	0.85	2.49
$\left(\frac{dy}{dx}\right)_a \Delta x$	0.10	0.29	0.85	2.49
y_b	1.20	1.58	2.70	6.08
$\left(\frac{dy}{dx}\right)_{av}^{(1)} = \frac{y_a + y_b^{(1)} - 2}{2}$	0.15	0.43	1.26	3.83
$\left(\frac{dy}{dx}\right)_{av}^{(1)} \Delta x$	0.15	0.43	1.26	3.83
$y_b^{(2)}$	1.25	1.72	3.11	7.32
$\left(\frac{dy}{dx}\right)_{av}^{(2)} = \frac{y_a + y_b^{(2)} - 2}{2}$	0.17	0.51	1.48	4.40
$\left(\frac{dy}{dx}\right)_{av}^{(2)} \Delta x$	0.17	0.51	1.48	4.40
$y_b^{(3)}$	1.27	1.80	3.33	7.89
$\left(\frac{dy}{dx}\right)_{av}^{(3)} = \frac{y_a + y_b^{(3)} - 2}{2}$	0.18	0.55	1.59	4.69
$\left(\frac{dy}{dx}\right)_{av}^{(3)} \Delta x$	0.18	0.55	1.59	4.69
$y_b^{(4)}$	1.28	1.84	3.44	8.18
$\left(\frac{dy}{dx}\right)_{av}^{(4)} = \frac{y_a + y_b^{(4)} - 2}{2}$	0.19	0.56	1.64	
$\left(\frac{dy}{dx}\right)_{av}^{(4)} \Delta x$	0.19	0.56	1.64	
$y_b^{(5)}$	1.29	1.85	3.49	

The most obvious method of obtaining a better average value for dy/dx is to repeat the calculation represented by Eq. (5-131) several times, using for dy/dx in each successive calculation its arithmetic average value obtained at the beginning and end of the interval. Such a calculation would proceed as follows for the interval $x_1 - x_0$:

$$y_1^{(1)} = y_0 + \left(\frac{dy}{dx}\right)_0 (x_1 - x_0) = y_0 + f(x_0,y_0)(x_1 - x_0) \quad (5\text{-}134)$$

$$y_1^{(2)} = y_0 + \left(\frac{dy}{dx}\right)_{av}^{(1)} (x_1 - x_0) \qquad \left(\frac{dy}{dx}\right)_{av}^{(1)} = \frac{f(x_0,y_0) + f(x_1,y_1^{(1)})}{2}$$

and the kth approximation becomes

$$y_1^{(k)} = y_0 + \left(\frac{dy}{dx}\right)_{\text{av}}^{(k-1)} (x_1 - x_0) \qquad \left(\frac{dy}{dx}\right)_{\text{av}}^{k-1} = \frac{f(x_0,y_0) + f(x_1,y_1^{(k-1)})}{2}$$

In practice, if the size of the interval is reasonably small, there is ordinarily but little difference between the successive values of y after the first few approximations.

As an illustration of how this process may be carried out systematically, an integral curve of Eq. (5-133) through the point (0,1.1) appears in Table 5-7, where subscripts a and b refer to the beginning and the end of an interval, respectively.

NOTATION FOR NUMERICAL SOLUTIONS

Point 0		Point 1		Point 2		Point 3		Point 4		Point 5
x_0		x_1		x_2		x_3		x_4		x_5
	Interval 1		Interval 2		Interval 3		Interval 4		Interval 5	
x_a		x_b		x_a		x_b		x_a		x_b
									etc.	
		x_a		x_b		x_a		x_b		

The accompanying tabulation is helpful in understanding the notation referring to points and intervals in the tables explaining solutions by the various numerical methods. Point 0 is the point at which calculation starts, and point 1 is at the end of interval 1. The column number in the tables is the interval number. Thus, x_b in column 3 refers to point 3; both x_b in column 3 and x_a in column 4 refer to point 3. The value of y_a at the top of the first column is the value of y at the fixed starting point, subsequent values of y_a are taken equal to $y_b^{(5)}$ from the preceding interval.

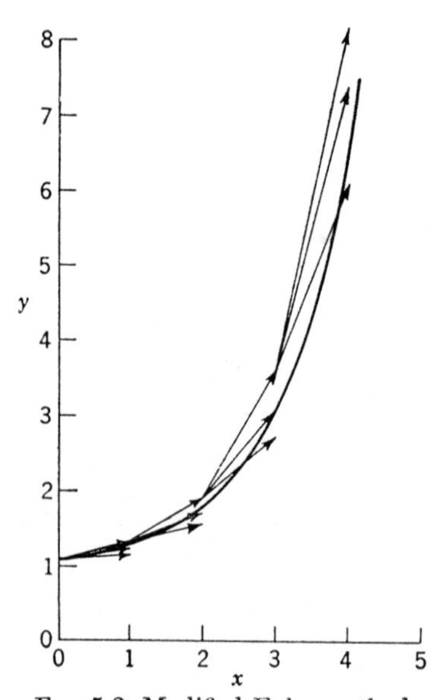

FIG. 5-3. Modified Euler method.

The arrowheads illustrated in Fig. 5-3 indicate the results of several successive approximations, the top arrow being $y_b^{(5)}$ for every interval but the fourth. For an interval of 1, this represents a better approximation than was obtained in curve b of Fig. 5-2. There is a bad divergence on the other side of the theoretical curve in this case, although the agreement could be considerably improved if the interval Δx were to be made smaller. Instead of making the interval smaller, however, an improved method of evaluating dy/dx over each interval will be used.

5-15. Method of W. E. Milne.[9] Consider again the general equation

$$\frac{dy}{dx} = f(x,y) \equiv f \tag{5-135}$$

and let it be specified that the integral curve shall pass through x_0, y_0. The first four points on the integral curve must be calculated by the

TABLE 5-8

	1	2	3	4	5	6	7	8
x_a	1.80	2.00	2.20	2.40	2.60	2.80	3.00	3.20
y_a	0.00	0.38	0.81	1.29	1.84	2.47	3.21	4.10
Δx	0.20	0.20	0.20	0.20	0.20	0.20	0.20	0.20
$\left(\frac{dy}{dx}\right)_a$	1.80	2.01	2.27	2.57	2.94	3.41	4.04	4.89
$\left(\frac{dy}{dx}\right)_a \Delta x$	0.36	0.40	0.45	0.51				
$y_b^{(1)}$	0.36	0.78	1.26	1.80	2.47	3.22	4.10	5.19
$\left(\frac{dy}{dx}\right)_b^{(1)}$	2.01	2.26	2.56	2.92				
$\left(\frac{dy}{dx}\right)_{av}^{(1)}$	1.91	2.13	2.41	2.75				
$\left(\frac{dy}{dx}\right)_{av}^{(1)} \Delta x$	0.38	0.43	0.48	0.55				
$y_b^{(2)}$	0.38	0.81	1.29	1.84	2.47	3.21	4.10	5.19
$\left(\frac{dy}{dx}\right)_b^{(2)}$	2.01	2.27	2.57	2.94	3.41	4.04	4.89	6.09

modified Euler method or by an alternative technique. The subsequent values of y are then obtained as follows:

1. $$y_n^{(1)} = y_{n-4} + \frac{4\Delta x}{3}(2f_{n-1} - f_{n-2} + 2f_{n-3}) \tag{5-136}$$

2. This value of $y_n^{(1)}$ is substituted into (5-135), together with x_n, to obtain the corresponding value of f_n.

3. The resulting value of f_n is now substituted into the formula

$$y_n^{(2)} = y_{n-2} + \frac{\Delta x}{3}(f_n + 4f_{n-1} + f_{n-2}) \tag{5-137}$$

If these two successive values of y_n agree to the desired number of significant figures, $y_n^{(2)}$ is taken to be the correct value of y_n and is substituted in (5-135) along with x_n to obtain the correct value of f_n. The process is now repeated for the next interval.

TABLE 5-9

$$y_4^{(1)} = 0 + \frac{0.8}{3}(2 \cdot 2.57 - 2.27 + 2 \cdot 2.02) = 1.84$$

$$\frac{dy}{dx} = 2.60 + 0.1(1.84)^2 = 2.94$$

$$y_4^{(2)} = 0.81 + \frac{0.20}{3}(2.94 + 4 \cdot 2.57 + 2.27) = 1.84$$

$$\frac{dy}{dx} = 2.60 + 0.1(1.84)^2 = 2.94$$

$$E = \frac{1.84 - 1.84}{29} = \frac{0}{29} = 0$$

Since this error does not affect the second decimal place, proceed to the next point.

$$y_5^{(1)} = 0.38 + \frac{0.8}{3}(2 \cdot 2.94 - 2.57 + 2 \cdot 2.27) = 2.47$$

$$\frac{dy}{dx} = 2.80 + 0.1(2.47)^2 = 3.40$$

$$y_5^{(2)} = 1.29 + \frac{0.20}{3}(3.40 + 4 \cdot 2.94 + 2.57) = 2.47$$

$$\frac{dy}{dx} = 2.80 + 0.1(2.47)^2 = 3.41$$

$$y_6^{(1)} = 0.81 + \frac{0.8}{3}(2 \cdot 3.41 - 2.94 + 2 \cdot 2.57) = 3.22$$

$$\frac{dy}{dx} = 3.00 + 0.1(3.22)^2 = 4.04$$

$$y_6^{(2)} = 1.84 + \frac{0.2}{3}(4.04 + 4 \cdot 3.41 + 2.94) = 3.22$$

$$\frac{dy}{dx} = 3.00 + 0.1(3.22)^2 = 4.04$$

$$y_7^{(1)} = 1.29 + \frac{0.8}{3}(2 \cdot 4.04 - 3.41 + 2 \cdot 2.94) = 4.11$$

$$\frac{dy}{dx} = 3.20 + 0.1(4.11)^2 = 4.89$$

$$y_7^{(2)} = 2.47 + \frac{0.2}{3}(4.89 + 4 \cdot 4.04 + 3.41) = 4.10$$

$$\frac{dy}{dx} = 3.20 + 0.1(4.10)^2 = 4.88$$

$$y_8^{(1)} = 1.84 + \frac{0.8}{3}(2 \cdot 4.88 - 4.04 + 2 \cdot 3.41) = 5.19$$

$$\frac{dy}{dx} = 3.41 + 0.1(5.19)^2 = 6.09$$

$$y_8^{(2)} = 3.22 + \frac{0.2}{3}(6.09 + 4 \cdot 4.88 + 4.04) = 5.19$$

On the other hand, if $y_n^{(1)}$ and $y_n^{(2)}$ do not agree, the error E due to approximate integration by formula (5-137) may be *estimated* by the relation

$$E \cong \frac{y_n^{(2)} - y_n^{(1)}}{29} \tag{5-138}$$

If this error is not large enough to affect the last significant figure in y_n, $y_n^{(2)}$ may be accepted and the process repeated for another interval. Otherwise, the only recourse is to repeat the entire calculation, using a smaller value of the interval Δx.

Although it is not necessary to be familiar with the detailed derivation of formulas (5-136) and (5-137) in order to use them successfully, it may be stated that both arise from approximating $f(x,y)$ by Newton's interpolation formulas, neglecting differences of the fourth order (see Chap. 1). Formula (5-137) turns out to be nothing but Simpson's rule applied over the interval $(n - 2)$, n.

Example 5-4. To illustrate a systematic application of the method, the following nonlinear equation of the first order will be solved:

$$\frac{dy}{dx} = x + 0.1y^2 \tag{5-139}$$

Let it be specified that the integral curve is to pass through the point $x = 1.8$, $y = 0$. The first four points on the curve are calculated by the modified Euler method, the values being tabulated in Table 5-8. The first point, that is, $x = 1.8$, $y = 0$, is designated point 0. The column headed 1 considers the interval between point 0 and point 1. As before, subscripts a and b refer to values of quantities at the beginning and end of an interval. The value of y at the end of the fourth interval, that is, y_4, is calculated by the Euler and the Milne methods, as shown in Table 5-9. Both methods agree, indicating that the first five points are correct to the accuracy of the present calculation. Note that for the calculation of $y_4^{(1)}$ by the Milne method, y_{n-4} refers to y at point 0. Successive values of y are calculated by the Milne method. Over the range of calculation ($x = 1.8$ to 3.4), an increment of 0.2 in x suffices to give accuracy to two decimal places.

5-16. Method of Runge-Kutta. A rather different method of procedure has been described by Runge[11] and elaborated by Kutta.[6] In this method, formulas are derived that enable the direct calculation of the increment in y corresponding to an increment in x. In application to the problem of calculating an integral curve of the equation $dy/dx = f(x,y)$ through the point x_0, y_0, these formulas are

$$k_1 = f(x_0,y_0)\,\Delta x \tag{5-140}$$

$$k_2 = f\left(x_0 + \frac{\Delta x}{2}, y_0 + \frac{k_1}{2}\right)\Delta x \tag{5-141}$$

$$k_3 = f\left(x_0 + \frac{\Delta x}{2}, y_0 + \frac{k_2}{2}\right)\Delta x \tag{5-142}$$

$$k_4 = f(x_0 + \Delta x,\ y_0 + k_3)\,\Delta x \tag{5-143}$$

$$\Delta y = \tfrac{1}{6}(k_1 + 2k_2 + 2k_3 + k_4) \tag{5-144}$$

$$x_1 = x_0 + \Delta x \tag{5-145}$$

$$y_1 = y_0 + \Delta y \tag{5-146}$$

To compute a second point on the integral curve, the same set of formulas may be applied in the order given, x_0 and y_0 being replaced by

TABLE 5-10

	$x_0 = 1.8,\ y_0 = 0$	$x_1 = 2.10,\ y_1 = 0.59$
Δx	0.30	0.30
k_1	$(1.80 + 0)0.30 = 0.54$	$[2.10 + 0.1(0.59)^2]0.30 = 0.64$
k_2	$\left[1.80 + \frac{0.30}{2} + 0.1\left(0 + \frac{0.54}{2}\right)^2\right] 0.30 = 0.58$	$\left[2.10 + \frac{0.30}{2} + 0.1\left(0.59 + \frac{0.64}{2}\right)^2\right] 0.30 = 0.70$
k_3	$\left[1.80 + \frac{0.30}{2} + 0.1\left(0 + \frac{0.58}{2}\right)^2\right] 0.30 = 0.59$	$\left[2.10 + \frac{0.30}{2} + 0.1\left(0.59 + \frac{0.70}{2}\right)^2\right] 0.30 = 0.70$
k_4	$[1.80 + 0.30 + 0.1(0 + 0.59)^2]0.30 = 0.64$	$[2.10 + 0.30 + 0.1(0.59 + 0.7)^2]0.30 = 0.77$
Δy	$\frac{1}{6}(0.54 + 2 \cdot 0.58 + 2 \cdot 0.59 + 0.64) = 0.59$	$\frac{1}{6}(0.64 + 2 \cdot 0.70 + 2 \cdot 0.70 + 0.77) = 0.70$
	$x_2 = 2.40,\ y_2 = 1.29$	$x_3 = 2.70,\ y_3 = 2.15$
Δx	0.30	0.30
k_1	$[2.40 + 0.1(1.29)^2]0.30 = 0.77$	$[2.70 + 0.1(2.15)^2]0.3 = 0.95$
k_2	$\left[2.40 + \frac{0.30}{2} + 0.1\left(1.29 + \frac{0.77}{2}\right)^2\right] 0.30 = 0.85$	$\left[2.70 + \frac{0.30}{2} + 0.1\left(2.15 + \frac{0.95}{2}\right)^2\right] 0.30 = 1.06$
k_3	$\left[2.40 + \frac{0.30}{2} + 0.1\left(1.29 + \frac{0.85}{2}\right)^2\right] 0.30 = 0.86$	$\left[2.70 + \frac{0.30}{2} + 0.1\left(2.15 + \frac{1.06}{2}\right)^2\right] 0.30 = 1.06$
k_4	$[2.40 + 0.30 + 0.1(1.29 + 0.86)^2]0.30 = 0.95$	$[2.70 + 0.30 + 0.1(2.15 + 1.06)^2]0.30 = 1.21$
Δy	$\frac{1}{6}(0.77 + 2 \cdot 0.85 + 2 \cdot 0.86 + 0.95) = 0.86$	$\frac{1}{6}(0.95 + 2 \cdot 1.06 + 2 \cdot 1.06 + 1.21) = 1.07$
	$x_4 = 3.00,\ y_4 = 3.22$	$x_5 = 3.30,\ y_4 = 4.62$
Δx	0.30	0.30
k_1	$[3.00 + 0.1(3.22)^2]0.30 = 1.21$	$[3.30 + 0.1(4.62)^2]0.30 = 1.63$
k_2	$\left[3.00 + \frac{0.30}{2} + 0.1\left(3.22 + \frac{1.21}{2}\right)^2\right] 0.30 = 1.39$	$\left[3.30 + \frac{0.30}{2} + 0.1\left(4.62 + \frac{1.63}{2}\right)^2\right] 0.30 = 1.92$
k_3	$\left[3.00 + \frac{0.30}{2} + 0.1\left(3.22 + \frac{1.39}{2}\right)^2\right] 0.30 = 1.41$	$\left[3.30 + \frac{0.30}{2} + 0.1\left(4.62 + \frac{1.92}{2}\right)^2\right] 0.30 = 1.97$
k_4	$[3.00 + 0.30 + 0.1(3.22 + 1.41)^2]0.30 = 1.63$	$[3.30 + 0.30 + 0.1(4.62 + 1.97)^2]0.30 = 2.38$
Δy	$\frac{1}{6}(1.21 + 2 \cdot 1.39 + 2 \cdot 1.41 + 1.63) = 1.40$	$\frac{1}{6}(1.63 + 2 \cdot 1.92 + 2 \cdot 1.97 + 2.38) = 1.96$

x_1 and y_1. The derivation of this set of equations is involved and will not be given here. The error in this method is not easy to estimate, but it is known to be of the same order as that in Simpson's rule; it is of interest to note that when $dy/dx = f(x)$ the method reduces to Simpson's rule.

Example 5-5. For illustration, consider again the equation

$$\frac{dy}{dx} = x + 0.1y^2$$

The particular integral curve is to pass through the point $x = 1.8$, $y = 0$. The x increment will be taken as 0.3. A systematic tabulation of the calculation appears in Table 5-10. The value of $y = 3.22$ at $x = 3.00$ is in good agreement with the value of $y = 3.21$ at $x = 3.00$ obtained earlier for the same problem by means of the Milne method.

5-17. Numerical Solution of Equations of Higher Order.

Any equation of higher order may be reduced to a set of first-order equations by the introduction of auxiliary variables. For example, the second-order equation

$$\frac{d^2y}{dx^2} + y^2\frac{dy}{dx} + y = 0 \tag{5-147}$$

may be reduced to a set of first-order equations by placing $dy/dx = z$, whereupon $d^2y/dx^2 = dz/dx$, and (5-147) becomes equivalent to two equations:

$$\frac{dz}{dx} + y^2z + y = 0 \tag{5-148}$$

$$\frac{dy}{dx} = z \tag{5-149}$$

These two first-order simultaneous equations may be solved by any of the previous methods.

Example 5-6. In order to illustrate the method of attack, a simple second-order equation will be solved. It is desired to calculate the integral curve of

$$\frac{d^2y}{dx^2} - y = 0 \tag{5-150}$$

passing through the point $x = 0$, $y = 1.00$ and having $dy/dx = 0$ at this point. Notice that to calculate an integral curve of this second-order equation it is necessary to specify two conditions corresponding to an assignment of values to the two arbitrary constants that must occur in its general solution. The necessity for these two conditions is emphasized as soon as (5-150) is replaced by its equivalent set of first-order equations:

$$\frac{dz}{dx} = y \tag{5-151}$$

$$\frac{dy}{dx} = z \tag{5-152}$$

These equations will be used to calculate curves of y vs. x and z vs. x, but, before it is possible to start either of the two curves, initial values of y and $z = dy/dx$ must be selected at a given value of x. Starting at $x = 0$, $y = 1.000$, and $z = 0$ (Fig. 5-4), it is assumed that dz/dx is constant over the short interval from $x = 0$ to $x = 0.050$. At $x = 0.050$,

$$z_1^{(1)} = z_0 + \frac{dz}{dx}\,\Delta x = 0.00 + 1 \cdot 0.050 = 0.050$$

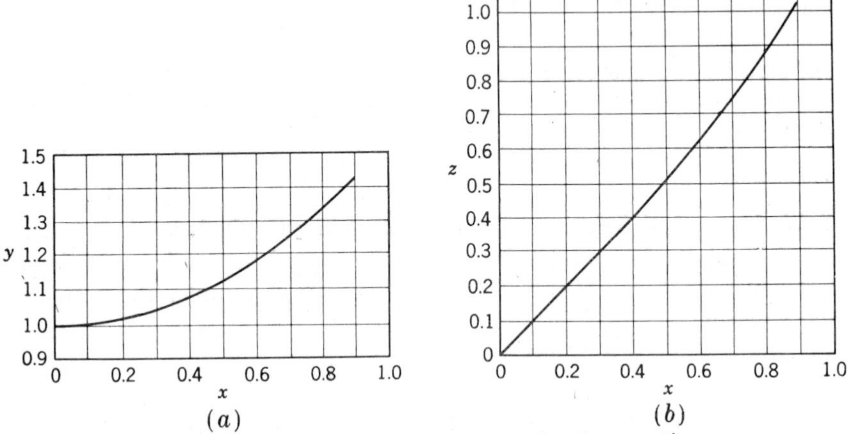

FIG. 5-4. Numerical solution of a second-order equation.

Using the arithmetic average value of z over the interval, compute

$$z_{\text{av}} = \frac{z_0 + z_1^{(1)}}{2} = \left(\frac{dy}{dx}\right)_{\text{av}}$$

$$y_b^{(1)} = y_a + \left(\frac{dy}{dx}\right)_{\text{av}} \Delta x = 1.000 + 0.001 = 1.001$$

In a similar manner, second approximations to y_1 and z_1 are computed by employing new arithmetic average values of dy/dx and dz/dx over the interval based on $z_1^{(2)}$ and $y_1^{(2)}$.

To continue the two curves, pass to the points x_1, y_1, and z_1; take a new increment in x equal to 0.050; and repeat the foregoing process. If a calculation of this nature is to be successful, it should be organized into a table. This has been done in Table 5-11, where subscripts a and b indicate values at the beginning and end of an interval. The exact solution of (5-150) is

$$y = C_1e^x + C_2e^{-x}$$

When $x = 0$, $y = 1$, and $C_1 + C_2 = 1$. Differentiating (5-146) gives, at $x = 0$,

$$\frac{dy}{dx} = C_1 - C_2 = 0$$

Therefore, $C_1 = C_2 = \frac{1}{2}$ and

$$y = \frac{e^x + e^{-x}}{2} = \cosh x$$

$$\frac{dy}{dx} = z = \sinh x$$

TABLE 5-11

Interval	1	2	3	4	5	6	7	8	9	10	11	12	13	14	15	16	17	18
x_a	0	0.050	0.100	0.150	0.200	0.250	0.300	0.350	0.400	0.450	0.500	0.550	0.600	0.650	0.700	0.750	0.800	0.850
Δx	0.050	0.050	0.050	0.050	0.050	0.050	0.050	0.050	0.050	0.050	0.050	0.050	0.050	0.050	0.050	0.050	0.050	0.050
x_b	0.050	0.100	0.150	0.200	0.250	0.300	0.350	0.400	0.450	0.500	0.550	0.600	0.650	0.700	0.750	0.800	0.850	0.900
$y_a = \left(\frac{dz}{dx}\right)_a$	1.000	1.001	1.004	1.010	1.018	1.030	1.044	1.060	1.079	1.101	1.125	1.152	1.182	1.215	1.251	1.290	1.332	1.378
$\frac{dz}{dx}\Delta x = \Delta z$	0.050	0.050	0.050	0.050	0.051	0.051	0.052	0.053	0.054	0.055	0.056	0.058	0.059	0.061	0.063	0.065	0.067	0.069
z_a	0.000	0.050	0.100	0.150	0.200	0.251	0.302	0.354	0.407	0.461	0.516	0.572	0.630	0.689	0.750	0.813	0.878	0.945
$z_b^{(1)}$	0.050	0.100	0.150	0.200	0.251	0.302	0.354	0.407	0.461	0.516	0.572	0.630	0.689	0.750	0.813	0.878	0.945	1.014
$z_{av}^{(1)} = \frac{z_a + z_b^{(1)}}{2} = \left(\frac{dy}{dx}\right)_{av}^{(1)}$	0.025	0.075	0.120	0.170	0.225	0.276	0.328	0.380	0.434	0.488	0.544	0.600	0.659	0.717	0.781	0.845	0.911	0.979
$\left(\frac{dy}{dx}\right)_{av}^{(1)} \Delta x = \Delta y$	0.001	0.003	0.006	0.009	0.012	0.014	0.016	0.019	0.022	0.024	0.027	0.030	0.033	0.036	0.039	0.042	0.045	0.049
$y_b^{(1)}$	1.001	1.004	1.010	1.018	1.030	1.044	1.060	1.079	1.101	1.125	1.152	1.182	1.215	1.251	1.290	1.332	1.377	1.427
$y_{av} = \frac{y_a + y_b^{(1)}}{2} = \left(\frac{dz}{dx}\right)_{av}^{(1)}$	1.000	1.003	1.007	1.014	1.024	1.037	1.052	1.069	1.090	1.113	1.138	1.167	1.198	1.233	1.270	1.311	1.354	1.402
$\left(\frac{dz}{dx}\right)_{av}^{(1)} \Delta x = \Delta z$	0.050	0.050	0.050	0.051	0.052	0.053	0.054	0.055	0.056	0.057	0.058	0.060	0.060	0.062	0.064	0.066	0.068	0.070
$z_b^{(2)}$	0.050	0.050	0.150	0.201	0.251	0.303	0.355	0.407	0.461	0.517	0.573	0.630	0.690	0.751	0.814	0.878	0.946	1.015
$z_{av}^{(2)} = \frac{z_a + z_b^{(2)}}{2} = \left(\frac{dy}{dz}\right)_{av}^{(2)}$	0.025	0.075	0.100	0.170	0.225	0.277	0.328	0.380	0.434	0.489	0.544	0.601	0.660	0.720	0.782	0.845	0.912	0.980
$\left(\frac{dy}{dx}\right)_{av}^{(2)} \Delta x = \Delta y$	0.001	0.003	0.005	0.008	0.012	0.014	0.016	0.019	0.022	0.024	0.027	0.030	0.033	0.036	0.039	0.042	0.046	0.049
$y_b^{(2)}$	1.001	1.004	1.010	1.018	1.030	1.044	1.060	1.079	1.101	1.125	1.152	1.182	1.215	1.251	1.290	1.332	1.378	1.427

The accuracy of the numerical method as employed in the present case may be judged from the fact that curves of $y = \cosh x$ and $z = \sinh x$ may be superimposed on the curves of Fig. 5-4.

It would be entirely possible to carry out this calculation graphically, increments of the two curves being drawn in with the proper slopes, by a process of approximation. There is usually no advantage to the graphical method, and the numerical procedure is more accurate.

5-18. Application of Runge-Kutta Method to Higher-order Equations. To integrate numerically an equation of higher order by this method, first obtain the equivalent system of first-order equations and then compute increments by means of a set of formulas similar to those used for a single equation (Sec. 5-16). Consider the simultaneous equations

$$\frac{dy}{dx} = f_1(x,y,z) \tag{5-153a}$$

$$\frac{dz}{dx} = f_2(x,y,z) \tag{5-153b}$$

Starting at x_0, y_0, z_0, the increments in y and z for the first increment in x are computed by means of the formulas

$$k_1 = f_1(x_0,y_0,z_0)\,\Delta x \tag{5-154}$$

$$k_2 = f_1\left(x_0 + \frac{\Delta x}{2}, y_0 + \frac{k_1}{2}, z_0 + \frac{l_1}{2}\right)\Delta x \tag{5-155}$$

$$k_3 = f_1\left(x_0 + \frac{\Delta x}{2}, y_0 + \frac{k_2}{2}, z_0 + \frac{l_2}{2}\right)\Delta x \tag{5-156}$$

$$k_4 = f_1(x_0 + \Delta x,\ y_0 + k_3,\ z_0 + l_3)\,\Delta x \tag{5-157}$$

$$\Delta y = \tfrac{1}{6}(k_1 + 2k_2 + 2k_3 + k_4) \tag{5-158}$$

$$l_1 = f_2(x_0,y_0,z_0)\,\Delta x \tag{5-159}$$

$$l_2 = f_2\left(x_0 + \frac{\Delta x}{2}, y_0 + \frac{k_1}{2}, z_0 + \frac{l_1}{2}\right)\Delta x \tag{5-160}$$

$$l_3 = f_2\left(x_0 + \frac{\Delta x}{2}, y_0 + \frac{k_2}{2}, z_0 + \frac{l_2}{2}\right)\Delta x \tag{5-161}$$

$$l_4 = f_2(x_0 + \Delta x,\ y_0 + k_3,\ z_0 + l_3)\,\Delta x \tag{5-162}$$

$$\Delta z = \tfrac{1}{6}(l_1 + 2l_2 + 2l_3 + l_4) \tag{5-163}$$

To compute the next increment, it is necessary only to replace x_0, y_0, and z_0 in the above formulas by x_1, y_1, and z_1. Extension of the method to four or more variables is effected by a set of equations analogous to those above.

5-19. Use of Taylor's Series. The Taylor-series form of the power-series representation of $y = f(x)$ was given in Sec. 5-4 as

$$y = y_0 + (x - x_0)f'(x_0) + \frac{(x - x_0)^2}{2!}f''(x_0) + \cdots$$

$$= \sum_{n=0}^{\infty} \frac{(x - x_0)^n}{n!} f^n(x_0) \tag{5-10}$$

where the notation $f^n(x_0)$ denotes the value of the nth derivative of (x) at the point x_0. If $f(x)$ is a continuous function with continuous derivatives of all orders, (5-10) converges to $f(x)$, provided that $x - x_0$ is less than the radius of convergence. Furthermore, the *difference* E between the true value of $y = f(x)$ and the value given by the first $n + 1$ terms of (5-10) is

$$E = \frac{(x - x_0)^{n+1} f^{n+1}(\xi)}{(n + 1)!} \tag{5-164}$$

where $x_0 \leqq \xi \leqq x$. Thus, although ξ is not known, the maximum possible value of the difference E can be computed. Equation (5-10) then provides a means for the numerical integration of a differential equation, and (5-164) provides a means for estimating the error caused by the use of a finite number of terms.

Example 5-7. Again consider the equation

$$\frac{dy}{dx} = x + 0.1y^2 \tag{5-139}$$

which includes the point $x = 1.8$, $y = 0$. Let $\Delta x = 0.2$ and $x = 1.8 = x_0$ and the corresponding value of y be y_0. The integration of (5-139) will be accomplished by use of the terms up to and including $n = 2$ in the Taylor series (5-10), and the error will be estimated.

$$f'(x) = \frac{dy}{dx} = x + 0.1y^2 \qquad n = 1$$

$$f''(x) = \frac{d^2y}{dx^2} = 1 + 0.2y\frac{dy}{dx} \qquad n = 2$$

$$f'''(x) = \frac{d^3y}{dx^3} = 0.2\left(\frac{dy}{dx}\right)^2 + 0.2y\frac{d^2y}{dx^2} \qquad n = 3$$

Then,

$$y_1^{(1)} \cong y_0 + (x - x_0)f'(x_0) + \frac{(x - x_0)^2}{2!}f''(x_0)$$

$$y_0 = 0$$

$$f'(x_0) = x_0 + 0.1y_0^2 = 1.8$$

$$f''(x_0) = 1 + 0.2y_0\left(\frac{dy}{dx}\right)_0 = 1$$

$$x - x_0 = 0.2$$

$$y_1^{(1)} \cong 0.2 \cdot 1.8 + \frac{0.04}{2}1 = 0.380$$

The exact difference between the true value of y_1 and the approximation $y_1^{(1)}$ could be evaluated by means of (5-164) if ξ were known:

$$E = \frac{(x - x_0)^3 f'''(\xi)}{3!}$$

Since ξ lies between x and x_0, use $\xi = x_0$ as an approximation. Then

$$f'''(\xi) \cong f'''(x_0) = 0.2(1.8)^2 = 0.647$$

$$E \cong \frac{(0.2)^3 0.647}{3!} = 0.00086$$

Consequently, the calculated value of y_1, $y_1^{(1)}$ of 0.380 does not differ from the true value by more than approximately 0.0009. If this precision is satisfactory, then $y_1 = 0.380$ is accepted, and the calculation is carried out for the next interval, taking $y_1^{(1)}$ to be the new y_0, etc. If the precision is not satisfactory, the process is repeated, using either more terms in the series or smaller increments. *Note that E is an error estimate only and should not be used to adjust $y_1^{(1)}$.*

The Taylor's-series method may be used to integrate higher-order equations by use of the reduction suggested in Sec. 5-17. It is also possible to develop a power-series approximation to the solution of the differential equation by means of Taylor's series.

BIBLIOGRAPHY

1. Bateman, H., and R. C. Archibald: A Guide to Tables of Bessel Functions, *Math. Tables Aids Comput.*, vol. 1, p. 7, 1944.
2. Bennett, A. A., W. E. Milne, H. Bateman, and L. E. Ford: Numerical Integration of Differential Equations, *Bull. Natl. Research Council*, 1933.
3. Fletcher, A., J. C. P. Miller, and L. Rosenhead: "An Index of Mathematical Tables," McGraw-Hill Book Company, Inc., New York, 1946.
4. Gray, A., G. B. Mathews, and J. M. MacRobert: "A Treatise on Bessel Functions and Their Applications to Physics," Macmillan & Co., Ltd., London, 1931.
5. Jahnke, E., and F. Emde: "Tables of Functions," Teubner Verlagsgesellschaft, Leipzig, 1933; also reprint by Dover Publications, New York, 1943.
6. Kutta, W.: *Z. Math. Phys.*, vol. 46, pp. 435–453, 1901.
7. Magnus, W., and F. Oberhettinger: "Formeln und Satze fur die speziellen Functionen der mathematischen Physik," Springer-Verlag OHG, Berlin, 1948.
8. McLachlan, N. W.: "Bessel Functions for Engineers," Oxford University Press, New York, 1934.
9. Milne, W. E.: Numerical Integration of Ordinary Differential Equations, *Am. Math. Monthly*, vol. 33, p. 455, 1926.
10. Peirce, B. O., and R. M. Foster: "A Short Table of Integrals," Ginn & Company, Boston, 1956.
11. Runge, C.: *Math. Ann.*, vol. 46, pp. 167–178, 1895.
12. Scarborough, J. B.: "Numerical Mathematical Analysis," Johns Hopkins Press, Baltimore, 1950.
13. Watson, G. N.: "Theory of Bessel Functions," Cambridge University Press, London, 1922.
14. Hildebrand, F. B.: "Advanced Calculus for Engineers," Prentice-Hall, Inc., Englewood Cliffs, N.J., 1949.
15. Kármán, T. v., and M. A. Biot: "Mathematical Methods in Engineering," McGraw-Hill Book Company, Inc., New York, 1940.
16. Marshall, W. R., Jr. and R. L. Pigford: "The Application of Differential Equations to Chemical Engineering Problems," University of Delaware, Newark, Del., 1947.
17. Wylie, C. R., Jr.: "Advanced Engineering Mathematics," McGraw-Hill Book Company, Inc., New York, 1951.

PROBLEMS

5-1. (*a*) Expand $\sqrt{1+x}$ about $x = 0$ by means of Taylor's series. If, as an approximation, it is assumed $\sqrt{1+x} = 1 + x/2$, what is the error when $x = 0.05$?

(*b*) Expand $\ln (c + x)$ about $x/c = 0$ by means of Taylor's series. If, as an approximation, it is assumed that $\ln (c + x) = x/c + \ln c$, what is the error when $x/c = 0.05$?

5-2. (*a*) Obtain a solution in power series for $dy/dx = x + y^2 + 1$ when $y = 0$ at $x = 0$. Evaluate the coefficients of the terms up to x^6.

(*b*) Compare the results of (*a*) with that obtained by solution in Taylor's series. Calculate the value of y at $x = -1$, and note that the series converges rapidly only when x is a fraction.

5-3. Locate and identify the singular points of the differential equation

$$x^2(1 - x^2)^2 \frac{d^2y}{dx^2} + 2x(1 - x) \frac{dy}{dx} + y = 0$$

5-4. Use the method of Frobenius to obtain the general solution of each of the following differential equations, valid near $x = 0$:

(*a*)
$$2x \frac{d^2y}{dx^2} + (1 - 2x) \frac{dy}{dx} - y = 0$$

(*b*)
$$x^2 \frac{d^2y}{dx^2} + x \frac{dy}{dx} + (x^2 - \tfrac{1}{4})y = 0$$

(*c*)
$$x \frac{d^2y}{dx^2} + 2 \frac{dy}{dx} + xy = 0$$

(*d*)
$$x(1 - x) \frac{d^2y}{dx^2} - 2 \frac{dy}{dx} + 2y = 0$$

5-5. Compute and compare the efficiencies of flat circular fins of uniform thickness t, where (1) the fin is stamped from a flat sheet of metal and (2) the fin is made from a rectangular sheet and is crimped around the tube. Heat transfer from the edge of the fins may be neglected. Clearly state all assumptions that are made in your derivation.

DATA: $h = 20$, $k = 25$ (steel), $t = 0.096$ in. (thickness), tube diameter $= 2$ in., total diameter of finned tube $= 6$ in.

5-6. A common practice in heterogeneous catalysis is the use of the catalyst in the form of grains. It is sometimes assumed that the reacting fluid in the interior of the grains has a composition equal to that of the bulk of the fluid bathing the grain at the time. However, the size of the grains cannot be increased indefinitely without reaching a point at which the reaction will yield products in the interior of the grain faster than diffusion can carry them away. The reaction will then tend to be confined to the outer layers of the grain, the interior being relatively inactive.

A mathematical analysis of the above problem was made by Thiele† on the following assumptions:

1. The fluid may be either liquid or gas but not a mixture of the two. Attention is fixed on an individual portion of catalyst, bathed in a fluid of constant composition.
2. The temperature is assumed to be uniform throughout the grain.
3. The greater part of the surface available for reaction is assumed to be on the walls of the pores in the catalyst. The actual external surface is assumed to be negligible in comparison.
4. Diffusion through a surface film is very fast compared with diffusion into the grain interior.
5. The pores in the catalyst grain are interconnecting, and the diffusion of reacting gases and products takes place through these pores and not through the solid catalyst.

† E. W. Thiele, *Ind. Eng. Chem.*, vol. 31, pp. 916–20, 1939.

6. The largest pores need not be straight or round, but it is assumed that the ratio of the periphery to the area of all cross sections is constant for each pore and the same from pore to pore. It is assumed that the length of the pores in pellets of different sizes is proportional to the shortest dimension of the grains.
7. There is no bulk flow through the catalyst grains, all transfer being by diffusion.
8. The reverse reaction rate is negligible.

Thiele treats various cases, including various shapes of pellets, first- and second-order reactions, with or without change in volume on reaction. The essential features of the results of all these cases are shown by the analysis for spherical pellets with a first-order reaction and no change in volume. Consider a spherical pellet of radius R surrounded by a binary gas mixture of A and B. The molal density of A in this mixture is C_0. Component A diffuses into the pores of the catalyst and undergoes the reaction

$$A \rightarrow B$$

on the surface of the catalyst. The reaction rate in terms of moles of A reaction per unit time is proportional to the catalyst surface area and the molal density of A at the surface. Define the effectiveness of the catalyst pellet as the ratio of the actual reaction rate divided by the rate at which the reaction would proceed if the molal density of A were C_0 everywhere inside the pellet. Determine the effectiveness as a function of catalyst pellet radius R and hydraulic radius of the pores in the pellet.

5-7. Compute an integral curve through the point $x = 1.0$, $y = 3.0$ and over the range $x = 1.0$ to $x = 2.0$ for the equation

$$\frac{dy}{dx} = 0.1(y + x)^{0.8}$$

5-8. Compute an integral curve passing through the point $x = 1.0$, $y = 4.0$, with a slope of ½ at this point and over the range $x = 1.0$ to $x = 2.0$ for the equation

$$\frac{d^2y}{dx^2} = y^2 + xy + x^2$$

5-9. In a physical problem the differential equation

$$\frac{d^2y}{dx^2} + e^{xy}\frac{dy}{dx} + y^2x = 0$$

with the boundary conditions

$$\begin{aligned} &\text{At } x = 0 \qquad && y = 0 \\ &\text{At } x = 0.5 \qquad && y = 1 \end{aligned}$$

arises. *Estimate y when $x = 1.0$.*

5-10. McCabe and Stevens† have measured the rate of growth of copper sulfate pentahydrate crystals. In a set of experiments in which an initially uniform set of crystals were grown in an agitated bath, they reported that the rate of growth was given by the equation

$$\frac{dL}{dt} = 0.00177L^{1.1}(\Delta C)^{1.8} \tag{1}$$

where L is a linear dimension measured in microns, t is time in minutes, and ΔC is the difference between the concentration of the salt in the solution and the saturation

† *Chem. Eng. Progr.*, vol. 47, no. 4, pp. 168–174, 1951.

concentration at the operating temperature. The concentration is reported in grams of salt per 100 g of "free" water. Free water designates the water in solution less that part which came from the pentahydrate. The characteristic length L is related to the weight of a crystal w by the relation

$$w = \rho b L^3 \tag{2}$$

b is a "shape factor" and is a function of L. The relation between ρb and L is given in the following table:

Shape Factor

L, μ	ρb, g/cm^3
250	0.778
290	0.781
375	0.685
514	0.599
535	0.555
585	0.540
778	0.489
877	0.455
950	0.457

It is proposed to reverse the experiment used to obtain Eq. (1) and to add to 100 g of pure free water 12.5 g of pentahydrate crystals each of length 950 μ and 12.6 g of pentahydrate crystals each of length 535 μ. At the temperature used, the saturation concentration is 0.25 g of pentahydrate crystals per gram of free water. How long would be required to obtain the first crystals of length 250 μ? What would be the length of the other crystals?

$$1\ \mu = 10^{-4}\ \text{cm}$$

CHAPTER 6

FORMULATION OF PARTIAL DIFFERENTIAL EQUATIONS

6-1. Introduction. The analysis of situations involving two or more independent variables frequently results in a partial differential equation. The mathematical techniques employed to formulate the partial differential equation are similar to those used to set up an ordinary differential equation. The details of the two processes and, in particular, the manipulations involved are, however, sufficiently different to warrant considerable discussion of the new case.

6-2. Partial Derivatives. The general function of n variables

$$w = f(x_1, x_2, \ldots, x_n) \tag{6-1}$$

may be reduced to a function of x_1 alone by holding the remaining variables $x_2, \ldots, x_n$ constant and allowing x_1 to vary. The function of x_1 thus created may have a derivative defined and computed by the ordinary methods applicable to functions of a single variable. This derivative is called the first partial derivative of f or w with respect to x_1, and its notation is the symbol f_{x_1}, or $(df/dx_1)_{x_2 \cdots x_n}$, or $\partial f/\partial x_1$.

This partial derivative is defined as the limit

$$\frac{\partial f(x_1, x_2, \ldots, x_n)}{\partial x_1} = \lim_{\Delta x_1 \to 0} \frac{f(x_1 + \Delta x_1, x_2, \ldots, x_n) - f(x_1, x_2, \ldots, x_n)}{\Delta x_1} \tag{6-2}$$

Partial derivatives are defined with respect to each of the variables $x_1, x_2, x_3, \ldots, x_n$ in an analogous manner.

Particular attention must be devoted to the notational aspects of partial differentiation, inasmuch as the customary symbols, carelessly employed, may become so nondescript with reference to the actual operations as to result in considerable ambiguity and confusion. The chief difficulty in this respect is with the common partial derivative symbol $(\partial f/\partial x_1)_{x_2 \cdots x_n}$.

The symbols ∂f and ∂x_1 should not be confused with the ordinary differentials dy and dx, which may be treated as algebraic quantities and whose ratio dy/dx is the ordinary derivative. In $(\partial f/\partial x_1)_{x_2 \cdots x_n}$,

∂f indicates that some function f of several variables is to be partially differentiated with respect to the one of these variables which is indicated in ∂x_1, the remaining variables in the function, as indicated in the subscript, being held constant. Clearly, ∂f standing alone has no definite meaning, for f may be partially differentiated with respect to any one of its arguments $x_1, x_2, \ldots, x_n$. Until it is indicated by ∂x_1 which of the n partial derivatives is to be taken, ∂f has only an indefinite operational significance.

As a symbol for a partial derivative, $(\partial f/\partial x_1)_{x_2 \cdots x_n}$ must indicate clearly three things:

1. The function that is to be differentiated, f
2. The variable x_1 with respect to which differentiation is to be performed
3. The variables that are to be held constant during the differentiation (usually indicated as subscripts)

Unfortunately, ambiguity is often present in 1 and 3. In the case of 3, the subscripts are often omitted where there may be some doubt as to which variables are really being held constant. In the case of 1, the functional symbol f is often replaced by the dependent variable w to give $\partial w/\partial x_1$, even though this dual notation may lead to great confusion when w is a function not only of $x_1, x_2, \ldots, x_n$ but also of some other group of variables.

The total differential of f is defined by the equation

$$df = \frac{\partial f}{\partial x_1}\, dx_1 + \frac{\partial f}{\partial x_2}\, dx_2 + \cdots + \frac{\partial f}{\partial x_n}\, dx_n \tag{6-3}$$

df is the net change in f resulting from differential changes in each of the variables $x_1, x_2, \ldots, x_n$. Equation (6-3) is the correct expression for the total differential even though one or more of the variables may be a function of the others, i.e., even though, say, x_3 were related to x_1 and x_2 by

$$x_3 = \phi(x_1,x_2) \tag{6-4}$$

The partial derivatives $\partial f/\partial x_1, \partial f/\partial x_2, \ldots, \partial f/\partial x_n$ of $f(x_1,x_2, \ldots ,x_n)$ are functions of $x_1, x_2, \ldots, x_n$ and in themselves may be partially differentiated with respect to one or all of the variables $x_1, x_2, \ldots, x_n$. If there are only two independent variables x and y, the *second* partial derivatives of $f(x,y)$ are

$$\frac{\partial}{\partial x}\left(\frac{\partial f}{\partial x}\right) \equiv \frac{\partial^2 f}{\partial x^2} \equiv f_{xx}$$

$$\frac{\partial}{\partial y}\left(\frac{\partial f}{\partial x}\right) \equiv \frac{\partial^2 f}{\partial y\, \partial x} \equiv f_{xy}$$

$$\frac{\partial}{\partial x}\left(\frac{\partial f}{\partial y}\right) \equiv \frac{\partial^2 f}{\partial y\, \partial x} \equiv f_{yx}$$

$$\frac{\partial}{\partial y}\left(\frac{\partial f}{\partial y}\right) \equiv \frac{\partial^2 f}{\partial y^2} \equiv f_{yy}$$

The notation $\partial^2 f/(\partial y\, \partial x)$ implies that $\partial f/\partial x$ is found first and then the operation $(\partial/\partial y)\, \partial f/\partial x$ is performed.

It can be proved that, if $\partial^2 f/(\partial y\, \partial x)$ and $\partial^2 f/(\partial x\, \partial y)$ are continuous functions of x and y, $\partial^2 f/(\partial y\, \partial x) = \partial^2 f/(\partial x\, \partial y)$, and the order of differentiation is immaterial. With the same restrictions regarding continuity, this statement applies to derivatives of all orders.

A function u of two variables, $u = f(x,y)$, has a partial differential with respect to each variable. These are written

$$d_x f\dagger = d_x u\dagger = \left(\frac{\partial u}{\partial x}\right)_y dx \tag{6-5}$$

$$d_y f = d_y u = \left(\frac{\partial u}{\partial y}\right)_x dy \tag{6-6}$$

Similar definitions hold for the n partial differentials of a function of n variables.

Upon dividing both sides of (6-5) by dx, there is obtained

$$\frac{d_x f}{dx} = \frac{d_x u}{dx} = \frac{(\partial u/\partial x)_y\, dx}{dx} = \left(\frac{\partial u}{\partial x}\right)_y \tag{6-7}$$

It therefore becomes proper to regard the partial derivative as the ratio of a partial differential $d_x u$ to the differential dx of the independent variable with respect to which the partial differentiation is performed.

Similarly, the partial differential of u with respect to y is

$$d_y f = d_y u = \left(\frac{\partial u}{\partial y}\right)_x dy$$

and

$$\frac{d_y f}{dy} = \frac{d_y u}{dy} = \frac{(\partial u/\partial y)_x\, dy}{dy} = \left(\frac{\partial u}{\partial y}\right)_x \tag{6-8}$$

Now solve the function $u = f(x,y)$ for y, obtaining $y = \phi(u,x)$. The partial differential of y with respect to x is

$$\frac{d_x \phi}{dx} = \frac{d_x y}{dx} = \frac{(\partial y/\partial x)_u\, dx}{dx} = \left(\frac{\partial y}{\partial x}\right)_u \tag{6-9}$$

† In this notation for partial differentials, the subscript denotes the variable with respect to which differentiation is performed and not the variable held constant.

Combining (6-8) and (6-9) by multiplication and employing the several notations, there results

$$\frac{d_y f}{dy}\frac{d_x \phi}{dx} = \frac{d_y u}{dy}\frac{d_x y}{dx} = \left(\frac{\partial u}{\partial y}\right)_x \left(\frac{\partial y}{\partial x}\right)_u \tag{6-10}$$

These equations emphasize the point that the terms ∂y in $(\partial u/\partial y)_x$ $(\partial y/\partial x)_u$ cannot be canceled to give $(\partial u/\partial x)_y$ and that ∂y in $(\partial u/\partial y)_x$ has an entirely different significance from ∂y in $(\partial y/\partial x)_u$. In the case of $(\partial u/\partial y)_x$, ∂y represents the differential of y considered as an independent variable in the equation $u = f(x,y)$, whereas ∂y in $(\partial y/\partial x)_u$ stands for the partial differential with respect to x of the function $y = \phi(u,x)$. It is possible to avoid confusion of this sort by utilizing the functional symbols f and ϕ and writing $(\partial f/\partial y)_x$ instead of $(\partial u/\partial y)_x$, and $(\partial \phi/\partial x)_u$ instead of $(\partial y/\partial x)_u$. The latter notation is widely used in spite of its ambiguity.

6-3. Differentiation of Composite Functions. The differential of a function of any number of variables

$$w = f(x_1, x_2, \ldots, x_n) \tag{6-1}$$

is given by

$$dw = df = \frac{\partial f}{\partial x_1}\,dx_1 + \frac{\partial f}{\partial x_2}\,dx_2 + \cdots + \frac{\partial f}{\partial x_n}\,dx_n \tag{6-3}$$

If $x_1, x_2, \ldots, x_n$ are each functions of a single independent variable t, as given by the equations

$$x_1 = f_1(t) \qquad x_2 = f_2(t) \qquad \cdots \qquad x_n = f_n(t)$$

the differentials $dx_1, dx_2, \ldots, dx_n$ are given by

$$\begin{aligned} dx_1 &= f_1'(t)\,dt = \frac{df_1(t)}{dt}\,dt = \frac{dx_1}{dt}\,dt \\ dx_2 &= f_2'(t)\,dt = \frac{df_2(t)}{dt}\,dt = \frac{dx_2}{dt}\,dt \\ dx_n &= f_n'(t)\,dt = \frac{df_n(t)}{dt}\,dt = \frac{dx_n}{dt}\,dt \end{aligned} \tag{6-11}$$

the three notations being used interchangeably.

Substituting in (6-3) results in

$$dw = \left(\frac{\partial f}{\partial x_1}\frac{dx_1}{dt} + \frac{\partial f}{\partial x_2}\frac{dx_2}{dt} + \cdots + \frac{\partial f}{\partial x_n}\frac{dx_n}{dt}\right)dt \tag{6-12}$$

Division by dt gives

$$\frac{dw}{dt} = \frac{\partial f}{\partial x_1}\frac{dx_1}{dt} + \frac{\partial f}{\partial x_2}\frac{dx_2}{dt} + \cdots + \frac{\partial f}{\partial x_n}\frac{dx_n}{dt} \tag{6-13}$$

Since the arguments $x_1, x_2, \ldots, x_n$ of the original functions are each functions of t, w is in reality a function of a single variable t, and the derivative of w with respect to t is an ordinary derivative. When the arguments of a function w are themselves functions of other variables, w is said to be a composite function.

Assume that in $w = f(x_1,x_2, \ldots ,x_n)$, $x_1, x_2, \ldots, x_n$ are each functions of two variables t and s, the functional relationships being given by the equations

$$\begin{aligned} x_1 &= \phi_1(t,s) \\ x_2 &= \phi_2(t,s) \\ x_n &= \phi_n(t,s) \end{aligned} \tag{6-14}$$

The differentials of $x_1, x_2, \ldots, x_n$ are

$$\begin{aligned} dx_1 &= \frac{\partial x_1}{\partial t}\,dt + \frac{\partial x_1}{\partial s}\,ds \\ dx_2 &= \frac{\partial x_2}{\partial t}\,dt + \frac{\partial x_2}{\partial s}\,ds \\ dx_n &= \frac{\partial x_n}{\partial t}\,dt + \frac{\partial x_n}{\partial s}\,ds \end{aligned} \tag{6-15}$$

Substituting in (6-3) gives

$$\begin{aligned} dw = df = {} & \left(\frac{\partial f}{\partial x_1}\frac{\partial x_1}{\partial t} + \frac{\partial f}{\partial x_2}\frac{\partial x_2}{\partial t} + \cdots + \frac{\partial f}{\partial x_n}\frac{\partial x_n}{\partial t}\right) dt \\ & + \left(\frac{\partial f}{\partial x_1}\frac{\partial x_1}{\partial s} + \frac{\partial f}{\partial x_2}\frac{\partial x_2}{\partial s} + \cdots + \frac{\partial f}{\partial x_n}\frac{\partial x_n}{\partial s}\right) ds \end{aligned} \tag{6-16}$$

But w is also a function of t and s which may be obtained by eliminating $x_1, x_2, \ldots, x_n$ in (6-1) by Eqs. (6-14). This may be written

$$\begin{aligned} w &= f[\phi_1(t,s),\phi_2(t,s), \ldots ,\phi_n(t,s)] \\ &= F(t,s) \end{aligned} \tag{6-17}$$

and
$$\begin{aligned} dw = dF = df &= \frac{\partial w}{\partial t}\,dt + \frac{\partial w}{\partial s}\,ds \\ &= \frac{\partial f}{\partial t}\,dt + \frac{\partial f}{\partial s}\,ds = \frac{\partial F}{\partial t}\,dt + \frac{\partial F}{\partial s}\,ds \end{aligned} \tag{6-18}$$

Comparing (6-16) and (6-18), there are obtained the partial derivatives of w with respect to t and s by equating the coefficients of dt and ds. In the case of t,

$$\frac{\partial w}{\partial t} = \frac{\partial f}{\partial x_1}\frac{\partial x_1}{\partial t} + \frac{\partial f}{\partial x_2}\frac{\partial x_2}{\partial t} + \cdots + \frac{\partial f}{\partial x_n}\frac{\partial x_n}{\partial t} \tag{6-19}$$

Attention is again drawn to the question of notation in Eq. (6-16). There is really no need to place subscripts on the quantities $\partial f/\partial x_1$,

$\partial f/\partial x_2$, $\partial f/\partial x_n$, because during the operation of differentiating the function f with respect to one of its arguments $x_1, x_2, \ldots, x_n$, the rest of these must be held constant. Similarly, there is no real need to employ subscripts on $\partial x_1/\partial t \cdots \partial x_n/\partial t$ and $\partial x_1/\partial s \cdots \partial x_n/\partial s$, again for the reason that differentiation of x_1 with respect to one of its arguments, t or s, necessitates constancy of the other. On the other hand, the function $f(x_1,x_2, \ldots ,x_n)$, being a function of t and s, as given by (6-17), may be differentiated with respect to t at constant s, with $x_1, x_2, \ldots, x_n$ permitted to vary in accordance with Eqs. (6-15). In this case, the nature of the differentiation would be better indicated by $\partial F/\partial t$ than by $\partial f/\partial t$, since t is an argument of F and not of f [see (6-17)]. It would be correct for $\partial w/\partial x_1$ to replace $\partial f/\partial x_1$ and $\partial w/\partial t$ to replace $\partial F/\partial t$, since w refers to both functions.

Equation (6-19) could have been obtained directly from (6-16) by holding s constant and dividing each side by dt. Under this circumstance, ds becomes equal to zero, and the last term on the right vanishes. It is important to note that no meaning is assigned to the symbol dw/dt, which would result from mere division of (6-16) by dt. If there are n independent variables, a derivative may be defined only by keeping $n - 1$ of these constant. Therefore, because there are two independent variables in (6-16) (t and s), division by dt must be at constant s to give $\partial w/\partial t$.

Equations (6-3), (6-13), and (6-19) are the basic formulas of partial differentiation. They will now be applied to several composite functions of frequent occurrence.

6-4. Differentiation Formulas

Case I. $w = f(x)$, *and* $x = F(u)$. Application of (6-13) yields the first derivative

$$\frac{dw}{du} = \frac{df}{dx}\frac{dx}{du} \tag{6-20}$$

The second derivative is

$$\frac{d}{du}\left(\frac{dw}{du}\right) = \frac{d^2w}{du^2} = \frac{df}{dx}\frac{d^2x}{du^2} + \frac{dx}{du}\frac{d}{du}\left(\frac{df}{dx}\right) \tag{6-21}$$

The first derivative df/dx is itself a function of x, usually written $f'(x)$. The term $d(df/dx)/du$ is then evaluated by application of (6-13) as

$$\frac{d^2f}{dx^2}\frac{dx}{du}$$

whence

$$\frac{d^2w}{du^2} = \frac{df}{dx}\frac{d^2x}{du^2} + \left(\frac{dx}{du}\right)^2\frac{d^2f}{dx^2} \tag{6-22}$$

Example 6-1. $w = x^n$, and $x = u^2 - 2$

$$\frac{df}{dx} = nx^{n-1} \qquad \frac{dx}{du} = 2u$$

$$\frac{dw}{du} = nx^{n-1}(2u) = n(u^2 - 2)^{n-1}2u$$

$$\frac{d^2f}{dx^2} = n(n-1)x^{n-2} \qquad \frac{d^2x}{du^2} = 2$$

$$\frac{d^2w}{du^2} = nx^{n-1}(2) + (2u)^2 n(n-1)x^{n-2}$$

Case II. $w = f(x)$, and $x = F(u,v)$. Application of (6-19) yields the two first partial derivatives

$$\frac{\partial w}{\partial u} = \frac{df}{dx}\frac{\partial x}{\partial u} \qquad \frac{\partial w}{\partial v} = \frac{df}{dx}\frac{\partial x}{\partial v} \tag{6-23}$$

$$\frac{\partial^2 w}{\partial u^2} = \frac{df}{dx}\frac{\partial}{\partial u}\left(\frac{\partial x}{\partial u}\right) + \frac{\partial x}{\partial u}\frac{\partial}{\partial u}\left(\frac{df}{dx}\right) \tag{6-24}$$

$\partial(df/dx)/\partial u$ is evaluated by noting that df/dx is itself a function of x, so that (6-19) may be applied to give

$$\frac{\partial}{\partial u}\left(\frac{df}{dx}\right) = \frac{d(df/dx)}{dx}\frac{\partial x}{\partial u} = \frac{d^2f}{dx^2}\frac{\partial x}{\partial u}$$

By substitution, (6-24) now becomes

$$\frac{\partial^2 w}{\partial u^2} = \frac{df}{dx}\frac{\partial^2 x}{\partial u^2} + \left(\frac{\partial x}{\partial u}\right)^2\frac{d^2f}{dx^2} \tag{6-25}$$

The other two second-order derivatives are obtained in similar fashion and are

$$\frac{\partial^2 w}{\partial u\,\partial v} = \frac{df}{dx}\frac{\partial^2 x}{\partial u\,\partial v} + \frac{\partial x}{\partial u}\frac{\partial x}{dv}\frac{d^2f}{dx^2} \tag{6-26}$$

$$\frac{\partial^2 w}{\partial v^2} = \frac{df}{dx}\frac{\partial^2 x}{\partial v^2} + \left(\frac{\partial x}{\partial v}\right)^2\frac{d^2f}{dx^2} \tag{6-27}$$

Example 6-2. $w = (x + 3)^2$, and $x = u^2 - v^2$

$$\frac{df}{dx} = 2(x+3) \qquad \frac{\partial x}{\partial u} = 2u \qquad \frac{\partial x}{\partial v} = -2v$$

$$\frac{d^2f}{dx^2} = 2 \qquad \frac{\partial^2 x}{\partial u^2} = 2 \qquad \frac{\partial^2 x}{\partial v^2} = -2$$

$$\frac{\partial^2 x}{\partial u\,\partial v} = \frac{\partial}{\partial v}\left(\frac{\partial x}{\partial u}\right) = \frac{\partial}{\partial u}\left(\frac{\partial x}{\partial v}\right) = 0$$

Substitution in (6-25) to (6-27) gives

$$\frac{\partial^2 w}{\partial u^2} = 2(x+3)2 + 2(2u)^2$$

$$\frac{\partial^2 w}{\partial u\,\partial v} = 2(x+3)0 + 2u(-2v)2 = -8uv$$

$$\frac{\partial^2 w}{\partial v^2} = 2(x+3)(-2) + (-2v)^2$$

Case III. $w = f(x,y)$, $x = F_1(u)$, $y = F_2(u)$. There is but one first derivative in this case, since w is a function of u only. By application of (6-13),

$$\frac{dw}{du} = \frac{\partial f}{\partial x}\frac{dx}{du} + \frac{\partial f}{\partial y}\frac{dy}{du} \tag{6-28}$$

Application of (6-13) to each term of (6-28) yields the second partial derivative

$$\frac{d^2w}{du^2} = \frac{\partial f}{\partial x}\frac{d^2x}{du^2} + \frac{dx}{du}\frac{d}{du}\left(\frac{\partial f}{\partial x}\right) + \frac{\partial f}{\partial y}\frac{d^2y}{du^2} + \frac{dy}{du}\frac{d}{du}\left(\frac{\partial f}{\partial y}\right) \tag{6-29}$$

But $\partial f/\partial x$ and $\partial f/\partial y$ are each functions of x and y, and (6-13) may be applied to each of them as follows:

$$\frac{d}{du}\left(\frac{\partial f}{\partial x}\right) = \frac{\partial(\partial f/\partial x)}{\partial x}\frac{dx}{du} + \frac{\partial(\partial f/\partial x)}{\partial y}\frac{dy}{du} = \frac{\partial^2 f}{\partial x^2}\frac{dx}{du} + \frac{\partial^2 f}{\partial y\,\partial x}\frac{dy}{du} \tag{6-30}$$

$$\frac{d}{du}\left(\frac{\partial f}{\partial y}\right) = \frac{\partial(\partial f/\partial y)}{\partial x}\frac{dx}{du} + \frac{\partial(\partial f/\partial y)}{\partial y}\frac{dy}{du} = \frac{\partial^2 f}{\partial x\,\partial y}\frac{dx}{du} + \frac{\partial^2 f}{\partial y^2}\frac{dy}{du} \tag{6-31}$$

Substituting in (6-29) gives

$$\frac{d^2w}{du^2} = \frac{\partial f}{\partial x}\frac{d^2x}{du^2} + \frac{\partial^2 f}{\partial x^2}\left(\frac{dx}{du}\right)^2 + 2\,\frac{\partial^2 f}{\partial x\,\partial y}\frac{dx}{du}\frac{dy}{du} + \frac{\partial f}{\partial y}\frac{d^2y}{du^2} + \frac{\partial^2 f}{\partial y^2}\left(\frac{dy}{du}\right)^2 \tag{6-32}$$

Example 6-3. $w = x^2 + y^2$, $x = u^3$, $y = u^2$

$$\frac{\partial f}{\partial x} = \frac{\partial(x^2 + y^2)}{\partial x} = 2x \qquad \frac{\partial f}{\partial y} = 2y \qquad \frac{dx}{du} = 3u^2 \qquad \frac{dy}{du} = 2u$$

$$\frac{dw}{du} = 2x(3u^2) + 2y(2u)$$

$$\frac{\partial^2 f}{\partial x^2} = 2 \qquad \frac{\partial^2 f}{\partial y^2} = 2 \qquad \frac{\partial^2 f}{\partial x\,\partial y} = 0 \qquad \frac{d^2x}{du^2} = 6u \qquad \frac{d^2y}{du^2} = 2$$

$$\frac{d^2w}{du^2} = 2x(6u) + 2(3u^2)^2 + 2\cdot 0(3u^2)2u + 2y(2) + 2(2u)^2$$

Case IV. $w = f(x,y)$, *where* $x = f_1(u)$, $y = f_2(v)$. By application of (6-19),

$$\frac{\partial w}{\partial u} = \frac{\partial f}{\partial x}\frac{\partial x}{\partial u} + \frac{\partial f}{\partial y}\frac{\partial y}{\partial u} \tag{6-33}$$

In this case, $y = f_2(v)$ is independent of u, and hence $\partial y/\partial u$ vanishes. Because x is a function of a single variable u, $\partial x/\partial u$ becomes the ordinary derivative dx/du. Equation (6-33) then becomes

$$\frac{\partial w}{\partial u} = \frac{\partial f}{\partial x}\frac{dx}{du} \tag{6-34}$$

Similarly, when x is held constant,

$$\frac{\partial w}{\partial v} = \frac{\partial f}{\partial y}\frac{dy}{dv} \tag{6-35}$$

$$\frac{\partial^2 w}{\partial u^2} = \frac{\partial f}{\partial x}\frac{d^2x}{du^2} + \frac{dx}{du}\frac{\partial}{\partial u}\left(\frac{\partial f}{\partial x}\right)$$

But $\partial f/\partial x$ is a function of x and y, so that (6-13) may be applied to give

$$\frac{\partial}{\partial u}\left(\frac{\partial f}{\partial x}\right) = \frac{\partial^2 f}{\partial x^2}\frac{dx}{du}$$

and

$$\frac{\partial^2 w}{\partial u^2} = \frac{\partial f}{\partial x}\frac{d^2x}{du^2} + \frac{\partial^2 f}{\partial x^2}\left(\frac{dx}{du}\right)^2 \tag{6-36}$$

Similarly,

$$\frac{\partial^2 w}{\partial v^2} = \frac{\partial f}{\partial y}\frac{d^2y}{dv^2} + \frac{\partial^2 f}{\partial y^2}\left(\frac{dy}{dv}\right)^2 \tag{6-37}$$

and

$$\frac{\partial^2 w}{\partial u\,\partial v} = \frac{dx}{du}\frac{dy}{dv}\frac{\partial^2 f}{\partial x\,\partial y} \tag{6-38}$$

This case is frequently of importance in effecting simplification in a formula by change of the independent variables.

Example 6-4. Simplify the equation $ax(\partial\phi/\partial x)_y + by(\partial\phi/\partial y)_x = 0$, where a and b are constants and ϕ is a function of x and y.

Place $x = u^a$, $y = v^b$.

$$\left(\frac{\partial\phi}{\partial u}\right)_v = \left(\frac{\partial\phi}{\partial x}\right)_y\frac{dx}{du} = \left(\frac{\partial\phi}{\partial x}\right)_y au^{a-1} = \left(\frac{\partial\phi}{\partial x}\right)_y\frac{ax}{u}$$

and

$$ax\left(\frac{\partial\phi}{\partial x}\right)_y = u\left(\frac{\partial\phi}{\partial u}\right)_v$$

Similarly,

$$by\left(\frac{\partial\phi}{\partial y}\right)_x = v\left(\frac{\partial\phi}{\partial v}\right)_u$$

The simplified equation is

$$u\left(\frac{\partial\phi}{\partial u}\right)_v + v\left(\frac{\partial\phi}{\partial v}\right)_u = 0$$

Case V. $w = f(x,y)$, $x = F_1(u,v)$, $y = F_2(u,v)$. Application of (6-19) gives the two first partial derivatives

$$\frac{\partial w}{\partial u} = \frac{\partial f}{\partial x}\frac{\partial x}{\partial u} + \frac{\partial f}{\partial y}\frac{\partial y}{\partial u} \tag{6-39}$$

$$\frac{\partial w}{\partial v} = \frac{\partial f}{\partial x}\frac{\partial x}{\partial v} + \frac{\partial f}{\partial y}\frac{\partial y}{\partial v} \tag{6-40}$$

The second partial derivatives are obtained by differentiating (6-39) and (6-40):

$$\frac{\partial^2 w}{\partial u^2} = \frac{\partial f}{\partial x}\frac{\partial^2 x}{\partial u^2} + \frac{\partial x}{\partial u}\frac{\partial}{\partial u}\left(\frac{\partial f}{\partial x}\right) + \frac{\partial f}{\partial y}\frac{\partial^2 y}{\partial u^2} + \frac{\partial y}{\partial u}\frac{\partial}{\partial u}\left(\frac{\partial f}{\partial y}\right)$$

But $\partial f/\partial x$ and $\partial f/\partial y$, being functions of x and y, may be treated by (6-19) to give

$$\frac{\partial}{\partial u}\left(\frac{\partial f}{\partial x}\right) = \frac{\partial^2 f}{\partial x^2}\frac{\partial x}{\partial u} + \frac{\partial^2 f}{\partial x\,\partial y}\frac{\partial y}{\partial u} \tag{6-41}$$

$$\frac{\partial}{\partial u}\left(\frac{\partial f}{\partial y}\right) = \frac{\partial^2 f}{\partial y\,\partial x}\frac{\partial x}{\partial u} + \frac{\partial^2 f}{\partial y^2}\frac{\partial y}{\partial u} \tag{6-42}$$

Substituting in (6-41) yields

$$\frac{\partial^2 w}{\partial u^2} = \frac{\partial f}{\partial x}\frac{\partial^2 x}{\partial u^2} + \frac{\partial^2 f}{\partial x^2}\left(\frac{\partial x}{\partial u}\right)^2 + 2\frac{\partial^2 f}{\partial x\,\partial y}\frac{\partial x}{\partial u}\frac{\partial y}{\partial u} + \frac{\partial f}{\partial y}\frac{\partial^2 y}{\partial u^2} + \frac{\partial^2 f}{\partial y^2}\left(\frac{\partial y}{\partial u}\right)^2 \tag{6-43}$$

Obviously the formula for $\partial^2 w/\partial v^2$ is the same as (6-43), with v substituted for u. The third partial derivative of the second order is obtained by applying (6-19) to either (6-39) or (6-40):

$$\frac{\partial^2 w}{\partial u\,\partial v} = \frac{\partial f}{\partial x}\frac{\partial^2 x}{\partial u\,\partial v} + \frac{\partial x}{\partial v}\frac{\partial}{\partial u}\left(\frac{\partial f}{\partial x}\right) + \frac{\partial f}{\partial y}\frac{\partial^2 y}{\partial u\,\partial v} + \frac{\partial y}{\partial v}\frac{\partial}{\partial u}\left(\frac{\partial f}{\partial y}\right)$$

Substituting from (6-41) and (6-42) results in

$$\begin{aligned}\frac{\partial^2 w}{\partial u\,\partial v} &= \frac{\partial f}{\partial x}\frac{\partial^2 x}{\partial u\,\partial v} + \frac{\partial^2 f}{\partial x^2}\frac{\partial x}{\partial u}\frac{\partial x}{\partial v} + \frac{\partial^2 f}{\partial x\,\partial y}\frac{\partial y}{\partial u}\frac{\partial x}{\partial v} + \frac{\partial f}{\partial y}\frac{\partial^2 y}{\partial u\,\partial v} + \frac{\partial y}{\partial v}\frac{\partial x}{\partial u}\frac{\partial^2 f}{\partial x\,\partial y} \\ &\quad + \frac{\partial^2 f}{\partial y^2}\frac{\partial y}{\partial v}\frac{\partial y}{\partial u} = \frac{\partial^2 x}{\partial u\,\partial v}\frac{\partial f}{\partial x} + \frac{\partial^2 y}{\partial u\,\partial v}\frac{\partial f}{\partial y} + \frac{\partial^2 f}{\partial x\,\partial y}\left(\frac{\partial y}{\partial v}\frac{\partial x}{\partial u} + \frac{\partial y}{\partial u}\frac{\partial x}{\partial v}\right) \\ &\quad + \frac{\partial^2 f}{\partial x^2}\frac{\partial x}{\partial u}\frac{\partial x}{\partial v} + \frac{\partial^2 f}{\partial y^2}\frac{\partial y}{\partial u}\frac{\partial y}{\partial v}\end{aligned} \tag{6-44}$$

An important application of case V occurs in the change of coordinate systems.

Example 6-5. Change from Cartesian to Cylindrical Coordinates. The relation between these two systems is shown in Fig. 6-1. Quantitatively the two coordinate systems are related as follows:

$$x = r\cos\theta \tag{6-45}$$
$$y = r\sin\theta \tag{6-46}$$
$$z = z \tag{6-47}$$
$$r = \sqrt{x^2 + y^2} \tag{6-48}$$

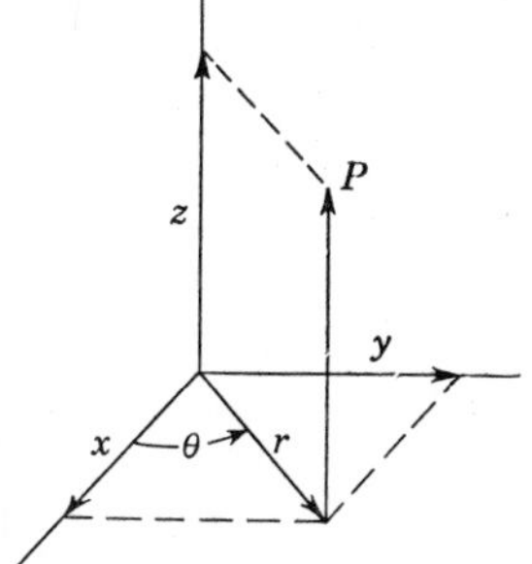

FIG. 6-1. Cylindrical-coordinate system.

Cylindrical coordinates are particularly well adapted to treating problems involving cylindrical symmetry. Consider the problem of heat transfer by conduction in a circular cylinder. The treatment of the steady-state case can start with the equation in cartesian coordinates:

$$\frac{\partial^2 T}{\partial x^2} + \frac{\partial^2 T}{\partial y^2} + \frac{\partial^2 T}{\partial z^2} = 0$$

and proceed to cylindrical coordinates for solution because of the resulting simplification of boundary conditions.

For a fixed value of z, T is some function of x and y which may be written $f(x,y)$. Since $x = F_1(r,\theta) = r \cos \theta$ and $y = F_2(r,\theta) = r \sin \theta$, T also is some function of r and θ, $T = f(x,y) = f(r \cos \theta, r \sin \theta) = \psi(r,\theta)$. Furthermore, r and θ are each functions of x and y and may be written $r = F_3(x,y)$ and $\theta = F_4(x,y)$. The application of case V is now evident, and the required transformation is effected by substitution into (6-39) and (6-43) with the changes in notation $w = T = \psi(r,\theta)$, $z = r$, $y = \theta$, $u = x$, $v = y$.

The following partial derivatives are used throughout the calculation:

$$\frac{\partial r}{\partial x} = \frac{1}{2\sqrt{x^2 + y^2}} 2x = \frac{r \cos \theta}{r} = \cos \theta \tag{6-49}$$

$$\frac{\partial r}{\partial y} = \sin \theta \tag{6-50}$$

$$\tan \theta = \frac{y}{x} \qquad \theta = \tan^{-1} \frac{y}{x} \tag{6-51}$$

$$\frac{\partial \theta}{\partial x} = \frac{1}{1 + (y/x)^2} y \left(-\frac{1}{x^2}\right) = -\frac{y}{x^2 + y^2} = -\frac{r \sin \theta}{r^2} = -\frac{\sin \theta}{r} \tag{6-52}$$

$$\frac{\partial \theta}{\partial y} = \frac{\cos \theta}{r} \tag{6-53}$$

$$\frac{\partial^2 r}{\partial x^2} = -\sin \theta \frac{\partial \theta}{\partial x} = \frac{\sin^2 \theta}{r} \tag{6-54}$$

$$\frac{\partial^2 r}{\partial y^2} = \cos \theta \frac{\partial \theta}{\partial y} = \frac{\cos^2 \theta}{r} \tag{6-55}$$

$$\frac{\partial^2 \theta}{\partial x^2} = -\frac{\partial}{\partial x}\left(\frac{y}{x^2 + y^2}\right) = \frac{2xy}{(x^2 + y^2)^2} \tag{6-56}$$

$$\frac{\partial^2 \theta}{\partial y^2} = \frac{\partial}{\partial y}\left(\frac{x}{x^2 + y^2}\right) = \frac{-2xy}{(x^2 + y^2)^2} \tag{6-57}$$

$$\frac{\partial^2 T}{\partial x^2} = \frac{\partial T}{\partial r}\frac{\partial^2 r}{\partial x^2} + \frac{\partial^2 T}{\partial r^2}\left(\frac{\partial r}{\partial x}\right)^2 + 2\frac{\partial^2 T}{\partial r\, \partial \theta}\frac{\partial r}{\partial x}\frac{\partial \theta}{\partial x} + \frac{\partial T}{\partial \theta}\frac{\partial^2 \theta}{\partial x^2} + \frac{\partial^2 T}{\partial \theta^2}\left(\frac{\partial \theta}{\partial x}\right)^2 \tag{6-58}$$

$$\frac{\partial^2 T}{\partial y^2} = \frac{\partial T}{\partial r}\frac{\partial^2 r}{\partial y^2} + \frac{\partial^2 T}{\partial r^2}\left(\frac{\partial r}{\partial y}\right)^2 + 2\frac{\partial^2 T}{\partial r\, \partial \theta}\frac{\partial r}{\partial y}\frac{\partial \theta}{\partial y} + \frac{\partial T}{\partial \theta}\frac{\partial^2 \theta}{\partial y^2} + \frac{\partial^2 T}{\partial \theta^2}\left(\frac{\partial \theta}{\partial y}\right)^2 \tag{6-59}$$

When (6-58) and (6-59) are added, there results

$$\begin{aligned}\frac{\partial^2 T}{\partial x^2} + \frac{\partial^2 T}{\partial y^2} &= \frac{\partial T}{\partial r}\left(\frac{\partial^2 r}{\partial x^2} + \frac{\partial^2 r}{\partial y^2}\right) + \frac{\partial^2 T}{\partial r^2}\left[\left(\frac{\partial r}{\partial x}\right)^2 + \left(\frac{\partial r}{\partial y}\right)^2\right] \\ &+ 2\frac{\partial^2 T}{\partial r\, \partial \theta}\left(\frac{\partial r}{\partial x}\frac{\partial \theta}{\partial x} + \frac{\partial r}{\partial y}\frac{\partial \theta}{\partial y}\right) + \frac{\partial T}{\partial \theta}\left(\frac{\partial^2 \theta}{\partial x^2} + \frac{\partial^2 \theta}{\partial y^2}\right) + \frac{\partial^2 T}{\partial \theta^2}\left[\left(\frac{\partial \theta}{\partial x}\right)^2 + \left(\frac{\partial \theta}{\partial y}\right)^2\right]\end{aligned} \tag{6-60}$$

The derivatives appearing in parentheses or brackets have already been computed and are given in Eqs. (6-49) to (6-57). Substitution gives

$$\begin{aligned}\frac{\partial^2 T}{\partial x^2} + \frac{\partial^2 T}{\partial y^2} &= \frac{\partial T}{\partial r}\frac{\sin^2 \theta + \cos^2 \theta}{r} + \frac{\partial^2 T}{\partial r^2}(\cos^2 \theta + \sin^2 \theta) \\ &+ 2\frac{\partial^2 T}{\partial r\, \partial \theta}\left(\frac{\sin \theta \cos \theta}{r} - \frac{\sin \theta \cos \theta}{r}\right) + \frac{\partial T}{\partial \theta}\left[\frac{2xy}{(x^2 + y^2)^2} - \frac{2xy}{(x^2 + y^2)^2}\right] \\ &+ \frac{\partial^2 T}{\partial \theta^2}\frac{\sin^2 \theta + \cos^2 \theta}{r^2}\end{aligned} \tag{6-61}$$

Since $\sin^2 \theta + \cos^2 \theta = 1$, (6-61) becomes

$$\frac{\partial^2 T}{\partial x^2} + \frac{\partial^2 T}{\partial y^2} = \frac{1}{r}\frac{\partial T}{\partial r} + \frac{\partial^2 T}{\partial r^2} + \frac{1}{r^2}\frac{\partial^2 T}{\partial \theta^2} \tag{6-62}$$

The final transformation is then

$$\frac{\partial^2 T}{\partial x^2} + \frac{\partial^2 T}{\partial y^2} + \frac{\partial^2 T}{\partial z^2} = \frac{1}{r}\frac{\partial T}{\partial r} + \frac{\partial^2 T}{\partial r^2} + \frac{1}{r^2}\frac{\partial^2 T}{\partial \theta^2} + \frac{\partial^2 T}{\partial z^2} \tag{6-63}$$

6-5. Differentiation of Implicit Functions. A relation of the form $f(x_1,x_2, \ldots ,x_n) = 0$ defines one of the variables, say x_1, as an implicit function of the remaining variables. Suppose that it is desired to evaluate the partial derivatives $\partial x_1/\partial x_2$, $\partial x_1/\partial x_3$, ... A straightforward approach would involve solving for x_1 explicitly to obtain a relation

$$x_1 = \phi(x_2,x_3, \ldots ,x_n) \tag{6-64}$$

and then to carry out the operations of partial differentiation. If the function $f(x_1,x_2, \ldots ,x_n)$ is complicated, it may prove impossible to obtain the explicit relation (6-64). In this situation, the desired relation may be obtained by application of the partial differentiation relations derived previously.

Consider the implicit function of two variables. The differential of $f(x,y) = 0$ is

$$df = \frac{\partial f}{\partial x} dx + \frac{\partial f}{\partial y} dy = 0 \tag{6-65}$$

Equation (6-65) yields for dy/dx the important relation

$$\frac{dy}{dx} = -\frac{\partial f/\partial x}{\partial f/\partial y} \tag{6-66}$$

provided that

$$\frac{\partial f}{\partial y} \neq 0$$

Example 6-6. Find dy/dx if $f(x,y) = 3x^3y^2 + x \cos y = 0$. Here

$$\frac{\partial f}{\partial x} = 9x^2y^2 + \cos y$$

$$\frac{\partial f}{\partial y} = 6x^3y - x \sin y$$

Use of (6-66) gives

$$\frac{dy}{dx} = -\frac{9x^2y^2 + \cos y}{6x^3y - x \sin y}$$

The relation $f(x,y,z) = 0$ may define any one of the variables as an implicit function of the other two. Let z be the dependent variable. Then

$$dz = \frac{\partial z}{\partial x} dx + \frac{\partial z}{\partial y} dy \tag{6-67}$$

But

$$df = \frac{\partial f}{\partial x} dx + \frac{\partial f}{\partial y} dy + \frac{\partial f}{\partial z} dz = 0 \tag{6-68}$$

When (6-67) is used to replace dz in (6-68), there results

$$\left(\frac{\partial f}{\partial x} + \frac{\partial f}{\partial z}\frac{\partial z}{\partial x}\right) dx + \left(\frac{\partial f}{\partial y} + \frac{\partial f}{\partial z}\frac{\partial z}{\partial y}\right) dy = 0 \tag{6-69}$$

Inasmuch as x and y are independent and (6-69) must hold for all values of dx and dy, it follows that the coefficients (in parentheses) of dx and dy must individually be zero. If $\partial f/\partial z \neq 0$, this condition gives

$$\begin{aligned} \frac{\partial z}{\partial x} &= -\frac{\partial f/\partial x}{\partial f/\partial z} \\ \frac{\partial z}{\partial y} &= -\frac{\partial f/\partial y}{\partial f/\partial z} \end{aligned} \tag{6-70}$$

The calculation of the derivatives of a function that is defined implicitly by a pair of simultaneous equations

$$\begin{aligned} f(x,y,z) &= 0 \\ \phi(x,y,z) &= 0 \end{aligned} \tag{6-71}$$

proceeds as follows: Differentiation of (6-71) gives

$$\begin{aligned} df &= \frac{\partial f}{\partial x}\,dx + \frac{\partial f}{\partial y}\,dy + \frac{\partial f}{\partial z}\,dz = 0 \\ d\phi &= \frac{\partial \phi}{\partial x}\,dx + \frac{\partial \phi}{\partial y}\,dy + \frac{\partial \phi}{\partial z}\,dz = 0 \end{aligned} \tag{6-72}$$

Algebraic manipulation of (6-72) yields the ratio of the differentials

$$dx\colon dy\colon dz = \begin{vmatrix} \frac{\partial f}{\partial y} & \frac{\partial f}{\partial z} \\ \frac{\partial \phi}{\partial y} & \frac{\partial \phi}{\partial z} \end{vmatrix} : \begin{vmatrix} \frac{\partial f}{\partial z} & \frac{\partial f}{\partial x} \\ \frac{\partial \phi}{\partial z} & \frac{\partial \phi}{\partial x} \end{vmatrix} : \begin{vmatrix} \frac{\partial f}{\partial x} & \frac{\partial f}{\partial y} \\ \frac{\partial \phi}{\partial x} & \frac{\partial \phi}{\partial y} \end{vmatrix} \tag{6-73}$$

Any desired derivative may be obtained from (6-73). For example,

$$\frac{dx}{dz} = \frac{\begin{vmatrix} \frac{\partial f}{\partial y} & \frac{\partial f}{\partial z} \\ \frac{\partial \phi}{\partial y} & \frac{\partial \phi}{\partial z} \end{vmatrix}}{\begin{vmatrix} \frac{\partial f}{\partial x} & \frac{\partial f}{\partial y} \\ \frac{\partial \phi}{\partial x} & \frac{\partial \phi}{\partial y} \end{vmatrix}} = \frac{(\partial f/\partial y)(\partial \phi/\partial z) - (\partial f/\partial z)(\partial \phi/\partial y)}{(\partial f/\partial x)(\partial \phi/\partial y) - (\partial f/\partial y)(\partial \phi/\partial x)}$$

The pair of simultaneous equations

$$\begin{aligned} f(x,y,u,v) &= 0 \\ \phi(x,y,u,v) &= 0 \end{aligned} \tag{6-74}$$

define u and v as implicit functions of x and y. The partial derivatives may be obtained as follows: Differentiation of (6-74) gives

$$\begin{aligned} df &= \frac{\partial f}{\partial x}\,dx + \frac{\partial f}{\partial y}\,dy + \frac{\partial f}{\partial u}\,du + \frac{\partial f}{\partial v}\,dv = 0 \\ d\phi &= \frac{\partial \phi}{\partial x}\,dx + \frac{\partial \phi}{\partial y}\,dy + \frac{\partial \phi}{\partial u}\,du + \frac{\partial \phi}{\partial v}\,dv = 0 \end{aligned} \tag{6-75}$$

However, u and v are functions of x and y. Hence,

$$\begin{aligned} du &= \frac{\partial u}{\partial x}\,dx + \frac{\partial u}{\partial y}\,dy \\ dv &= \frac{\partial v}{\partial x}\,dx + \frac{\partial v}{\partial y}\,dy \end{aligned} \tag{6-76}$$

When du and dv are replaced in (6-75) by (6-76), there results

$$\begin{aligned} \left(\frac{\partial f}{\partial x} + \frac{\partial f}{\partial u}\frac{\partial u}{\partial x} + \frac{\partial f}{\partial v}\frac{\partial v}{\partial x}\right)dx + \left(\frac{\partial f}{\partial y} + \frac{\partial f}{\partial u}\frac{\partial u}{\partial y} + \frac{\partial f}{\partial v}\frac{\partial v}{\partial y}\right)dy &= 0 \\ \left(\frac{\partial \phi}{\partial x} + \frac{\partial \phi}{\partial u}\frac{\partial u}{\partial x} + \frac{\partial \phi}{\partial v}\frac{\partial v}{\partial x}\right)dx + \left(\frac{\partial \phi}{\partial y} + \frac{\partial \phi}{\partial u}\frac{\partial u}{\partial x} + \frac{\partial \phi}{\partial v}\frac{\partial v}{\partial y}\right)dy &= 0 \end{aligned} \tag{6-77}$$

Inasmuch as x and y are independent variables, the coefficients of dx and dy in (6-77) must equal zero individually. This provides four equations for the determination of $\partial u/\partial x$, $\partial u/\partial y$, $\partial v/\partial x$, and $\partial v/\partial y$. A typical expression is

$$\frac{\partial u}{\partial x} = -\frac{\begin{vmatrix} \dfrac{\partial f}{\partial x} & \dfrac{\partial f}{\partial v} \\ \dfrac{\partial \phi}{\partial x} & \dfrac{\partial \phi}{\partial v} \end{vmatrix}}{\begin{vmatrix} \dfrac{\partial f}{\partial u} & \dfrac{\partial f}{\partial v} \\ \dfrac{\partial \phi}{\partial u} & \dfrac{\partial \phi}{\partial v} \end{vmatrix}} \tag{6-78}$$

6-6. Directional Derivatives. The relation

$$u = f(x,y) \tag{6-79}$$

defines the value of u in the x, y plane. In a practical problem, u may be temperature, concentration, etc. Use of (6-79) enables lines of constant u to be plotted in the x, y plane, as shown in Fig. 6-2. The change in u corresponding to a change in x and y is

$$du = \frac{\partial f}{\partial x}\,dx + \frac{\partial f}{\partial y}\,dy \tag{6-80}$$

Suppose that x and y are connected by a relation of the form

$$\phi\,(x,y) = 0 \tag{6-81}$$

The equation $\phi(x,y)$ defines a line in the x, y plane which, in general, crosses the $u =$ const lines (Fig. 6-2). Denote the length of this line between the point where it crosses the $u = u_1$ curve and the point where

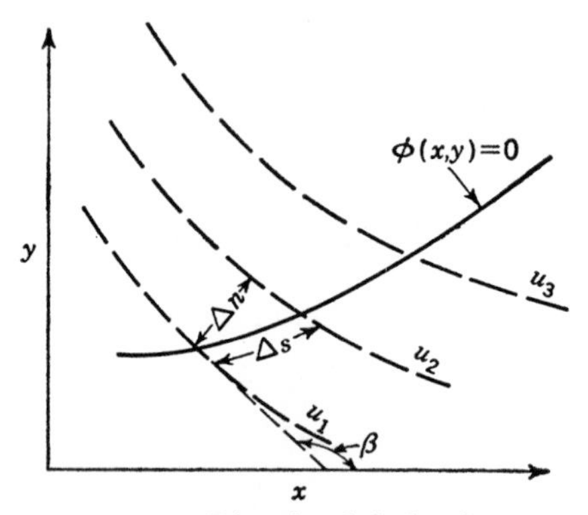

FIG. 6-2. Directional derivative.

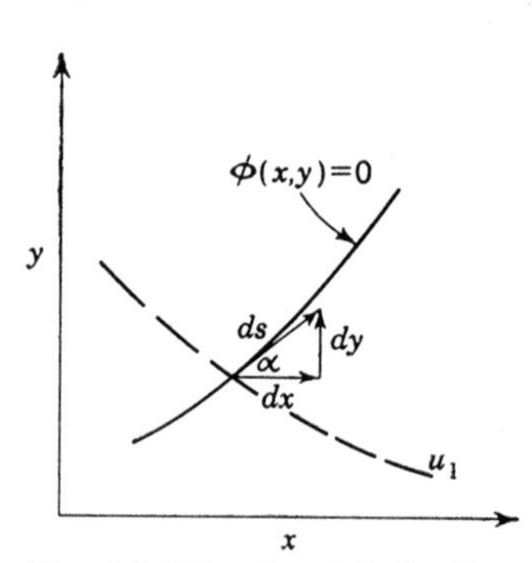

FIG. 6-3. Directional derivative.

it crosses the $u = u_2$ curve by Δs. Then the average rate at which u changes along the line $\phi(x,y) = 0$ is

$$\frac{u_2 - u_1}{\Delta s} = \frac{\Delta u}{\Delta s} \tag{6-82}$$

The instantaneous rate at which u changes with distance along the curve is

$$\lim_{\Delta s \to 0} \frac{\Delta u}{\Delta s} = \frac{du}{ds} = \frac{\partial f}{\partial x}\frac{dx}{ds} + \frac{\partial f}{\partial y}\frac{dy}{ds} \tag{6-83}$$

However,

$$\frac{dx}{ds} = \cos\alpha \qquad \frac{dy}{ds} = \sin\alpha$$

where $\tan\alpha$ is the slope of the curve $\phi(x,y) = 0$ (see Fig. 6-3). Hence,

$$\frac{du}{ds} = \frac{\partial f}{\partial x}\cos\alpha + \frac{\partial f}{\partial y}\sin\alpha \tag{6-84}$$

It is clear that du/ds depends upon the direction of the curve $\phi(x,y) = 0$. For this reason, du/ds is called the "directional derivative" and represents the rate of change of u in the direction of the tangent to the curve $\phi(x,y) = 0$.

The maximum value of du/ds is found by differentiating (6-84) with respect to α and setting the derivative equal to zero:

$$\tan \alpha = \frac{\partial f/\partial y}{\partial f/\partial x} \tag{6-85}$$

Since the slope of the curve $u =$ const is (see Fig. 6-2)

$$\tan \beta = -\frac{\partial f/\partial x}{\partial f/\partial y}$$

the maximum value of du/ds is in the direction *normal* to the line $u =$ const. If distance along a line normal to the line $u =$ const is denoted by n, then the maximum value of du/ds is du/dn. It is easy to show that

$$\frac{du}{dn} = \sqrt{\left(\frac{\partial f}{\partial x}\right)^2 + \left(\frac{\partial f}{\partial y}\right)^2} \tag{6-86}$$

and

$$\frac{du}{ds} = \frac{du}{dn} \cos \psi \tag{6-87}$$

where ψ is the angle between Δn and Δs (Fig. 6-2).

In three-dimensional space, if

$$u = f(x,y,z) \tag{6-88}$$

it may be shown that

$$\frac{du}{dn} = \sqrt{\left(\frac{\partial f}{\partial x}\right)^2 + \left(\frac{\partial f}{\partial y}\right)^2 + \left(\frac{\partial f}{\partial z}\right)^2} \tag{6-89}$$

6-7. Maxima and Minima. Maximum and minimum values of $u = f(x,y,z)$ occur at conditions for which

$$du = \frac{\partial f}{\partial x} dx + \frac{\partial f}{\partial y} dy + \frac{\partial f}{\partial z} dz = 0 \tag{6-90}$$

If x, y, and z are *independent* variables, then the following must be satisfied:

$$\frac{\partial f}{\partial x} = 0 \qquad \frac{\partial f}{\partial y} = 0 \qquad \frac{\partial f}{\partial z} = 0 \tag{6-91}$$

Frequently the maximum or minimum of the function $u = f(x,y,z)$ is desired when x, y, and z are related by a relation $\phi(x,y,z) = 0$. If a straightforward approach proves awkward, a technique due to Lagrange may be used. The total differential of $\phi(x,y,z) = 0$ is

$$\frac{\partial \phi}{\partial x} dx + \frac{\partial \phi}{\partial y} dy + \frac{\partial \phi}{\partial z} dz = 0 \tag{6-92}$$

Multiply (6-92) by an as yet undetermined number λ, and add the result to (6-90). The result is

$$\left(\frac{\partial f}{\partial x} + \lambda \frac{\partial \phi}{\partial x}\right) dx + \left(\frac{\partial f}{\partial y} + \lambda \frac{\partial \phi}{\partial y}\right) dy + \left(\frac{\partial f}{\partial z} + \lambda \frac{\partial \phi}{\partial z}\right) dz = 0 \tag{6-93}$$

If λ is assigned a value which makes

$$\begin{aligned} \frac{\partial f}{\partial x} + \lambda \frac{\partial \phi}{\partial x} &= 0 \\ \frac{\partial f}{\partial y} + \lambda \frac{\partial \phi}{\partial y} &= 0 \\ \frac{\partial f}{\partial z} + \lambda \frac{\partial \phi}{\partial z} &= 0 \\ \phi(x,y,z) &= 0 \end{aligned} \tag{6-94}$$

the condition for a maximum or minimum is satisfied. The numerical values of x, y, and z which satisfy (6-94) are the values at which $f(x,y,z)$ is a maximum or minimum.

6-8. Differentiation of Integrals. Definite integrals may appear in the form

$$F(b) - F(a) = \int_a^b f(x,c)\,dx \tag{6-95}$$

where $F(x)$ is the integral function of $f(x,c)$ and a and b are constant values of x; c is an arbitrary constant which remains constant during any one integration but which may vary to form different integrals all taken between the same two limits. In such a case, the integral is a function of c and may be written

$$\phi(c) = \int_a^b f(x,c)\,dx \tag{6-96}$$

The function $\phi(c)$ is represented by the area $acdb$ in Fig. 6-4. When c receives an increment Δc, $\phi(c + \Delta c)$ is the area $aefb$ under the curve

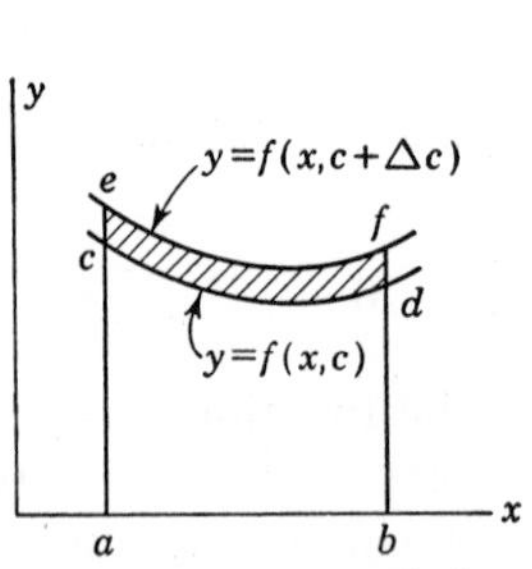

FIG. 6-4. Differentiation of integrals.

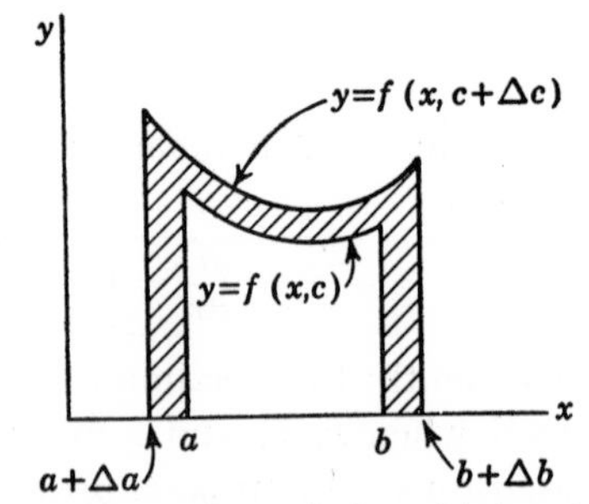

FIG. 6-5. Differentiation of integrals.

$y = f(x, c + \Delta c)$, and $\Delta\phi$ is represented by the shaded area $cefd$. It is often desirable to compute the rate of change of ϕ with respect to c, that is, $d\phi/dc$. This is done by the formula

$$\frac{d\phi}{dc} = \int_a^b \frac{\partial f(x,c)}{\partial c}\,dx \tag{6-97}$$

When the limits of integration a and b are functions of c, an increment in c produces an increment in ϕ represented by the shaded area in Fig. 6-5. In this case, it may be shown that

$$\frac{d\phi}{dc} = \int_a^b \frac{\partial f(x,c)}{\partial c}\,dx + f(b,c)\frac{db}{dc} - f(a,c)\frac{da}{dc} \tag{6-98}$$

This type of differentiation has many possible uses in determining maxima and minima in economic-balance problems.

6-9. Exact Differentials. The total differential of a function is often known as a "complete" or an "exact differential." It has already been shown that the exact differential of the function $u = f(x,y)$ is

$$du = \frac{\partial f}{\partial x}\,dx + \frac{\partial f}{\partial y}\,dy \tag{6-99}$$

where $\partial f/\partial x$ and $\partial f/\partial y$ are, in general, functions of x and y. It is very common to encounter expressions of the form

$$du = M\,dx + N\,dy \tag{6-100}$$

in which du is not an exact differential even though M and N are functions of x and y. The necessary and sufficient conditions for du to be an exact differential are easily stated. If du is an exact differential, there must exist a function $f(x, y)$ such that, by comparison of (6-99) and (6-100), $\partial f/\partial x = M$ and $\partial f/\partial y = N$. Furthermore, since the order of differentiation is immaterial, it is necessary that

$$\frac{\partial}{\partial y}\left(\frac{\partial f}{\partial x}\right) = \frac{\partial M}{\partial y} = \frac{\partial}{\partial x}\left(\frac{\partial f}{\partial y}\right) = \frac{\partial N}{\partial x} \tag{6-101}$$

As an example of the application of this criterion, consider the equation

$$du = 3x^2y\,dx + y^2x\,dy \tag{6-102}$$

where

$$\frac{\partial M}{\partial y} = \frac{\partial(3x^2y)}{\partial y} = 3x^2 \tag{6-103}$$

and

$$\frac{\partial N}{\partial x} = \frac{\partial(y^2x)}{\partial x} = y^2 \tag{6-104}$$

The two derivatives are not equal, and therefore du is not an exact differential; it is impossible to find a function f such that $\partial f/\partial x = 3x^2y$ and $\partial f/\partial y = y^2x$.

On the other hand, in the case of

$$du = 3x^2y^2\,dx + 2x^3y\,dy \tag{6-105}$$

$\partial M/\partial y = 6x^2y$, and $\partial N/\partial x = 6x^2y$, indicating that du in this case is an exact differential. This function is easily determined to be

$$u = x^3y^2 + c \tag{6-106}$$

There is a further property relating to exact differentials that is of particular importance in the field of thermodynamics. If (6-102) is integrated between limits, there results

$$\int_{u_a}^{u_b} du = u_b - u_a = \int_{x_a}^{x_b} 3x^2y\,dx + \int_{y_a}^{y_b} y^2x\,dy \tag{6-107}$$

In order to evaluate these integrals, it is necessary to have a functional relation between x and y; that is, it is necessary to specify the path of the integration.† In general, the value of $u_b - u_a$ will depend on the nature of the relation between y and x.

Consider now the integration of (6-105):

$$\int_{u_a}^{u_b} du = u_b - u_a = \int_{x_a}^{x_b} 3x^2y^2\,dx + \int_{y_a}^{y_b} 2x^3y\,dy \tag{6-108}$$

Again, it would appear as though the value of $u_b - u_a$ might depend upon the relation between x and y. It will be found, however, that, regardless of the nature of the functional relationship chosen between x and y, the value of $u_b - u_a$ remains the same. That this must be the case is evident upon reference to (6-106), where it is seen that

$$\begin{aligned} du &= d(x^3y^2) = 3x^2y^2\,dx + 2x^3y\,dy \\ \int_{u_b}^{u_a} du &= \int_a^b d(x^3y^2) = x_b{}^3y_b{}^2 - x_a{}^3y_a{}^2 \end{aligned} \tag{6-109}$$

Evidently, $u_b - u_a$ may be evaluated from (6-109), in which no functional relation between x and y is involved. Since the value of $\int_{u_b}^{u_a} du$ in (6-109) depends only upon the limits of the integration or upon the two points x_a, y_a and x_b, y_b, u is said to be a "point function," and du is an exact differential.

On the other hand, du in (6-102) is said to be an "inexact differential," because its integral depends not only upon the limits but also upon the path of integration, i.e., upon the nature of the function of x and y.

In the case of an equation in three independent variables,

$$du = M\,dx + N\,dy + P\,dz \tag{6-110}$$

where M, N, and P are functions of x, y, and z, the necessary and sufficient condition that du be an exact differential is that

$$\frac{\partial M}{\partial y} = \frac{\partial N}{\partial x} \qquad \frac{\partial M}{\partial z} = \frac{\partial P}{\partial x} \qquad \frac{\partial N}{\partial z} = \frac{\partial P}{\partial y} \tag{6-111}$$

If du is known to be an exact differential, it follows that $u = f(x,y,z)$ and $M = \partial f/\partial x$, $N = \partial f/\partial y$, $P = \partial f/\partial z$.

Similar relations hold for a function of any number of variables.

† Since the path of integration is along some curve, in the plane xy, (6-107) is called a "line integral."

6-10. Formulation of Partial Differential Equations. The mathematical formulation of a physical problem involving two or more independent variables proceeds in a fashion analogous to that used in Chap. 3 to formulate a problem involving only one independent variable. The resulting equation for the multiple variable system will be in terms of partial differentials rather than total differentials and is called a "partial differential equation." The procedure details are illustrated in the following examples.

Example 6-7. The Continuity Equation. Consider the unsteady flow of a compressible fluid. It is desired to apply the law of conservation of mass to this flowing fluid as an initial step in the formulation of the equations which govern its motion. Adopt the Eulerian point of view and examine an infinitesimal element of volume, fixed in space, and situated within the fluid (Fig. 6-6). Let the volume element be a

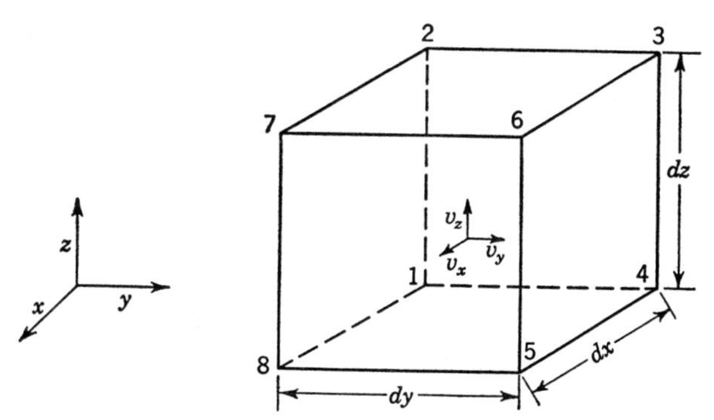

FIG. 6-6. Control volume.

rectangular parallelepiped whose sides are of length dx, dy, and dz, parallel, respectively, to the x, y, and z axes of a rectangular-coordinate system. At time t the conditions at a point located at the *center* of the element are fluid density ρ and velocity $\mathbf{V}$ with components V_x in the x direction, V_y in the y direction, and V_z in the z direction. Now consider the face 1234 of Fig. 6-6. The rate of input of matter across this face and into the element will be the product of the area $dz\,dy$ and the mean value of the product of the density and velocity in the x direction, ρV_x, across the surface. This mean value can be taken as the average of the values at the four corners. The corner values are:

At 1: $$\rho V_x - \frac{\partial \rho V_x}{\partial x}\frac{dx}{2} - \frac{\partial \rho V_x}{\partial y}\frac{dy}{2} - \frac{\partial \rho V_x}{\partial z}\frac{dz}{2}$$

At 2: $$\rho V_x - \frac{\partial \rho V_x}{\partial x}\frac{dx}{2} - \frac{\partial \rho V_x}{\partial y}\frac{dy}{2} + \frac{\partial \rho V_x}{\partial y}\frac{dz}{2}$$

At 3: $$\rho V_x - \frac{\partial \rho V_x}{\partial x}\frac{dx}{2} + \frac{\partial \rho V_x}{\partial y}\frac{dy}{2} + \frac{\partial \rho V_x}{\partial z}\frac{dz}{2}$$

At 4: $$\rho V_x - \frac{\partial \rho V_x}{\partial x}\frac{dx}{2} + \frac{\partial \rho V_x}{\partial y}\frac{dy}{2} - \frac{\partial \rho V_x}{\partial z}\frac{dz}{2}$$

The mean value across the face 1234 is then the average of the corner values and is

$$\rho V_x - \frac{\partial \rho V_x}{\partial x}\frac{dx}{2}$$

Note that this mean value is the same as the value of the mass flux at the *center* of the face 1234. This result will be found to hold in all cases. *Hence, the area average value of a continuous quantity may be evaluated at the center of the area, provided that the area is infinitesimal in extent.*

The rate of *input* of mass across face 1234 into the volume element is then

$$\left(\rho V_x - \frac{\partial \rho V_x}{\partial x}\frac{dx}{2}\right) dy\, dz$$

The rate of *input* of mass across face 1278 is

$$\left(\rho V_y - \frac{\partial \rho V_y}{\partial y}\frac{dy}{2}\right) dx\, dz$$

The rate of *input* of mass across face 1458 is

$$\left(\rho V_z - \frac{\partial \rho V_z}{\partial z}\frac{dz}{2}\right) dx\, dy$$

The rate of *output* of mass across face 5678 is

$$\left(\rho V_x + \frac{\partial \rho V_x}{\partial x}\frac{dx}{2}\right) dy\, dz$$

The rate of *output* of mass across face 3456 is

$$\left(\rho V_y + \frac{\partial \rho V_y}{\partial y}\frac{dy}{2}\right) dx\, dz$$

The rate of *output* of mass across face 2367 is

$$\left(\rho V_z + \frac{\partial \rho V_z}{\partial z}\frac{dz}{2}\right) dx\, dy$$

The rate of *accumulation* of mass within the volume element is

$$\frac{\partial}{\partial t}(\rho\, dx\, dy\, dz) = \frac{\partial \rho}{\partial t}\, dx\, dy\, dz$$

Since input rate − output rate = rate of accumulation, the application of the law of conservation of mass to an infinitesimal volume element within the fluid yields

$$\frac{\partial \rho V_x}{\partial x} + \frac{\partial \rho V_y}{\partial y} + \frac{\partial \rho V_z}{\partial z} = -\frac{\partial \rho}{\partial t} \tag{6-112}$$

Equation (6-112) is called the "continuity equation" and is used extensively in fluid mechanics. It may be written in alternative forms. From the Eulerian point of view, the density is a function of x, y, z, and t, and the total derivative of ρ is

$$d\rho = \frac{\partial \rho}{\partial x}\, dx + \frac{\partial \rho}{\partial y}\, dy + \frac{\partial \rho}{\partial z}\, dz + \frac{\partial \rho}{\partial t}\, dt \tag{6-113}$$

From the Lagrangian point of view, ρ is a function of elapsed time τ only, and the total derivative of ρ is

$$d\rho = \frac{d\rho}{d\tau}\, d\tau \tag{6-114}$$

Since $\qquad d\tau = dt$

and

$$\frac{dx}{d\tau} = V_x \qquad \frac{dy}{d\tau} = V_y \qquad \frac{dz}{d\tau} = V_z \tag{6-115}$$

combination of (6-113) and (6-114) yields

$$\frac{d\rho}{d\tau} = \frac{D\rho}{dt} = \frac{\partial \rho}{\partial t} + V_x \frac{\partial \rho}{\partial x} + V_y \frac{\partial \rho}{\partial y} + V_z \frac{\partial \rho}{\partial z} \tag{6-116}$$

The notation D/dt, frequently termed the "substantive derivative," is ordinarily used to replace the notation $d/d\tau$.

Expansion of (6-112) coupled with (6-116) yields an alternative form of the continuity equation

$$\frac{1}{\rho}\frac{D\rho}{dt} + \frac{\partial V_x}{\partial x} + \frac{\partial V_y}{\partial y} + \frac{\partial V_z}{\partial z} = 0 \tag{6-117}$$

The vector notation

$$\text{div } \mathbf{V} \equiv \frac{\partial V_x}{\partial x} + \frac{\partial V_y}{\partial y} + \frac{\partial V_z}{\partial z} \tag{6-118}$$

where div **V** is read "the divergence of the velocity vector," is commonly employed. With this symbolism, (6-117) becomes

$$\frac{1}{\rho}\frac{D\rho}{dt} + \text{div } \mathbf{V} = 0 \tag{6-119}$$

For an *incompressible* fluid, ρ is a constant, and the continuity equation reduces to

$$\text{div } \mathbf{V} = \frac{\partial V_x}{\partial x} + \frac{\partial V_y}{\partial y} + \frac{\partial V_z}{\partial z} = 0 \tag{6-120}$$

Example 6-8. Heat Conduction, Rectangular Coordinates. Consider a solid through which heat is flowing by conduction only. Let the solid be isotropic; i.e., the solid thermal conductivity k, specific heat c, and the density ρ are functions of temperature T only. Examine a rectangular parallelepiped volume element similar to that shown in Fig. 6-6. The application of the first law of thermodynamics to this system gives

Rate of input of heat − rate of output of heat = rate of accumulation of heat

The rate at which heat flows across an infinitesimal area element δA in a direction n normal to the area element is given by the Fourier relation

$$q_n = -k \frac{\partial T}{\partial n} \delta A$$

Let the conditions at time t at the center of the volume element be thermal conductivity k, specific heat c, density ρ, temperature T, temperature gradient in the x direction $\partial T/\partial x$, in the y direction $\partial T/\partial y$, and in the z direction $\partial T/\partial z$. Then the rate at which heat flows in the direction of x across a plane of area $dy\,dz$ normal to the axis of x and passing through the mid-point of the volume element is

$$q_{xx} = -k \frac{\partial T}{\partial x} dy\,dz$$

The rate of heat input across plane area 1234 is then

$$q_{xx} - \frac{\partial q_{xx}}{\partial x}\frac{dx}{2} = -k\frac{\partial T}{\partial x} dy\,dz - \frac{\partial}{\partial x}\left(-k\frac{\partial T}{\partial x} dy\,dz\right)\frac{dx}{2}$$
$$= \left[-k\frac{\partial T}{\partial x} + \frac{\partial}{\partial x}\left(k\frac{\partial T}{\partial x}\right)\frac{dx}{2}\right] dy\,dz$$

Similar expressions obtain for the rate at which heat flows across the other infinitesimal plane areas forming the boundaries of the volume element. The difference between the rate of heat input and the rate of heat output is found to be

$$\left[\frac{\partial}{\partial x}\left(k\frac{\partial T}{\partial x}\right) + \frac{\partial}{\partial y}\left(k\frac{\partial T}{\partial y}\right) + \frac{\partial}{\partial z}\left(k\frac{\partial T}{\partial z}\right)\right] dx\,dy\,dz$$

The rate of accumulation is

$$\frac{\partial}{\partial t}\left[\left(\int_{T_R}^{T} c\rho\,dT\right) dx\,dy\,dz\right] = c\rho\,dx\,dy\,dz\,\frac{\partial T}{\partial t}$$

The energy balance on the infinitesimal element then is

$$\frac{\partial}{\partial x}\left(k\frac{\partial T}{\partial x}\right) + \frac{\partial}{\partial y}\left(k\frac{\partial T}{\partial y}\right) + \frac{\partial}{\partial z}\left(k\frac{\partial T}{\partial z}\right) = c\rho\,\frac{\partial T}{\partial t} \tag{6-121}$$

If the thermal properties of the solid are invariant, (6-121) becomes

$$\nabla^2 T = \frac{\partial^2 T}{\partial x^2} + \frac{\partial^2 T}{\partial y^2} + \frac{\partial^2 T}{\partial z^2} = \frac{1}{\alpha}\frac{\partial T}{\partial t} \tag{6-122}$$

where $\alpha = k/c\rho$ is termed the "thermal diffusivity," and ∇^2, termed the "Laplacian," denotes the operation

$$\nabla^2 = \frac{\partial^2}{\partial x^2} + \frac{\partial^2}{\partial y^2} + \frac{\partial^2}{\partial z^2}$$

Example 6-9. Heat Conduction, Spherical Coordinates. The heat-conduction equation in rectangular coordinates (6-121) may be transformed into spherical-coordinate form by formal differentiation techniques. This procedure will not be used; instead, an energy balance will be written for a differential element of volume in spherical coordinates.

A spherical-coordinate system is shown in Fig. 6-7. The three coordinates are r, θ, ϕ, where r is the distance from the origin, θ is the polar angle measured from the xz plane, and ϕ is the "cone angle" measured from the z axis. The relation between a spherical- and a rectangular-coordinate system is

$$\begin{aligned} x &= r\cos\theta\sin\phi \\ y &= r\sin\theta\sin\phi \\ z &= r\cos\phi \end{aligned} \tag{6-123}$$

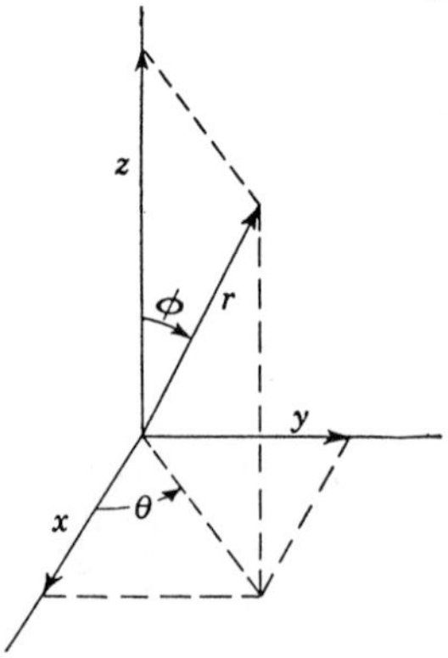

FIG. 6-7. Spherical-coordinate system.

Figure 6-8 shows an infinitesimal volume element in spherical coordinates. A spherical-coordinate system is orthogonal because lines of constant r and θ intersect lines of constant r and ϕ and lines of constant θ and ϕ at right angles. In this sense, a rectangular-coordinate system is orthogonal and a cylindrical-coordinate system is orthogonal.

Infinitesimal changes in the coordinate result in the distance changes shown in Table 6-1. The area elements normal to the three coordinate directions are given in Table 6-2. The volume of an infinitesimal element in spherical coordinates is

$$dv = r^2 \sin\phi\,d\phi\,d\theta\,dr \tag{6-124}$$

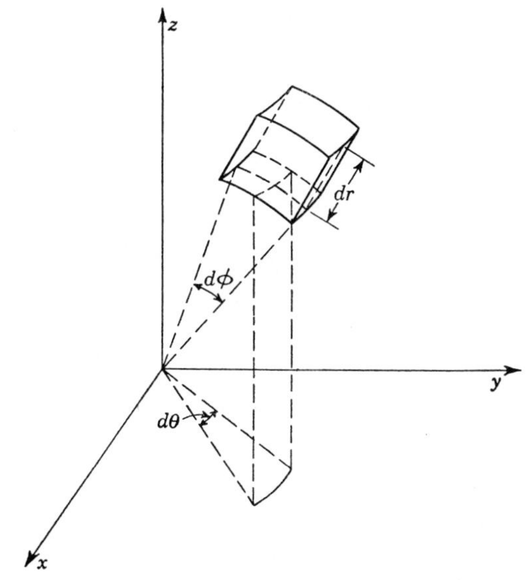

FIG. 6-8. Volume element in spherical coordinates.

TABLE 6-1. SPHERICAL-COORDINATE SYSTEM

Coordinate change	Coordinates held constant	Change in arc length
dr	θ,ϕ	dr
$d\theta$	r,ϕ	$r \sin \phi \, d\theta$
$d\phi$	r,θ	$r \, d\phi$

TABLE 6-2. SPHERICAL-COORDINATE SYSTEM

Symbol	Normal to coordinate	Area element
dA_r	r	$r^2 \sin \phi \, d\theta \, d\phi$
dA_θ	θ	$r \, d\phi \, dr$
dA_ϕ	ϕ	$r \sin \phi \, d\theta \, dr$

The reader may verify all the above relations from elementary geometrical considerations.

Now consider a volume element in spherical coordinates (Fig. 6-8). Let the coordinates of the center of the element be r, θ, ϕ. The rate at which heat flows by conduction through each of the three mutually perpendicular area elements intersecting at r, θ, ϕ is

$$\begin{aligned} q_r &= -k \frac{\partial T}{\partial r} dA_r = -k \frac{\partial T}{\partial r} r^2 \sin \phi \, d\theta \, d\phi \\ q_\theta &= -k \frac{1}{r \sin \phi} \frac{\partial T}{\partial \theta} dA_\theta = -k \frac{\partial T}{\partial \theta} \frac{d\phi \, dr}{\sin \phi} \\ q_\phi &= -k \frac{1}{r} \frac{\partial T}{\partial \phi} dA_\phi = -k \frac{\partial T}{\partial \phi} \sin \phi \, d\theta \, dr \end{aligned} \tag{6-125}$$

Note that in (6-125) the derivative of T is taken with respect to arc length in all cases. This is because the physical law invoked (Fourier's equation) states that the rate of heat flow in the n direction is proportional to the rate of change of T with *distance* in the n direction.

The rate at which heat crosses the surfaces forming the boundaries of the volume element is:

Rate of heat input

$$q_r - \frac{\partial q_r}{\partial r} \frac{dr}{2}$$

$$q_\theta - \frac{\partial q_\theta}{r \sin \phi \, \partial \theta} \frac{r \sin \phi \, d\theta}{2} = q_\theta - \frac{\partial q_\theta}{\partial \theta} \frac{d\theta}{2}$$

$$q_\phi - \frac{\partial q_\phi}{r \, \partial \phi} \frac{r \, d\phi}{2} = q_\phi - \frac{\partial q_\phi}{\partial \phi} \frac{d\phi}{2}$$

Rate of heat output

$$q_r + \frac{\partial q_r}{\partial r} \frac{dr}{2}$$

$$q_\theta + \frac{\partial q_\theta}{\partial \theta} \frac{d\theta}{2}$$

$$q_\phi + \frac{\partial q_\phi}{\partial \phi} \frac{d\phi}{2}$$

Rate of energy accumulation

$$\frac{\partial}{\partial t} \int_{T_R}^{T} \rho c d\vartheta \, dT = c\rho r^2 \sin \phi \, d\phi \, d\theta \, dr \frac{\partial T}{\partial t}$$

Then, since rate of input − rate of output = rate of accumulation, there results, following use of (6-125) and some rearrangement,

$$\frac{1}{r^2} \frac{\partial}{\partial r} \left(k r^2 \frac{\partial T}{\partial r} \right) + \frac{1}{r^2 \sin^2 \phi} \frac{\partial}{\partial \theta} \left(k \frac{\partial T}{\partial \theta} \right) + \frac{1}{r^2 \sin \phi} \frac{\partial}{\partial \phi} \left(k \sin \phi \frac{\partial T}{\partial \phi} \right) = \rho c \frac{\partial T}{\partial t} \tag{6-126}$$

Equation (6-126) is the spherical-coordinate-system equivalent of (6-121).

6-11. Vectors. Quantities which require specification of magnitude and direction for complete characterization are called "vectors." Quantities characterized by magnitude alone are termed "scalars." Velocity is a vector quantity; speed is a scalar quantity. Operations involving vectors can always be handled by first decomposing the vector quantity into its components along the coordinate axes. This is not the most efficient procedure, however, and a technique termed "vector analysis" for handling the undecomposed vector itself has been developed. In the following the introductory concepts of vector analysis are presented. Boldface roman type is used to denote a vector.

Vector Addition and Subtraction. Addition of the vector **B** to the vector **A** to form the vector **C**

$$\mathbf{A} + \mathbf{B} = \mathbf{C} \tag{6-127}$$

is accomplished by placing the initial point of **B** with the terminal point of **A**. The vector **C** which joins the initial point of **A** with the terminal point of **B** is the vector sum of **A** and **B**. (See Fig. 6-9.)

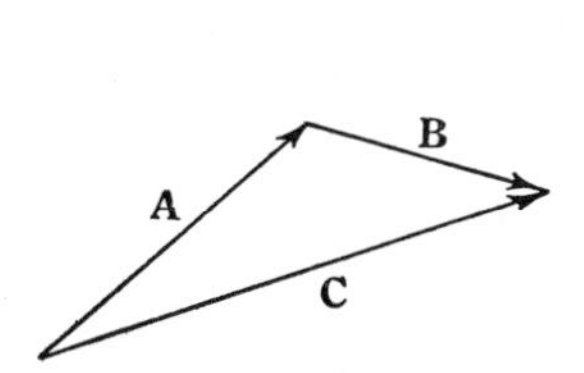

FIG. 6-9. Addition of vectors.

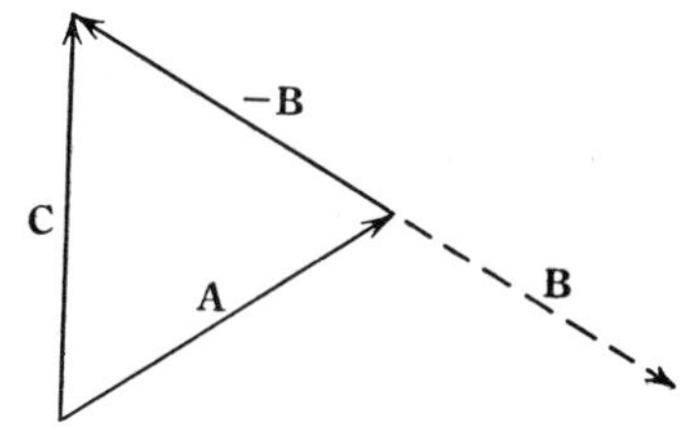

FIG. 6-10. Subtraction of vectors.

Subtraction of **B** from **A** is defined as the "addition of the negative vector" −**B** to **A**. (See Fig. 6-10.)

Unit Vectors. If m is a scalar, then $m\mathbf{A}$ is defined as the vector **B** whose magnitude is $B = |m|A$ and whose direction is that of **A** if m is positive and opposite to that of **A** if m is negative.

A vector whose magnitude is unity is called a "unit vector." A unit vector **a** that is directed along a vector **A** can be written as

$$\mathbf{a} = \frac{\mathbf{A}}{A}$$

where A is the absolute magnitude of **A**.

Consider a right-handed system of rectangular-coordinate axes [i.e., a right-handed screw directed along the positive z axis will advance in the positive direction when it is rotated from the positive x axis toward the positive y axis through the smaller (90°) angle, as shown in Fig. 6-11].

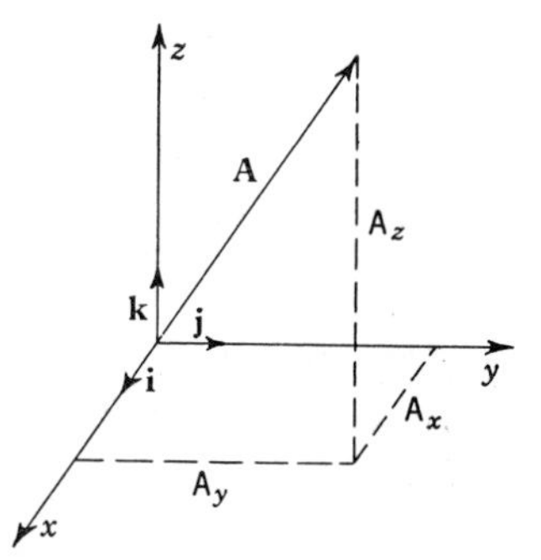

FIG. 6-11. Representation of a vector in a right-handed coordinate system.

Let **i**, **j**, **k** denote unit vectors directed along the positive directions of the x, y, and z axes, respectively. Then any vector **A** can be represented uniquely in the form

$$\mathbf{A} = A_x\mathbf{i} + A_y\mathbf{j} + A_z\mathbf{k} \tag{6-128}$$

where A_x, A_y, and A_z are the coordinates of the terminal point of **A**. The magnitude of **A** is

$$A = \sqrt{A_x^2 + A_y^2 + A_z^2} \tag{6-129}$$

Let the angles made by **A** and the positive x, y, and z axes, respectively, be (A,x), (A,y), and (A,z). Then

$$\begin{aligned} A_x &= A \cos (A,x) \\ A_y &= A \cos (A,y) \\ A_z &= A \cos (A,z) \end{aligned} \tag{6-130}$$

and

$$\cos^2 (A,x) + \cos^2 (A,y) + \cos^2 (A,z) = 1 \tag{6-131}$$

6-12. Scalar, or Dot, Product. The scalar, or dot, product of two vectors **A** and **B** is the scalar quantity representing the product of the length of one of the vectors and the scalar projection of the other vector upon the first. The scalar, or dot-product, operation is symbolized by

$$\mathbf{A} \cdot \mathbf{B} = AB \cos (A,B) = \mathbf{B} \cdot \mathbf{A} \tag{6-132}$$

where (A,B) is the angle between the two vectors. Since $\mathbf{A} \cdot \mathbf{B} = \mathbf{B} \cdot \mathbf{A}$, scalar multiplication is *commutative*. It is easily shown that

$$\mathbf{A} \cdot (\mathbf{B} + \mathbf{C}) = \mathbf{A} \cdot \mathbf{B} + \mathbf{A} \cdot \mathbf{C} \tag{6-133}$$

i.e., that the operation is also *distributive*. If **A** and **B** are perpendicular, $\cos (A,B)$ is zero, and conversely, if the scalar product is zero, one of the vectors is zero, or else the vectors are perpendicular. Since the scalar product may be zero when neither factor is zero, *division by a vector is not possible.* For example, if

$$\mathbf{A} \cdot \mathbf{B} = \mathbf{A} \cdot \mathbf{C}$$

it does *not* follow that $\mathbf{B} = \mathbf{C}$ but merely that

$$\mathbf{A} \cdot (\mathbf{B} - \mathbf{C}) = 0$$

and $\mathbf{B} - \mathbf{C}$ is either zero or perpendicular to **A**.

In view of (6-132), the scalar products formed by the unit vectors **i**, **j**, **k** are

$$\begin{aligned} \mathbf{i} \cdot \mathbf{i} = \mathbf{j} \cdot \mathbf{j} = \mathbf{k} \cdot \mathbf{k} = 1 \\ \mathbf{i} \cdot \mathbf{j} = \mathbf{i} \cdot \mathbf{k} = \mathbf{j} \cdot \mathbf{k} = 0 \end{aligned} \tag{6-134}$$

Since

$$\mathbf{A} \cdot \mathbf{B} = (A_x\mathbf{i} + A_y\mathbf{j} + A_z\mathbf{k}) \cdot (B_x\mathbf{i} + B_y\mathbf{j} + B_z\mathbf{k})$$

use of (6-133) and (6-134) gives

$$\mathbf{A} \cdot \mathbf{B} = A_xB_x + A_yB_y + A_zB_z \tag{6-135}$$

6-13. Vector, or Cross, Product. The vector, or cross, product of **A** and **B** is a vector **C** which is normal to the plane formed by **A** and **B** and directed so that the vectors **A**, **B**, and **C** form a right-handed system (Fig. 6-12). The notation for vector, or cross-product, multiplication is

$$\mathbf{A} \times \mathbf{B} = \mathbf{C} \tag{6-136a}$$

and the magnitude of **C** is defined as

$$C = AB|\sin(A,B)| \tag{6-136b}$$

i.e., the product of magnitude of **A**, the magnitude of **B**, and the absolute value of the sine of the angle between **A** and **B**. The magnitude of **C** is equal to the area of the parallelogram constructed with **A** and **B** as sides.

Rotation from **B** to **A** is opposite to rotation from **A** to **B**. Consequently

$$\mathbf{A} \times \mathbf{B} = -\mathbf{B} \times \mathbf{A} \tag{6-137}$$

and *the commutative law does not hold for vector products. The order of vectors in a vector cross product must not be changed unless the necessary sign changes are also made.*

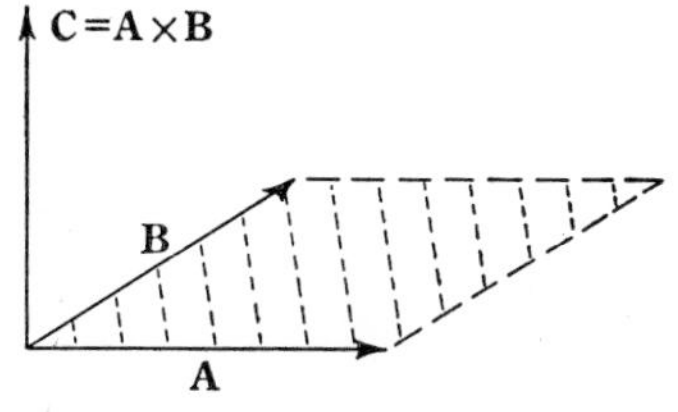

FIG. 6-12. Vector, or cross, product.

By contrast, the *distributive law applies unrestrictedly to vector products:*

$$(\mathbf{A} + \mathbf{B}) \times \mathbf{C} = \mathbf{A} \times \mathbf{C} + \mathbf{B} \times \mathbf{C} \tag{6-138}$$

There follows from the definition of unit vectors and the definition of the cross product

$$\begin{aligned} \mathbf{i} \times \mathbf{i} &= \mathbf{j} \times \mathbf{j} = \mathbf{k} \times \mathbf{k} = 0 \\ \mathbf{i} \times \mathbf{j} &= -\mathbf{j} \times \mathbf{i} = \mathbf{k} \\ \mathbf{j} \times \mathbf{k} &= -\mathbf{k} \times \mathbf{j} = \mathbf{i} \\ \mathbf{k} \times \mathbf{i} &= -\mathbf{i} \times \mathbf{k} = \mathbf{j} \end{aligned} \tag{6-139}$$

Expansion of **A** and **B** in terms of their cartesian components followed by use of (6-138) and (6-139) gives

$$\mathbf{A} \times \mathbf{B} = \begin{vmatrix} \mathbf{i} & \mathbf{j} & \mathbf{k} \\ A_x & A_y & A_z \\ B_x & B_y & B_z \end{vmatrix} \tag{6-140}$$

6-14. Multiple Products. The *triple scalar* product

$$\mathbf{A} \cdot (\mathbf{B} \times \mathbf{C}) = \mathbf{B} \cdot (\mathbf{C} \times \mathbf{A}) = \mathbf{C} \cdot (\mathbf{A} \times \mathbf{B}) \tag{6-141}$$

is numerically equal to the volume of the parallelepiped formed with **A**, **B**, and **C** as coterminous edges. The sign of the product depends upon the relative orientation of the three vectors and is positive if and only if **A**, **B**, and **C** form a right-handed system in the sense that **A** × **B** and **C** lie on the same side of the plane formed by **A** and **B**. Equation (6-141) discloses that the triple scalar product is not changed by a cyclic permutation of **A**, **B**, and **C** and that the dot and cross can be interchanged. Consequently, the notation (**ABC**) is often used to represent (6-141). Straightforward manipulation gives

$$(\mathbf{ABC}) = \mathbf{A} \cdot (\mathbf{B} \times \mathbf{C}) = \begin{vmatrix} A_x & A_y & A_z \\ B_x & B_y & B_z \\ C_x & C_y & C_z \end{vmatrix} \tag{6-142}$$

It also follows that the product vanishes if two of the vectors are parallel.

The product $(\mathbf{A} \times \mathbf{B}) \times \mathbf{C}$ is a vector which is in the plane of **A** and **B** and is perpendicular to **C**. Relatively involved manipulation gives

$$(\vec{\mathbf{A}} \times \vec{\mathbf{B}}) \times \vec{\mathbf{C}} = (\vec{\mathbf{A}} \cdot \vec{\mathbf{C}})\vec{\mathbf{B}} - (\vec{\mathbf{B}} \cdot \vec{\mathbf{C}})\vec{\mathbf{A}} \tag{6-143}$$

6-15. Differentiation of Vectors. A vector $\mathbf{A} = A_x\mathbf{i} + A_y\mathbf{j} + A_z\mathbf{k}$ may be differentiated. If **A** is solely a function of a parameter t, then

$$\frac{d\mathbf{A}}{dt} = \frac{dA_x}{dt}\mathbf{i} + \frac{dA_y}{dt}\mathbf{j} + \frac{dA_z}{dt}\mathbf{k} \tag{6-144}$$

and $d\mathbf{A}/dt$ is also a vector. The partial derivative is also defined:

$$\frac{\partial \mathbf{A}}{\partial x} = \frac{\partial A_x}{\partial x}\mathbf{i} + \frac{\partial A_y}{\partial x}\mathbf{j} + \frac{\partial A_z}{\partial x}\mathbf{k} \tag{6-145}$$

If a denotes a scalar quantity, then

$$\frac{d(a\mathbf{A})}{dt} = a\frac{d\mathbf{A}}{dt} + \mathbf{A}\frac{da}{dt} \tag{6-146}$$

The derivatives of vector products correspond to the scalar rules, *provided that the order of factors is retained:*

$$\frac{d}{dt}(\mathbf{A} \cdot \mathbf{B} \times \mathbf{C}) = \frac{d\mathbf{A}}{dt} \cdot \mathbf{B} \times \mathbf{C} + \mathbf{A} \cdot \frac{d\mathbf{B}}{dt} \times \mathbf{C} + \mathbf{A} \cdot \mathbf{B} \times \frac{d\mathbf{C}}{dt} \tag{6-147}$$

The derivative of a vector of constant length but changing directions is perpendicular to the vector. This conclusion results from the following: Consider the equivalent expressions

$$\frac{d}{dt}(\mathbf{A} \cdot \mathbf{A}) = 2\mathbf{A} \cdot \frac{d\mathbf{A}}{dt} = \frac{d}{dt}(A^2) = 0$$

where A denotes the magnitude of **A**. These expressions are compatible only if $d\mathbf{A}/dt = 0$ or if $d\mathbf{A}/dt$ is perpendicular to **A**.

6-16. The Position Vector R. Let the vector **R** denote the position vector from a fixed origin of coordinates to a point $P(x,y,z)$ in three-dimensional space (see Fig. 6-13). Then

$$\mathbf{R} = x\mathbf{i} + y\mathbf{j} + z\mathbf{k} \tag{6-148}$$

Suppose that the point $P(x,y,z)$ lies on a space curve $x = x(t)$, $y = y(t)$, $z = z(t)$, and t is given the increment Δt. Then

$$\Delta\mathbf{R} = \Delta x\mathbf{i} + \Delta y\mathbf{j} + \Delta z\mathbf{k} \tag{6-149}$$

When Δt is the infinitesimal dt,

$$d\mathbf{R} = dx\mathbf{i} + dy\mathbf{j} + dz\mathbf{k} \tag{6-150}$$

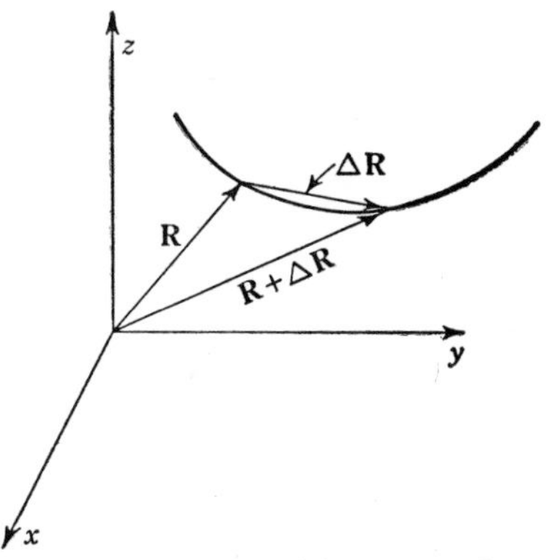

FIG. 6-13. The position vector **R**.

Let dl denote arc length along the space curve. Then

$$dl = \sqrt{(dx)^2 + (dy)^2 + (dz)^2} \tag{6-151}$$

and

$$\frac{d\mathbf{R}}{dl} = \frac{dx}{dl}\mathbf{i} + \frac{dy}{dl}\mathbf{j} + \frac{dz}{dl}\mathbf{k} \tag{6-152}$$

Since

$$\left|\frac{d\mathbf{R}}{dl}\right| = \sqrt{\left(\frac{dx}{dl}\right)^2 + \left(\frac{dy}{dl}\right)^2 + \left(\frac{dz}{dl}\right)^2} = 1 \tag{6-153}$$

$d\mathbf{R}/dl$ is a unit vector and is tangent to the space curve. Consequently, the *derivative of the position vector* **R** *to a space curve, with respect to arc length l along the curve, is a unit vector tangent to the space curve and points in the direction of increasing arc length.* Since $d\mathbf{R}/dl$ is a unit vector,

$$\frac{d}{dl}\left(\frac{d\mathbf{R}}{dl}\right) = \frac{d^2\mathbf{R}}{dl^2}$$

is perpendicular to $d\mathbf{R}/dl$ and hence normal to the space curve. The unit vector normal to a space curve is then

$$\mathbf{n} = \frac{d^2\mathbf{R}/dl^2}{|d^2\mathbf{R}/dl^2|} \tag{6-154}$$

where $1/|d^2\mathbf{R}/dl^2|$ is the radius of curvature of the space curve.

6-17. Scalar and Vector Fields. If a function is defined at every point in a region in space, that region is called a field. If the function represents a scalar quantity, the field is termed a "scalar field"; if the function represents a vector quantity, the field is termed a "vector field." Each scalar field can be related to an associated vector field, a circumstance of considerable utility.

In Sec. 6-6 the directional derivative du/ds was defined and described (see also Fig. 6-2). In terms of vector notation, du/ds is the scalar rate of change of the scalar quantity u with distance s measured in the direction of the vector $\mathbf{s}$. In three-dimensional space, the extension of Eq. (6-83) is

$$\frac{du}{ds} = \frac{\partial u}{\partial x}\frac{dx}{ds} + \frac{\partial u}{\partial y}\frac{dy}{ds} + \frac{\partial u}{\partial z}\frac{dz}{ds} = \frac{\partial u}{\partial x}\cos(s,x) + \frac{\partial u}{\partial y}\cos(s,y) + \frac{\partial u}{\partial z}\cos(s,z) \qquad (6\text{-}155)$$

In the above, (s,x) denotes the angle made by a vector having the direction of $\mathbf{s}$ with the positive x axis.

Now a vector $\mathbf{A} = A_x\mathbf{i} + A_y\mathbf{j} + A_z\mathbf{k}$ has a component in the direction of $\mathbf{s}$ which is equal to the scalar product of $\mathbf{A}$ and the unit vector $\mathbf{s}/s$:

$$\frac{\mathbf{A}\cdot\mathbf{s}}{s} = A\cos(A,s) = A_x\cos(s,x) + A_y\cos(s,y) + A_z\cos(s,z) \qquad (6\text{-}156)$$

In view of the similarity of Eqs. (6-155) and (6-156), it is evident that the directional derivative can be considered to be the scalar component in the direction of $\mathbf{s}$ of the vector

$$\nabla u = \frac{\partial u}{\partial x}\mathbf{i} + \frac{\partial u}{\partial y}\mathbf{j} + \frac{\partial u}{\partial z}\mathbf{k} \qquad (6\text{-}157)$$

that is,

$$\frac{du}{ds} = \frac{\nabla u\cdot\mathbf{s}}{s} = |\nabla u|\cos(\nabla u,s) \qquad (6\text{-}158)$$

Now $(\nabla u\cdot\mathbf{s}/s)$ is a maximum when ∇u and $\mathbf{s}$ are parallel, and it was shown in Sec. 6-6 that the maximum value of du/ds occurs when the direction of s is normal to the surface $u =$ const, i.e., in the direction of the surface normal n. Consequently, in view of (6-89) and (6-158),

$$\frac{du}{dn} = |\nabla u| = \sqrt{\left(\frac{\partial u}{\partial x}\right)^2 + \left(\frac{\partial u}{\partial y}\right)^2 + \left(\frac{\partial u}{\partial z}\right)^2} \qquad (6\text{-}159)$$

It is now seen that *the vector* ∇u *formed from the scalar function* u *has the direction of the normal to the surface* $u =$ *const and has the magnitude of* du/dn. The vector ∇u is called the "gradient" of u, and the symbol ∇u is read "nabla u" or "del u." ∇u is also written "grad u."

The symbol

$$\nabla = \mathbf{i}\frac{\partial}{\partial x} + \mathbf{j}\frac{\partial}{\partial y} + \mathbf{k}\frac{\partial}{\partial z} \qquad (6\text{-}160)$$

denotes the vector differential operator "del" or "nabla" but does not itself represent a vector quantity. The symbol ∇ indicates that the operations expressed by (6-160) are to be performed on the *scalar* function which follows. The operation ∇u generates a vector field which is associated with the scalar field of u.

It is evident that when the direction **s** denotes the direction of the tangent to a space curve,

$$\frac{d\mathbf{R}}{dl} = \frac{\mathbf{s}}{s}$$

and

$$\frac{du}{ds} = \nabla u \cdot \frac{d\mathbf{R}}{dl} = \frac{du}{dl} \tag{6-161}$$

$$du = \nabla u \cdot d\mathbf{R} \tag{6-162}$$

As an example of the utility of the operation associated with ∇, consider the problem of three-dimensional heat flow by conduction. The solid temperature T forms a scalar field, the vector $-\text{grad } T \equiv -\nabla T$ has the direction of the resultant heat flow, and $|\nabla T|$ is directly proportional to the heat-flow rate.

6-18. Vector Differential Operator. The vector differential operator ∇ is also used to operate on vectors. Several operations are *defined*. Consider the vector $\mathbf{A} = A_x\mathbf{i} + A_y\mathbf{j} + A_z\mathbf{k}$. Since

$$\nabla \equiv \mathbf{i}\frac{\partial}{\partial x} + \mathbf{j}\frac{\partial}{\partial y} + \mathbf{k}\frac{\partial}{\partial z} \tag{6-160}$$

it is natural to define the operation $\nabla \cdot \mathbf{A}$ as

$$\nabla \cdot \mathbf{A} = \mathbf{i} \cdot \frac{\partial \mathbf{A}}{\partial x} + \mathbf{j} \cdot \frac{\partial \mathbf{A}}{\partial y} + \mathbf{k} \cdot \frac{\partial \mathbf{A}}{\partial z} \tag{6-163a}$$

Since $\mathbf{i} \cdot \mathbf{i} = 1$ and $\mathbf{i} \cdot \mathbf{j} = 0$, etc., expansion of (6-163*a*) gives

$$\nabla \cdot \mathbf{A} = \frac{\partial A_x}{\partial x} + \frac{\partial A_y}{\partial y} + \frac{\partial A_z}{\partial z} = \text{div } \mathbf{A} \tag{6-163b}$$

$\nabla \cdot \mathbf{A}$ is called the "divergence of **A**" and is a scalar. The physical significance of the divergence will be discussed shortly.

The symbolism $\mathbf{A} \cdot \nabla$ defines a *scalar operator* and *is not* equal to $\nabla \cdot \mathbf{A}$. Formal expansion gives

$$\mathbf{A} \cdot \nabla = \mathbf{A} \cdot \left(\mathbf{i}\frac{\partial}{\partial x} + \mathbf{j}\frac{\partial}{\partial y} + \mathbf{k}\frac{\partial}{\partial z}\right) = (\mathbf{i}A_x + \mathbf{j}A_y + \mathbf{k}A_z) \cdot \left(\mathbf{i}\frac{\partial}{\partial x} + \mathbf{j}\frac{\partial}{\partial y} + \mathbf{k}\frac{\partial}{\partial z}\right) = A_x\frac{\partial}{\partial x} + A_y\frac{\partial}{\partial y} + A_z\frac{\partial}{\partial z} \tag{6-164}$$

Then

$$(\mathbf{A} \cdot \nabla)\,\mathbf{B} \equiv \mathbf{A} \cdot \nabla\mathbf{B} = A_x\frac{\partial \mathbf{B}}{\partial x} + A_y\frac{\partial \mathbf{B}}{\partial y} + A_z\frac{\partial \mathbf{B}}{\partial z} \tag{6-165}$$

where the parentheses surrounding the operator $\mathbf{A} \cdot \nabla$ are conventionally omitted since $\nabla\mathbf{B}$ is undefined and $\mathbf{A} \cdot \nabla\mathbf{B}$ can only mean $\mathbf{A} \cdot \nabla$ operating on $\mathbf{B}$.

Since $d\mathbf{R} = \mathbf{i}\,dx + \mathbf{j}\,dy + \mathbf{k}\,dz$,

$$d\mathbf{B} = d\mathbf{R} \cdot \nabla\mathbf{B} \tag{6-166a}$$

and the derivative of $\mathbf{B}$ in the direction of $\overrightarrow{d\mathbf{R}}$ is

$$\frac{d\mathbf{B}}{dl} = \frac{d\mathbf{R}}{dl} \cdot \nabla\mathbf{B} \tag{6-166b}$$

Quite generally, if $\mathbf{s}/s$ is any unit vector, the derivative of $\mathbf{B}$ in the direction of $\mathbf{s}$ is

$$\frac{d\mathbf{B}}{ds} = \frac{d\mathbf{s}}{ds} \cdot \nabla\mathbf{B} \tag{6-167}$$

The scalar operator known as the "Laplacian operator" is defined as

$$\nabla^2 \equiv \nabla \cdot \nabla = \frac{\partial^2}{\partial x^2} + \frac{\partial^2}{\partial y^2} + \frac{\partial^2}{\partial z^2} \tag{6-168}$$

The cross-product operation $\nabla \times \mathbf{A}$ is called the "curl" of $\mathbf{A}$ and is defined as

$$\nabla \times \vec{\mathbf{A}} = \text{curl } \vec{\mathbf{A}} = \mathbf{i} \times \frac{\partial \mathbf{A}}{\partial x} + \mathbf{j} \times \frac{\partial \mathbf{A}}{\partial y} + \vec{\mathbf{k}} \times \frac{\partial \vec{\mathbf{A}}}{\partial z}$$

$$= \begin{vmatrix} \mathbf{i} & \mathbf{j} & \mathbf{k} \\ \frac{\partial}{\partial x} & \frac{\partial}{\partial y} & \frac{\partial}{\partial z} \\ A_x & A_y & A_z \end{vmatrix}$$

$$= \mathbf{i}\left(\frac{\partial A_z}{\partial y} - \frac{\partial A_y}{\partial z}\right) + \mathbf{j}\left(\frac{\partial A_x}{\partial z} - \frac{\partial A_z}{\partial x}\right) + \mathbf{k}\left(\frac{\partial A_y}{\partial x} - \frac{\partial A_x}{\partial y}\right) \tag{6-169}$$

The physical significance of curl $\mathbf{A}$ will be discussed shortly.

If $\mathbf{A}$ and $\mathbf{B}$ are arbitrary vectors and ϕ represents an arbitrary scalar, the following relations may be verified by direct expansion, *provided* that the indicated derivatives exist and that in relations (6-170*f*) and (6-170*g*) the second-order partial derivatives of $\mathbf{A}$ and ϕ are continuous. This latter condition requires that $\partial^2\phi/(\partial x\,\partial y) = \partial^2\phi/(\partial y\,\partial x)$, that is, that the order of differentiation be immaterial.

$$\nabla \cdot \phi\mathbf{A} = \phi\nabla \cdot \mathbf{A} + \mathbf{A} \cdot \nabla\phi \tag{6-170a}$$

$$\nabla \times \phi\mathbf{A} = \phi\nabla \times \mathbf{A} + \nabla\phi \times \mathbf{A} \tag{6-170b}$$

$$\nabla \cdot \mathbf{A} \times \mathbf{B} = \mathbf{B} \cdot \nabla \times \mathbf{A} - \mathbf{A} \cdot \nabla \times \mathbf{B} \tag{6-170c}$$

$$\nabla \times (\mathbf{A} \times \mathbf{B}) = \mathbf{B} \cdot \nabla\mathbf{A} - \mathbf{A} \cdot \nabla\mathbf{B} + \mathbf{A}(\nabla \cdot \mathbf{B}) - \mathbf{B}(\nabla \cdot \mathbf{A}) \tag{6-170d}$$

$$\nabla(\mathbf{A} \cdot \mathbf{B}) = \mathbf{A} \cdot \nabla\mathbf{B} + \mathbf{B} \cdot \nabla\mathbf{A} + \mathbf{A} \times (\nabla \times \mathbf{B}) + \mathbf{B} \times (\nabla \times \mathbf{A}) \tag{6-170e}$$

$$\nabla \times \nabla\phi = \text{curl grad } \phi = 0 \tag{6-170f}$$

$$\nabla \cdot (\nabla \times \mathbf{A}) = \text{div curl } \mathbf{A} = 0 \tag{6-170g}$$

$$\begin{aligned} \nabla \times (\nabla \times \mathbf{A}) &= \text{curl curl } \mathbf{A} = \nabla(\nabla \cdot \mathbf{A}) - \nabla \cdot \nabla\mathbf{A} \\ &= \text{grad div } \mathbf{A} - \nabla^2\mathbf{A} \end{aligned} \tag{6-170h}$$

$$\nabla \cdot (\nabla\phi_1 \times \nabla\phi_2) = 0 \tag{6-170i}$$

6-19. Line Integrals. Consider a space curve for which each point on the curve $P(x,y,z)$ is specified by the corresponding position vector $\mathbf{R}$ taken from the fixed origin of coordinates to P. Inasmuch as $d\mathbf{R}$ is tangent to the curve at P and has the length of the arc dl, $\mathbf{A} \cdot d\mathbf{R}$ is the product of the component of $\mathbf{A}$ in the direction of the space curve times the differential arc length dl. Then the integral

$$\int_{C\,P_0}^{P_1} \mathbf{A} \cdot d\mathbf{R} = \int_{C\,P_0}^{P_1} (A_x\,dx + A_y\,dy + A_z\,dz) \tag{6-171}$$

where P_0 and P_1 are points on the space curve C, is a *line integral.* In general, the integration of (6-171) demands specification of the space curve in the form $x = x(t)$, $y = y(t)$, $z = z(t)$; that is, the value of the integral depends upon the specific curve chosen as well as upon the limits P_0 and P_1. However, such information is not always required. If a single-valued scalar function ϕ exists for which

$$d\phi = A_x\,dx + A_y\,dy + A_z\,dz$$

then

$$\int_{C\,P_0}^{P_1} \mathbf{A} \cdot d\mathbf{R} = \int_{P_0}^{P_1} d\phi = \phi(P_1) - \phi(P_0)$$

and the value of the line integral depends only on P_0 and P_1 and *not* on the curve (path). A scalar function ϕ for which $d\phi = \mathbf{A} \cdot d\mathbf{R}$ is said to be the potential function of the vector $\mathbf{A}$. The *potential* functions are an important class and include the thermodynamic point functions. If the scalar function ϕ is to be the potential function of the vector $\mathbf{A}$, it is necessary that $\mathbf{A} \cdot d\mathbf{R}$ be a perfect differential. However, this requires that

$$\frac{\partial A_x}{\partial y} = \frac{\partial A_y}{\partial x} \qquad \frac{\partial A_x}{\partial z} = \frac{\partial A_z}{\partial x} \qquad \frac{\partial A_y}{\partial z} = \frac{\partial A_z}{\partial y}$$

which will be true if $\nabla \times \mathbf{A} = 0$ and if $\mathbf{A}$ is continuously differentiable.

When the above conditions are fulfilled, a scalar function ϕ exists for which

$$d\phi = \mathbf{A} \cdot d\mathbf{R}$$

However, in view of Eq. (6-162),

$$d\phi = \nabla\phi \cdot d\mathbf{R}$$

also. Subtraction yields

$$(\mathbf{A} - \nabla\phi) \cdot d\mathbf{R} = 0$$

which is true only if one of two conditions is fulfilled: $\mathbf{A} - \nabla\phi$ is either normal to $d\mathbf{R}$, or $\mathbf{A} = \nabla\phi$. Since $d\mathbf{R}$ can assume any direction, $\mathbf{A} - \nabla\phi$ cannot always be normal to $d\mathbf{R}$ and hence $\mathbf{A} = \nabla\phi$ is required.

Consequently, *when the vector* **A** *is both continuously differentiable and* $\nabla \times \mathbf{A} = 0$ *in a region, a scalar function exists for which*

$$\mathbf{A} = \nabla\phi \tag{6-172}$$

and the value of the line integral

$$\int_{C\ P_1}^{P_2} \mathbf{A} \cdot d\mathbf{R} = \int_{C\ P_1}^{P_2} \nabla\phi \cdot d\mathbf{R} = \int_{P_1}^{P_2} d\phi$$

will be independent of the path, provided that ϕ is single-valued; i.e., the value approached by ϕ at a point P must be independent of the manner of approach.

The line integral around a closed path

$$\oint \mathbf{A} \cdot d\mathbf{R} \tag{6-173}$$

is called the *circulation* of the vector **A**. The sign of the circulation is set by the convention that the direction of integration is such that motion along the curve keeps the enclosed area on the left. In general, the circulation of **A** is not zero, but when $\mathbf{A} = \nabla\phi$ and ϕ is single-valued the integral vanishes.

6-20. Surface Integrals. Consider an infinitesimal element of area dS located on a surface S in three-dimensional space. This infinitesimal area can be considered to define a direction, namely, that of the unit vector normal to dS, **n**. Consequently, it is possible to define a *differential surface-area vector* $d\mathbf{S}$ as the vector of magnitude dS and of direction **n**. Clearly then, the *surface integral*

$$\iint_s \mathbf{A} \cdot d\mathbf{S} = \iint_s \mathbf{A} \cdot \mathbf{n}\, dS \tag{6-174a}$$

is the integrated product of the component of **A** normal to the surface times the scalar magnitude of the surface-area element dS. For example, if **V** denotes fluid velocity and ρ fluid density,

$$\iint_s \rho\mathbf{V} \cdot \mathbf{n}\, dS$$

is the rate at which mass flows across the surface S in the fluid.

Since the equation of a surface is of the form

$$f(x,y,z) = \text{const}$$

the vector ∇f is normal to the surface and

$$\mathbf{n} = \pm \frac{\nabla f}{|\nabla f|} \tag{6-174b}$$

is the unit vector normal to dS. For a closed surface, $\mathbf{n}$ is taken to be positive when it points *outward* from the enclosed volume.

In Example 6-7, it was shown that the net rate at which matter crossed the boundaries of an infinitesimal volume $dx\,dy\,dz = d\nu$ situated within the fluid is

$$\left(\frac{\partial \rho V_x}{\partial x} + \frac{\partial \rho V_y}{\partial y} + \frac{\partial \rho V_z}{\partial z}\right) d\nu = (\nabla \cdot \rho V)\, d\nu = (\text{div}\ \rho\mathbf{V})\, d\nu$$

Now let the surface dS enclose the infinitesimal element of volume $d\nu$. Then the net rate at which matter crosses the surface boundary of the infinitesimal volume element is

$$\rho\mathbf{V} \cdot \mathbf{n}\, dS = \nabla \cdot \rho\mathbf{V}\, d\nu \tag{6-175}$$

If (6-175) is written in the form

$$\nabla \cdot \rho\mathbf{V} = (\rho\mathbf{V} \cdot \mathbf{n}) \frac{dS}{d\nu}$$

the physical interpretation of the divergence is evident; divergence $\rho\mathbf{V}$ is the net rate of flux of matter across a surface *enclosing* unit volume.

Equation (6-175) may be integrated to give a relation known as the "divergence theorem":

$$\oiint_S \rho\mathbf{V} \cdot \mathbf{n}\, dS = \iiint_\nu \nabla \cdot \rho\mathbf{V}\, d\nu \tag{6-176}$$

in which the symbol $\oiint$ indicates integration over the surface S enclosing ν.

Examination of the methods used to obtain (6-175) and (6-176) will show that these relations are not dependent upon the physical situation invoked here and that $\rho\mathbf{V}$ *may be replaced by any vector* $\mathbf{A}$, *provided only that* $\mathbf{A}$ *and its partial derivatives are continuous in* ν *and on* S*:*

$$\iint_S \mathbf{A} \cdot \mathbf{n}\, dS = \iiint_\nu \nabla \cdot \mathbf{A}\, d\nu \tag{6-177}$$

The divergence of a vector $\mathbf{A}$ is then the net flux of $\mathbf{A}$ across a surface enclosing unit volume.

6-21. Coordinate Transformation. One of the useful properties of vector notation results from the fact that the vector equations are independent of the coordinate system employed. Consequently, if the vector operations are worked out for alternative coordinate systems, the transformation from one system to another is easy. The transformations for cylindrical- and spherical-coordinate systems are given below.

Cylindrical Coordinates (r, θ, z) (Fig. 6-1)

$$x = r\cos\theta \qquad y = r\sin\theta \qquad z = z \tag{6-178a}$$

Unit vectors in r, θ, z directions:

$$\begin{aligned} \mathbf{v}_r &= \mathbf{i}\cos\theta + \mathbf{j}\sin\theta \\ \mathbf{v}_\theta &= -\mathbf{i}\sin\theta + \mathbf{j}\cos\theta \\ \mathbf{v}_z &= \mathbf{k} \end{aligned} \tag{6-178b}$$

Position vector:

$$\mathbf{R} = \mathbf{v}_r r + \mathbf{k}z \tag{6-178c}$$

Arc length:

$$dl = \sqrt{dr^2 + r^2\,d\theta^2 + dz^2} \tag{6-178d}$$

Element of volume:

$$dv = r\,d\theta\,dr\,dz \tag{6-178e}$$

Area elements *normal* to r, θ, z directions:

$$\begin{aligned} dA_r &= r\,d\theta\,dz \\ dA_\theta &= dr\,dz \\ dA_z &= r\,d\theta\,dr \end{aligned} \tag{6-178f}$$

$$\nabla f = \mathbf{v}_r \frac{\partial f}{\partial r} + \mathbf{v}_\theta \frac{1}{r}\frac{\partial f}{\partial \theta} + \mathbf{v}_z \frac{\partial f}{\partial z} \tag{6-178g}$$

$$\nabla \cdot \mathbf{B} = \frac{1}{r}\frac{\partial}{\partial r}(rB_r) + \frac{1}{r}\frac{\partial B_\theta}{\partial \theta} + \frac{\partial B_z}{\partial z} \tag{6-178h}$$

$$\nabla^2 f = \frac{1}{r}\frac{\partial}{\partial r}\left(r\frac{\partial f}{\partial r}\right) + \frac{1}{r^2}\frac{\partial^2 f}{\partial \theta^2} + \frac{\partial^2 f}{\partial z^2} \tag{6-178i}$$

$$\nabla \times \mathbf{B} = \frac{1}{r}\begin{vmatrix} \mathbf{v}_r & r\mathbf{v}_\theta & \mathbf{v}_z \\ \dfrac{\partial}{\partial r} & \dfrac{\partial}{\partial \theta} & \dfrac{\partial}{\partial z} \\ B_r & rB_\theta & B_z \end{vmatrix} \tag{6-178j}$$

Spherical Coordinates (r, ϕ, θ) (Fig. 6-7)

$$x = r\cos\theta\sin\phi \qquad y = r\sin\theta\sin\phi \qquad z = r\cos\phi \tag{6-179a}$$

$$\begin{aligned} \mathbf{v}_r &= \mathbf{i}\cos\theta\sin\phi + \mathbf{j}\sin\theta\sin\phi + \mathbf{k}\cos\phi \\ \mathbf{v}_\phi &= \mathbf{i}\cos\theta\cos\phi + \mathbf{j}\sin\theta\cos\phi - \mathbf{k}\sin\phi \\ \mathbf{v}_\theta &= -\mathbf{i}\sin\theta + \mathbf{j}\cos\theta \end{aligned} \tag{6-179b}$$

$$\mathbf{R} = \mathbf{v}_r r \tag{6-179c}$$

$$dl = \sqrt{dr^2 + r^2\,d\phi^2 + r^2\sin^2\phi\,d\theta^2} \tag{6-179d}$$

$$dv = r^2\sin\phi\,dr\,d\phi\,d\theta \tag{6-179e}$$

$$dA_r = r^2\sin\phi\,d\phi\,d\theta$$

$$dA_\phi = r \sin \phi \, dr \, d\theta \tag{6-179f}$$
$$dA_\theta = r \, dr \, d\phi$$

$$\nabla f = \mathbf{v}_r \frac{\partial f}{\partial r} + \mathbf{v}_\phi \frac{1}{r} \frac{\partial f}{\partial \phi} + \mathbf{v}_\theta \frac{1}{r \sin \phi} \frac{\partial f}{\partial \theta} \tag{6-179g}$$

$$\nabla \cdot \mathbf{B} = \frac{1}{r^2} \frac{\partial}{\partial r} (r^2 B_r) + \frac{1}{r \sin \phi} \frac{\partial}{\partial \phi} (B_\phi \sin \phi) + \frac{1}{r \sin \phi} \frac{\partial B_\theta}{\partial \theta} \tag{6-179h}$$

$$\nabla^2 f = \frac{1}{r^2} \frac{\partial}{\partial r} \left(r^2 \frac{\partial f}{\partial r} \right) + \frac{1}{r^2 \sin \phi} \frac{\partial}{\partial \phi} \left(\sin \phi \frac{\partial f}{\partial \phi} \right) + \frac{1}{r^2 \sin^2 \phi} \frac{\partial^2 f}{\partial \theta^2} \tag{6-179i}$$

$$\nabla \times \mathbf{B} = \frac{1}{r^2 \sin \phi} \begin{vmatrix} \mathbf{v}_r & r\mathbf{v}_\phi & r \sin \phi \mathbf{v}_\theta \\ \dfrac{\partial}{\partial r} & \dfrac{\partial}{\partial \phi} & \dfrac{\partial}{\partial \theta} \\ B_r & rB_\phi & r \sin \phi B_\theta \end{vmatrix} \tag{6-179j}$$

6-22. Mass Transfer in a Binary Gas Mixture. Consider a binary perfect-gas mixture at constant temperature and pressure, but let the velocity and composition of the mixture vary both with time and position. It is desired to set up the partial differential equation which controls the mass-transfer process taking place. Let

M_A = molecular weight of component A
M_B = molecular weight of component B
Y_A = mole fraction of A in mixture
$Y_B = 1 - Y_A$ = mole fraction of B in mixture
ρ_A = *mass* density of A in mixture
ρ_B = *mass* density of B in mixture
$\rho = \rho_A + \rho_B$ = *mass* density of mixture
$M = Y_A M_A + (1 - Y_A) M_B$ = mean molecular weight of mixture
$\mathfrak{D}$ = molecular diffusivity
$\mathbf{V}_A$ = velocity of component A

If neither component A nor B is created or destroyed within the system, a material balance written for component A for an infinitesimal volume fixed in space gives

$$-\rho \mathbf{V}_A \cdot \mathbf{n} \, dS = \frac{\partial \rho_A}{\partial t} \, dv \tag{6-180}$$

In view of the divergence theorem in the form (6-175), (6-180) may be written

$$\nabla \cdot \rho_A \mathbf{V}_A + \frac{\partial \rho_A}{\partial t} = 0 \tag{6-181}$$

Similarly, a balance on component B gives

$$\nabla \cdot \rho_B \mathbf{V}_B + \frac{\partial \rho_B}{\partial t} = 0 \tag{6-182}$$

Now divide (6-181) by M_A and (6-182) by M_B and add the result:

$$\nabla \cdot \left(\frac{\rho_A \mathbf{V}_A}{M_A} + \frac{\rho_B \mathbf{V}_B}{M_B} \right) + \frac{\partial}{\partial t} \left(\frac{\rho_A}{M_A} + \frac{\rho_B}{M_B} \right) = 0 \tag{6-183}$$

But

$$\frac{\rho}{M} = \frac{\rho_A}{M_A} + \frac{\rho_B}{M_B} = \frac{P}{RT} = \text{const} \tag{6-184}$$

for a perfect-gas system at constant temperature T and pressure P. Hence, (6-183) becomes

$$\nabla \cdot \left(\frac{\rho_A \mathbf{V}_A}{M_A} + \frac{\rho_B \mathbf{V}_B}{M_B} \right) = 0 \tag{6-185}$$

Now define the *molal lineal velocity* of the *system* $\mathbf{U}$ by the relation

$$\mathbf{U} = \frac{\rho_A \mathbf{V}_A / M_A + \rho_B \mathbf{V}_B / M_B}{\rho / M} \tag{6-186}$$

Since ρ/M is a constant, (6-185) may be written

$$\nabla \cdot \mathbf{U} = 0 \tag{6-187}$$

Despite the fact that the *mass* density of the system is not a constant, a characteristic velocity of the system has been found whose divergence is zero. It is evident that the molal lineal velocity $\mathbf{U}$ differs from the customary *mass lineal velocity* $\mathbf{V}$ assigned to a system. The mass lineal velocity is defined by the relation

$$\mathbf{V} = \frac{\rho_A \mathbf{V}_A + \rho_B \mathbf{V}_B}{\rho} \tag{6-188}$$

The mass lineal velocity $\mathbf{V}$ is seen to be proportional to the rate at which *mass* flows across an infinitesimal area oriented normal to $\mathbf{V}$. Examination of (6-186) shows that the molal lineal velocity is proportional to the rate at which moles (or molecules or volume) flow across an infinitesimal area oriented normal to $\mathbf{U}$. In a system of varying composition, $\mathbf{V}$ and $\mathbf{U}$ generally will differ in *both magnitude and direction.*

In their present form, neither Eq. (6-181) nor Eq. (6-182) is particularly useful. It is desired to relate the velocities of individual components to the composition and mean velocity of the system. This is done by means of the diffusion equation

$$\frac{\rho_A}{M_A} (\mathbf{V}_A - \mathbf{U}) = -\mathfrak{D} \nabla Y_A \tag{6-189}$$

The significance of the diffusion equation (6-189) is as follows: $(\rho_A/M_A)\mathbf{V}_A$ is the rate at which moles of component A actually move inside the system, and $(\rho_A/M_A)\mathbf{U}$ is the rate at which moles of component A would move if the lineal velocity of A were equal to the molal lineal velocity of the system. The kinetic theory shows that in a perfect-gas mixture at constant temperature and pressure the difference between these two rates is proportional to the gradient ∇Y_A. The constant of proportionality $\mathfrak{D}$ is called the molecular diffusivity, and its engineering units are pound moles per foot per hour (unit ∇Y). Equation (6-189) in effect defines $\mathfrak{D}$. The *molal lineal velocity* of the system $\mathbf{U}$ is used in the defining equation (6-189) because this results in a molecular diffusivity which is *substantially independent* of *the total pressure and relative composition* of the binary perfect-gas mixture. These desirable attributes would not be obtained if the mass lineal velocity were employed.

Combination of (6-181) and (6-189) gives

$$\nabla \cdot \frac{\rho_A \mathbf{U}}{M_A} + \frac{\partial \rho_A / M_A}{\partial t} = \nabla \cdot \mathfrak{D} \nabla Y_A \tag{6-190}$$

But

$$\nabla \cdot \rho_A \mathbf{U} = \rho_A \nabla \cdot \mathbf{U} + U \cdot \nabla \rho_A$$

and $\nabla \cdot \mathbf{U} = 0$. Furthermore, $\mathfrak{D}$ can be considered a constant, ρ/M is a constant, and

$$Y_A = \frac{\rho_A / M_A}{\rho / M} \tag{6-191}$$

Hence, (6-190) may be written in the form

$$\begin{aligned} \left(\frac{DY_A}{dt}\right)_{\mathbf{U}} &= \frac{\partial Y_A}{\partial t} + \mathbf{U} \cdot \nabla Y_A \\ &= \frac{\partial Y_A}{\partial t} + U_x \frac{\partial Y_A}{\partial x} + U_y \frac{\partial Y_A}{\partial y} + U_z \frac{\partial Y_A}{\partial z} \\ &= \frac{M}{\rho} \mathfrak{D} \nabla^2 Y_A \\ &= \frac{M}{\rho} \mathfrak{D} \left(\frac{\partial^2 Y_A}{\partial x^2} + \frac{\partial^2 Y_A}{\partial y^2} + \frac{\partial^2 Y_A}{\partial z^2}\right) \end{aligned} \tag{6-192}$$

where U_x, U_y, U_z denote the components of $\mathbf{U}$ parallel to the x, y, and z axes, respectively. Equation (6-192) is the "diffusion equation" for a binary perfect-gas mixture at constant temperature and pressure.

Although special cases may be treated by means of (6-192) and suitable boundary conditions only, in the general case the solution of (6-192) requires simultaneous solution of the "momentum equation." The momentum equation is derived from Newton's laws of motion and is based upon the mass lineal velocity $\mathbf{V}$. Consequently, it is desirable to replace $\mathbf{U}$ by $\mathbf{V}$ in (6-192). This may be done as follows:

If Eq. (6-188) is solved for $\mathbf{V}_B$ and the resulting expression introduced into (6-186), there results

$$\frac{\rho_A \mathbf{V}_A}{M_A} = \frac{\rho M_B}{M_B - M_A}\left(\frac{\mathbf{U}}{M} - \frac{\mathbf{V}}{M_B}\right) \tag{6-193}$$

Now combine (6-189) and (6-193) and solve for $\mathbf{U}$:

$$\mathbf{U} = \mathbf{V} - \frac{\mathfrak{D}(M_B - M_A)}{\rho/M}\frac{1}{M}\nabla Y_A \tag{6-194}$$

Combination of (6-192) and (6-194) yields

$$\frac{\partial Y_A}{\partial t} + \mathbf{V}\cdot\nabla Y_A = \frac{M}{\rho}\mathfrak{D}\left(\nabla^2 Y_A + \frac{M_B - M_A}{M}\nabla Y_A \cdot \nabla Y_A\right) \tag{6-195}$$

Equation (6-195) may be transformed as follows: In view of the definition of M,

$$Y_A = \frac{M - M_B}{M_A - M_B} \tag{6-196}$$

$$\nabla Y_A = \frac{\nabla M}{M_A - M_B} = \frac{M}{M_A - M_B}\nabla \ln M \tag{6-197}$$

$$\frac{\partial Y_A}{\partial t} = \frac{1}{M_A - M_B}\frac{\partial M}{\partial t} = \frac{M}{M_A - M_B}\frac{\partial \ln M}{\partial t} \tag{6-198}$$

Consequently,

$$\frac{\partial Y_A}{\partial t} + \mathbf{V}\cdot\nabla Y_A = \frac{M}{M_A - M_B}\left(\frac{\partial \ln M}{\partial t} + \mathbf{V}\cdot\nabla \ln M\right) \tag{6-199}$$

Now

$$\nabla^2 Y_A = \nabla\cdot\nabla Y_A = \nabla\cdot\frac{M}{M_A - M_B}\nabla \ln M = \frac{M}{M_A - M_B}\nabla\cdot\nabla \ln M + (\nabla \ln M)\cdot\nabla\frac{M}{M_A - M_B} \tag{6-200}$$

and

$$\nabla \ln M \cdot \nabla\frac{M}{M_A - M_B} = \frac{M_A - M_B}{M}\nabla Y_A \cdot \nabla Y_A \tag{6-201}$$

Consequently,

$$\nabla^2 Y_A + \frac{M_A - M_B}{M}\nabla Y_A \cdot \nabla Y_A = \frac{M}{M_A - M_B}\nabla^2 \ln M \tag{6-202}$$

and Eq. (6-195) becomes, after introduction of (6-199) and (6-202),

$$\frac{\partial \ln M}{\partial t} + \mathbf{V}\cdot\nabla \ln M = \frac{M}{\rho}\mathfrak{D}\nabla^2 \ln M \tag{6-203}$$

Equation (6-203) is the diffusion equation expressed in terms of the mass lineal velocity. Note that the dependent variable is the logarithm of the mean molecular weight of the system. It is in this form that comparison with the momentum and energy equations leads to the so-called "analogies" between momentum, heat, and mass transfer.

6-23. The Equations of Motion of a Perfect Fluid. Consider the flow of a perfect fluid, i.e., one without viscosity. Although no such fluid exists, it is found that, except for an important region termed the "boundary layer," most fluids may be treated as perfect. Fortunately, within the boundary layer where shear stresses resulting from viscous effects must be considered, certain other simplifications are often possible. Consequently, in modern fluid mechanics, it is customary to attack a complicated situation by dividing the flow into two regions—the ideal-flow region and the boundary-layer region. The flow characteristics of these regions are obtained by quite different methods, and the total flow pattern is obtained by joining the two solutions at the outer edge of the boundary layer. Consequently, the theory of an ideal fluid, which formerly was regarded as purely a mathematical exercise, has assumed great engineering importance.

The equations of motion may be derived from either the Eulerian or Lagrangian point of view. Here the Lagrangian approach is used. Examine a small element of volume which always encloses a fixed mass of fluid, δM. At time t let the volume element be a rectangle with sides of length δa, δb, δc parallel to the x, y, and z axes, respectively. It is assumed that no sources or sinks are present. Hence, the condition of constant mass requires

$$\frac{D}{dt}\rho\,\delta a\,\delta b\,\delta c = \delta a\,\delta b\,\delta c\,\frac{D\rho}{dt} + \rho\frac{D}{dt}(\delta a\,\delta b\,\delta c) = 0 \tag{6-204}$$

However, in Example 3-9, it was shown that $(D/dt)\,\delta a = (\partial V_x/\partial x)\,\delta a$, etc. Hence, (6-204) becomes

$$\frac{1}{\rho}\frac{D\rho}{dt} + \nabla\cdot\mathbf{V} = 0 \tag{6-119}$$

Now Newton's law of motion requires that the resultant of the external forces applied to a body be equal to the time rate of change of the momentum of the body. In a perfect fluid only normal stresses exist, and these are due to the fluid pressure. The resultant of the normal stresses is easily shown to be

$$-\left(\mathbf{i}\frac{\partial p}{\partial x} + \mathbf{j}\frac{\partial p}{\partial y} + \mathbf{k}\frac{\partial p}{\partial z}\right)\delta a\,\delta b\,\delta c = -\delta a\,\delta b\,\delta c\,\nabla p$$

The minus sign appears because the pressure is a compressive force and acts inward on a surface surrounding a fluid element.

The remaining external forces will be due to gravitational, magnetic, electric, etc., fields and are called "body forces." Denote the resultant of these combined body forces by

$$\rho \mathbf{F}\ \delta a\ \delta b\ \delta c$$

The time rate of change of momentum of the fluid element is, using (6-204),

$$\frac{D}{dt}\rho\ \delta a\ \delta b\ \delta c\ \mathbf{V} = \rho\ \delta a\ \delta b\ \delta c \frac{D\mathbf{V}}{dt}$$

Then, as a consequence of Newton's law of motion, there results

$$\frac{D\mathbf{V}}{dt} = \mathbf{V}\cdot\nabla\mathbf{V} + \frac{\partial \mathbf{V}}{\partial t} = \mathbf{F} - \frac{\nabla p}{\rho} \tag{6-205}$$

Equation (6-205) is frequently called the "momentum equation." If the vector equation is decomposed into its components, three equations result:

$$\begin{aligned}
\frac{DV_x}{dt} &= \frac{\partial V_x}{\partial t} + V_x\frac{\partial V_x}{\partial x} + V_y\frac{\partial V_x}{\partial y} + V_z\frac{\partial V_x}{\partial z} = F_x - \frac{1}{\rho}\frac{\partial p}{\partial x}\\
\frac{DV_y}{dt} &= \frac{\partial V_y}{\partial t} + V_x\frac{\partial V_y}{\partial x} + V_y\frac{\partial V_y}{\partial y} + V_z\frac{\partial V_y}{\partial z} = F_y - \frac{1}{\rho}\frac{\partial p}{\partial y}\\
\frac{DV_z}{dt} &= \frac{\partial V_z}{\partial t} + V_x\frac{\partial V_z}{\partial x} + V_y\frac{\partial V_z}{\partial y} + V_z\frac{\partial V_z}{\partial z} = F_z - \frac{1}{\rho}\frac{\partial p}{\partial z}
\end{aligned} \tag{6-206}$$

If the *fluid density is constant,* the equation of motion can be greatly simplified. In most cases the body force will be due to gravity and will be conservative; i.e., the net work done by moving a fluid element around a closed path will be zero. In mathematical terms, this statement may be written as

$$\oint \mathbf{F}\cdot d\mathbf{R} = 0$$

In view of the discussion in Sec. 6-19, the fact that the circulation of $\mathbf{F}$ vanishes permits $\mathbf{F}$ to be expressed as the gradient of a scalar function which, in this case, will be proportional to *minus the potential energy*

$$\mathbf{F} = \nabla\gamma$$

Then the term $(-1/\rho)\nabla p + \mathbf{F}$ appearing in (6-205) can be written as

$$\nabla\lambda = \nabla\left(\gamma - \frac{1}{\rho}p\right) = -\frac{1}{\rho}\nabla p - \mathbf{F} \tag{6-207}$$

and Eq. (6-205) becomes

$$\mathbf{V}\cdot\nabla\mathbf{V} + \frac{\partial\mathbf{V}}{\partial t} = \nabla\lambda \tag{6-208}$$

or, when broken down into its components, the three equations equivalent to (6-208) are

$$\frac{\partial V_x}{\partial t} + V_x \frac{\partial V_x}{\partial x} + V_y \frac{\partial V_x}{\partial y} + V_z \frac{\partial V_x}{\partial z} = \frac{\partial \lambda}{\partial x} \tag{6-209a}$$

$$\frac{\partial V_y}{\partial t} + V_x \frac{\partial V_y}{\partial x} + V_y \frac{\partial V_y}{\partial y} + V_z \frac{\partial V_y}{\partial z} = \frac{\partial \lambda}{\partial y} \tag{6-209b}$$

$$\frac{\partial V_z}{\partial t} + V_x \frac{\partial V_z}{\partial x} + V_y \frac{\partial V_z}{\partial y} + V_z \frac{\partial V_z}{\partial z} = \frac{\partial \lambda}{\partial z} \tag{6-209c}$$

For the constant-density case, the continuity equation (6-119) becomes

$$\nabla \cdot \mathbf{V} = \frac{\partial V_x}{\partial x} + \frac{\partial V_y}{\partial y} + \frac{\partial V_z}{\partial z} = 0 \tag{6-210}$$

Note that even in the case of an ideal fluid of constant density the determination of the velocity and force field of the fluid involves the solution of four simultaneous partial differential equations! In order to attack the problem, it is necessary to reduce the system of equations involving *four* dependent variables (V_x, V_y, V_z, and λ) to *one* partial differential equation involving but *one* dependent variable. A straightforward approach might proceed as follows: Differentiate (6-209*a*) *twice*, first with respect to y and then with respect to z, so that the term on the right-hand side becomes $\partial^3\lambda/(\partial z\,\partial y\,\partial x)$. A similar operation is performed on (6-209*b*) (but differentiation is with respect to x and then z) and on (6-209*c*) (but differentiation is with respect to x and then y). The resulting equations may then be used to eliminate λ, giving two new third-order equations involving only V_x, V_y, and V_z. These new equations, together with the continuity equation (6-210), are further manipulated to eliminate, say, V_y and V_z. Although this process will work, it is evident that it is extremely cumbersome. A much more efficient approach is as follows: Suppose that $\nabla \times \mathbf{V} = 0$; that is, the curl of the velocity vector is zero. The physical significance of this will be discussed shortly. The development of Sec. 6-19 then permits $\mathbf{V}$ to be taken as the gradient of a scalar, provided that $\mathbf{V}$ is continuously differentiable:

$$\mathbf{V} = \nabla\Phi \tag{6-211}$$

where Φ is termed the "velocity potential." Clearly

$$V_x = \frac{\partial \Phi}{\partial x} \qquad V_y = \frac{\partial \Phi}{\partial y} \qquad V_z = \frac{\partial \Phi}{\partial z} \tag{6-212}$$

Now substitute (6-211) into the continuity equation (6-210):

$$\nabla \cdot \mathbf{V} = \nabla \cdot \nabla\Phi = \nabla^2\Phi = 0 \tag{6-213}$$

that is, the continuity equation is satisfied by Φ, provided that Φ satisfies LaPlace's equation. Since only one dependent variable Φ appears in (6-213), it may be solved directly without recourse to the momentum equation and, by virtue of (6-212), the velocity components obtained at once. The momentum equation is needed to determine the force potential λ. The condition $\nabla \times \mathbf{V} = 0$ guarantees that the momentum relation will not place demands on Φ in conflict with (6-213). The condition that $\nabla \times \mathbf{V} = 0$, that is, that the flow be irrotational, may seem unduly restrictive. This is not the case. It is easily shown that the curl of the velocity is equal to twice the angular velocity ω of the fluid

$$\nabla \times \mathbf{V} = 2\omega$$

Consider a small spherical element of fluid. As the viscosity is zero (perfect fluid), no tangential stresses may be applied to the surface. The pressure stresses act normal to the surface, and hence their line of action is through the center. The body forces act through the mass center, and for constant density the mass center and geometric center are identical. Consequently, no torque may be applied to the fluid element, and the angular *acceleration* of the element must always be zero. Therefore, if the fluid was originally at rest (as is usually the case), the angular velocity of the fluid must remain zero, and hence $\nabla \times \mathbf{V} = 0$ is always satisfied. Clearly these remarks apply only to a perfect fluid.

BIBLIOGRAPHY

1. Wylie, C. R., Jr.: "Advanced Engineering Mathematics," McGraw-Hill Book Company, Inc., New York, 1951.
2. Sokolnikoff, I. S., and E. S. Sokolnikoff: "Higher Mathematics for Engineers and Physicists," McGraw-Hill Book Company, Inc., New York, 1941.
3. Hildebrand, F. B.: "Advanced Calculus for Engineers," Prentice-Hall, Inc., Englewood Cliffs, N.J., 1949.
4. Phillips, H. B.: "Vector Analysis," John Wiley & Sons, Inc., New York, 1933.
5. Lass, H.: "Vector and Tensor Analysis," McGraw-Hill Book Company, Inc., New York, 1950.
6. Lamb, Sir Horace: "Hydrodynamics," Dover Publications, New York, 1945.
7. Streeter, V. L.: "Fluid Dynamics," McGraw-Hill Book Company, Inc., New York, 1948.
8. Bedingfield, C. H., and T. B. Drew: Analogy between Heat Transfer and Mass Transfer, *Ind. Eng. Chem.*, vol. 42, no. 6, pp. 1164–1173, 1950.
9. Marshall, W. R., Jr. and R. L. Pigford: "The Application of Differential Equations to Chemical Engineering Problems," University of Delaware, Newark, Del., 1947.

PROBLEMS

6-1. If $Z = f(x,y)$ and

$$x = e^u \cos v$$
$$y = e^u \sin v$$

find $\partial^2 Z/(\partial u\ \partial v)$.

6-2. Find $(\partial/\partial x) I_0(2\sqrt{xy})$, where x and y are independent variables.

6-3. The following equation often is obtained in the analysis of unsteady-state transfer processes:

$$-Gc_p\left(\frac{\partial t}{\partial x} + \frac{1}{V}\frac{\partial t}{\partial \theta}\right) = h_c A(t - t') = cj\,\frac{\partial t'}{\partial \theta}$$

t and t' are dependent variables, and x and θ are independent variables.

(*a*) Replace t, t', x, and θ in the equation by the new variables

$$T = \frac{t - t_0}{t_0' - t_0}$$
$$T' = \frac{t' - t_0}{t_0' - t_0} \qquad t_0 \text{ and } t_0' \text{ are constants}$$
$$Z = \frac{h_c A}{c_p G}\,x$$
$$\beta = \frac{Ah_c}{cj}\left(\theta - \frac{x}{V}\right)$$

(*b*) Eliminate T' from the resulting equation, and obtain a partial differential equation in terms of T, Z, and β.

6-4. The partial specific volume of component k in a single-phase mixture of j components is defined by the following equation:

$$\bar{V}_k = \left(\frac{\partial V}{\partial m_k}\right)_{T,P,m_j}$$

Calculate the partial specific volume of methane and pentane as a function of composition at a pressure of 1,000 psia and a temperature of 460°F from the following data. The use of residual methods to improve the accuracy of the calculation is suggested.

Behavior of Methane-n Pentane System
At 460°F and 1,000 psia

Composition, weight fraction CH_4	*Specific volume,* ft^3/lb
0.92476	0.58112
0.7783	0.50740
0.08450	0.12485
0.3857	0.30597
0.031144	0.079812
0.161191	0.17612
1.0	0.61755
0.0	0.05233

6-5. Given:

$$u^3 + v^3 + x^3 - 3y = 0$$
$$u^2 + v^2 + y^2 + 2x = 0$$

Find $(\partial u/\partial x)_y$.

6-6. P is a function of both x and y as given by

$$P = Ax + B\frac{xy - a}{y} + \frac{c}{\sqrt{xy}}$$

where A, B, C, and a are constants. Obtain expressions for the values of x and y (in terms of the constants) corresponding to maximum or minimum values of P.

6-7. Reduce the following integral equation to a differential equation involving only x, and integrate the expression. Be sure to evaluate any constants of integration.

$$f(x) = x + \int_0^x (z - x)f(z)\, dz$$

The variable z is independent of x. The functions $f(x)$ and $f(z)$ are formally the same; i.e., *if, for instance,* $f(t) = t^2$, then $f(x) = x^2$ and $f(z) = z^2$.

6-8. From the fact that

$$\int_0^1 x^N\, dx = \frac{1}{N + 1}$$

show that

$$\int_0^1 x^N (\log x)^m\, dx = (-1)^m \frac{m!}{(N + 1)^{m+1}}$$

6-9. Determine the value of

$$\int_0^1 \frac{x^t - 1}{\ln x}\, dx$$

without resource to the tabulated integral.

6-10. Verify Eqs. (6-170*a*) to (6-170*i*).

6-11. Find the circulation about the square enclosed by the lines $x = \pm 1$, $y = \pm 1$ for the flow given by $V_x = x + y$, $V_y = x^2 - y$, $V_z = 0$. Does this flow satisfy $\nabla^2\Phi = 0$?

6-12. In Sec. 6-22 it was shown that $\nabla \cdot \mathbf{U} = 0$. What is $\nabla \cdot \mathbf{V}$ for this case?

6-13. Develop a technique for carrying out the transformation of Eq. (6-192) to the form (6-203) without the use of vector notation.

6-14. Show that for a constant-density system with $\nabla \times \mathbf{V} = 0$ the momentum equation (6-208) does not impose conditions which are incompatible with $\nabla^2\Phi = 0$.

6-15. Show that for the case of constant density, irrotational flow, the momentum equation (6-208) may be integrated to give Bernoulli's equation

$$\frac{1}{2}V^2 + \frac{\partial\Phi}{\partial t} - \lambda = F(t)$$

where

$$V^2 = V_x^2 + V_y^2 + V_z^2$$

HINT: Decompose (6-208) into its components. Use $\nabla \times \mathbf{V} = 0$ to show that $V_y(\partial V_x/\partial y) = V_y(\partial V_y/\partial x) = (\partial/\partial x)V_y^2/2$, etc.

6-16. Consider the two-dimensional motion of an ideal fluid of constant density.

(*a*) Show that the "stream function" ψ defined by

$$\frac{\partial\psi}{\partial x} = -V_y$$
$$\frac{\partial\psi}{\partial y} = V_x$$

satisfies momentum and continuity equations if $\nabla^2\psi = 0$. Is it necessary that the flow be irrotational?

(*b*) Show that the line $\psi = \text{const}$ is a streamline.

(*c*) Show that lines of constant Φ are normal to lines of constant ψ.

6-17. Show that $\Phi = xy$ satisfies $\nabla^2\Phi = 0$ for two-dimensional flow. Determine the stream function ψ. Plot lines of constant Φ and ψ on rectangular coordinates. Discuss the flow pattern.

6-18. The Navier-Stokes equation is the "momentum equation" for a real fluid. For a constant-density system this equation has the form

$$\frac{D\mathbf{V}}{dt} = \mathbf{V} \cdot \nabla\mathbf{V} + \frac{\partial\mathbf{V}}{\partial t} = \mathbf{F} - \frac{\nabla p}{\rho} + \frac{\mu}{\rho}\nabla^2\mathbf{V}$$

where μ is the fluid viscosity. Is this equation compatible with the concept of a velocity potential Φ? In the two-dimensional case is it compatible with the concept of a stream function ψ?

CHAPTER 7

SOLUTION OF PARTIAL DIFFERENTIAL EQUATIONS

7-1. Nature of a Partial Differential Equation. An equation involving partial derivatives is a partial differential equation. The equation

$$\frac{\partial^2 u}{\partial x\, \partial y} = 0 \tag{7-1}$$

is a partial differential equation. Such equations are of importance in engineering because the significant variables are so frequently functions of more than one independent variable, and the basic differential expressions for the natural laws are, therefore, partial differential equations. The usual problem is to determine a particular relation between u, x, and y, expressed as $u = f(x,y)$, that satisfies the basic differential equation and also satisfies some particular conditions specified by the practical problem at hand.

Integrating (7-1) with respect to x gives

$$\frac{\partial u}{\partial y} = f_1(y) \tag{7-2}$$

and integrating (7-2) with respect to y gives

$$u = \int f_1(y)\, dy + f_2(x) = f_3(y) + f_2(x) \tag{7-3}$$

which may be checked by differentiating with respect to x and y separately and comparing with (7-1). It will be observed that (7-3) is a solution of (7-1), no matter what may be the nature of the functions $f_3(y)$ and $f_2(x)$, which are arbitrary functions. *Whereas the solution of ordinary differential equations involves the introduction of arbitrary constants, the solution of partial differential equations involves the introduction of arbitrary functions.*

7-2. Types of Partial Differential Equations. The "order" of a partial differential equation is the order of the highest derivative appearing after the equation is rationalized and cleared of fractions. The equation

$$\frac{\partial^3 w}{\partial x\, \partial y^2} + \frac{\partial^2 w}{\partial x^2} + w^2 = 0$$

is third-order.

A partial differential equation is said to be "linear" if, after rationalization and clearing of fractions, no powers or products of the dependent variable or its partial derivatives are present. The second-order equation

$$P(x,y)\,\frac{\partial^2 w}{\partial x^2} + Q(x,y)\,\frac{\partial w}{\partial y} = w\,R_1(x,y) + R_2(x,y)$$

is linear. If, however, one of the coefficients, say P, were a function of the dependent variable w, then the equation would be nonlinear.

A "homogeneous" partial differential equation is one in which all terms contain derivatives of the same order.

7-3. Boundary Conditions. The solution of a partial differential equation must satisfy both the original differential equation and the boundary conditions. Whereas in a total differential equation the boundary conditions permit definite numbers to be assigned to the constants of integration, in a partial differential equation the boundary conditions demand that the arbitrary functions resulting from integration assume specific functional forms. However, a solution procedure which first determines the arbitrary functions and then specializes them to fit the boundary conditions is usually not feasible. A more fruitful attack is to determine directly a set of specialized ("particular") solutions and then to combine them in such a way that the boundary conditions are satisfied. The following property of a *linear, homogeneous* partial differential equation is of extreme importance in this respect:

If each of the functions $v_1, v_2, \ldots, v_n \ldots$ is a solution of a linear, homogeneous partial differential equation, then the function

$$v = \sum_{1}^{\infty} v_n$$

is also a solution, provided that the infinite series converges and is termwise differentiable as far as the highest derivatives appearing in the original differential equation. If the series is termwise differentiable, the sum of the series of partial derivatives resulting from termwise differentiation of the infinite series must converge to the corresponding partial derivatives of v, the function defined by the series.

A discussion of the number and type of boundary conditions necessary and sufficient to ensure a specialized solution of a general partial differential equation is beyond the scope of this discussion. Fortunately, however, in most engineering applications the physical situation enables the necessary boundary conditions to be established relatively easily.

7-4. Method of Solution of Partial Differential Equations. No general, formalized analytical procedure for the solution of an arbitrary partial differential equation is known. The solution of a partial differential equation is essentially a guessing game. The object of this game is to guess a form of the specialized solution which will reduce the partial differential equation to one or more total differential equations. The solutions of the total differential equations are then combined (if possible) in such a way that the boundary conditions as well as the original partial differential equation are satisfied. If this proves to be impossible, the analyst has made an incorrect guess and must start over. Fortunately, for a limited number of frequently occurring types of partial differential equations, methods of getting the game off to a good start are available. This is particularly true for linear, homogeneous partial differential equations with constant coefficients.

The following example illustrates one method of attack.

Example 7-1. Heat Transfer in a Flowing Fluid. An infinitely wide flat plate is maintained at a constant temperature T_0. The plate is immersed in an infinitely wide and thick stream of constant-density fluid originally at temperature T_1. If the origin of coordinates is taken at the leading edge of the plate, a *rough approximation* to the true velocity distribution is

$$V_x = \beta y \qquad (\beta \text{ is a constant}) \qquad V_y = 0 \qquad V_z = 0 \tag{7-4}$$

Turbulent heat transfer is assumed negligible, and molecular transport of heat is assumed important only in the y direction. The thermal conductivity of the fluid, k, is assumed constant. These assumptions are clearly so drastic that the results will only approximate the true situation. However, the method used here is a simplified version of the procedure used when boundary-layer theory is invoked (see Prob. 7-1 and reference 5).

It is desired to determine the temperature distribution within the fluid and the heat-transfer coefficient between the fluid and the plate.

Apply the first law of thermodynamics to an infinitesimal volume element of length dx and height dy situated in the fluid. If the dissipation resulting from fluid shear is neglected and the fluid properties are considered constant, there result

$$\text{Input energy rate} = V_x\rho CT\,dy - k\,dx\,\frac{\partial T}{\partial y}$$

$$\text{Output energy rate} = \left[V_x\rho CT + \frac{\partial}{\partial x}(V_x\rho CT)\,dx\right]dy - \left[k\frac{\partial T}{\partial y} + \frac{\partial}{\partial y}\left(k\frac{\partial T}{\partial y}\right)dy\right]dx$$

$$\text{Accumulation} = 0 \qquad \text{steady state assumed}$$

Then
$$\frac{\partial V_x\rho CT}{\partial x} = \frac{\partial}{\partial y}\left(k\frac{\partial T}{\partial y}\right)$$

and with constant properties and using (7-4) there results

$$\frac{\partial T}{\partial x} = \frac{k}{\beta\rho C}\frac{1}{y}\frac{\partial^2 T}{\partial y^2} \tag{7-5}$$

with the boundary conditions

$$T = T_1 \qquad \text{at } x = 0,\ y > 0 \tag{7-6a}$$
$$T = T_1 \qquad \text{at } y = \infty,\ x > 0 \tag{7-6b}$$
$$T = T_0 \qquad \text{at } y = 0,\ x > 0 \tag{7-6c}$$

Now let
$$\frac{k}{\beta\rho C} = A \tag{7-7}$$

$$\theta = \frac{T - T_1}{T_0 - T_1} \tag{7-8}$$

Then (7-5) becomes

$$\frac{\partial \theta}{\partial x} = \frac{A}{y} \frac{\partial^2 \theta}{\partial y^2} \tag{7-9}$$

with the boundary conditions

$$\theta = 0 \qquad \text{at } x = 0,\ y > 0 \tag{7-10a}$$
$$\theta = 0 \qquad \text{at } y = \infty,\ x > 0 \tag{7-10b}$$
$$\theta = 1 \qquad \text{at } y = 0,\ x > 0 \tag{7-10c}$$

It is now *assumed* that (7-9) and (7-10) can be satisfied by a solution of the form

$$\theta = f\left(\frac{y}{x^n}\right) = f(\eta) \tag{7-11}$$

where now

$$\theta = 0 \qquad \text{at } \eta = \infty \tag{7-12a}$$
$$\theta = 1 \qquad \text{at } \eta = 0 \tag{7-12b}$$

Note that the assumption (7-11) is not a wild guess. The purpose here is to replace the independent variables x and y by a single variable η. In view of (7-10a) and (7-10b), the form of the new function of x and y *must* be such that it has the same value when $x = 0$ as it does when $y = \infty$. This condition is satisfied by (7-11). As a matter of fact, the boundary conditions (7-10a) and (7-10b) suggest the trial form (7-11).

The next step is to replace y and x in the partial differential equation (7-9) by η. It is hoped that a value of n exists which will completely eliminate x and y in the equation. This process is straightforward. Since θ is assumed a function of η only and η is a function of x and y, $\eta = y/x^n$,

$$d\theta = \frac{d\theta}{d\eta}\,d\eta = \frac{d\theta}{d\eta}\left(\frac{\partial \eta}{\partial x}\,dx + \frac{\partial \eta}{\partial y}\,dy\right)$$

and

$$\frac{\partial \theta}{\partial x} = \frac{d\theta}{d\eta}\frac{\partial \eta}{\partial x} = -\frac{ny}{x^{n+1}}\frac{d\theta}{d\eta} = -\frac{n\eta}{x}\frac{d\theta}{d\eta}$$

$$\frac{\partial \theta}{\partial y} = \frac{d\theta}{d\eta}\frac{\partial \eta}{\partial y} = \frac{1}{x^n}\frac{d\theta}{d\eta}$$

$$\begin{aligned}\frac{\partial^2 \theta}{\partial y^2} &= \frac{\partial}{\partial y}\left(\frac{\partial \theta}{\partial y}\right) = \frac{\partial}{\partial y}\left(\frac{1}{x^n}\frac{d\theta}{d\eta}\right)\\ &= \frac{1}{x^n}\frac{\partial}{\partial y}\left(\frac{d\theta}{d\eta}\right) = \frac{1}{x^n}\frac{d^2\theta}{d\eta^2}\frac{d\eta}{dy}\\ &= \frac{1}{x^{2n}}\frac{d^2\theta}{d\eta^2}\end{aligned}$$

Substituting the above expressions into (7-9) yields

$$-\frac{1}{x}n\eta\frac{d\theta}{d\eta} = \frac{1}{x^{3n}}\frac{A}{\eta}\frac{d^2\theta}{d\eta^2} \tag{7-13}$$

If (7-13) is to be a function of η only,

$$n = \tfrac{1}{3} \tag{7-14}$$

and (7-13) becomes

$$\frac{d^2\theta}{d\eta^2} + \frac{\eta^2}{3A}\frac{d\theta}{d\eta} = 0 \tag{7-15}$$

The first integration of (7-15) gives

$$\frac{d\theta}{d\eta} = B \exp\left(-\frac{\eta^3}{9A}\right) \tag{7-16}$$

Use of (7-12*a*) gives

$$\int_0^\theta d\theta = \theta = B \int_\infty^\eta \left[\exp\left(-\frac{\eta^3}{9A}\right)\right] d\eta = \frac{T - T_1}{T_0 - T_1} \tag{7-17}$$

Since $\theta = 1$ when $\eta = 0$ (7-12*b*)

$$B = \frac{-1}{\int_0^\infty [\exp(-\eta^3/9A)]\, d\eta}$$

and

$$\frac{T - T_1}{T_0 - T_1} = \frac{\int_\eta^\infty [\exp(-\eta^3/9A)]\, d\eta}{\int_0^\infty [\exp(-\eta^3/9A)]\, d\eta} \tag{7-18}$$

Now the *local* heat-transfer coefficient is defined by the relation

$$h(T_0 - T_1) = -k\left(\frac{\partial T}{\partial y}\right)_0 \tag{7-19}$$

Since

$$\frac{\partial T}{\partial y} = (T_0 - T_1)\frac{\partial \theta}{\partial y} = \frac{T_0 - T_1}{x^{1/3}}\frac{d\theta}{d\eta} = \frac{T_0 - T_1}{x^{1/3}} \frac{-\exp(-\eta^3/9A)}{\int_0^\infty [\exp(-\eta^3/9A)]\, d\eta}$$

and $\eta = 0$ at $y = 0$,

$$h = \frac{k}{x^{1/3}\int_0^\infty [\exp(-\eta^3/9A)]\, d\eta}$$

If the integral is evaluated, the final result is

$$h = 0.43k\left(\frac{\beta\rho C}{kx}\right)^{1/3} \tag{7-20}$$

Equation (7-20) predicts that the local coefficient will decrease as x increases, which is qualitatively correct.

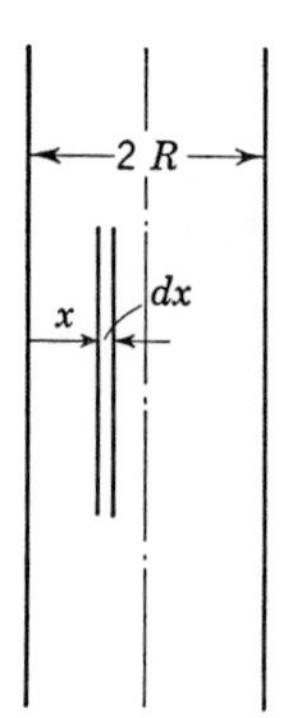

Fig. 7-1. Infinite slab.

7-5. Separation of Variables.

A technique termed "separation of variables" often can be used to determine the solution of a linear partial differential equation. The method is best illustrated by an example.

Example 7-2. Heat Conduction in a Slab. Suppose that a slab (extending indefinitely in the y and z directions) at an initial temperature T_1 has its two faces suddenly cooled to T_0. What is the relation between temperature, time after quenching, and position within the slab?

The origin is placed at one face, with the other face at $x = 2R$, as shown in Fig. 7-1. (By experience, it is found that the problem is more difficult if the origin is placed in the center of the slab.) Heat is lost at the two faces, and, since the solid extends indefinitely in the y and z directions, it is apparent that heat flows only in the x direction. Consequently, $\partial^2 t/\partial y^2$ and $\partial^2 t/\partial z^2$ are zero, and the heat-conduction equation (6-122) becomes

$$\frac{\partial T}{\partial t} = \alpha \frac{\partial^2 T}{\partial x^2} \tag{7-21}$$

with the boundary conditions

$$T = T_1 \qquad \text{at } t = 0,\ x > 0 \tag{7-22a}$$
$$T = T_0 \qquad \text{at } t = \infty,\ x > 0 \tag{7-22b}$$
$$T = T_0 \qquad \text{at } x = 0,\ t > 0 \tag{7-22c}$$
$$T = T_0 \qquad \text{at } x = 2R,\ t > 0 \tag{7-22d}$$

In order to simplify the boundary conditions, the change of variable

$$\theta = \frac{T - T_0}{T_1 - T_0} \tag{7-23}$$

is made. Then (7-21) and (7-22) become

$$\frac{\partial \theta}{\partial t} = \alpha \frac{\partial^2 \theta}{\partial x^2} \tag{7-24}$$

with the boundary conditions

$$\theta = 1 \qquad \text{at } t = 0,\ x > 0 \tag{7-25a}$$
$$\theta = 0 \qquad \text{at } t = \infty,\ x > 0 \tag{7-25b}$$
$$\theta = 0 \qquad \text{at } x = 0,\ t > 0 \tag{7-25c}$$
$$\theta = 0 \qquad \text{at } x = 2R,\ t > 0 \tag{7-25d}$$

Now *assume* that the solution of (7-24) is of the form

$$\theta = X(x)\tau(t) \tag{7-26}$$

where *$X(x)$ is a function of x only* and *$\tau(t)$ is a function of t only*. Note that this form of solution permits the boundary conditions (7-25) to be satisfied, although it does not guarantee that this is possible. When (7-26) is replaced in (7-24), there results

$$X \frac{\partial \tau}{\partial t} = \alpha\tau \frac{\partial^2 X}{\partial x^2}$$

or

$$\frac{1}{X}\frac{\partial^2 X}{\partial x^2} = \frac{1}{\alpha\tau}\frac{\partial \tau}{\partial t} \tag{7-27}$$

Now, by definition, τ is independent of x, and so as x varies, the right-hand side of (7-27) remains constant. Similarly, the left-hand side of (7-27) remains constant as t varies. *Consequently,* (7-27) *must equal a constant:*

$$\frac{1}{X}\frac{\partial^2 X}{\partial x^2} = \frac{1}{\alpha\tau}\frac{\partial \tau}{\partial t} = -a^2 \tag{7-28}$$

Equation (7-28) is equivalent to two total differential equations:

$$\frac{d\tau}{dt} + a^2\alpha\tau = 0 \tag{7-29}$$

$$\frac{d^2 X}{dx^2} + a^2 X = 0 \tag{7-30}$$

The argument leading to (7-28) requires only that (7-27) equal a constant. This constant could be real and either positive or negative, zero, imaginary, or complex.

A decision regarding the nature of the constant depends upon the boundary conditions. Now except when $a^2 = 0$, the solution of (7-29) and (7-30) will always be of the form

$$\tau = c_1 \exp(-a^2\alpha t) \tag{7-31a}$$
$$X = c_2 \cos ax + c_3 \sin ax \tag{7-31b}$$

and when $a^2 = 0$,

$$\tau = c_4 \tag{7-32a}$$
$$X = c_5 + c_6 x \tag{7-32b}$$

and the solution will be of the form

$$\theta = c_7 + c_8 x + \exp(-a^2\alpha t)(c_9 \cos ax + c_{10} \sin ax) \tag{7-33}$$

Now the boundary condition (7-25*b*) requires $\theta \to 0$ as $t \to \infty$, all x. This will be true only if a^2 is *real* and *positive;* for any other condition on a^2 the exponential term in (7-33) will not vanish as $t \to \infty$. Furthermore, c_7 and c_8 must be zero. These conditions reduce (7-33) to

$$\theta = \exp(-a^2\alpha t)(c_9 \cos ax + c_{10} \sin ax) \tag{7-34}$$

where a^2 is real and positive and consequently a is real. The condition (7-25*c*), $\theta = 0$ at $x = 0$, requires c_9 to be zero, since $\cos 0 = 1$. Then (7-34) becomes

$$\theta = A\,[\exp(-a^2\alpha t)] \sin ax \tag{7-35}$$

The condition (7-25*d*), $\theta = 0$ at $x = 2R$, can be satisfied if

$$a = \frac{n\pi}{2R} \tag{7-36}$$

where *n is a nonzero integer*. The solution then has the form

$$\theta = A\left[\exp\left(-\alpha\frac{n^2\pi^2}{4R^2}t\right)\right]\sin\frac{n\pi x}{2R} \tag{7-37}$$

and it remains to determine A such that (7-25*a*), $\theta = 1$ at $t = 0$, is satisfied for all values of x. It is evident that no single value of A will satisfy (7-25*a*). However, as the original differential equation (7-24) was linear, it follows that the partial differential equation and the boundary conditions (7-25*b, c, d*) will be satisfied by a solution which consists of an infinite sum of terms having the form of (7-37):

$$\begin{aligned}\theta = {} & A_1\left[\exp\left(-\alpha\frac{\pi^2 t}{4R^2}\right)\right]\sin\frac{\pi x}{2R} + A_2\left[\exp\left(-\alpha\frac{4\pi^2 t}{4R^2}\right)\right]\sin\frac{2\pi x}{2R} \\ & + \cdots + A_n\left[\exp\left(-\alpha\frac{n^2\pi^2 t}{4R^2}\right)\right]\sin\frac{n\pi x}{2R} + \cdots \\ = {} & \sum_{n=1}^{n=\infty} A_n\left[\exp\left(-\frac{\alpha n^2\pi^2 t}{4R^2}\right)\right]\sin\frac{n\pi x}{2R}\end{aligned} \tag{7-38}$$

The boundary condition (7-25*a*) then requires

$$1 = \sum_{n=1}^{\infty} A_n \sin\frac{n\pi x}{2R} \tag{7-39}$$

The problem now is to determine the values of A_n which make (7-39) true. To do this, (7-39) is multiplied by $[\sin(m\pi x/2R)]\,dx$, where m is an integer, and the result integrated between 0 and $2R$:

$$\int_0^{2R} \left(\sin \frac{n\pi x}{2R}\right) dx = \int_0^{2R} \left(\sin \frac{m\pi x}{2R}\right) \left(\sum_{n=1}^{\infty} A_n \sin \frac{n\pi x}{2R}\right) dx \qquad (7\text{-}40)$$

If the integral of the sum may be taken equal to the sum of the integrals, (7-40) may be written in the form

$$\int_0^{2R} \left(\sin \frac{m\pi x}{2R}\right) dx = \sum_{n=1}^{\infty} A_n \int_0^{2R} \left(\sin \frac{m\pi x}{2R}\right) \left(\sin \frac{n\pi x}{2R}\right) dx \qquad (7\text{-}41)$$

Now the integral of each term in the right-hand side of (7-41) is zero *except* when $m = n$ and

$$\int_0^{2R} \left(\sin \frac{n\pi x}{2R}\right)^2 dx = R\dagger$$

Consequently,

$$A_n = \frac{1}{R} \int_0^{2R} \left(\sin \frac{n\pi x}{2R}\right) dx = [1 - (-1)^n] \frac{2}{n\pi} \qquad (7\text{-}42)$$

The formal solution of the problem is then given by (7-38) and (7-42). When these two relations are combined, the solution may be written in the form

$$\begin{aligned} \frac{T - T_1}{T_0 - T_1} &= \frac{4}{\pi} \left\{ \left[\exp\left(-\alpha \frac{\pi^2 t}{4R^2}\right) \right] \sin \frac{\pi x}{2R} + \frac{1}{3} \left[\exp\left(-\frac{9\alpha\pi^2 t}{4R^2}\right) \right] \right. \\ &\left. \sin \frac{3\pi x}{2R} + \frac{1}{5} \left[\exp\left(-\frac{25\alpha\pi^2 t}{4R^2}\right) \right] \sin \frac{5\pi x}{2R} + \cdots \right\} \\ &= \frac{2}{\pi} \sum_{n=1}^{\infty} \frac{1 - (-1)^n}{n} \left[\exp\left(-\frac{\alpha n^2 \pi^2 t}{4R^2}\right) \right] \sin \frac{n\pi x}{2R} \end{aligned} \qquad (7\text{-}43)$$

The analytical solution is then an infinite series.

The representation of a function by means of an infinite series of sine functions as in (7-39) is known as a "Fourier sine series" and represents a special case of a very useful technique. The following section discusses the procedure in more detail.

7-6. Orthogonal Functions. Two functions $\phi_m(x)$ and $\phi_n(x)$ are said to be "orthogonal" with respect to the *weighting function* $r(x)$ over the interval a, b if

$$\int_a^b r(x)\phi_m(x)\,\phi_n(x)\,dx = 0 \qquad (7\text{-}44)$$

In Example 7-2 it was demonstrated that $\sin(m\pi x/2R)$ and $\sin(n\pi x/2R)$, where m and n are distinct integers, were orthogonal in the sense of (7-44) with respect to the weighting function unity over the interval 0, $2R$. Reexamination of the solution of Example 7-2 will show that the function $\sin n\pi x/2R$ resulted from the solution of the differential equation

$$\frac{d^2X}{dx^2} + a^2X = 0 \qquad (7\text{-}30)$$

† Peirce Nos. 488 and 489 (see Chap. 1, reference 3).

subject to the boundary conditions

$$X = 0 \qquad \text{at } x = 0$$
$$X = 0 \qquad \text{at } x = 2R$$

The solution of (7-30) which satisfies the condition $X = 0$ at $x = 0$ is of the form

$$X = c \sin ax$$

and the condition $X = 0$ at $x = 2R$ requires

$$c \sin 2aR = 0$$

This condition is satisfied only if a takes on definite discrete values: $a_n = n\pi/2R$, where n is a nonzero integer. These values of a, a_n, are called the "characteristic values," or "eigenvalues," of the equation, and the corresponding solutions, $\sin a_n x$, are called the "characteristic functions," or "eigenfunctions."

Now it can be shown that the linear, homogeneous, second-order equation

$$\frac{d}{dx}\left[p(x)\frac{dy}{dx}\right] + [q(x) + \lambda r(x)]y = 0 \tag{7-45}$$

with boundary conditions at *each* end point $x = a$ and $x = b$ which satisfy *one* of the following forms:

$$\begin{aligned} y &= 0 \\ \frac{dy}{dx} &= 0 \qquad x = a \qquad \text{or} \qquad x = b \\ y + \alpha\frac{dy}{dx} &= 0 \qquad \alpha = \text{const} \end{aligned} \tag{7-46}$$

has as solutions the characteristic functions $\phi_m(x)$ and $\phi_n(x)$ which are orthogonal in the sense of (7-44), provided that the *characteristic values* λ_n and λ_m *are different.*

The restrictive boundary conditions (7-46) may be made less severe if the following is true:

If $p(x) = 0$ when $x = a$ or $x = b$, then the only requirement is that y be finite at the point where $p(x) = 0$.

If $p(a) = p(b)$, the characteristic solutions will be orthogonal if $y(b) = y(a)$ and $y'(b) = y'(a)$. This latter condition is satisfied if the characteristic solutions are required to be *periodic* of period $b - a$.

A second-order differential equation of the form

$$g_0(x)\frac{d^2y}{dx^2} + g_1(x)\frac{dy}{dx} + [g_2(x) + \lambda g_3(x)]y = 0 \tag{7-47}$$

may be transformed into the form of (7-45) by means of the relations

$$\begin{aligned} p(x) &= \exp \int \frac{g_1(x)}{g_0(x)}\,dx \\ q(x) &= \frac{g_2(x)}{g_0(x)}\,p(x) \\ r(x) &= \frac{g_3(x)}{g_0(x)}\,p(x) \end{aligned} \tag{7-48}$$

An equation of the form and with the boundary conditions of (7-45) is known as a "Sturm-Liouville equation."

7-7. Expansion in a Series of Orthogonal Functions. Let the characteristic functions $\phi(x)$ result from an equation of the form of (7-45) and hence be orthogonal. The formal expansion of a function $f(x)$ can be written in the form

$$f(x) = A_0\phi_0(x) + A_1\phi_1(x) + A_2\phi_2(x) + \cdots = \sum_{n=0}^{\infty} A_n\phi_n(x) \tag{7-49}$$

The values of the constants A_n may be obtained by making use of the orthogonal properties of the functions $\phi(x)$:

$$\int_a^b r(x)\phi_m(x)f(x)\,dx = \int_a^b r(x)\phi_m(x)\left[\sum_{n=0}^{\infty} A_n\phi_n(x)\right]dx \tag{7-50}$$

If the integral of the sum is equal to the sum of the integrals, (7-50) becomes

$$\int_a^b r(x)\phi_m(x)f(x)\,dx = \sum_{n=0}^{\infty} A_n \int_a^b r(x)\phi_m(x)\phi_n(x)\,dx \tag{7-51}$$

However, the terms of the sum appearing on the right-hand side of (7-51) are zero *except for the case where* $m = n$. When $m = n$, there results

$$A_n \int_a^b r(x)[\phi_n(x)]^2\,dx = \int_a^b r(x)\phi_n(x)f(x)\,dx$$

or

$$A_n = \frac{\int_a^b r(x)\phi_n(x)f(x)\,dx}{\int_a^b r(x)[\phi_n(x)]^2\,dx} \tag{7-52}$$

Equation (7-52) permits the constants A_n of the series expansion (7-49) to be evaluated.

It can be shown that the series (7-49) represents $f(x)$ if the following conditions are fulfilled: $f(x)$ is bounded and piecewise differentiable; $p(x)$, $q(x)$, and $r(x)$ are regular in the interval a, b; and $p(x)$ and $r(x)$ are positive throughout the interval a, b, including the points a and b. Under these circumstances, (7-49) converges to $f(x)$ inside the interval a, b at all points where $f(x)$ is continuous and converges to the value $\frac{1}{2}[f(x+) + f(x-)]$ at points where the function $f(x)$ makes finite jumps. The series (7-49) may or may not converge to $f(x)$ at $x = a$ and $x = b$. The series is extremely versatile; $f(x)$ may exhibit a finite number of finite discontinuities and may even be defined by different analytical expressions over different parts of the interval of representation.

Expansion in a series of orthogonal functions was used in the solution of Example 7-2. The equation (7-30) and its boundary conditions satisfy the Sturm-Liouville conditions with $\phi_n(x) = \sin a_n x = \sin (n\pi x/2R)$. The end points are $a = 0$, $b = 2R$, and Eq. (7-30) may be written (since $a_n = n\pi/2R$)

$$\frac{d^2X}{dx^2} + \frac{n^2\pi^2}{4R^2} X = 0$$

Comparison with (7-47) gives $g_0(x) = 1; g_1(x) = 0; g_2(x) = 0; g_3(x) = 1;$ $\lambda = a_n^2 = n^2\pi^2/4R^2$. Use of (7-48) gives $p(x) = 1$; $q(x) = 0$; $r(x) = 1$. The series (7-39) then represents an expansion in a series of orthogonal functions, and the steps leading to (7-42) follow from the discussion of this section.

Example 7-3. Unsteady-state Heat Transfer to a Sphere. A sphere, initially at a uniform temperature T_0, is suddenly placed in a fluid medium whose temperature is maintained constant at a value T_1. The heat-transfer coefficient between the medium and the sphere is constant at a value h. The sphere is isotropic, and the temperature variation of the physical properties of the material forming the sphere may be neglected. Derive the equation relating the temperature of the sphere to the radius r and time t.

The differential equation covering this situation was derived in Chap. 6:

$$\nabla^2 T = \frac{\rho}{k}\frac{\partial T}{\partial t} = \frac{1}{\alpha}\frac{\partial T}{\partial t} \tag{7-53a}$$

In the present case, the temperature is independent of the coordinates ϕ and θ, and (7-53a) may be written as

$$\frac{\partial^2 T}{\partial r^2} + \frac{2}{r}\frac{\partial T}{\partial r} = \frac{1}{\alpha}\frac{\partial T}{\partial t} \tag{7-53b}$$

with the boundary conditions

$$T = T_0 \qquad \text{at } t = 0, r > 0 \tag{7-54a}$$
$$T = T_1 \qquad \text{at } t = \infty, r > 0 \tag{7-54b}$$
$$T = \text{finite} \qquad \text{at } r = 0, t > 0 \qquad \text{or, alternatively} \qquad \frac{\partial T}{\partial r} = 0 \qquad \text{at } r = 0 \tag{7-54c}$$

which is a consequence of the symmetry of the temperature distribution.

$$q = 4\pi r_0^2 h(T_s - T_1) = -k4\pi r_0^2 \left(\frac{\partial T}{\partial r}\right)_{r=r_0} \tag{7-54d}$$

or

$$\left(\frac{\partial T}{\partial r}\right)_{r=r_0} = -\frac{h}{k}(T_s - T_1)$$

The last boundary condition (7-54d) is derived from the following considerations: Examine a small spherical shell bounded by $r = r_0$ and $r = r_0 - \epsilon$, where r_0 is the outer radius of the sphere. The rate at which heat enters the shell is

$$-k4\pi(r_0 - \epsilon)^2 \left(\frac{\partial T}{\partial r}\right)_{r=r_0-\epsilon}$$

The rate at which heat leaves the shell is $4\pi r_0^2 h(T_s - T_1)$, where T_s is the temperature of the sphere at $r = r_0$. The rate of accumulation of heat within the shell is

$$\frac{\partial}{\partial t}\left[\rho c 4\pi \left(r_0 - \frac{\epsilon}{2}\right)^2 \epsilon T\right]$$

Hence,

$$-k4\pi(r_0 - \epsilon)^2 \left(\frac{\partial T}{\partial r}\right)_{r=r_0-\epsilon} - 4\pi r_0^2 h(T_s - T_1) = \left[\rho c 4\pi \left(r_0 - \frac{\epsilon}{2}\right)^2\right] \epsilon \frac{\partial T}{\partial t} \tag{7-55}$$

Now let the shell thickness ϵ approach zero. The right-hand side of (7-55) vanishes as $\epsilon \to 0$ and $r_0 - \epsilon \to r_0$, leading to the boundary condition (7-54d).

The solution of (7-53) is somewhat facilitated by a change of variable, but this is not necessary and will not be done here. It is now assumed that the solution of (7-53) is of the form

$$T = R(r)\tau(t) \tag{7-56}$$

Substituting (7-56) into (7-53) and using the arguments employed in Example 7-2 lead to

$$\frac{R''}{R} + \frac{2R'}{rR} = \frac{1}{\alpha}\frac{\tau'}{\tau} = -a^2 \tag{7-57}$$

where a is a constant. Two equations result from (7-57):

$$r^2R'' + 2rR' + a^2r^2R = 0 \tag{7-58}$$

$$\tau' + \alpha a^2\tau = 0 \tag{7-59}$$

Equation (7-58) is one form of Bessel's equation and has solutions of the form

$$R = c_1 r^{-1/2} J_{1/2}ar + c_2 r^{-1/2} J_{-1/2}ar \qquad \text{if } a \neq 0 \tag{7-60a}$$

$$R = c_3 + c_4/r \qquad \text{if } a = 0 \tag{7-60b}$$

The solutions of (7-59) are of the form

$$\tau = c_5 \exp(-\alpha a^2 t) \qquad \text{if } a \neq 0 \tag{7-61a}$$

$$\tau = c_6 \qquad \text{if } a = 0 \tag{7-61b}$$

However, the Bessel functions of half order may be expressed in terms of sines and cosines. Consequently, (7-60a) may be written in the form

$$R = \frac{1}{r}\sqrt{\frac{2}{a\pi}}\,(c_1 \sin ar + c_2 \cos ar) \tag{7-60c}$$

Use of (7-60) and (7-61) in (7-56) gives

$$T = \left(\frac{1}{r}\sqrt{\frac{2}{a\pi}}\right)[\exp\,(-\alpha a^2 t)](A \sin ar + B \cos ar) + \frac{C}{r} + D \tag{7-62}$$

as the form of the assumed solution. The boundary condition (7-54c) requires $B = C = 0$.

The boundary condition (7-54*b*) requires

$$D = T_1$$

Equation (7-62) then becomes

$$T - T_1 = \sqrt{\frac{2}{a\pi}}\,[\exp(-\alpha a^2 t)]\,\frac{1}{r}\,(A \sin ar) \tag{7-63}$$

Differentiation of (7-63) gives

$$\frac{\partial T}{\partial r} = \sqrt{\frac{2}{a\pi}}\,[\exp(-\alpha a^2 t)]\left(\frac{Aa \cos ar}{r} - \frac{A \sin ar}{r^2}\right) \tag{7-64}$$

Substitution of (7-63) and (7-64) into the boundary condition (7-54*d*) gives, after simplification,

$$\tan ar_0 = \frac{ar_0 k}{k - r_0 h} \tag{7-65}$$

Equation (7-65) is a transcendental equation, and, for specific values of r_0, k, and h, an infinite number of values of a: a_0, a_1, a_2, . . . satisfy (7-65). Denote the values of a which satisfy (7-65) by the generic term a_n. Then (7-63) becomes

$$T - T_1 = \sqrt{\frac{2}{a_n\pi}}\,[\exp(-\alpha a_n^2 t)]\,\frac{1}{r}\,(A \sin a_n r) \tag{7-66}$$

The boundary condition (7-54*a*) requires that

$$T_0 - T_1 = \sqrt{\frac{2}{a_n\pi}}\,\frac{A \sin a_n r}{r} \tag{7-67}$$

Again it is clear that no single value of A will satisfy (7-67) for all values of r, and Eq. (7-66) only represents the form of the final solution. The original differential equation is satisfied by a solution which consists of a sum of terms each of the form of (7-66). Consequently, the solution is taken as

$$T - T_1 = \sum_n \sqrt{\frac{2}{a_n\pi}}\,[\exp(-\alpha a_n^2 t)]\,\frac{A_n \sin a_n r}{r} \tag{7-68}$$

when the summation is made over the infinite series of values of a_n that satisfy (7-65). The boundary condition (7-54*a*) provides a possible means to evaluate the value of A_n corresponding to a_n:

$$T_0 - T_1 = \sum_n \sqrt{\frac{2}{a_n\pi}}\,\frac{A_n \sin a_n r}{r} \tag{7-69}$$

It remains to be determined if the constants A_n can be obtained by making use of the properties of orthogonal functions.

The differential equation (7-58) has solutions of the form $\phi_n(r) = \sqrt{2/a_n\pi}\,[(\sin a_n r)/r]$ [see (7-60*a*)]. Furthermore, it only is desired that the series expansion of $f(x) = f(r) = T_0 - T_1$ represented by (7-69) hold over the range $0 \leqq r \leqq r_0$. At $r = 0$, the boundary condition (7-54*c*) in the form $\partial T/\partial r = 0$ is equivalent to $dR/dr = 0$ which satisfies (7-46). At $r = r_0$, (7-46) is satisfied in the form $R + b(dR/dr)$ by virtue of (7-63) and (7-64) and is equivalent to (7-65).

The differential equation (7-58) may be compared with (7-47) to give $g_0(x) = r^2$;

$g_1(x) = 2r$; $g_2(x) = 0$; $\lambda = a^2$; $g_c(x) = r^2$. Use of (7-48) then gives

$$p(x) = \exp \int \frac{2r}{r^2}\,dr = \exp \ln r^2 = r^2$$
$$q(x) = 0$$
$$r(x) = \frac{r^2}{r^2} r^2 = r^2$$

Consequently, $p(x)$, $q(x)$, and $r(x)$ are regular in the interval 0, r_0, and $p(x)$ and $r(x)$ are positive. Since $T_0 - T_1$ is bounded and piecewise differentiable, the functions

$$\phi_n(r) = \sqrt{\frac{2}{a_n\pi}}\,\frac{\sin a_n r}{r}$$

are orthogonal with respect to the weighting function r^2, and expansion of $T_0 - T_1$ in the series (7-69) is valid. Note also that since $p(x) = r^2$ is zero at the end point $r = 0$, the boundary-condition requirement discussion was not necessary.

Equation (7-69) corresponds then to (7-49), and the constants A_n of (7-69) may be obtained by use of the relation (7-52)

$$A_n = \frac{\int_0^{r_0} (r^2)\sqrt{2/a_n\pi}\,[(\sin a_n r)/r](T_0 - T_1)\,dr}{\int_0^{r_0} (r^2)(2/a_n\pi)[(\sin^2 a_n r)/r^2]\,dr}$$
$$= (T_0 - T_1)\frac{\sqrt{2a_n\pi}}{r_0}\left(\frac{\sin a_n r_0}{a_n^2} - r_0\frac{\cos a_n r_0}{a_n}\right) \tag{7-70}$$

The reader is urged to convince himself that the functions $\phi_n(r)$ of this example are truly orthogonal by carrying out the integrations indicated in Sec. 7-6.

The solution of this problem is then given by (7-68), with a_n specified by (7-65) and A_n by (7-70).

7-8. Expansion in a Double Series of Orthogonal Functions.

The following example illustrates the type of problem that gives rise to a solution involving expansion of a function in a double series of orthogonal functions and the method of handling such a situation.

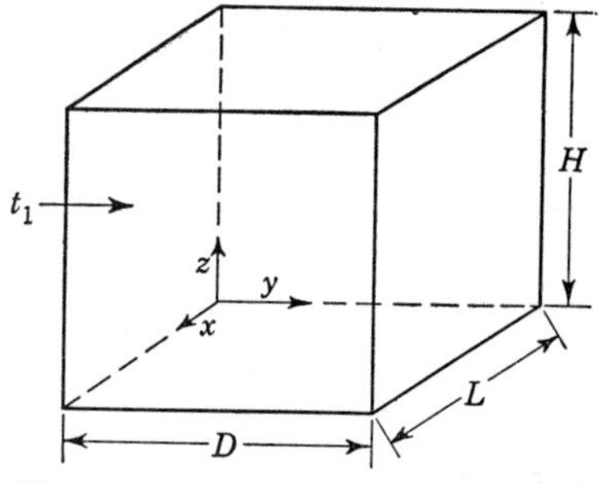

FIG. 7-2. Rectangular parallelepiped.

Example 7-4. Temperature Distribution in a Rectangular Parallelepiped. It is desired to determine the equation relating the *steady-state* temperature to position in an isotropic rectangular parallelepiped. The dimensions of the parallelepiped are $0 \leqq x \leqq L$, $0 \leqq y \leqq D$, $0 \leqq z \leqq H$ (see Fig. 7-2). Five sides are maintained at the temperature T_0, and the remaining face (corresponding to $z = H$) is maintained at the temperature T_1.

The applicable partial differential equation is

$$\nabla^2 T = \frac{\partial^2 T}{\partial x^2} + \frac{\partial^2 T}{\partial y^2} + \frac{\partial^2 T}{\partial z^2} = 0 \tag{7-71}$$

With the change of variables

$$\theta = \frac{T - T_0}{T_1 - T_0} \tag{7-72}$$

Eq. (7-71) becomes

$$\frac{\partial^2 \theta}{\partial x^2} + \frac{\partial^2 \theta}{\partial y^2} + \frac{\partial^2 \theta}{\partial z^2} = 0 \tag{7-73}$$

with the boundary conditions

$$\theta = 0 \quad \text{for } x = 0, \text{ all } y \text{ and } z \tag{7-74a}$$
$$\theta = 0 \quad \text{for } x = L, \text{ all } y \text{ and } z \tag{7-74b}$$
$$\theta = 0 \quad \text{for } y = 0, \text{ all } x \text{ and } z \tag{7-74c}$$
$$\theta = 0 \quad \text{for } y = D, \text{ all } x \text{ and } z \tag{7-74d}$$
$$\theta = 0 \quad \text{for } z = 0, \text{ all } x \text{ and } y \tag{7-74e}$$
$$\theta = 1 \quad \text{for } z = H, \text{ all } x \text{ and } y \tag{7-74f}$$

Assume a solution of the form

$$\theta = X(x)Y(y)Z(z) \tag{7-75}$$

which reduces (7-73) to

$$\frac{X''}{X} + \frac{Y''}{Y} + \frac{Z''}{Z} = 0 \tag{7-76}$$

By virtue of the arguments used previously,

$$\frac{X''}{X} + \frac{Y''}{Y} = -\frac{Z''}{Z} = -a^2 \tag{7-77}$$

and

$$-\frac{X''}{X} = a^2 + \frac{Y''}{Y} = b^2 \tag{7-78}$$

where a and b are constants. The three total differential equations to be solved are then

$$X'' + b^2 X = 0 \tag{7-79a}$$
$$Y'' + (a^2 - b^2)Y = 0 \tag{7-79b}$$
$$Z'' - a^2 Z = 0 \tag{7-79c}$$

The solutions of (7-79) are of the following form:

For $a \neq b \neq 0$,

$$\begin{aligned} X &= c_1 \cos bx + c_2 \sin bx \\ Y &= c_3 \cos \sqrt{a^2 - b^2}\, y + c_4 \sin \sqrt{a^2 - b^2}\, y \\ Z &= c_5 \cosh az + c_6 \sinh az \end{aligned} \tag{7-80}$$

For $a = b \neq 0$,

$$\begin{aligned} X &= c_1 \cos bx + c_2 \sin bx \\ Y &= c_7 + c_8 y \\ Z &= c_5 \cosh az + c_6 \sinh az \end{aligned} \tag{7-81}$$

For $a = b = 0$,

$$\begin{aligned} X &= c_9 + c_{10} x \\ Y &= c_7 + c_8 y \\ Z &= c_{11} + c_{12} z \end{aligned} \tag{7-82}$$

For $a = 0$, $b \neq 0$,

$$\begin{aligned} X &= c_1 \cos bx + c_2 \sin bx \\ Y &= c_{13} \cosh by + c_{14} \sinh by \\ Z &= c_{11} + c_{12}z \end{aligned} \tag{7-83}$$

For $a \neq 0$, $b = 0$,

$$\begin{aligned} X &= c_9 + c_{10}x \\ Y &= c_{15} \cos ay + c_{16} \sin ay \\ Z &= c_5 \cosh az + c_6 \sinh az \end{aligned} \tag{7-84}$$

The boundary conditions permit some immediate simplification.

From (7-74*a*), $\theta = 0$ at $x = 0$, it is required that $X(0) = 0$ and hence $c_1 = 0$, $c_9 = 0$. From (7-74*c*), $\theta = 0$ at $y = 0$, it is required that $Y(0) = 0$ and hence $c_3 = 0$, $c_7 = 0$, $c_{13} = 0$, $c_{15} = 0$. From (7-74*e*), $\theta = 0$ at $z = 0$, it is required that $Z(0) = 0$ and hence $c_5 = 0$, $c_{11} = 0$.

From (7-74*b*), $X(L) = 0$ and $c_{10} = 0$, $b = n\pi/L$, where $n = 1, 2, 3, \ldots$, but not zero.

From (7-74*d*), $Y(D) = 0$ and $c_7 = 0$, $c_{14} = 0$, $\sqrt{a^2 - b^2} = m\pi/D$, where $m = 1, 2, 3, \ldots$, but not zero.

Then

$$a = \pm\pi \sqrt{\frac{n^2}{L^2} + \frac{m^2}{D^2}}$$

and the assumed solution is of the form

$$\theta = A \left(\sin \frac{n\pi x}{L}\right) \left(\sin \frac{m\pi y}{D}\right) \left(\sinh \pi \sqrt{\frac{n^2}{L^2} + \frac{m^2}{D^2}}\, z\right) \tag{7-85}$$

Since $\sinh z = (e^z - e^{-z})/2$, the positive root of a is taken, because θ cannot be negative.

The final boundary condition (7-74*f*) requires

$$1 = A \left(\sinh \pi \sqrt{\frac{n^2}{L^2} + \frac{m^2}{D^2}}\, H\right) \left(\sin \frac{n\pi x}{L}\right) \left(\sin \frac{m\pi y}{L}\right) \tag{7-86}$$

Again, (7-86) cannot be satisfied for all values of x and y by a single value of A. Consequently, a double infinite sum is assumed as the solution, each term having the form of (7-85):

$$\theta = \sum_{m=1}^{\infty} \sum_{n=1}^{\infty} A_{mn} \left(\sin \frac{n\pi x}{L}\right) \left(\sin \frac{m\pi y}{D}\right) \left(\sinh \pi \sqrt{\frac{n^2}{L^2} + \frac{m^2}{D^2}}\, z\right) \tag{7-87}$$

Equation (7-87) satisfies the original differential equation (7-73), and the problem is to satisfy the boundary condition (7-74*f*) which requires

$$1 = \sum_{n=1}^{\infty} \sum_{m=1}^{\infty} A_{mn} \left(\sinh \pi \sqrt{\frac{n^2}{L^2} + \frac{m^2}{D^2}}\, H\right) \left(\sin \frac{n\pi x}{L}\right) \left(\sin \frac{m\pi y}{D}\right) \tag{7-88}$$

Now the differential equations and associated boundary conditions which give rise to the terms $\sin (n\pi x/L)$ and $\sin (m\pi x/D)$ are such that each forms an orthogonal set with a weighting function of unity. Consequently, it is expected that the constants

A_{mn} of (7-88) can be evaluated as follows:

$$\iint_{00}^{LD} \left(\sin \frac{p\pi x}{L}\right)\left(\sin \frac{q\pi y}{D}\right) dy\, dx = \int_0^L \left(\sin \frac{p\pi x}{L}\right) dx \int_0^D \left(\sin \frac{q\pi y}{D}\right) dy$$

$$= \iint_{00}^{LD} \left(\sin \frac{p\pi x}{L}\right)\left(\sin \frac{q\pi y}{D}\right)\left[\sum_{m=1}^{\infty}\sum_{n=1}^{\infty} A_{mn}\left(\sinh \pi \sqrt{\frac{n^2}{L^2}+\frac{m^2}{D^2}}\, H\right)\left(\sin \frac{n\pi x}{L}\right)\right.$$

$$\left.\left(\sin \frac{m\pi y}{D}\right)\right] dy\, dx = \sum_{m=1}^{\infty}\sum_{n=1}^{\infty} A_{mn}\left(\sinh \pi \sqrt{\frac{n^2}{L^2}+\frac{m^2}{D^2}}\, H\right)$$

$$\left[\int_0^L \left(\sin \frac{p\pi x}{L}\right)\left(\sin \frac{n\pi x}{L}\right) dx\right]\left[\int_0^D \left(\sin \frac{q\pi y}{D}\right)\left(\sin \frac{m\pi y}{D}\right) dy\right] \quad (7\text{-}89)$$

The right-hand side of (7-89) is equal to zero unless $p = n$ and $q = m$. For this case, the right-hand side equals A_{mn} $(\sinh \pi \sqrt{n^2/L^2 + m^2/D^2}\, H)LD/4$. Consequently,

$$A_{mn} = \frac{4}{LD \sinh \pi \sqrt{n^2/L^2 + m^2/D^2}\, H} \int_0^L \left(\sin \frac{n\pi x}{L}\right) dx \int_0^D \left(\sin \frac{m\pi y}{L}\right) dy \quad (7\text{-}90)$$

The solution of the problem then consists of Eq. (7-87) with A_{mn} given by (7-90). The extension of this procedure to more complicated situations follows easily.

7-9. Fourier Series. A special but widely used case of expansion in a series of orthogonal functions is the "Fourier series." In the *symmetrical* interval $-L, L$, the Fourier-series representation of the function $f(x)$ is defined by

$$f(x) = \sum_{n=0}^{\infty} A_n \cos \frac{n\pi x}{L} + \sum_{n=0}^{\infty} B_n \sin \frac{n\pi x}{L} \quad (7\text{-}91)$$

where n is a positive integer. The justification for this form follows from the discussion of Sec. 7-7. The limitation to the symmetrical interval $-L, L$ forces the representation to be periodic in agreement with the condition of Sec. 7-6. If (7-91) is multiplied by $[\cos (m\pi x/L)]\, dx$ and the result integrated between $-L$ and L, the coefficients A_n are found. A similar procedure using $[\sin (m\pi x/L)]\, dx$ gives B_n:

$$\begin{aligned} A_0 &= \frac{1}{2L}\int_{-L}^{L} f(x)\, dx \\ A_n &= \frac{1}{L}\int_{-L}^{L} f(x) \cos \frac{n\pi x}{L}\, dx \\ B_0 &= 0 \\ B_n &= \frac{1}{L}\int_{-L}^{L} f(x) \sin \frac{n\pi x}{L}\, dx \end{aligned} \quad (7\text{-}92)$$

Whether or not the Fourier-series expansion of the function $f(x)$ consists only of sine terms or only of cosine terms or of both depends upon whether the function is *even* or *odd* or neither.

An even function, illustrated in Fig. 7-3, is one for which $f(x) = f(-x)$. x^2, $\cos nx$, $x \sin mx$, and a pure number are all even functions. If the function $f(x)$ is *even*,

$$A_0 = \frac{1}{L}\int_0^L f(x)\,dx$$
$$A_n = \frac{2}{L}\int_0^L f(x)\cos\frac{n\pi x}{L}\,dx \qquad (7\text{-}93)$$
$$B_n = 0$$

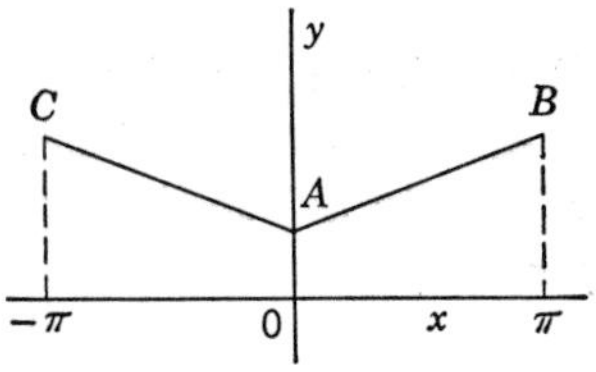

FIG. 7-3. Even function.

This results from (7-92) and the fact that

$$f(x) = f(-x)$$
$$f(x)\cos\left(\frac{n\pi x}{L}\right) = f(-x)\cos\left(-\frac{n\pi x}{L}\right)$$

and

$$f(x)\sin\frac{n\pi x}{L} = -f(-x)\sin -\frac{n\pi x}{L}$$

Consequently, in the evaluation of A_n, the integral from $-L$ to 0 adds to the integral from 0 to L, whereas in the evaluation of B_n these integrals cancel. This may be shown formally as follows: Let $f(x)$ be even. Then

$$A_0 = \frac{1}{2L}\int_{-L}^{L} f(x)\,dx = \frac{1}{2L}\int_0^L f(x)\,dx + \int_{-L}^{0} f(x)\,dx$$
$$= \frac{1}{2L}\int_0^L f(x)\,dx - \int_0^{-L} f(x)\,dx$$

Now let $x = -w$. Then

$$\int_0^{-L} f(x)\,dx = -\int_0^L f(-w)\,dw = -\int_0^L f(w)\,dw = -\int_0^L f(x)\,dx$$

whereupon

$$A_0 = \frac{1}{L}\int_0^L f(x)\,dx$$

The expression for A_n follows in the same fashion.

B_n is found as follows:

$$B_n = \frac{1}{L}\int_{-L}^{L} f(x)\sin\frac{n\pi x}{L}\,dx$$
$$= \frac{1}{L}\int_0^L f(x)\sin\frac{n\pi x}{L}\,dx - \int_0^{-L} f(x)\sin\frac{n\pi x}{L}\,dx$$

But, for an even function,

$$-\int_0^{-L} f(x) \sin \frac{n\pi x}{L}\,dx = +\int_0^{L} f(-w) \sin -\frac{n\pi w}{L}\,dw$$
$$= -\int_0^{L} f(x) \sin \frac{n\pi x}{L}\,dx$$

Hence for an *even* function $B_n = 0$.

An odd function, illustrated in Fig. 7-4, is one for which $f(x) = -f(-x)$. x, x^3, $\sin nx$, $x^2 \sin mx$, $x \cos nx$ are odd functions. If the function $f(x)$ is *odd*,

$$A_0 = 0$$
$$A_n = 0 \tag{7-94}$$
$$B_n = \frac{2}{L}\int_0^L f(x) \sin \frac{n\pi x}{L}\,dx$$

If only the interval 0 to L is of interest, then $f(x)$ may be expanded in either a pure sine series by use of (7-94), or a pure cosine series by use of (7-93). This may be done when $f(x)$ is even, odd, or neither. When this is done, the expansion in terms of sines generates an odd function which, in general, will not represent $f(x)$ outside 0, L, and expansion in terms of cosines generates an even function which, in general, will not represent $f(x)$ outside 0, L. Just this technique was used in Example 7-2. There, it was necessary to represent $f(x) = 1$ by means of a sine series in the interval 0, $2R$. $f(x) = 1$ is actually an even function, but over the interval 0, $2R$ it was represented by a sine series.

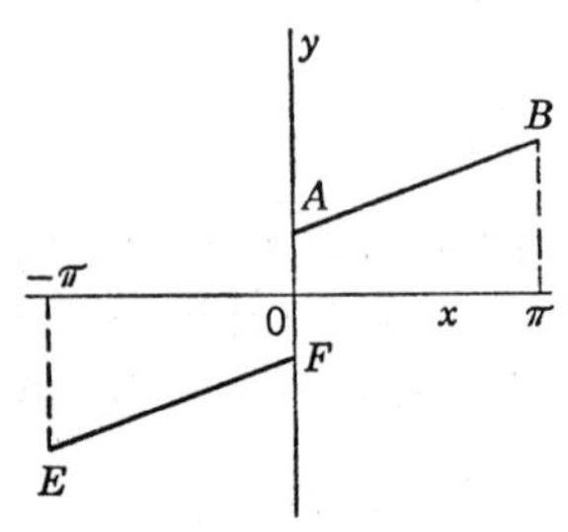

FIG. 7-4. Odd function.

Example 7-5. Fourier-series Representation of $f(x) = 1$. In order to illustrate the manner in which the terms of the series combine to represent $f(x)$, the function $f(x) = 1$ will be expressed as a sine series in the interval 0, π. Use of (7-94) gives

$$A_0 = 0$$
$$A_n = 0$$
$$B_n = \frac{2}{n\pi}[1 - (-1)^n]$$

Substitution into (7-91) gives

$$f(x) = 1 = \frac{4}{\pi}\left(\sin x + \frac{1}{3}\sin 3x + \frac{1}{5}\sin 5x + \cdots\right)$$

Figure 7-5 indicates the manner in which the function is built up by the addition of successive terms of the series. The single sine term represents the function poorly, and even five terms do not show good agreement. The series converges slowly in the region 0 to π. If an infinite number of terms are added, the result is $\pi/4 = 0.7854$, which is represented by the solid horizontal line of Fig. 7-5.

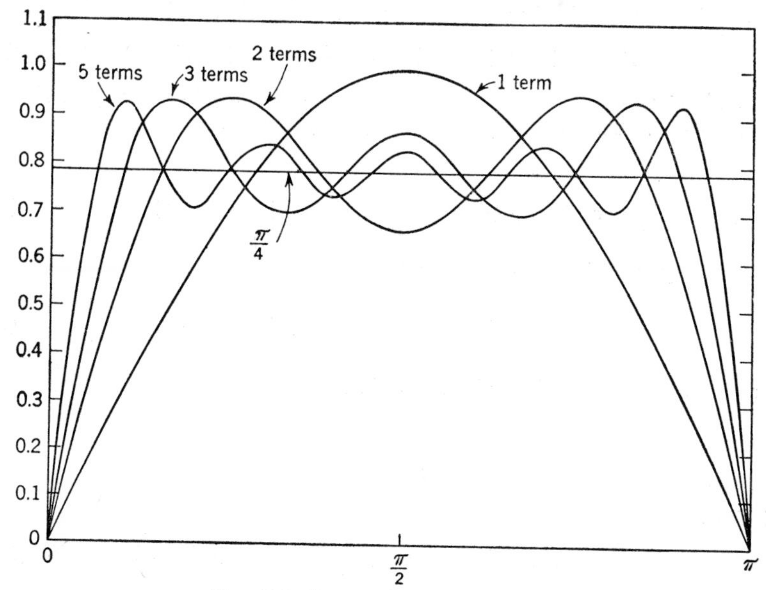

FIG. 7-5. Summation of sine terms.

7-10. Representation over an Infinite Interval. The previous examples have illustrated the utility of representation of a function $f(x)$ over a finite interval a, b by means of an infinite series of orthogonal functions. In certain cases, however, the interval becomes infinite, that is, $b \to \infty$. The treatment of this situation differs from the finite-interval technique. The following example illustrates such a case.

Example 7-6. Heat Conduction in a Slab of Infinite Thickness. Imagine a slab which extends indefinitely in the x direction and so arranged that heat flows in the x direction only. At time $t = 0$, suppose that the temperature distribution inside the slab is given by $T = f(x)$. Now expose the slab face at $x = 0$ to a constant-temperature medium at T_1. The heat-transfer coefficient between the medium and the slab surface is constant at a value h. The temperature variation of the physical properties of the material forming the slab may be neglected. Determine the temperature distribution within the slab.

The heat-conduction equation for this case becomes

$$\frac{\partial T}{\partial t} = \alpha \frac{\partial^2 T}{\partial x^2} \tag{7-95}$$

with the boundary conditions

$$T = f(x) \qquad \text{at } t = 0 \tag{7-96a}$$

$$\frac{\partial T}{\partial x} = \frac{h}{k}(T - T_1) \qquad \text{at } x = 0,\ t > 0 \tag{7-96b}$$

$$T \to T_1 \qquad \text{at } x = 0,\ t \to \infty \tag{7-96c}$$

If a solution of the form

$$T = X(x)\tau(t)$$

is assumed, steps completely paralleling those of Example 7-2 give

$$T = c_1 + c_2 x + [\exp(-a^2\alpha t)](c_3 \cos ax + c_4 \sin ax) \tag{7-97}$$

Where a is a nonzero, real constant, a must be real in order to ensure that T remain finite as $t \to \infty$. The boundary condition (7-96c) requires

$$c_1 = T_1$$

$$\frac{\partial T}{\partial x} = c_2 + [\exp(-a^2\alpha t)](-ac_3 \sin ax + ac_4 \cos ax) \tag{7-98}$$

The boundary condition (7-96b) demands

$$c_2 + [\exp(-a^2\alpha t)]ac_4 = +[\exp(-a^2\alpha t)]c_3 \frac{h}{k} \tag{7-99}$$

Equation (7-99) must hold for all values of $t > 0$. Hence

$$c_2 = 0$$

$$c_3 = ac_4 \frac{k}{h}$$

and (7-97) becomes

$$T = T_1 + c_4 [\exp(-a^2\alpha t)] \left(\frac{k}{h} a \cos ax + \sin ax\right) \tag{7-100}$$

The remaining boundary condition (7-96a) must be satisfied. Now (7-95) is satisfied by (7-100) for all real values of a. Hence, in a formal manner, a solution of the type

$$T = T_1 + \sum_a A_a [\exp(-a^2\alpha t)] \left(\frac{k}{h} a \cos ax + \sin ax\right)$$

or

$$T = T_1 + \int_0^\infty A_a [\exp(-a^2\alpha t)] \left(\frac{k}{h} a \cos ax + \sin ax\right) da \tag{7-101}$$

is tried. The integral replaces the sum because the only restriction on a is that it be real. Furthermore, a may be permitted to take on the value of zero in the integration, since the kernel of the integral is zero at $a = 0$. The limits of the integral are taken to be zero and infinity rather than $-\infty$ to $+\infty$, because the functions forming the kernel are odd functions of a and with the limits of $-\infty$ to ∞ the integral would vanish.

The boundary condition (7-96a) requires

$$f(x) = T_1 + \int_0^\infty A_a \left(\frac{k}{h} a \cos ax + \sin ax\right) da \tag{7-102}$$

Now proceed as though the values of A_a were to be obtained by the methods used for expansion in a series of orthogonal functions. In this case the weighting function is unity. For the moment, let the interval over which it is desired to obtain the expansion be 0, L. Shortly, L will be permitted to approach infinity in order to obtain an expansion valid over the infinite depth of the slab. Then (7-102) is operated on to give

$$\int_0^L [f(x) - T_1] \sin bx \, dx = \int_0^L \sin bx \left[\int_0^\infty A_a \left(\frac{k}{h} a \cos ax + \sin ax\right) da\right] dx$$
$$= \int_0^\infty A_a \left[\int_0^L \left(\frac{k}{h} a \cos ax \sin bx + \sin ax \sin bx\right) dx\right] da \tag{7-103}$$

In (7-103) b denotes any positive value of a, and it is assumed that the order of integration may be interchanged.

Now (using Peirce Nos. 359 and 360)

$$\int_0^L a \cos ax \sin bx \, dx = \frac{a}{2}\left[\frac{\cos L(a-b)}{a-b} - \frac{\cos L(a+b)}{a+b}\right]$$

$$\int_0^L \sin ax \sin bx \, dx = \frac{1}{2}\left[\frac{\sin L(a-b)}{a-b} - \frac{\sin L(a+b)}{a+b}\right]$$

Let

$$L(a-b) = z \tag{7-104}$$

$$L(a+b) = u \tag{7-105}$$

and the right-hand side of (7-103) becomes

$$R_H = \frac{1}{2}\int_{-bL}^{\infty} A_{\frac{z}{L}+b}\left[\frac{k}{h}\left(\frac{z}{L}+b\right)\frac{\cos z}{z} + \frac{\sin z}{z}\right] dz - \frac{1}{2}\int_{bL}^{\infty} A_{(u/L)-b}\left[\frac{k}{h}\left(\frac{u}{L}-b\right)\frac{\cos u}{u} + \frac{\sin u}{u}\right] du \tag{7-106}$$

Now let $L \to \infty$. Then the second integral on the right-hand side of (7-106) formally vanishes and there results

$$\lim_{L\to\infty} R_H = \frac{1}{2}\int_{-\infty}^{\infty} A_b\left(\frac{kb}{h}\frac{\cos z}{z} + \frac{\sin z}{z}\right) dz = \frac{\pi}{2} A_b \tag{7-107}$$

The integration of (7-107) makes use of the fact that the cosine term vanishes since it is odd and the sine term is integrated, using Peirce No. 484. When (7-107) is replaced in (7-103) with $L \to \infty$, there results

$$A_b = \frac{2}{\pi}\int_0^{\infty} [f(x) - T_1] \sin bx \, dx$$

However, b represents a particular value of a. Consequently, the expression for A_a is

$$A_a = \frac{2}{\pi}\int_0^{\infty} [f(v) - T_1] (\sin av) \, dv \tag{7-108}$$

Equation (7-101) now becomes

$$T - T_1 = \int_0^{\infty}\left\{[\exp(-a^2\alpha t)]\left(\frac{k}{h} a \cos ax + \sin ax\right)\frac{2}{\pi}\int_0^{\infty}[f(v) - T_1](\sin av) \, dv\right\} da \tag{7-109}$$

which is the solution to the problem.

The procedure just used depended upon the development of a method for expanding a function $F(x)$ in terms of sines and cosines over a semi-infinite interval. Thus, notice that combining (7-108) and (7-102) gives

$$f(x) - T_1 = F(x) = \int_0^{\infty}\left[\left(\frac{k}{h} a \cos ax + \sin ax\right)\int_0^{\infty}\frac{2}{\pi}F(v) \sin av \, dv\right] da \quad \text{for } 0 < x < \infty \tag{7-110}$$

It can be shown that the expansion represents $F(x)$, provided that $F(x)$ meets the conditions of Sec. 7-7 and if $\int_0^{\infty} |F(x)| \, dx$ exists.

Generally speaking, any function $F(x)$ which meets the conditions of Sec. 7-7 and for which $\int_0^\infty r(x)|F(x)|\,dx$ exists may be represented over the interval 0, ∞ by use of an orthogonal-function integral representation which may be developed in a manner analogous to that used in Example 7-6. The following particular cases are frequently encountered.

Fourier Sine Integral

$$F(x) = \frac{2}{\pi}\int_0^\infty \sin ax \int_0^\infty F(v) \sin av\,dv\,da \qquad 0 < x < \infty \qquad (7\text{-}111)$$

Fourier Cosine Integral

$$F(x) = \frac{2}{\pi}\int_0^\infty \cos ax \int_0^\infty F(v) \cos av\,dv\,da \qquad 0 < x < \infty \qquad (7\text{-}112)$$

Fourier-Bessel Integral

$$F(x) = \int_0^\infty \int_0^\infty avF(v)J_p(ax)\,dv\,da \qquad \begin{matrix} 0 < x < \infty \\ p > -1 \end{matrix} \qquad (7\text{-}113)$$

Complete Fourier Integral

$$F(x) = \frac{1}{2\pi}\int_{-\infty}^\infty \int_{-\infty}^\infty F(v) \cos a(v - x)\,dv\,da \qquad -\infty < x < \infty \qquad (7\text{-}114)$$

7-11. Potential Flow around a Sphere

Example 7-7. A solid sphere of radius R is placed in an incompressible, inviscid fluid of infinite extent. The flow was initially uniform and parallel to the z axis, flowing with a speed V_0 in the *negative* z direction. (See Fig. 7-6.) It is desired to determine the velocity of the fluid after the sphere is placed in the flow and steady-state has been achieved.

In Sec. 6-22 it was shown that the equations of motion of an inviscid irrotational flow are satisfied simultaneously by Laplace's equation

$$\nabla^2 P = 0 \qquad (7\text{-}115)$$

where P is the velocity potential and

$$\nabla P = \mathbf{V} \qquad (7\text{-}116)$$

The boundary conditions are

$$\left.\begin{aligned} V_x &= 0 \\ V_y &= 0 \\ V_z &= -V_0 \end{aligned}\right\} \quad z = \infty, \text{ all } x \text{ and } y \qquad (7\text{-}117a)$$

$$V_r = 0 \qquad \text{at } r = R \qquad (7\text{-}117b)$$

The last boundary condition results from the fact that the fluid cannot penetrate the solid sphere.

The problem is best handled by means of spherical coordinates. Use of the transformation equation (6-179i), together with the fact that the velocity is independent of θ, gives

$$\nabla^2 P = \frac{\partial}{\partial r}\left(r^2 \frac{\partial P}{\partial r}\right) + \frac{1}{\sin \phi}\frac{\partial}{\partial \phi}\left(\sin \phi \frac{\partial P}{\partial \phi}\right) = 0 \tag{7-118}$$

The transformation of the boundary conditions to a spherical-coordinate system is readily accomplished by use of vector notation:

$$\begin{aligned}\nabla P = \mathbf{V} &= \mathbf{i}\frac{\partial P}{\partial x} + \mathbf{j}\frac{\partial P}{\partial y} + \mathbf{k}\frac{\partial P}{\partial z} = \mathbf{i}V_x + \mathbf{j}V_y + \mathbf{k}V_z \\ &= \mathbf{v}_r \frac{\partial P}{\partial r} + \mathbf{v}_\phi \frac{1}{r}\frac{\partial P}{\partial \phi} \\ &= \mathbf{v}_r V_r + \mathbf{v}_\phi V_\phi\end{aligned}$$

where $\mathbf{v}_r$ is a *unit vector* in the r direction and $\mathbf{v}_\phi$ is a *unit vector* in the ϕ direction and

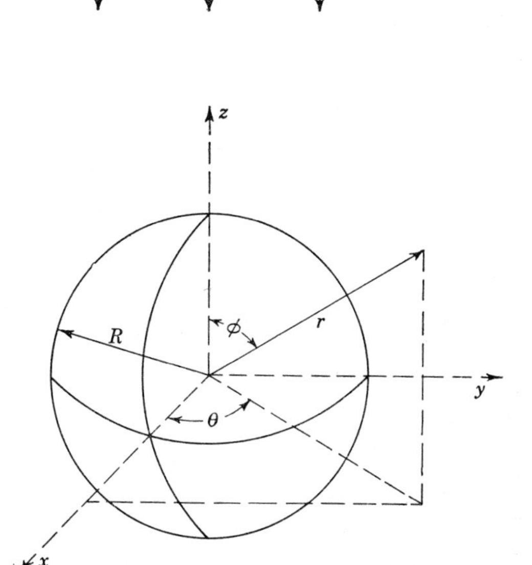

FIG. 7-6. Potential flow around a sphere.

the relation (6-179g) has been used. At $z \to \infty$, (7-117a) gives $\mathbf{V} = -\mathbf{k}V_0$. The component of this vector in the r direction is V_r. Hence, using (6-179b),

$$\begin{aligned}V_r &= \frac{\partial P}{\partial r} = \mathbf{V} \cdot \mathbf{v}_r = -\mathbf{k}V_0 \cdot (\mathbf{i} \cos \theta \sin \phi + \mathbf{j} \sin \theta \sin \phi + \mathbf{k} \cos \phi) \\ &= -V_0 \cos \phi \qquad \text{at } z = r \cos \phi \to \infty \\ V_\phi &= \frac{1}{r}\frac{\partial P}{\partial \phi} = \mathbf{V} \cdot \mathbf{v}_\phi = V_0 \sin \phi \qquad \text{at } z = r \cos \phi \to \infty\end{aligned}$$

The boundary conditions to be satisfied by (7-118) are then

$$\frac{\partial P}{\partial r} = -V_0 \cos \phi \qquad \text{at } r \cos \phi \to \infty \tag{7-119a}$$

$$\frac{1}{r}\frac{\partial P}{\partial \phi} = V_0 \sin \phi \qquad \text{at } r \cos \phi \to \infty \tag{7-119b}$$

$$\frac{\partial P}{\partial r} = 0 \qquad \text{at } r = R \tag{7-119c}$$

A solution of the form

$$P = \lambda(r)\Phi(\phi) \tag{7-120}$$

will be tried. With the assumption (7-120), (7-118) becomes

$$\frac{d}{dr}\left(r^2 \frac{d\lambda}{dr}\right) - a^2\lambda = 0 \tag{7-121}$$

$$\frac{d}{d\phi}\left(\sin \phi \frac{d\Phi}{d\phi}\right) + a^2 \sin \phi\Phi = 0 \tag{7-122}$$

where a is the separation constant. If

$$a^2 = n(n+1) \tag{7-123}$$

where n is real and positive or zero, Eq. (7-122) is in the form of Legendre's equation [see Eq. (5-107)]. The solution of (7-122) is then of the form

$$\Phi = c_1 P_n(\cos \phi) + c_2 Q_n(\cos \phi) \tag{7-124}$$

where $P_n(\cos \phi)$ and $Q_n(\cos \phi)$ are Legendre polynomials and are defined in Sec. 5-12.

With the substitution $r = \exp t$, the solution of (7-121) is found to be of the form

$$\begin{aligned} \lambda &= c_3 \exp nt + c_4 \exp[-(n+1)t] \\ &= c_3 r^n + c_4 r^{-(n+1)} \end{aligned} \tag{7-125}$$

Then (7-120) may be written in the form

$$P = \sum_n [c_3 r^n + c_4 r^{-(n+1)}][c_1 P_n(\cos \phi) + c_2 Q_n(\cos \phi)] \tag{7-126}$$

$$\frac{\partial P}{\partial r} = \sum_n [c_3 n r^{n-1} - c_4(n+1)r^{-(n+2)}][c_1 P_n(\cos \phi) + c_2 Q_n(\cos \phi)] \tag{7-127}$$

Since $\partial P/\partial r$ represents the magnitude of the velocity in the r direction, V_r, it must be finite. This requires that P_n and Q_n be finite at all values of ϕ. However, at $\phi = 0$, $\cos \phi = 1$ and at $\phi = \pi$, $\cos \phi = -1$. Only P_n is finite at these points and then only if *n is integer*. Hence

$$\begin{aligned} c_2 &= 0 \\ n &= \text{integer} \end{aligned} \tag{7-128}$$

Furthermore, in order to avoid infinite velocities as $r \to \infty$, *n must be limited to the values* 0 *and* 1.

The boundary condition (7-119c) demands that $(\partial P/\partial r)_{r=R} = 0$. Consequently,

$$c_3 n R^{n-1} - c_4(n+1)R^{-(n+2)} = 0$$

and

$$c_4 = c_3 \frac{n}{n+1} R^{2n+1} \tag{7-129}$$

Equations (7-126) and (7-127) then take the form

$$P = A_0 + A_1 \left(r + \frac{R^3}{2} r^{-2} \right) \cos \phi \tag{7-130}$$

$$\frac{\partial P}{\partial r} = A_1 \left(1 - \frac{R^3}{r^3} \right) \cos \phi \tag{7-131}$$

since $P_0(x) = 1$ and $P_1(x) = x$. Application of the boundary condition (7-119*a*) to (7-131) gives $A_1 = -V_0$ and

$$P = A_0 - V_0 \cos \phi \left(r + \frac{R^3}{2r^2} \right) \tag{7-132}$$

Equation (7-132) satisfies the boundary condition (7-119*b*) identically. The constant A_0 is arbitrary, which is, of course, satisfactory since its value has no influence on the velocity itself. Then the velocity is given by

$$\nabla P = \mathbf{V} = \mathbf{v}_r \left[-V_0 \left(1 - \frac{R^3}{r^3} \right) \cos \phi \right] + \mathbf{v}_\phi \left[V_0 \left(1 + \frac{R^3}{r^3} \right) \sin \phi \right] \tag{7-133}$$

where $\mathbf{v}_r$ and $\mathbf{v}_\phi$ are unit vectors in the r and ϕ directions, respectively.

7-12. Conduction of Heat and Diffusion of Matter. Many of the previous examples have dealt with the solution of the heat-conduction equation

$$\frac{\partial T}{\partial t} = \alpha \nabla^2 T \tag{7-134}$$

for various physical cases. Equation (7-134) is a special form of the general temperature-distribution equation in that it applies only to the case where heat is transferred solely by conduction and without convection or radiation.

In special situations, the general equation for the diffusion of matter reduces to a form which is formally identical to (7-134). These situations arise as follows:

In Sec. 6-22 the diffusion equation for a binary perfect-gas mixture was derived. The results were

$$\frac{\partial Y_A}{\partial t} + \mathbf{U} \cdot \nabla Y_A = \frac{M}{\rho} \mathfrak{D} \nabla^2 Y_A \tag{6-192}$$

and

$$\frac{\partial \ln M}{\partial t} + \mathbf{V} \cdot \nabla \ln M = \frac{M \mathfrak{D}}{\rho} \nabla^2 \ln M \tag{6-203}$$

Now if the molal mean velocity $\mathbf{U}$ is everywhere zero, as would be the case if the velocities of the two components were equal in magnitude but oppositely directed, *i.e., if equal-molal counterflow took place*, then (6-192) becomes

$$\frac{\partial Y_A}{\partial t} = \frac{M \mathfrak{D}}{\rho} \nabla^2 Y_A \tag{7-135}$$

which is formally identical with (7-134).

If, on the other hand, the mass mean velocity **V** is everywhere zero, then (6-203) becomes

$$\frac{\partial \ln M}{\partial t} = \frac{M\mathfrak{D}}{\rho} \nabla^2 \ln M \tag{7-136}$$

which is formally identical with both (7-134) and (7-135).

When these special situations occur, the diffusion problem may be handled like the conduction problem. Indeed, if the boundary conditions are the same, as sometimes occurs, the formal solutions of the diffusion and pure conduction problems are identical.

Actually, the formal analogy between heat transport and mass transport extends into the region where convection is important and forms the basis for the so-called "analogies" between heat and mass transfer. The interested reader will find an informative discussion of this topic in reference 6.

BIBLIOGRAPHY

1. Churchill, R. V.: "Fourier Series and Boundary Value Problems," McGraw-Hill Book Company, Inc., New York, 1941.
2. Carslaw, H. S., and J. C. Jaeger: "Conduction of Heat in Solids," Clarendon Press, Oxford, 1947.
3. Kellogg, O. D.: "Foundations of Potential Theory," Springer-Verlag OHG, Berlin, 1929.
4. Bateman, H.: "Partial Differential Equations of Mathematical Physics," Dover Publications, New York, 1944.
5. Fluid Motion Panel of the Aeronautical Research Committee and Others, "Modern Developments in Fluid Dynamics," Clarendon Press, Oxford, 1938.
6. Bedingfield, C. H., and T. B. Drew: Analogy between Heat Transfer and Mass Transfer, *Ind. Eng. Chem.*, vol. 42, no. 6, pp. 1164–1173, 1950.
7. Hildebrand, F. B.: "Advanced Calculus for Engineers," Prentice-Hall, Inc., Englewood Cliffs, N.J., 1949.
8. Marshall, W. R., Jr. and R. L. Pigford: "The Application of Differential Equations to Chemical Engineering Problems," University of Delaware, Newark, Del., 1947.

PROBLEMS

7-1. In the study of boundary-layer theory, the easiest case to treat analytically is that of a flat plate immersed in a stream of unlimited extent. The results obtained will approximate the flow over airplane wings, turbine blades, etc.

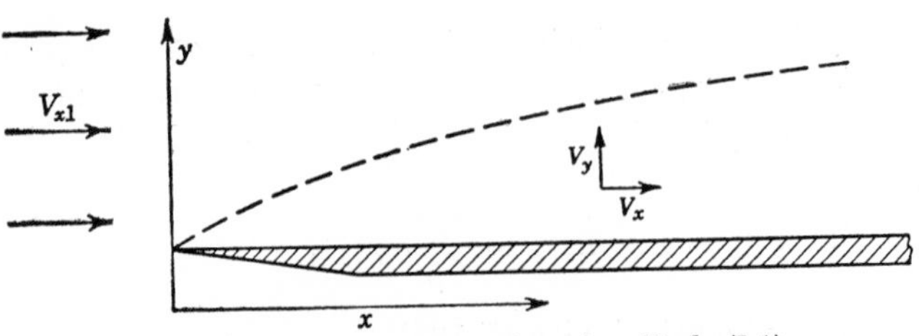

FIG. 7-7. Flow over a flat plate (see Prob. 7-1).

Consider a flat plate immersed in a stream of unlimited extent, as shown in Fig. 7-7

The main stream has a constant velocity V_{x1} parallel to the plate. Under these circumstances, it can be shown that $\partial p/\partial y = \partial p/\partial x = 0$. It is desired to determine the friction factor f, defined by the relation

$$f = \frac{2\tau_w}{\rho V_{x1}^2}$$

which exists at any point. τ_w denotes the local shear stress exerted on the fluid by the plate. Since the flow is laminar at the wall (even if the main portion of the boundary layer were turbulent),

$$\tau_w = \mu \left(\frac{\partial V_x}{\partial y}\right)_{y=0} \qquad \mu = \text{viscosity}$$

It may be assumed that the fluid is incompressible, that the plate extends to plus and minus infinity in the z direction and to plus infinity in the x direction, and that the flow in the boundary layer is laminar. In the laminar boundary layer the following equations apply:

Continuity,

$$\frac{\partial V_x}{\partial x} + \frac{\partial V_y}{\partial y} = 0 \tag{1}$$

Force balance,

$$V_x \frac{\partial V_x}{\partial x} + V_y \frac{\partial V_x}{\partial y} = \frac{\mu}{\rho}\frac{\partial^2 V_x}{\partial y^2} \tag{2}$$

with the boundary conditions

(1) $V_x = V_y = 0$ at $y = 0$, for $x > 0$
(2) $V_x = V_{x1}$ at $y = \infty$, all values of x
(3) $V_x = V_{x1}$ at $x = 0$, all values of y

(*a*) By use of the stream function ψ, reduce Eqs. (1) and (2) to a single partial differential equation in ψ, x, and y.

$$\frac{\partial \psi}{\partial x} = -V_y \qquad \frac{\partial \psi}{\partial y} = +V_x$$

(*b*) Into the resulting equation make the substitution

$$\psi = (ax)^p \phi(\eta)$$

where $\eta = (bx)^q y$, and show that for specific values of p and q, the resultant equation reduces to a total differential equation which can satisfy the boundary conditions. Find these values of p and q.

(*c*) Taking

$$a = \frac{\mu V_{x1}}{\rho}$$

$$b = \frac{4\mu}{\rho V_{x1}}$$

write the resulting differential equation. What are the boundary conditions? Discuss the techniques available for solution and, in particular, the one to use to obtain the friction factor as a function of x.

7-2. The diffusivity $\mathfrak{D}$ for a binary perfect-gas mixture is to be obtained experimentally by measuring the rate of interdiffusion of two gases originally confined in the two ends of a hollow cylinder. A thin diaphragm separating the gases divides the cylinder into two sections of equal volume. The diaphragm is suddenly removed and the gases allowed to diffuse for a measured time. The diaphragm is then replaced, and the gas in one-half the cylinder is well mixed and analyzed. It is assumed that convection effects are not important. Pure gas A is originally confined in one-half the cylinder and pure gas B in the other half, and the pressure and temperature of the system are kept uniform and constant. Derive the expression which permits the calculation of the diffusivity from the experimental data.

7-3. A solid cylinder of infinite length and radius R has an initial temperature distribution $T = f(r)$, where r is the local radius. The surface at $r = R$ is suddenly brought to and maintained at the temperature T_1. Determine the interior temperature as a function of time. Compute the temperature distribution if $T = f(r) = T_0 = 70°F$, $T_1 = 1500°F$, $R = 2$ ft, and the solid is steel.

7-4. The cylinder of Prob. 7-3, with initial temperature distribution $T = f(r)$, is placed in a constant-temperature medium at T_1. The heat-transfer coefficient between the medium and the cylinder is finite and equal to h. Determine the interior temperature as a function of time. Compute the temperature distribution if $T = f(r) = T_0 = 70°F$, $T_1 = 1500°F$, $R = 2$ ft, $h = 10$ Btu/(hr)(ft²)(°F), and the cylinder is made of steel.

7-5. A slab of porous material is to be dried from both faces at a constant rate of W lb/(hr)(ft²). The initial moisture content of the slab is uniform at C_0 pcf. The movement of moisture through the slab is in the x direction only, and the *rate* of movement is proportional to the concentration gradient of the moisture in the slab, dc/dx. If the slab is of thickness L, determine the concentration of moisture within the slab as a function of time and distance. Show that, as the drying time increases, the concentration distribution becomes parabolic.

7-6. A hollow cylinder has the inner face ($r = R_0$) maintained at $T = f_0(\theta)$ and the outer face ($r = R_1$) maintained at $T = f_1(\theta)$. Determine the steady-state temperature distribution within the cylinder. θ denotes a coordinate in a cylindrical coordinate system.

7-7. The void volume of a sphere of porosity P and radius R is filled with a liquid containing a concentration C_0 of a dissolved organic compound. The sphere is placed in a small volume ν of the same liquid containing a concentration Y_0 of the organic compound. As a result of the concentration difference, some of the organic compound is transferred from the inside of the sphere to the surrounding liquid. Within the sphere, diffusion controls the transport of the compound. The external liquid is mechanically mixed. Determine the rate at which the organic compound is transferred from the sphere to the surrounding liquid.

CHAPTER 8

THE LAPLACE TRANSFORM

8-1. Introduction. The Laplace-transform method reduces the solution of a *linear* total or partial differential equation to essentially an algebraic procedure. In addition, the relevant boundary conditions are introduced early in the analysis, and the "constants of integration" are automatically evaluated. An excellent and complete treatment of the Laplace-transform method which both covers the mathematics more thoroughly and extends the technique to more complicated situations than given here will be found in R. V. Churchill, "Modern Operational Mathematics in Engineering." Marshall and Pigford, "The Application of Differential Equations to Chemical Engineering Problems,"[9] pioneered the application of the Laplace transform to typical chemical engineering problems.

8-2. Illustration of the Transform Method

Example 8-1. Consider the differential equation

$$\frac{dy}{dt} - y = e^{at} \tag{8-1}$$

with the boundary condition

$$y = -1 \qquad \text{at } t = 0 \tag{8-2}$$

A solution is desired which is valid for positive values of t. The Laplace-transform method proceeds by multiplying both sides of (8-1) by $e^{-pt}\,dt$ and integrating the result from zero to infinity:

$$\int_0^\infty e^{-pt}\frac{dy}{dt}\,dt - \int_0^\infty e^{-pt}y\,dt = \int_0^\infty e^{-pt}\,e^{at}\,dt \tag{8-3}$$

$$\int_0^\infty e^{-pt}\,e^{at}\,dt = -\left[\frac{e^{-(p-a)t}}{p-a}\right]_0^\infty = \frac{1}{p-a} \tag{8-4}$$

provided that $p > a$. The first integral in (8-3) may be integrated by parts:

$$\begin{aligned}\int_0^\infty e^{-pt}\frac{dy}{dt}\,dt &= \Big[e^{-pt}y\Big]_0^\infty + p\int_0^\infty e^{-pt}y\,dt \\ &= 1 + p\int_0^\infty e^{-pt}y\,dt \end{aligned} \tag{8-5}$$

provided that $e^{-pt}y \to 0$ as $t \to \infty$ and following the introduction of the boundary condition (8-2). When (8-4) and (8-5) are combined with (8-3) there results

$$(p - 1) \int_0^\infty e^{-pt} y \, dt = \frac{1}{p - a} - 1$$

or

$$\int_0^\infty e^{-pt} y \, dt = \frac{a + 1 - p}{(p - 1)(p - a)} \tag{8-6}$$

The problem now is to determine the function y whose Laplace transform is given by the right-hand side of (8-6). In the present case, this may be done by expanding the right-hand side by the method of partial fractions to give

$$\int_0^\infty e^{-pt} y \, dt = \frac{a + 1 - p}{(p - 1)(p - a)} = \frac{1}{a - 1} \frac{1}{p - a} - \frac{a}{a - 1} \frac{1}{p - 1} \tag{8-7}$$

In view of (8-4), there follows

$$y = \frac{1}{a - 1} e^{at} - \frac{a}{a - 1} e^t \tag{8-8}$$

Direct substitution into (8-1) and comparison with (8-2) demonstrate that (8-8) is the solution of the original problem.

8-3. Properties of the Transformation. The preceding section illustrated the use of the Laplace transform, but it left many questions unanswered. This section will develop the procedure in more detail.

Definition of the Laplace Transform. The Laplace transform of a function $f(t)$ is defined for positive values of t as *a function of the new variable* p by the integral

$$\mathcal{L}\{f(t)\} \equiv \int_0^\infty e^{-pt} f(t) \, dt = \bar{f}(p) \tag{8-9}\dagger$$

The transform exists if $f(t)$ satisfies the following conditions:

1. *$f(t)$ is continuous or piecewise continuous in any interval $t_1 \leqq t \leqq t_2$, where $t_1 > 0$.*
2. *$t^n|f(t)|$ is bounded near $t = 0$ when approached from positive values of t for some number n, where $n < 1$.*
3. *$e^{-p_0 t}|f(t)|$ is bounded for large values of t for some number p_0.*

The function $f(t)$ is *piecewise continuous* in the range $t_1 \leqq t \leqq t_2$ if it is possible to divide the range into a finite number of intervals in such a way that $f(t)$ is continuous within each interval and approaches finite values as either end of the interval is approached from within the interval. Thus, a piecewise continuous function may have a number of *finite* discontinuities. At a typical discontinuity, say at $t = t_0$, $f(t)$ approaches one value as t approaches t_0 from larger values of t—the *right-hand limit*—and a different value as t approaches t_0 from smaller values of t—the

† Some writers use the definition

$$\mathcal{L}\{f(t)\} = p \int_0^\infty e^{-pt} f(t) \, dt$$

left-hand limit. The notation used for the right-hand limit is

$$\lim_{t \to t_0+} f(t) = f(t_0+)$$

and the notation used for the left-hand limit is $\lim_{t \to t_0-} f(t) = f(t_0-)$.

If $e^{-p_0 t}|f(t)|$ is bounded for large values of t, the function $f(t)$ is said to be of "exponential order." If $f(t)$ is such a function, it may well approach infinity as $t \to \infty$, but it does not grow as rapidly as $e^{p_0 t}$, and hence $e^{-p_0 t}|f(t)|$ will be bounded.

If $f(t)$ is piecewise continuous and of exponential order, then $\int_0^t f(t)\,dt$ may be shown to be continuous and of exponential order. Consequently, if $\mathcal{L}\{f(t)\}$ exists, then $\mathcal{L}\left\{\int_0^t f(t)\,dt\right\}$ also exists. The above statement is not generally true when reversed; i.e., if $\mathcal{L}\{f(t)\}$ exists, it does not always follow that $\mathcal{L}\{d^n f(t)/dt^n\}$ exists. However, the Laplacian of the derivatives of $f(t)$ exists in most practical problems.

Linear Property of the Transform. Direct application of the definition gives

$$\mathcal{L}\{af(t) + bg(t)\} = a\bar{f}(p) + b\bar{g}(p) \tag{8-10}$$

Transform of Derivatives of $f(t)$. Integration by parts in a fashion similar to that used in Sec. 8-2 gives

$$\mathcal{L}\left\{\frac{d^n f(t)}{dt^n}\right\} = p^n \bar{f}(p) - \left[p^{n-1} f(0+) + p^{n-2}\frac{df(0+)}{dt} + p^{n-3}\frac{d^2 f(0+)}{dt^2} + \cdots + \frac{d^{n-1} f(0+)}{dt^{n-1}}\right] \tag{8-11}$$

Equation (8-11) is one of the most useful properties of the transform. It relates the transform of the nth derivative of $f(t)$ to the transform of $f(t)$ itself and to the *numerical* values approached by the lower-order derivatives of $f(t)$ as $t \to 0$ from positive values. When the Laplace-transform method is used to solve a differential equation, it is by virtue of the relation (8-11) that the boundary conditions are introduced. This is a distinct advantage when the boundary conditions are specified at $t = 0$ but may prove awkward when the boundary conditions are specified for nonzero values of the independent variable.

The conditions under which Eq. (8-11) is valid are more stringent than those imposed upon the transform of $f(t)$ itself:

Equation (8-11) *is valid if $f(t)$ and its first $n - 1$ derivatives are continuous over the interval $0 \to t_2$, if $d^n f(t)/dt^n$ is at least piecewise continuous over the interval $0 \to t_2$, and if $f(t)$ and its first n derivatives are of exponential order.*

Transform of $\int_0^t f(t)\,dt$. Integration by parts gives

$$\mathcal{L}\left\{\int_0^t f(t)\,dt\right\} = \frac{1}{p}\bar{f}(p) \tag{8-12}$$

and

$$\mathcal{L}\left\{\int_a^t f(t)\,dt\right\} = \frac{1}{p}\bar{f}(p) - \frac{1}{p}\int_0^a f(t)\,dt \tag{8-13}$$

Equations (8-12) and (8-13) are valid if $\mathcal{L}\{f(t)\}$ exists.

Translation Properties of the Transform

$$\mathcal{L}\{e^{at}f(t)\} = \bar{f}(p - a) \tag{8-14}$$

Equation (8-14) states that the transform of the product $e^{at}f(t)$ is obtained by replacing p by $p - a$ in the transform of $f(t)$. If

$$f(t) = \begin{cases} 0 & t < a \\ g(t - a) & t \geqq a \end{cases}$$

then

$$\bar{f}(p) = e^{-ap}\bar{g}(p) \tag{8-15}$$

The utility of (8-15) is as follows: Suppose that the function $f(t)$ is such that it is zero for all values of t less than some positive number a and of the form $g(t - a)$ for $t \geqq a$. The transform of this function is then found as the product of e^{-ap} and the transform of $g(t)$.

8-4. The Inverse Transform. In practical applications the most difficult problem is the determination of the function which corresponds to a known transform. Thus, in the course of the solution of a typical differential equation, $\bar{f}(p)$ is found relatively easily, and the final step is the determination of the *inverse Laplace transform* $f(t)$. The following sections deal with this step.

A short table of transforms is given in Table 8-1 at the end of this chapter. This table, together with the additional techniques to be described, will handle most problems. A very complete set of tables has been compiled by Campbell and Foster.[3] When all other methods of finding the inverse transform fail, numerical-integration techniques may be employed. These more advanced procedures are described in references 4, 5, and 7.

Fortunately, the inversion process is unique; i.e., *if one function* $f(t)$ *corresponding to the known transform* $\bar{f}(p)$ *can be found, it is the correct one.* This statement is known as "Lerch's theorem."†

Not all functions of p are transforms, since continuity and other considerations must be satisfied. However, *if* $\bar{f}(p) \to 0$ *as* $p \to \infty$ *and*

† Actually, Lerch's theorem states that two functions having the same transform cannot differ over any finite positive interval. This more precise conclusion seldom contradicts the above statement.

$p\bar{f}(p)$ is bounded as $p \to \infty$, then $\bar{f}(p)$ is the transform of some function $f(t)$ which is at least piecewise continuous in some interval $0 \leqq t \leqq t_2$ and is of exponential order.

When the initial value of $f(t)$ is needed and $\bar{f}(p)$ is known, the following is very useful:

$$\lim_{p \to \infty} p\bar{f}(p) = f(0+) \tag{8-16}$$

provided that $f(t)$ is continuous and $df(t)/dt$ is piecewise continuous in every finite interval $0 \leqq t \leqq t_2$ and if $f(t)$ and $df(t)/dt$ are of exponential order.

An obvious modification of Eq. (8-16) may be used as an approximation to find the form taken by $f(t)$ as $t \to 0$ and hence to evaluate $f(t)$ for small values of t.

8-5. Convolution. Often the transform $\bar{f}(p)$ is known but the inverse transform $f(t)$ is not given by a simple table of transforms nor is it readily obtained by the methods already discussed. In such a case, it may be possible to break down $\bar{f}(p)$ into the product of two transforms

$$\bar{f}(p) = \bar{g}(p)\bar{h}(p) \tag{8-17}$$

where $\bar{g}(p)$ is known to be the transform of the function $g(t)$ and $\bar{h}(p)$ is known to be the transform of the function $h(t)$. The *convolution* integral then enables $f(t)$ to be determined:

$$\begin{aligned} f(t) &= \mathfrak{L}^{-1}\{\bar{f}(p)\} = \int_0^t g(t-s)h(s)\,ds \\ &= \int_0^t g(s)h(t-s)\,ds = g_*h = h_*g \end{aligned} \tag{8-18}$$

Example 8-2. As a simple illustration of the use of convolution, let $\bar{f}(p) = 1/p^3$. This may be taken as the product $\bar{g}(p)\bar{h}(p)$, where $\bar{g}(p) = 1/p$ and $\bar{h}(p) = 1/p^2$. Relation 1 of Table 8-1 gives $g(t) = 1$, and relation 2 gives $h(t) = t$. Then, using (8-18),

$$f(t) = \int_0^t g(s)h(t-s)\,ds = \int_0^t 1(t-s)\,ds = \left[ts - \frac{s^2}{2}\right]_0^t = \frac{t^2}{2}$$

which agrees with relation 3 of Table 8-1.

8-6. Solution of a Partial Differential Equation

Example 8-3. Heating of a Diathermanous Solid. Considerable practical interest has recently arisen in situations where moderately transparent (or diathermanous) materials are exposed to a time-invariant thermal flux. At the heating rates and irradiation periods in question, the samples may be assumed to behave as one-dimensional semi-infinite solids (i.e., the depth of penetration for the "heat wave" may be assumed slight, relative to the dimensions of the exposed area), and any cooling losses from the irradiated surface may be neglected. It is desired to develop an expression for the temperature-time-space relation in these irradiated slabs subject to the following assumptions:

1. The materials in question are homogeneous, isotropic, and have physical properties (that is, k, ρ, c_p, etc.) that are independent of temperature.

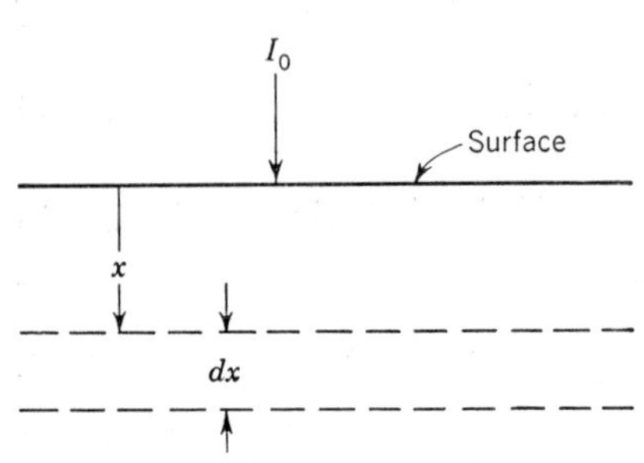

FIG. 8-1. Heating of a diathermanous solid.

2. The absorption of radiant energy below the surface may be characterized by a Lambert's-decay-law expression of the form

$$I_x = I_0 e^{-\gamma x} = \text{Btu}/(\text{hr})(\text{ft}^2)$$

where I_x = intensity at any depth x
I_0 = intensity of unreflected radiation at surface
γ = "extinction coefficient"

3. The initial temperature of the solid is uniform at T_0.

Solution. Consider a section of thickness dx and of unit area located a distance x from the surface of the slab of Fig. 8-1. Apply the first law of thermodynamics.

$$\text{Rate of energy input} = -k\frac{\partial T}{\partial x} + I_x$$

$$\text{Rate of energy output} = -k\left(\frac{\partial T}{\partial x} + \frac{\partial^2 T}{\partial x^2}dx\right) + \left(I_x + \frac{\partial I_x}{\partial x}dx\right)$$

$$\text{Rate of energy accumulation} = \frac{\partial}{\partial t}(\rho\, cT\, dx)$$

The applicable differential equation is then

$$\alpha\frac{\partial^2\theta}{\partial x^2} + \beta e^{-\gamma x} = \frac{\partial\theta}{\partial t} \tag{8-19}$$

where

$$\alpha = \frac{k}{\rho c} \tag{8-20}$$

$$\beta = \frac{\gamma I_0}{\rho c} \tag{8-21}$$

$$\theta = T - T_0 \tag{8-22}$$

The relevant boundary conditions are

$$\begin{aligned} \theta &= 0 \quad \text{at } t = 0, \text{ all } x \\ \theta &= 0 \quad \text{at } x = \infty, \text{ all } t \\ \frac{\partial\theta}{\partial x} &= 0 \quad \text{at } x = 0, \text{ all } t \end{aligned} \tag{8-23}$$

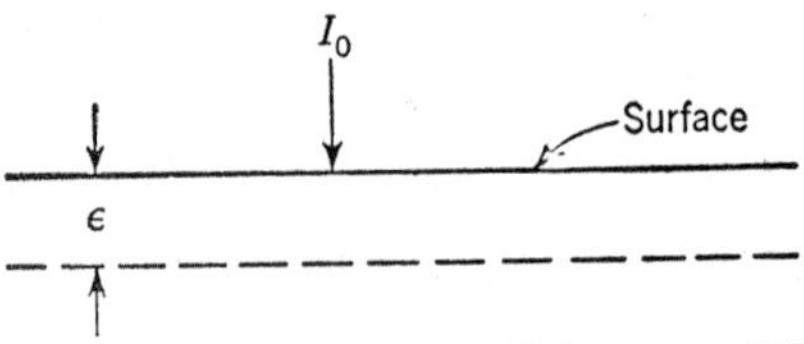

FIG. 8-2. Boundary condition at solid surface.

The third boundary condition results from the following considerations: Examine a slice of material of infinitesimal thickness ϵ at the surface of the slab (Fig. 8-2).

$$\text{Rate of energy input} = I_0$$

$$\text{Rate of energy output} = -k\left(\frac{\partial T}{\partial x}\right)_{x=\epsilon} + I_0 e^{-\gamma\epsilon}$$

$$\text{Rate of energy accumulation} = \frac{\partial}{\partial t}(\rho\epsilon cT)$$

Hence

$$\left(\frac{\partial T}{\partial x}\right)_{x=\epsilon} = I_0(e^{-\gamma\epsilon} - 1) + \rho c\epsilon \frac{\partial T}{\partial t} \tag{8-24}$$

Now let $\epsilon \to 0$, corresponding to $x \to 0$. Then the right-hand side of (8-24) vanishes, and there results

$$\left(\frac{\partial T}{\partial x}\right)_{x=0} = 0$$

It is now desired to take the Laplace transform of the partial differential equation (8-19):

$$\alpha \int_0^\infty e^{-pt} \frac{\partial^2 \theta}{\partial x^2}\, dt + \beta \int_0^\infty e^{-pt} e^{-\gamma x}\, dt = \int_0^\infty e^{-pt} \frac{\partial \theta}{dt}\, dt \tag{8-25}$$

The indicated integrations may be performed if it is noted that θ is a function of two *independent* variables x and t, that the integrations are with respect to t, and that the limits of integration are independent of x. Then,

$$\int_0^\infty e^{-pt} \frac{\partial^2 \theta}{\partial x^2}\, dt = \frac{\partial^2}{\partial x^2} \int_0^\infty e^{-pt}\theta\, dt = \frac{\partial^2 \bar{\theta}(x,p)}{\partial x^2}$$

$$\int_0^\infty e^{-pt} e^{-\gamma x}\, dt = \frac{e^{-\gamma x}}{p}$$

$$\int_0^\infty e^{-pt} \frac{\partial \theta}{\partial t}\, dt = p\bar{\theta}(x,p) - \theta(x,0+) = p\bar{\theta}(x,p)$$

[using (8-11)] when the first boundary condition of (8-23) is used. Equation (8-25) is then

$$\frac{\partial^2 \bar{\theta}}{\partial x^2} - \frac{p}{\alpha}\bar{\theta} = -\frac{\beta}{\alpha}\frac{e^{\gamma x}}{p} \tag{8-26a}$$

Inasmuch as p appears in (8-26a) as a parameter only, (8-26a) may be regarded as a total differential equation and written

$$\frac{d^2 \bar{\theta}}{dx^2} - \frac{p}{\alpha}\bar{\theta} = -\frac{\beta e^{-\gamma x}}{\alpha p} \tag{8-26b}$$

The boundary conditions which apply to (8-26b) are obtained by transformation of the boundary conditions (8-23): The condition $\theta = 0$ at $x = \infty$, all t, becomes

$$\bar{\theta} = 0 \qquad \text{at } x = \infty \tag{8-27a}$$

The condition $\partial\theta/\partial x = 0$ at $x = 0$, all t, becomes

$$\frac{d\bar{\theta}}{dx} = 0 \qquad \text{at } x = 0 \tag{8-27b}$$

The differential equation (8-26b) may now be solved. Since the boundary condition (8-27a) involves $\bar{\theta}$ at $x = \infty$ rather than at $x = 0$, it is convenient to use a conventional method rather than to apply the Laplace-transformation procedure a second time. The solution of (8-26b) is readily found to be

$$\bar{\theta} = Ae^{(\sqrt{p/\alpha})x} + Be^{-(\sqrt{p/\alpha})x} - \frac{\beta e^{-\gamma x}}{\alpha p(\gamma^2 - p/\alpha)} \tag{8-28}$$

Since $\bar{\theta} \to 0$ as $x \to \infty$, A must equal zero. Furthermore,

$$\frac{d\bar{\theta}}{dx} = -B\sqrt{\frac{p}{\alpha}}\,e^{-(\sqrt{p/\alpha})x} + \frac{\beta\gamma e^{-\gamma x}}{\alpha p(\gamma^2 - p/\alpha)}$$

and

$$\left(\frac{d\bar{\theta}}{dx}\right)_{x=0} = 0 = -B\sqrt{\frac{p}{\alpha}} + \frac{\beta\gamma}{\alpha p(\gamma^2 - p/\alpha)}$$

$$B = \frac{\beta\gamma}{(\sqrt{p/\alpha})\alpha p(\gamma^2 - p/\alpha)} \tag{8-29}$$

Equation (8-28) then becomes

$$\bar{\theta} = \frac{\beta}{\alpha p(\gamma^2 - p/\alpha)}\left(\frac{\gamma e^{-(\sqrt{p/\alpha})x}}{\sqrt{p/\alpha}} - e^{-\gamma x}\right) \tag{8-30}$$

It is now necessary to determine the inverse transform of (8-30). The inversion of the term $-[\beta e^{-\gamma x}/\alpha p(\gamma^2 - p/\alpha)]$ appearing as part of (8-30) is given by relation 12 of Table 8-1 as

$$\frac{\beta e^{-\gamma x}(e^{\gamma^2 \alpha t} - 1)}{\gamma^2\alpha} \tag{8-31}$$

The remaining term in (8-30) is

$$\frac{\beta\gamma}{\alpha^2}\,\frac{e^{-(\sqrt{p/\alpha})x}}{(p/\alpha)^{3/2}(\gamma^2 - p/\alpha)}$$

The corresponding inverse transformation is not given directly in Table 8-1. However, if this term is expanded:

$$\frac{\beta\gamma}{\alpha^2}\,\frac{e^{-(\sqrt{p/\alpha})x}}{(p/\alpha)^{3/2}(\gamma^2 - p/\alpha)} = \beta\gamma\sqrt{\alpha}\,\frac{-1}{p(p - \gamma^2\alpha)}\,\frac{e^{-(\sqrt{p})x/\sqrt{\alpha}}}{p^{1/2}} = \bar{f}(p)$$

the inverse transform of $-1/p(p - \gamma^2\alpha) = \bar{g}(p)$ is given by 12 of Table 8-1 as $-(1/\gamma^2\alpha)(e^{\gamma^2\alpha t} - 1) = g(t)$, and the inverse transform of $e^{(-\sqrt{p})x/\sqrt{\alpha}}/p^{1/2} = \bar{h}(p)$ is given by 84 of Table 8-1 as $(1/\sqrt{\pi t})e^{-(x^2/4\alpha t)} = h(t)$. The inverse transform of $\bar{f}(p)$ may now be found by convolution:

$$f(t) = \beta\gamma\sqrt{\alpha}\int_0^t g(t - s)h(s)\,ds = \frac{-\beta}{\gamma\alpha}\int_0^t \left(e^{\gamma^2\alpha(t-s)} - 1\right)\frac{1}{\sqrt{\pi s}}\,e^{-(x^2/4\alpha s)}\,ds \tag{8-32}$$

The integration of (8-32) involves considerable manipulation. The final result is

$$f(t) = \frac{2\beta\sqrt{\alpha t}}{\alpha\gamma}\,ierfc\left(\frac{x}{2\sqrt{\alpha t}}\right) - \frac{\beta e^{\alpha\gamma^2 t}}{2\alpha\gamma}\left[e^{-\gamma x}erfc\left(\frac{x}{2\sqrt{\alpha t}} - \gamma\sqrt{\alpha t}\right) - e^{\gamma x}erfc\left(\frac{x}{2\sqrt{\alpha t}} + \gamma\sqrt{\alpha t}\right)\right] \tag{8-33}$$

In (8-33)

$$ierfc(z) = \int_0^z erfc(z)\,dz$$

$$erfc(z) = 1 - erf(z)$$

$$erf(z) = \frac{2}{\sqrt{\pi}}\int_0^z e^{-z^2}\,dz = \text{“probability integral”}$$

The probability integral is a tabulated function. The sum of the relations (8-31) and (8-33) constitutes the result of the inversion of (8-30). The final result is consequently

$$T - T_0 = \frac{I_0 e^{-\gamma x}}{k\gamma}(e^{\alpha\gamma^2 t} - 1) + \frac{2I_0\sqrt{\alpha t}}{k}\, ierfc\left(\frac{x}{2\sqrt{\alpha t}}\right)$$
$$- \frac{I_0 e^{\alpha\gamma^2 t}}{2k\gamma}\left[e^{-\gamma x} erfc\left(\frac{x}{2\sqrt{\alpha t}} - \gamma\sqrt{\alpha t}\right) - e^{\gamma x} erfc\left(\frac{x}{2\sqrt{\alpha t}} + \gamma\sqrt{\alpha t}\right)\right] \quad (8\text{-}34)$$

8-7. Inversion by the Method of Residues. The most powerful method of carrying out the inversion of a Laplace transform is by means of a line integral in the complex plane, commonly called the "complex inversion integral" or the "Bromwich-Wagner integral." Most of the complicated inversion transforms which are tabulated have been obtained by this method. The successful use of the complex inversion integral itself requires considerable background in the theory of a complex variable. However, it is possible to carry out most inversion processes with a minimum background in complex-variable theory if the analyst is willing to accept certain consequences of the more general methods without proof. This is the procedure which will be followed here. A procedure, termed the "method of residues," will be presented which is applicable to a large class of functions. The combination of the residue method, a simple table of transforms, and the use of the technique of convolution will solve most inversion problems. The successful use of the residue method demands some background in the properties of a complex variable.

Complex Numbers. A complex number is conventionally represented by the symbol Z:

$$Z = x + iy \quad (8\text{-}35)$$

where x and y are both real numbers and $i = \sqrt{-1}$. The number represented by x is termed the "real part of Z,"

$$x = R(Z) \quad (8\text{-}36)$$

and the number represented by y is termed the "imaginary part of Z,"

$$y = I(Z) \quad (8\text{-}37)$$

The symbol $\bar{Z}$ or Z^* is used to denote the complex conjugate which is obtained by replacing the symbol i by $-i$:

$$\bar{Z} = x - iy \quad (8\text{-}38)$$

Two complex numbers are equal if, and only if, both the real parts and the imaginary parts are individually equal. Thus,

$$Z_1 = x_1 + iy_1 = Z_2 = x_2 + iy_2 \quad (8\text{-}39)$$

if $x_1 = x_2$ and $y_1 = y_2$. This leads to the following rules of addition, subtraction, multiplication, and division:

$$Z_1 \pm Z_2 = (x_1 + iy_1) \pm (x_2 + iy_2) = (x_1 \pm x_2) + i(y_1 \pm y_2) \tag{8-40}$$

$$Z_1 Z_2 = (x_1 + iy_1)(x_2 + iy_2) = (x_1x_2 - y_1y_2) + i(x_2y_1 + x_1y_2) \tag{8-41}$$

$$\frac{Z_1}{Z_2} = \frac{x_1 + iy_1}{x_2 + iy_2} = \frac{(x_1 + iy_1)(x_2 - iy_2)}{x_2^2 + y_2^2} = \frac{(x_1x_2 + y_1y_2) + i(x_2y_1 - x_1y_2)}{x_2^2 + y_2^2} \tag{8-42}$$

The complex number Z has properties which closely resemble those of a vector. For example, in the so-called "complex" or "Argand plane" of Fig. 8-3 let the distance along the x axis measure the magnitude of the real part of the complex number and the distance along the y axis measure the imaginary part of the complex number. Then the complex number Z_1 with real part x_1 and imaginary part y_1 will be represented by a point in the complex plane, as shown in Fig. 8-3. If a line is drawn from the origin to the point Z_1, the length of that line will be equal to the absolute magnitude of the complex number:

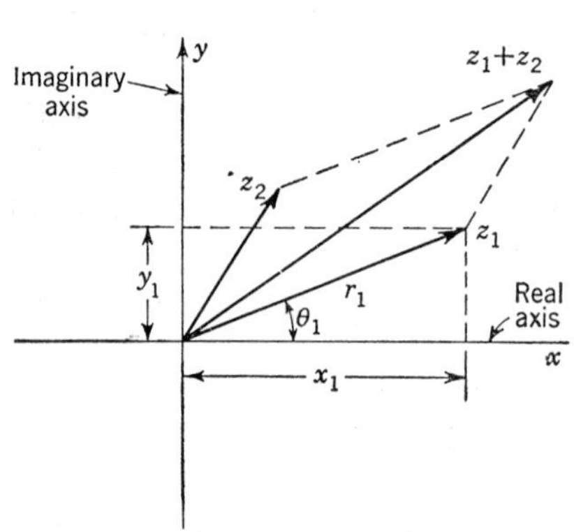

FIG. 8-3. The complex plane.

$$|Z| = \sqrt{x^2 + y^2} = \sqrt{Z\bar{Z}} = r \tag{8-43}$$

If r denotes the absolute magnitude of Z and θ denotes the angle between the x axis and the point Z, then

$$\begin{aligned} r &= \sqrt{x^2 + y^2} \\ x &= r \cos \theta \\ y &= r \sin \theta \end{aligned} \tag{8-44}$$

$$Z = r(\cos \theta + i \sin \theta) \tag{8-45}$$

$$Z_1 Z_2 = r_1 r_2[\cos (\theta_1 + \theta_2) + i \sin (\theta_1 + \theta_2)] \tag{8-46}$$

$$\frac{Z_1}{Z_2} = \frac{r_1}{r_2}[\cos (\theta_1 - \theta_2) + i \sin (\theta_1 - \theta_2)] \tag{8-47}$$

$$Z^n = r^n(\cos n\theta + i \sin n\theta) \tag{8-48}$$

Analytic Functions. The utility of complex numbers is largely a result of the properties of what are known as "analytic functions" of a complex variable. Thus, let W denote a function of the complex number Z:

$$W = f(Z) = U(x,y) + iV(x,y)$$
$$U = R(W) \qquad V = I(W) \tag{8-49}$$

W is said to be an analytic function of the complex variable if it fits certain rather stringent criteria. First, it must be a single-valued function of Z; that is, only one value of W can correspond to a given value of Z. Second, the derivative dW/dZ must have the same value, regardless of the manner in which ΔZ approaches zero. This latter requirement is not fulfilled by all functions, but it is found to be true for a very important class of functions which arise frequently during the course of an analytic solution of linear partial differential equations. When the function $W = f(Z)$ is analytic, the so-called "Cauchy-Riemann conditions"

$$\frac{\partial U}{\partial x} = \frac{\partial V}{\partial y} \qquad \frac{\partial V}{\partial x} = -\frac{\partial U}{\partial y} \tag{8-50}$$

are satisfied, and, conversely, when the Cauchy-Riemann conditions are satisfied, the function is analytic. The Cauchy-Riemann conditions often are a convenient way of testing a function for analytic properties.

It can be shown that every polynomial

$$W = \sum_n A_n Z^n \qquad n = \text{integer, positive} \tag{8-51}$$

is analytic except at $Z = \infty$. The function

$$W = f(Z) = \frac{1}{Z} = \frac{1}{x + iy} = \frac{x}{x^2 + y^2} - \frac{iy}{x^2 + y^2} \tag{8-52}$$

is analytic except at $Z = 0$, where U and V and their derivatives do not exist. It may also be shown that an analytic function of an analytic function is analytic. For example, $\sin Z$ is analytic, and $1/Z$ is analytic except at $Z = 0$; it then follows that $\sin 1/Z$ is analytic except at $Z = 0$.

Exponential and Trigonometric Functions. Formal power-series expansions give the following results which are used to define the exponential and trigonometric functions of the complex variable:

$$e^{iy} = \cos y + i \sin y \tag{8-53}$$
$$e^Z = e^{x+iy} = e^x(\cos y + i \sin y) \tag{8-54}$$

As a result of (8-45) and (8-53),

$$Z = re^{i\theta} \tag{8-55}$$
$$e^{-iy} = \cos y - i \sin y \tag{8-56}$$

$$\cos Z = \frac{e^{iZ} + e^{-iZ}}{2} = \cosh iZ \tag{8-57}$$

$$\sin Z = \frac{e^{iZ} - e^{-iZ}}{2i} = i \sinh iZ \tag{8-58}$$

$$\cosh Z = \frac{e^{Z} + e^{-Z}}{2} \tag{8-59}$$

$$\sinh Z = \frac{e^{Z} - e^{-Z}}{2} \tag{8-60}$$

$$\cosh Z = \cosh x \cos y + i \sinh x \sin y \tag{8-61}$$

$$\sinh Z = \sinh x \cos y + i \cosh x \sin y \tag{8-62}$$

All circular and hyperbolic functions are analytic for all finite values of Z except where the denominator vanishes. All trigonometric identities and differentiation rules which are valid for real variables apply without change to the complex variable.

It is frequently necessary to determine those points at which the denominator of an otherwise analytic function vanishes. In the case of the circular and hyperbolic functions, the above relations are of considerable help in this regard.

Example 8-4. Suppose that the values of Z at which $\sin Z$ is 0 are desired.

$$\sin Z = \frac{e^{iZ} - e^{iZ}}{2i} = \frac{e^{ix-y} - e^{-ix+y}}{2i}$$

$$= \frac{1}{2i} [e^{-y} (\cos x + i \sin x) - e^{y} (\cos x - i \sin x)] \tag{8-63}$$

For the above to equal zero, the real and imaginary parts both must vanish:

$$\begin{aligned} R(\sin Z) &= \sin x \frac{e^{y} + e^{-y}}{2} = \sin x \cosh y \\ I(\sin Z) &= -\cos x \frac{e^{y} - e^{-y}}{2i} = -\frac{1}{i} \cos x \sinh y \end{aligned} \tag{8-64}$$

Both terms vanish if $x = \pm n\pi$ $(n = 0, 1, \ldots)$ and $y = 0$ or if $x = 0$ and $y = \pm in\pi$, and both cases are covered if $Z = \pm n\pi$ $(n = 0, 1, \ldots)$.

Singularities. Every function [with the exception of $f(Z)$ equals a constant] fails to be analytic at one or more values of Z. If there are no such finite values of Z, then the function fails to be analytic at Z = infinity. This is the case with power series. Those points at which an otherwise analytic function fails to prove analytic are termed "singular points." Singular points are divided into three classes: branch points; removable singular points, or poles; and essential singular points.

Branch Points. Consider the real function of the real variable

$$q = (x + a)^n \tag{8-65}$$

As long as n is integer, the function q will have one, and only one, value corresponding to a given value of x. However, this is not the case if n is not integer. For example, suppose that n is $\frac{1}{2}$, then (as may be seen from Fig. 8-4) each value of x will correspond to two values of q. The curve $q = +(x + a)^{\frac{1}{2}}$ is called the "upper branch" of the function; curve $q = -(x + a)^{\frac{1}{2}}$ is called the "lower branch" of the function. The point $x = -a$, where the two branches of the curve join, is called a "branch point."

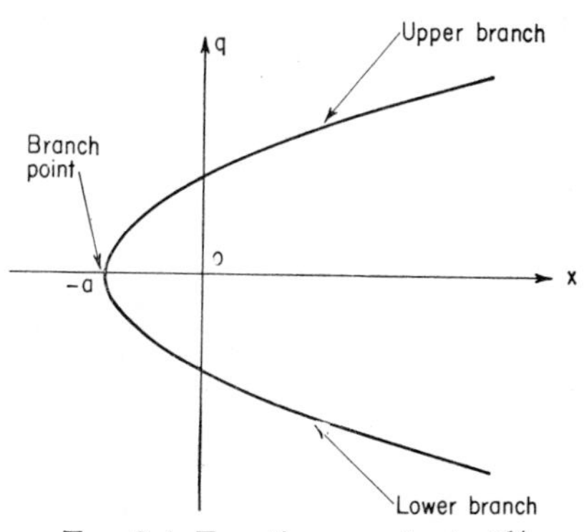

FIG. 8-4. Function $q = (x + a)^{\frac{1}{2}}$.

An analogous situation occurs with functions of the complex variable. Consider the function

$$Z^{1/n} = r^{1/n}\left(\cos\frac{\theta}{n} + i\sin\frac{\theta}{n}\right) \tag{8-66}$$

where n is integer ($n \geqq 0$) and $r^{1/n}$ is the positive nth root of r. Now

$$Z = r(\cos\theta + i\sin\theta) = r[\cos(\theta + 2\pi m) + i\sin(\theta + 2\pi m)] \tag{8-67}$$

where m is any positive integer. Then

$$Z^{1/n} = r^{1/n}\left(\cos\frac{\theta + 2\pi m}{n} + i\sin\frac{\theta + 2\pi m}{n}\right) \tag{8-68}$$

and $Z^{1/n}$ has n distinct values, or branches, corresponding to the same value of Z. However, each branch corresponds to a specific value of m, that is, $m = 0, 1, 2, \ldots, n - 1$. As a specific case, let $n = 2$. Then

$$f_1(Z) = w_1 = r^{\frac{1}{2}}\left(\cos\frac{\theta}{2} + i\sin\frac{\theta}{2}\right) \tag{8-69}$$

$$f_2(Z) = w_2 = r^{\frac{1}{2}}\left[\cos\left(\frac{\theta}{2} + \pi\right) + i\sin\left(\frac{\theta}{2} + \pi\right)\right] = -f_1(Z) \tag{8-70}$$

w_1 corresponds to one branch of the function and w_2 to the second branch. If the range of θ is limited to, say, $0 \leqq \theta < 2\pi$, w_1 lies in the first and second quadrants of the complex plane, since

$$0 \leqq \frac{\theta}{2} < \pi$$

and w_2 lies in the third and fourth quadrants of the complex plane. The functions w_1 and w_2 are called the "branches" of $w = \sqrt{Z}$. These two branches join at $Z = 0$, and consequently this point is called a "branch point" of the function $w = \sqrt{Z}$.

A function with more than one branch is not analytic since it is not single-valued. *However, if the range covered by the angle θ is restricted so that only one branch of the function can be traversed, then the restricted function is single-valued.* For example, if θ is restricted so that

$$\theta_0 \leqq \theta < \theta_0 + 2\pi$$

the function

$$w = Z^{1/n}$$
$$= r^{1/n}\left(\cos\frac{\theta}{n} + i\sin\frac{\theta}{n}\right) \qquad (8\text{-}71)$$

with r required to be positive, will be analytic in the defined region. This restriction on the range of the angle is frequently indicated in the complex plane, as shown in Fig. 8-5.

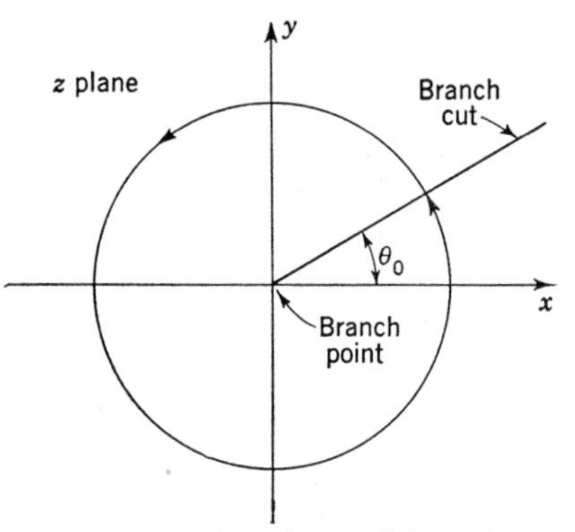

FIG. 8-5. Branch point and branch cut in complex plane.

A line extending from the origin to infinity and making the angle θ_0 with the x axis is called a "branch cut." If the function is to be analytic, values of w cannot be used which involve crossing the branch cut.

In general,

$$w = f(Z + a)^\nu \qquad (8\text{-}72)$$

where ν is noninteger, is multiple-valued with a branch point at $Z = -a$. Similarly,

$$\ln Z = \ln r + \ln e^{i\theta} = \ln r + i\theta$$
$$= \ln r + i(\theta + 2n\pi) \qquad (8\text{-}73)$$

is multiple-valued with a branch point at $Z = 0$. The inverse trigonometric and inverse hyperbolic functions are also multiple-valued. All three of these types of functions may be made analytic if the range of θ is suitably restricted and only one branch of the function is used.

An important aspect of functions which possess branch points is that they fail to be analytic on the branch cut and hence are not analytic at every point on a closed curve surrounding the branch point.

Poles. Frequently a function fails to be analytic at a point, say Z_0, because the denominator vanishes at that point. When this is the case, the point Z_0 is a singular point of the function. If, however, the function $f(Z)$ becomes infinite at $Z = Z_0$ in such a way that the new function

$$\phi(Z) = (Z - Z_0)^m f(Z) \tag{8-74}$$

(where m is some positive integer) is finite at $Z = Z_0$, Z_0 is said to be a "removable singular point," or a "pole," of order m of the function $f(Z)$. Here, m is the *smallest* positive integer which will make $\phi(Z)$ analytic at $Z = Z_0$.

If no finite value of m will make $\phi(Z)$ analytic at Z_0, Z_0 is said to be an "essential singular point" of the function $f(Z)$.

The quotients of polynomials are examples of functions with poles.

$$f(Z) = \frac{Z^2 + a^2}{(Z - a)^2(Z + a)} \tag{8-75}$$

has a simple pole at $Z = -a$ and a double pole at $Z = a$.

The function

$$\coth Z = \frac{\cosh Z}{\sinh Z} \tag{8-76}$$

is analytic except at the simple poles $Z = 0, \pm i\pi, \pm 2i\pi, \ldots, \pm ni\pi$. This may be seen as follows:

1. Expand the analytic function $\sinh Z$ in Taylor's series about $Z = 0$. Then

$$\sinh Z = Z + \frac{Z^3}{3!} + \frac{Z^5}{5!} + \cdots \tag{8-77}$$

$$\coth Z = \frac{\cosh Z}{Z + Z^3/3! + Z^5/5! + \cdots} \tag{8-78}$$

$$\phi(Z) = Z \coth Z = \frac{\cosh Z}{1 + Z^2/3! + Z^4/5! + \cdots} \tag{8-79}$$

Since $\phi(Z)$ is analytic at $Z = 0$, $Z = 0$ is a pole of $\coth Z$.

2. Expand the analytic function $\sinh Z$ in Taylor's series about $Z = i\pi$:

$$\begin{aligned}\sinh Z &= \sinh i\pi + (Z - i\pi)\cosh i\pi + \frac{(Z - i\pi)^2}{2!}\sinh i\pi \cdots \\ &= i \sin \pi - (Z - i\pi)\cos \pi + \frac{i(Z - i\pi)^2}{2!}\sin \pi \\ &\qquad - \frac{(Z - i\pi)^3}{3!}\cos \pi + \cdots \\ &= -(Z - i\pi) - \frac{1}{3!}(Z - i\pi)^3 - \frac{1}{5!}(Z - i\pi)^5 \cdots \end{aligned} \tag{8-80}$$

Then $\phi(Z) = (Z - i\pi) \coth Z$ is analytic at $Z = i\pi$, and $Z = i\pi$ is a pole of $\coth Z$.

3. Expansions carried out in the same fashion as (2) above will show that $Z = \pm ni\pi$ are also simple poles of $\coth Z$.

The function

$$e^{1/Z} = 1 + \frac{1}{Z} + \frac{1}{2!Z^2} + \frac{1}{3!Z^3} + \cdots \tag{8-81}$$

has an essential singularity at $Z = 0$.

Every analytic function may be expanded in Taylor's series. The expansion is valid within a region enclosed by a circle that does not include a singularity. The radius of the circle is termed the "radius of convergence."

Inversion. It can be shown that in the transform relation

$$\bar{f}(p) = \int_0^\infty e^{-pt} f(t)\, dt \tag{8-9}$$

p may be interpreted as a complex number and that, except for singularities and branches, the function $\bar{f}(p)$ is ordinarily analytic. *In particular, the transform of a solution to a Sturm-Liouville equation is analytic for all finite p except for poles which correspond to the characteristic numbers, or eigenvalues, of the system.*

When $\bar{f}(p)$ is analytic [or is made analytic by suitably restricting θ so that only one branch of $\bar{f}(p)$ is traversed] except for singularities which are *poles* only, the inverse transform of $\bar{f}(p)$ is given by the expression

$$f(t) = \sum_1^\infty \rho_n(t) \tag{8-82}$$

where $\rho_n(t)$ is termed the "residue" of $\bar{f}(p)$ at the pole p_n. The residues of $\bar{f}(p)$ may be determined as follows: Let $\bar{f}(p)$ be represented as the ratio of two functions

$$\bar{f}(p) = \frac{j(p)}{l(p)} \tag{8-83}$$

If, when expressed in the form of power series in terms of p, the degree of $l(p)$ is at least one greater than that of $j(p)$, then we have the following:

When p_n is a simple pole of $\bar{f}(p)$,

$$\rho_n(t) = \frac{j(p_n)}{l'(p_n)} e^{p_n t} \tag{8-84}$$

where $l'(p_n)$ denotes the value of $dl(p)/dp$ at $p = p_n$.

When p_n is a multiple pole of order m of $\bar{f}(p)$, then we have the following:

$$\rho_n(t) = e^{p_n t}\left[A_1 + tA_2 + \frac{t^2}{2!}A_3 + \cdots + \frac{t^{m-1}A_m}{(m-1)!}\right]$$

$$= e^{p_n t} \sum_{s=1}^{s=m} A_s \frac{t^{s-1}}{(s-1)!} \tag{8-85}$$

where the A's are given by

$$A_s = \frac{1}{(m-s)!}\,\phi_n^{m-s}(p_n) \qquad s = 1, 2, \ldots, m \tag{8-86}$$

$$\phi_n(p) = (p - p_n)^m \bar{f}(p) \tag{8-87}$$

$$\phi_n^{m-s}(p_n) = \left[\frac{d^{m-s}}{dp^{m-s}}\,\phi_n(p)\right]_{p_n} \tag{8-88}$$

evaluated at $p = p_n$. For instance,

$$\begin{aligned} \phi_n{}^0(p_n) &= \phi_n(p_n) & m - s &= 0 \\ \phi_n{}^1(p_n) &= \left[\frac{d}{dp}\,\phi_n(p)\right]_{p_n} & m - s &= 1 \end{aligned} \tag{8-89}$$

evaluated at p_n. The following examples illustrate the use of the residue method.

Example 8-5. Let $\bar{f}(p) = 1/p(p-a)^2$ which has a simple pole at $p = 0$ and a double pole at $p = a$. Then $j(p) = 1$, and $l(p) = p(p-a)^2$. The residue at $p = 0$ is given by (8-84) and is

$$\rho_0(t) = \frac{1}{a^2}$$

since $\qquad l'(p) = (p-a)^2 + 2p(p-a) \qquad$ and $\qquad l'(0) = a^2$

The residue at the double pole $p = a$ is given by (8-85) as

$$\rho_a(t) = e^{at}(A_1 + tA_2) \qquad \text{since } m = 2$$

$$\phi(p) = (p-a)^2\bar{f}(p) = \frac{1}{p} \qquad \phi'(p) = -\frac{1}{p^2}$$

$$A_1 = \frac{1}{1}\,\phi_a'(a) = -\frac{1}{a^2}$$

$$A_2 = 1\phi(a) = \frac{1}{a}$$

Then

$$\rho_a(t) = e^{at}\left(-\frac{1}{a^2} + \frac{t}{a}\right)$$

The complete inversion is then

$$f(t) = +\frac{1}{a^2}(1 - e^{at}) + \frac{te^{at}}{a}$$

Example 8-6. Let $\bar{f}(p) = (\sinh a\sqrt{p})/(p \sinh \sqrt{p})$; find $f(t)$. The function $\sinh \sqrt{p}$ has a branch at $p = 0$ arising from the $\sqrt{p}$ term. At first glance, then, it would appear that $\bar{f}(p)$ has a branch point also. Closer examination discloses that this is not the case. This is shown as follows: Consider p to be the complex variable. Then limit the range of both $\sinh a\sqrt{p}$ and $\sinh \sqrt{p}$ to $r > 0$, $0 \leqq \theta < 2\pi$; that is, choose corresponding branches of $\sinh a\sqrt{p}$ and $\sinh \sqrt{p}$. With this limitation, both $\sinh a\sqrt{p}$ and $\sinh \sqrt{p}$ are analytic for all finite values of the argument and may be expanded in Taylor series:

$$\sinh a\sqrt{p} = a\sqrt{p} + \frac{(a\sqrt{p})^3}{3!} + \frac{(a\sqrt{p})^5}{5!} + \cdots$$

$$\sinh \sqrt{p} = \sqrt{p} + \frac{(\sqrt{p})^3}{3!} + \frac{(\sqrt{p})^5}{5!} + \cdots$$

When the ratio $\sinh a\sqrt{p}/\sinh \sqrt{p}$ is taken, $\sqrt{p}$ disappears:

$$\bar{f}(p) = \frac{a + a^3p/3! + a^5p^2/5! + \cdots}{p(1 + p/3! + p^2/5! + \cdots)}$$

Hence, $\bar{f}(p)$ does not have a branch point. It has poles, however. The poles are most easily found by use of the form

$$\bar{f}(p) = \frac{\sinh a\sqrt{p}}{p \sinh \sqrt{p}} = \frac{(\sinh a\sqrt{p})/\sqrt{p}}{p[(\sinh \sqrt{p})/\sqrt{p}]} = \frac{(\sinh a\sqrt{p})/\sqrt{p}}{\sqrt{p}\sinh \sqrt{p}}$$

which has simple poles at $p = 0$, $p = -n^2\pi^2$; $n = 1, 2, 3, \ldots$. Since only simple poles are involved,

$$j(p) = \frac{\sinh a\sqrt{p}}{\sqrt{p}}$$

$$l(p) = \sqrt{p}\sinh \sqrt{p}$$

$$l'(p) = \frac{1}{2}\frac{\sinh \sqrt{p}}{\sqrt{p}} + \sqrt{p}(\cosh \sqrt{p})\frac{1}{2\sqrt{p}}$$

$$\frac{j(p)}{l'(p)} = \frac{2 \sinh a\sqrt{p}}{\sinh \sqrt{p} + \sqrt{p}\cosh \sqrt{p}}$$

$$f(t) = \sum_{n=0}^{\infty} \frac{2 \sinh ian\pi e^{-n^2\pi^2t}}{\sinh in\pi + in\pi \cosh in\pi}$$

$$= \sum_{0}^{\infty} \frac{2 \sin an\pi e^{-n^2\pi^2t}}{\sin n\pi + n\pi \cos n\pi}$$

$$= a + \sum_{n=1}^{\infty} \frac{2 \sin an\pi e^{-n^2\pi^2t}}{n\pi \cos n\pi}$$

$$= a + \frac{2}{\pi}\sum_{n=1}^{\infty} \frac{(-1)^n}{n}(\sin an\pi)e^{-n^2\pi^2t}$$

8-8. Applications to Chemical Engineering

Example 8-7. Catalytic Reaction in a Flow System. The irreversible reaction

$$A \rightarrow B$$

occurs at an appreciable rate only at the surface of a catalyst. It is believed that the surface reaction is first-order and is governed by an equation of the form

$$\frac{-dN_A}{dt} = KaC_s \tag{8-90}$$

where $-dN_A/dt$ denotes the moles of A reacting per unit time; K is a reaction-rate constant, a is the projected surface area of the catalyst in contact with the reacting gas; and C_s is the concentration of A (moles per unit volume) in the immediate vicinity of the catalyst surface. It is proposed to determine the reaction-rate constant K by the following experiment:

The walls of a round tube of radius R and length L are to be coated with catalyst. The walls of the tube are to be maintained at a constant temperature T_w, and the gas passing through the tube will be at the same constant temperature. Pure gas A (initial concentration C_i) will be passed through the tube with a constant velocity V. After steady state is achieved, the exit gas concentration C_m (after complete mixing) of A will be measured and used to calculate the rate constant. The tube is short, and the gas flows as a "slug" of uniform velocity *without radial velocity gradient* through the tube. Consequently, the radial transport of A to the catalyst is entirely by *molecular diffusion*. Determine the integrated expression which will permit the reaction-rate constant to be determined from the following known quantities:

C_i	Moles/ft³
C_m	Moles/ft³
R	Ft
L	Ft
D, molecular diffusivity	Ft²/sec†
V, constant gas velocity	Ft/sec

† D, the diffusivity in units of square feet per second, is obtained from $\mathfrak{D}$, the diffusivity in moles per second per foot, by the relation

$$D = \frac{\mathfrak{D}}{\rho}$$

where ρ is the *molal* density of the fluid.

Solution. Consider the differential volume element shown in Fig. 8-6. Write a material balance on component A, neglecting molecular diffusion in the axial (x) direction.

$$\text{Rate of input} = V2\pi r\,dr\,C - D2\pi r\,dx\,\frac{\partial C}{\partial r}$$

$$\text{Rate of output} = V2\pi r\,dr\,C + \frac{\partial}{\partial x}(V2\pi r\,dr\,C)\,dx - \left[D2\pi r\,dx\,\frac{\partial C}{\partial r} + \frac{\partial}{\partial r}\left(D2\pi r\,dx\,\frac{\partial C}{\partial r}\right)dr\right]$$

$$\text{Rate of accumulation} = 0 \qquad \text{at steady state}$$

The material balance is then

$$\frac{\partial C}{\partial x} = \frac{D}{V}\left(\frac{\partial^2 C}{\partial r^2} + \frac{1}{r}\frac{\partial C}{\partial r}\right) \tag{8-91}$$

The boundary conditions are

$$C = C_i \qquad \text{at } x = 0, \text{ all } r \tag{8-92a}$$

$$\frac{\partial C}{\partial r} = 0 \qquad \text{at } r = 0, \text{ all } x \text{ (because of symmetry)} \tag{8-92b}$$

$$\frac{\partial C}{\partial r} = -\frac{K}{D}C \qquad \text{at } r = R, \text{ all } 0 < x < L \text{ (because of chemical reaction at } r = R) \tag{8-92c}$$

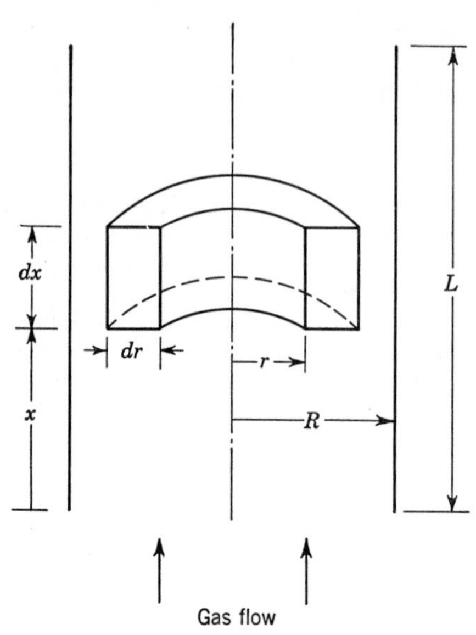

FIG. 8-6. Catalytic reaction in a flow system.

The Laplace transform, with respect to x, of (8-91) is

$$p\bar{C}(p,r) - C_i = \frac{D}{V}\left(\frac{d^2\bar{C}}{dr^2} + \frac{1}{r}\frac{d\bar{C}}{dr}\right) \tag{8-93}$$

The boundary conditions become

$$\frac{d\bar{C}}{dr} = 0 \qquad \text{at } r = 0 \tag{8-94}$$

$$\frac{d\bar{C}}{dr} = -\frac{K}{D}\bar{C} \qquad \text{at } r = R \tag{8-95}$$

If (8-93) is written in the form

$$r^2\frac{d^2\bar{C}}{dr^2} + r\frac{d\bar{C}}{dr} - \frac{Vp}{D}r^2\bar{C} = -\frac{Vr^2}{D}C_i \tag{8-96}$$

the solution may be obtained in terms of Bessel functions by use of Eq. (5-65). The result is

$$\bar{C} - \frac{C_i}{p} = AI_0\sqrt{\frac{Vp}{D}}\,r + BK_0\sqrt{\frac{Vp}{D}}\,r \tag{8-97}$$

Differentiation of (8-97), using relation (5-83), gives

$$\frac{d\bar{C}}{dr} = A\sqrt{\frac{Vp}{D}}\,I_1\sqrt{\frac{Vp}{D}}\,r - B\sqrt{\frac{Vp}{D}}\,K_1\sqrt{\frac{Vp}{D}}\,r \tag{8-98}$$

The boundary condition (8-94) requires B to be zero, a result which could also be concluded from (8-97), together with the fact that $\lim_{r\to 0} K_0(r) \to \infty$. Use of the boundary condition (8-95) gives

$$-\frac{K}{D}\bar{C}_R = A\sqrt{\frac{Vp}{D}}\, I_1\sqrt{\frac{Vp}{D}}\, R = -\frac{K}{D}\left(AI_0\sqrt{\frac{Vp}{D}}\, R + \frac{C_i}{p}\right) \tag{8-99}$$

The value of the constant A is obtained from (8-99) as

$$A = \frac{-(KC_i/Dp)}{(K/D)I_0(\sqrt{Vp/D})R + (\sqrt{Vp/D})I_1(\sqrt{Vp/D})R} \tag{8-100}$$

The solution of (8-93) is then

$$\bar{C} = C_i\left[\frac{1}{p} - \frac{(K/Dp)I_0(\sqrt{Vp/D})r}{(K/D)I_0(\sqrt{Vp/D})R + (Vp/D)I_1(\sqrt{Vp/D})R}\right] \tag{8-101}$$

The inversion of (8-101) would result in a relation between C and x and r. This relation could then be integrated over the cross section to give an equation between the mixed or bulk concentration C_m and x. It is easier, however, to find the transform of C_m, $\bar{C}_m$, and then to carry out the inversion. The justification of this procedure is shown as follows:

$$C_m = \frac{\int_0^R rC\, dr}{\int_0^R r\, dr} \tag{8-102}$$

Hence,

$$\begin{aligned}\bar{C}_m &= \int_0^\infty e^{-px}C_m\, dx = \frac{\int_0^\infty e^{-px}\, dx \int_0^R rC\, dr}{\int_0^R r\, dr} \\ &= \frac{\int_0^R r\, dr \int_0^\infty e^{-px}C\, dx}{\int_0^R r\, dr} = \frac{\int_0^R r\bar{C}\, dr}{\int_0^R r\, dr}\end{aligned} \tag{8-103}$$

Then

$$\begin{aligned}\bar{C}_m &= C_i\left\{\frac{1}{p} - \frac{2K/Dp\int_0^R [I_0(\sqrt{Vp/D})r]r\, dr}{R^2[(K/D)I_0(\sqrt{Vp/D})R + (\sqrt{Vp/D})I_1(\sqrt{Vp/D})R]}\right\} \\ &= C_i\left\{\frac{1}{p} - \frac{(2K/DRp)[\sqrt{(1/Vp)/D}]I_1(\sqrt{Vp/D})R}{(K/D)I_0(\sqrt{Vp/D})R + (\sqrt{Vp/D})I_1(\sqrt{Vp/D})R}\right\}\end{aligned} \tag{8-104}$$

Let

$$\alpha = R\sqrt{\frac{V}{D}} \qquad \beta = \frac{KR}{D} \tag{8-105}$$

Then (8-104) becomes

$$\frac{\bar{C}_m}{C_i} = \frac{1}{p} - \frac{(2\beta/\alpha\sqrt{p})I_1(\alpha\sqrt{p})}{p[\beta I_0(\alpha\sqrt{p}) + \alpha\sqrt{p}I_1(\alpha\sqrt{p})]} \tag{8-106}$$

The inversion of (8-106) may be carried out by means of the residue method. Let

$$\bar{f}(p) = \frac{(1/\alpha\sqrt{p})I_1(\alpha\sqrt{p})}{p[\beta I_0(\alpha\sqrt{p}) + \alpha\sqrt{p}I_1(\alpha\sqrt{p})]} = \frac{j(p)}{l(p)} \tag{8-107}$$

The series definition of $I_n(x)$ is

$$I_n(x) = \sum_{m=0}^{\infty} \frac{(x/2)^{2m+n}}{m!(m+n)!} \tag{5-63}$$

Then

$$I_1(\alpha\sqrt{p}) = \frac{\alpha\sqrt{p}}{2} \sum_{m=0}^{\infty} \frac{(\alpha^2 p/4)^m}{(m+1)m!m!}$$

$$I_0(\alpha\sqrt{p}) = \sum_{m=0}^{\infty} \frac{(\alpha^2 p/4)^m}{m!m!}$$

$$j(p) = \frac{1}{2} \sum_{m=0}^{\infty} \frac{(\alpha^2 p/4)^m}{(m+1)m!m!} \tag{8-108}$$

$$l(p) = p \sum_{m=0}^{\infty} \left(\beta + \frac{\alpha^2 p}{2(m+1)}\right) \frac{(\alpha^2 p/4)^m}{m!m!} \tag{8-109}$$

The degree of $l(p)$ is higher than that of $j(p)$. Furthermore, $l(p)$ is a polynomial in p, and the singularities of $\bar{f}(p)$ all are poles. The poles occur at those values of p which make (8-109) vanish. Inspection of (8-109) shows that only simple poles are involved; one pole occurs at $p = 0$, and the remainder occur at *negative* values of p. The numerical values of these poles are most easily found by returning to the Bessel-function form of (8-109):

$$l(p) = p[\beta I_0(\alpha\sqrt{p}) + \alpha\sqrt{p}\, I_1(\alpha\sqrt{p})] \tag{8-110}$$

The poles occur at $p = 0$ and at values of p for which

$$\beta I_0(\alpha\sqrt{p}) + \alpha\sqrt{p}\, I_1(\alpha\sqrt{p}) = 0 \tag{8-111}$$

Since

$$I_n(x) = i^{-n} J_n(ix) \tag{5-63}$$

(8-111) may be put in the form

$$\beta J_0(i\alpha\sqrt{p}) - i\alpha\sqrt{p}\, J_1(i\alpha\sqrt{p}) = 0 \tag{8-112}$$

Since (8-112) is only satisfied for negative values of p, all

$$\lambda_j = i\alpha\sqrt{p_j} \tag{8-113}$$

which satisfy

$$\beta J_0(\lambda_j) - \lambda_j J_1(\lambda_j) = 0 \tag{8-114}$$

will be real, will form an infinite sequence, and will define the poles (other than the pole at $p = 0$) of $\bar{f}(p)$, Eq. (8-107). With the new notation, (8-107) becomes

$$\bar{f}(p) = \frac{(J_1\lambda)/\lambda}{p[\beta J_0(\lambda) - J_1(\lambda)]} \tag{8-115}$$

$$\begin{aligned}\frac{d}{dp}[l(p)] &= \frac{d}{dp}\{p[\beta J_0(\lambda) - \lambda J_1(\lambda)]\} \\ &= \beta J_0(\lambda) - \lambda J_1(\lambda) + p\left(\frac{d}{d\lambda}\{\beta[J_0(\lambda) - \lambda J_1(\lambda)]\}\right)\frac{d\lambda}{dp} \\ &= \beta J_0(\lambda) - \lambda J_1(\lambda) - \frac{\lambda}{2}[\beta J_1(\lambda) + \lambda J_0(\lambda)]\end{aligned} \tag{8-116}$$

The residue at the pole $p = 0$ is

$$\rho_0(x) = \frac{j(0)}{l'(0)} = \lim_{\lambda\to 0} \frac{[J_1(\lambda)]/\lambda}{\beta J_0(\lambda) - \lambda J_1(\lambda) - (\lambda/2)[\beta J_1(\lambda) + \lambda J_0(\lambda)]} = \frac{1}{2\beta}$$

The residues at the poles $p = p_j$, where p_j satisfies (8-113) and (8-114), are

$$\sum_j \frac{-2\beta e^{-(\lambda_j^2 x/\alpha^2)}}{\lambda_j^2(\beta^2 + \lambda_j^2)}$$

Consequently,

$$f(x) = \frac{1}{2\beta} - \sum_j \frac{2\beta e^{-(\lambda_j^2 x/\alpha^2)}}{\lambda_j^2(\beta^2 + \lambda_j^2)} \tag{8-117}$$

The inversion of (8-106) is then

$$\frac{C_m}{C_i} = \sum_j \frac{4\beta^2 e^{-(\lambda_j^2 x/\alpha^2)}}{\lambda_j^2(\lambda_j^2 + \beta^2)} \tag{8-118}$$

where λ_j satisfies

$$\lambda_j = \beta \frac{J_0(\lambda_j)}{J_1(\lambda_j)} \tag{8-119}$$

and

$$\beta = \frac{KR}{D}$$

$$\alpha = R\sqrt{\frac{V}{D}} \tag{8-105}$$

Equations (8-118), (8-119), and (8-105) represent the formal solution of the problem. In a given experiment, the reaction-rate constant K is determined by comparing experimental data with curves representing Eq. (8-118) plotted as a function of β at constant values of α with $C_m = C_0$ at $x = L$.

Baron, Manning, and Johnstone[2] used the procedure described in this example to study the oxidation of sulfur dioxide in the presence of a vanadium pentoxide catalyst.

Example 8-8. Unsteady-state Operation of a Packed Bed. A number of chemical-engineering processes utilize the unsteady-state operation of a packed bed. Common examples include a heat regenerator, an adsorption column, an ion-exchange column, a chromatographic column. Typically, the equipment consists of a column filled with a loosely packed solid. A fluid is passed through the column, and the fluid and solid exchange heat or mass or both. Since the solid remains fixed in place, the temperature and/or composition of the solid change with time, and the operation of the column is an unsteady-state process.

Consider an adsorption column in which a component is transferred from the fluid phase to the solid phase. With reference to Fig. 8-7, let

a = area for mass transfer per unit tower volume, ft²/ft³
h = fluid holdup per unit tower volume, moles/ft³
H = solid holdup per unit tower volume, moles/ft³
k_g = mass-transfer coefficient, moles/(hr)(ft³)(mole/ft³)
ρ = fluid density, moles/ft³
t = time since start of process, hr
V = fluid velocity, ft/hr
x = mole fraction of adsorbable component on solid
y = mole fraction of adsorbable component in fluid
y^* = mole fraction of adsorbable component at fluid-solid interface
z = height of section in tower with reference to fluid inlet, ft

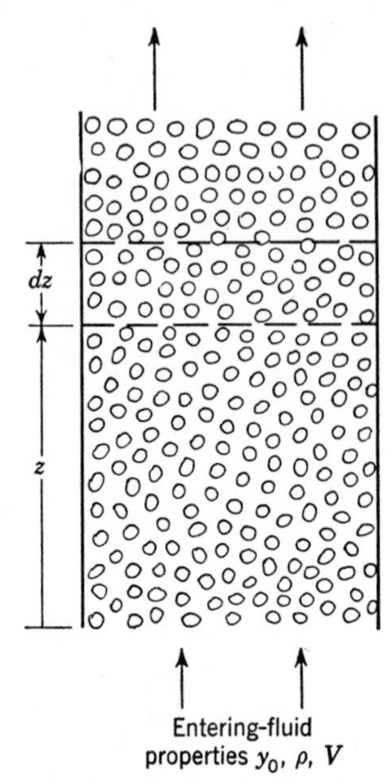

FIG. 8-7. Unsteady-state operation of a packed bed.

It will be assumed that a, h, H, k_g, ρ, and V are constant. Composition gradients within a solid granule will be neglected, and mass transferred by diffusion as the result of granule-to-granule contact will be ignored.

A material balance on the fluid phase contained in an element of unit cross section and of height dz is

$$\text{Rate of input} = V\rho y$$

$$\text{Rate of output} = V\rho\left(y + \frac{\partial y}{\partial z}\,dz\right) + k_g a\,dz\,(y - y^*)$$

$$\text{Rate of accumulation} = \frac{\partial}{\partial t}\,(h\,dz\,y)$$

The fluid-phase material balance is then

$$-V\rho\,\frac{\partial y}{\partial z} - k_g a(y - y^*) = h\,\frac{\partial y}{\partial t} \tag{8-120}$$

Similarly, the solid-phase material balance is

$$k_g a(y - y^*) = H\,\frac{\partial x}{\partial t} \tag{8-121}$$

The equilibrium relation which relates y^* and x must be specified. The analytical difficulties encountered in the solution process are strongly dependent upon the form of this relation. Here, the simple linear form

$$y^* = mx + b = w \tag{8-122}$$

will be assumed. When Eq. (8-122), together with the transformation variables

$$N = \frac{k_g a}{V\rho} z \tag{8-123}$$

$$\theta = \frac{k_g a m}{H}\left(t - \frac{h}{V\rho} z\right) \tag{8-124}$$

is introduced into (8-120) and (8-121), the material-balance expressions become

$$-\left(\frac{\partial y}{\partial N}\right)_\theta = y - w \tag{8-125}$$

$$\left(\frac{\partial w}{\partial \theta}\right)_N = y - w \tag{8-126}$$

The transformation variable N is simply the column height measured in terms of the number of transfer units. The quantity $(h/V\rho)z$ is the time required to displace the fluid held up in a column of height z. Consequently, $t - (h/V\rho)z$ is the time during which a bed element at height z has been contacted by fluid other than that originally present in the column.

It is assumed that the composition of the initial solid is uniform and that the composition of the entering fluid is constant. The boundary conditions are then

$$\begin{array}{lll} \text{At } t = 0 & x = x_0, \text{ all } z & \\ \text{or} \quad \text{At } \theta = 0 & w = w_0, \text{ all } N & (8\text{-}127) \\ \text{At } z = 0 & y = y_0, \text{ all } t & \\ \text{or} \quad \text{At } N = 0 & y = y_0, \text{ all } \theta & (8\text{-}128) \end{array}$$

Now carry out the Laplace transformation of (8-125) and (8-126), replacing the variable θ. Equation (8-125) becomes

$$-\frac{d\bar{y}}{dN} = \bar{y} - \bar{w} \tag{8-129}$$

and (8-126) becomes

$$p\bar{w} - w_0 = \bar{y} - \bar{w} \tag{8-130}$$

$$\bar{w} = \bar{w}(N,p) = \int_0^\infty e^{-p\theta} w(N,\theta)\, d\theta \tag{8-131}$$

If Eq. (8-130) is solved for $\bar{w}$,

$$\bar{w} = \frac{\bar{y} + w_0}{p + 1} \tag{8-132}$$

and the result substituted into (8-129), there obtains

$$\frac{d\bar{y}}{dN} = -\frac{p}{p+1}\left(\bar{y} - \frac{w_0}{p}\right) \tag{8-133}$$

Integration of (8-133) gives

$$\bar{y} = \frac{w_0}{p} + Ae^{-[p/(p+1)]N} \tag{8-134}$$

The integration constant A in (8-134) is evaluated by transformation of the boundary condition (8-128):

$$\text{At } N = 0 \qquad \bar{y} = \frac{y_0}{p}, \text{ all } \theta \tag{8-135}$$

Hence,

$$A = \frac{y_0 - w_0}{p} \tag{8-136}$$

and
$$y = \frac{w_0}{p} + \frac{y_0 - w_0}{p} e^{-[p/(p+1)]N} \tag{8-137}$$

Since
$$\frac{p}{p+1} = 1 - \frac{1}{p+1} \tag{8-138}$$

(8-137) may be written

$$\bar{y} = \frac{w_0}{p} + \frac{y_0 - w_0}{p} e^{-N} e^{N/(p+1)} \tag{8-139}$$

The direct inversion of (8-139) is difficult. However, combining (8-139) and (8-132) gives

$$\bar{w} = \frac{w_0}{p} + \frac{y_0 - w_0}{p} e^{-N} \frac{e^{N/(p+1)}}{p+1} \tag{8-140}$$

Now, transform relation 75 of Table 8-1 is

$$\mathcal{L}^{-1}\left\{\frac{e^{a/p}}{p}\right\} = J_0(2i\sqrt{at})$$

Use of Eq. (8-14) then gives

$$\mathcal{L}^{-1}\left\{\frac{e^{N/(p+1)}}{p+1}\right\} = e^{-\theta} J_0(2i\sqrt{N\theta}) \tag{8-141}$$

The inverse transform of (8-140) is then obtained by convolution as

$$\frac{w - w_0}{y_0 - w_0} = e^{-N} \int_0^\theta e^{-s} J_0(2i\sqrt{Ns})\, ds \tag{8-142}$$

Combination of (8-125), (8-126), and (8-142) gives

$$\frac{\partial y}{\partial N} = -\frac{\partial w}{\partial \theta} = (w_0 - y_0) e^{-(N+\theta)} J_0(2i\sqrt{N\theta}) \tag{8-143}$$

$$y = (w_0 - y_0) e^{-\theta} \int_0^N e^{-s} J_0(2i\sqrt{s\theta})\, ds + f(\theta) \tag{8-144}$$

The arbitrary function $f(\theta)$ appearing in (8-144) is evaluated by means of the boundary condition (8-128). There results

$$\frac{y - y_0}{y_0 - w_0} = -e^{-\theta} \int_0^N e^{-s} J_0(2i\sqrt{s\theta})\, ds \tag{8-145}$$

Equations (8-142) and (8-145), together with the definitions of w [Eq. (8-122)], N [Eq. (8-123)], and θ [Eq. (8-124)], constitute the solution of the problem. Note that the same formal solution applies to the case of a packed-bed heat regenerator if the following interpretation of terms is made:

Mass transfer	*Heat transfer*
y	T_F, temperature of fluid
w	T_B, temperature of bed
θ	$\frac{Ua}{Hc_B}\left(t - \frac{h}{V\rho c_F} z\right)$
N	$\frac{Ua}{V\rho c_F} z$

where a = area for heat transfer per unit tower volume
c_B = heat capacity of solid, Btu/(lb mole)(°F)
c_F = heat capacity of fluid, Btu/(lb mole)(°F)
U = heat-transfer coefficient between solid and fluid, Btu/(ft²)(hr)(°F)

The heat-transfer case was originally solved by Anzelius[1] by a different analytical technique. Curves representing the numerical values of the integral relations given formally by (8-142) and (8-145) have been published by Furnas.[6]

8-9. Utility of the Transform Method. The transform method is not of universal applicability. It is of particular value in the solution of linear total or partial differential equations with linear boundary conditions. The solution in such cases generally proceeds smoothly if the coefficients of the original equation are constant. If the coefficients are variable, success is sometimes obtained if the variability of the coefficients is due to functions of the untransformed variable; i.e., if x and t are the two independent variables and the transformation is made with respect to t, the problem is sometimes solvable if the coefficients are functions of x only. In other cases, the Laplace-transform procedure ordinarily does not simplify the analytical problems.

BIBLIOGRAPHY

1. Anzelius, A.: *Z. angew. Math. u. Mech.*, vol. 6, pp. 291–294, 1926.
2. Baron, T., W. R. Manning, and H. F. Johnstone: Reaction Kinetics in a Tubular Reactor, *Chem. Eng. Progr.*, vol. 48, no. 3, pp. 125–132, 1952.
3. Campbell, G. A., and R. M. Foster: "Fourier Integrals for Practical Applications," D. Van Nostrand Company, Inc., Princeton N.J., 1948.
4. Carslaw, H. S., and J. C. Jaeger: "Operational Methods in Applied Mathematics," Oxford University Press, New York, 1941.
5. Churchill, R. V.: "Modern Operational Mathematics in Engineering," McGraw-Hill Book Company, Inc., New York, 1944.
6. Furnas, C. C.: *Trans. Am. Inst. Chem. Engrs.*, vol. 24, pp. 142–193, 1930.
7. McLachlan, N. W.: "Modern Operational Calculus, with Applications in Technical Mathematics," Macmillan & Co., Ltd., London, 1948.
8. Jahnke, E., and F. Emde: "Tables of Functions," Dover Publications, New York, 1943.
9. Marshall, W. R., Jr. and R. L. Pigford: "The Application of Differential Equations to Chemical Engineering Problems," University of Delaware, Newark, Del., 1947.

PROBLEMS

8-1. A solid is bounded *internally* by the sphere $r = R$. The solid extends to infinity otherwise. At zero time the solid temperature is everywhere T_0. Immediately thereafter, the spherical surface at $r = R$ is brought to and maintained at the temperature T_1. Find the temperature in the solid as a function of time and position.

8-2. With reference to Example 8-7, suppose that the reaction

$$A \rightleftharpoons B$$

is reversible, with the forward-rate constant K_1 and the backward-rate constant K_2. If conditions are otherwise the same as in Example 8-7, can the experimental method proposed there be used to find K_1 and K_2? Discuss quantitatively.

8-3. One method of determining the integral diffusivity of a liquid system is as follows: A straight capillary tube of constant bore and known internal length and volume, but with one end sealed off, is filled with a binary solution of known composition. The capillary is held in a vertical position with the open end up, and a very slow stream of pure solvent is allowed to flow at right angles to the longitudinal axis of the capillary. After an elapsed time t, the capillary is removed and the amount and composition of the solution now inside the capillary determined. Generally, several such experiments are performed and the data presented as E (per cent of extractable material remaining unextracted) vs. time of extraction t.

In a liquid a satisfactory method of defining the diffusivity so that it will be independent of relative composition has yet to be devised. However, define the diffusivity of a binary liquid system, D_L, by the relation which corresponds to the gaseous case

$$C_A(\mathbf{V}_A - \mathbf{U}) = -D_L \nabla C_A \tag{1}$$

where C_A is the solute concentration in moles per unit volume. In many liquid systems it is a permissible approximation to assume that the solvent concentration is constant, i.e., that the partial molal volume of the solute is zero.

(*a*) Assume the solvent concentration to be constant and show that the partial differential equation which applies to the capillary experiment can be taken as

$$\frac{\partial C_A}{\partial t} = \frac{\partial}{\partial x}\left(D_L \frac{\partial C_A}{\partial x}\right) \tag{2}$$

with the boundary conditions

$$C_A = C_0 \qquad \text{at } t = 0,\ 0 = x = L \tag{3a}$$

$$C_A = 0 \qquad \text{at } x = L,\ t > 0 \tag{3b}$$

$$+D_L \frac{\partial C_A}{\partial x} = \frac{\partial}{\partial t}\int_0^L C_A\, dx \qquad \text{at } x = L,\ t > 0 \tag{3c}$$

(*b*) Assume that D_L is constant and show that the solution of Eq. (2) is then

$$E = \frac{800}{\pi^2}\sum_{n=1}^{\infty} \frac{1}{(2n-1)^2} \exp \frac{-(2n-1)^2\pi^2 D_L t}{4L^2} \tag{4}$$

(*c*) Show that for small values of $D_L t/L^2$, (4) may be approximated by

$$E = 100 - 200\sqrt{\frac{D_L t}{L^2 \pi}} \tag{5}$$

(*d*) Piret et al.† used the capillary-tube method to determine the mean or integral diffusivity for a number of systems. Their data for KCl-H_2O solutions ($C_0 = 4N$) are given below:

t/L^2, *sec/cm*^{2}	$100 - E$, %
$51.84 \cdot 10^3$	95.1
36.50	85.8
28.97	80.4
17.03	63.2
8.805	45.9
3.689	28.8
2.838	24.9
1.242	15.9
1.160	15.8
0.741	12.2

Use these data to obtain D_L.

† E. L. Piret, R. A. Ebel, C. T. Kiang, and W. P. Armstrong, *Chem. Eng. Progr.*, vol. 47, no. 8, pp. 405–414, 1951.

8-4. Two liquid phases are brought into contact at time $t = 0$. Component A only is transferred across the interface by diffusion only. Each liquid phase extends indefinitely in a direction normal to the interface. At the interface the mole fraction of solute A in one phase is related to that in the second by the expression

$$y_i = mx_i \tag{1}$$

Let the liquid diffusion coefficient be defined by

$$y_A\rho(\mathbf{V}_A - \mathbf{U}) = -D_L\nabla y_A\rho \tag{2}$$

where ρ is the molal density of the liquid phase. If the over-all mass-transfer coefficient K_L is defined by

$$N_A = \text{moles/(ft}^2\text{)(hr) of } A \text{ transferred between phases} = K_L(y_\infty - mx_\infty) \tag{3}$$

find K_L as a function of the time of contact. Show that K_L can be considered to be made up of the contributions of the individual coefficients for each phase,

$$N_A = k_1(y_\infty - y_i) = k_2(x_i - x_\infty) \tag{4}$$

in a manner which formally resembles the expression customarily used for the over-all rate coefficient in terms of the individual coefficients when it is assumed that the transfer process takes place across two "stagnant" films in which the accumulation of the diffusing component is negligible. Assume that the molal density of the solvent in a given phase is constant and that the respective diffusivities are constant.

8-5. Molino and Hougen† measured the "thermal conductivity" of a granular solid bed through which air was flowing as follows: Celite pellets were packed into a cylindrical tube. The walls of the tube were maintained at a constant temperature. Air was passed through the tube at a constant rate, and the temperature of the entering and leaving air was measured after steady-state had been achieved. In one set of runs the following data were observed:

Data	Run 1	Run 2	Run 3
Pellets	⅛-in. celite cylinders		
Void fraction of celite-packed section, F	0.388	0.388	0.388
Depth of packed section, L, in.	6.43	6.43	6.43
Diameter of tube, D, in.	1.37	1.37	1.37
Mass velocity of air based on empty tube, lb/(hr)(ft²), G	4,209.8	1,926.3	2,666.1
T_S = average surface temperature of tube, °F	286.4	290.0	120.9
T_i = inlet air temperature, °F	108.6	117.6	120.9
T_0 = mixed (bulk) temperature of leaving air, °F	218.0	234.7	230.9

If the following assumptions are taken as correct, calculate the "thermal conductivity" of the granular solid bed:

1. The thermal conductivity k of the bed is defined by the expression

$$q_r = -k(2\pi r\, dL)\frac{dT}{dr} = \text{Btu/hr}$$

† D. F. Molino and J. O. Hougen, *Chem. Eng. Progr.*, vol. 48, no. 3, p. 147, 1952.

where q_r is the heat transported *radially* by both conduction and convection.

2. k is assumed independent of position.
3. Resistance to heat transfer at and through container walls is negligible.
4. Axial mixing of the gas is negligible.
5. The velocity distribution of the gas is uniform over the cross section of the tube.

8-6. Repeat the analysis of Prob. 8-5 but use the assumption that the heat-transfer coefficient between the inside surface of the container and the packing, h, is finite, to replace assumption 3 of Prob. 8-5. Should the experimental measurements of Molino and Hougen be amplified in order to obtain h?†

8.7. Porous spheres of radius R containing a solution of uniform concentration C_0 enter an extraction tower with a velocity V_1 and fall through an upward-flowing solvent. Solvent of concentration y_0 enters at the bottom of the tower with a velocity V_2. The ratio a of the volumetric flow rate of solution in the external (solvent) phase to the volumetric flow rate of solution contained inside the spheres is known. In addition, the diffusivity D_L of the system and the pore-shape factor K^2 are known. The pore-shape factor takes into account (in an approximate manner) the fact that the passages inside the porous sphere are tortuous and constricted. Thus, the actual sphere of radius R is considered to be equivalent to an idealized sphere of radius KR. The idealized sphere is considered to contain only liquid.

If diffusion within the spheres controls the extraction process, determine the expression relating the tower height and the leaving concentration of the extractant solvent. (See Piret et al., Prob. 8.3.)

† See C. A. Coberly and W. R. Marshall, Jr., *Chem. Eng. Progr.*, vol. 47, no. 3, pp. 141–150, 1951, for a complete discussion and data.

TABLE 8-1. LAPLACE TRANSFORMS[a]

No.	Transform $\bar{f}(p) = \int_0^\infty e^{-pt} f(t)\,dt$	Function $f(t)$
1	$\frac{1}{p}$	1
2	$\frac{1}{p^2}$	t
3	$\frac{1}{p^n} \qquad n = 1, 2, 3, \ldots$	$\frac{t^{n-1}}{(n-1)!}$
4	$\frac{1}{\sqrt{p}}$	$\frac{1}{\sqrt{\pi t}}$
5	$\frac{1}{p^{3/2}}$	$2\sqrt{\frac{t}{\pi}}$
6	$\frac{1}{p^{n+1/2}} \qquad n = 1, 2, 3, \ldots$	$\frac{2^n t^{n-1/2}}{[1 \cdot 3 \cdot 5 \cdots (2n-1)]\sqrt{\pi}}$
7	$\frac{\Gamma(k)}{p^k} \qquad k > 0$	t^{k-1}
8	$\frac{1}{p-a}$	e^{at}
9	$\frac{1}{(p-a)^2}$	te^{at}
10	$\frac{1}{(p-a)^n} \qquad n = 1, 2, 3, \ldots$	$\frac{1}{(n-1)!} t^{n-1} e^{at}$
11	$\frac{\Gamma(k)}{(p-a)^k} \qquad k > 0$	$t^{k-1} e^{at}$
12	$\frac{1}{(p-a)(p-b)} \qquad a \neq b$	$\frac{1}{a-b}(e^{at} - e^{bt})$
13	$\frac{p}{(p-a)(p-b)} \qquad a \neq b$	$\frac{1}{a-b}(ae^{at} - be^{bt})$
14	$\frac{1}{(p-a)(p-b)(p-c)} \qquad a \neq b \neq c$	$-\frac{(b-c)e^{at} + (c-a)e^{bt} + (a-b)e^{ct}}{(a-b)(b-c)(c-a)}$

[a] By permission from "Modern Operational Mathematics in Engineering," by R. V. Churchill. Copyright 1944. McGraw-Hill Book Company, Inc., New York. A more comprehensive set of transforms, particularly those involving Bessel functions, forms a part of G. A. Campbell and R. M. Foster, "Fourier Integrals for Practical Applications," D. Van Nostrand Company, Inc., Princeton N.J., 1948. In their compilation only those relations for which $0 < g$ or $k < g$, where g corresponds to the t of Table 8-1, are Laplace transforms.

TABLE 8-1. LAPLACE TRANSFORMS (*Continued*)

No.	Transform $\bar{f}(p) = \int_0^\infty e^{-pt} f(t)\, dt$	Function $f(t)$
15	$\dfrac{1}{p^2 + a^2}$	$\dfrac{1}{a} \sin at$
16	$\dfrac{p}{p^2 + a^2}$	$\cos at$
17	$\dfrac{1}{p^2 - a^2}$	$\dfrac{1}{a} \sinh at$
18	$\dfrac{p}{p^2 - a^2}$	$\cosh at$
19	$\dfrac{1}{p(p^2 + a^2)}$	$\dfrac{1}{a^2} (1 - \cos at)$
20	$\dfrac{1}{p^2(p^2 + a^2)}$	$\dfrac{1}{a^3} (at - \sin at)$
21	$\dfrac{1}{(p^2 + a^2)^2}$	$\dfrac{1}{2a^3} (\sin at - at \cos at)$
22	$\dfrac{p}{(p^2 + a^2)^2}$	$\dfrac{t}{2a} \sin at$
23	$\dfrac{p^2}{(p^2 + a^2)^2}$	$\dfrac{1}{2a} (\sin at + at \cos at)$
24	$\dfrac{p^2 - a^2}{(p^2 + a^2)^2}$	$t \cos at$
25	$\dfrac{p}{(p^2 + a^2)(p^2 + b^2)} \qquad a^2 \neq b^2$	$\dfrac{\cos at - \cos bt}{b^2 - a^2}$
26	$\dfrac{1}{(p - a)^2 + b^2}$	$\dfrac{1}{b} e^{at} \sin bt$
27	$\dfrac{p - a}{(p - a)^2 + b^2}$	$e^{at} \cos bt$
28	$\dfrac{3a^2}{p^3 + a^3}$	$e^{-at} - e^{at/2} \left(\cos \dfrac{at\sqrt{3}}{2} - \sqrt{3} \sin \dfrac{at\sqrt{3}}{2} \right)$
29	$\dfrac{4a^3}{p^4 + 4a^4}$	$\sin at \cosh at - \cos at \sinh at$
30	$\dfrac{p}{p^4 + 4a^4}$	$\dfrac{1}{2a^2} \sin at \sinh at$

TABLE 8-1. LAPLACE TRANSFORMS (*Continued*)

No.	Transform $\bar{f}(p) = \int_0^\infty e^{-pt} f(t)\, dt$	Function $f(t)$
31	$\dfrac{1}{p^4 - a^4}$	$\dfrac{1}{2a^3} (\sinh at - \sin at)$
32	$\dfrac{p}{p^4 - a^4}$	$\dfrac{1}{2a^2} (\cosh at - \cos at)$
33	$\dfrac{8a^3p^2}{(p^2 + a^2)^3}$	$(1 + a^2t^2) \sin at - at \cos at$
34	$\dfrac{1}{p}\left(\dfrac{p-1}{p}\right)^n$	$\dfrac{e^t}{n!}\dfrac{d^n}{dt^n}(t^n e^{-t})$ = Laguerre polynomial of degree n
35	$\dfrac{p}{(p - a)^{3/2}}$	$\dfrac{1}{\sqrt{\pi t}} e^{at}(1 + 2at)$
36	$\sqrt{p - a} - \sqrt{p - b}$	$\dfrac{1}{2\sqrt{\pi t^3}} (e^{bt} - e^{at})$
37	$\dfrac{1}{\sqrt{p} + a}$	$\dfrac{1}{\sqrt{\pi t}} - ae^{a^2t} erfc(a\sqrt{t})$
38	$\dfrac{\sqrt{p}}{p - a^2}$	$\dfrac{1}{\sqrt{\pi t}} + ae^{a^2t} erf(a\sqrt{t})$
39	$\dfrac{\sqrt{p}}{p + a^2}$	$\dfrac{1}{\sqrt{\pi t}} - \dfrac{2ae^{-a^2t}}{\sqrt{\pi}} \int_0^{a\sqrt{t}} e^{\lambda^2}\, d\lambda$
40	$\dfrac{1}{\sqrt{p}(p - a^2)}$	$\dfrac{1}{a} e^{a^2t} erf(a\sqrt{t})$
41	$\dfrac{1}{\sqrt{p}\,(p + a^2)}$	$\dfrac{2e^{-a^2t}}{a\sqrt{\pi}} \int_0^{a\sqrt{t}} e^{\lambda^2}\, d\lambda$
42	$\dfrac{b^2 - a^2}{(p - a^2)(b + \sqrt{p})}$	$e^{a^2t}b - a[erf(a\sqrt{t})] - be^{b^2t} erfc(b\sqrt{t})$
43	$\dfrac{1}{\sqrt{p}\,(\sqrt{p} + a)}$	$e^{a^2t} erfc(a\sqrt{t})$
44	$\dfrac{1}{(p + a)\sqrt{p + b}}$	$\dfrac{1}{\sqrt{b - a}} e^{-at} erf(\sqrt{b - a}\sqrt{t})$
45	$\dfrac{b^2 - a^2}{\sqrt{p}\,(p - a^2)(\sqrt{p} + b)}$	$e^{a^2t}\left[\dfrac{b}{a} erf(a\sqrt{t}) - 1\right] + e^{b^2t} erfc(b\sqrt{t})$

TABLE 8-1. LAPLACE TRANSFORMS (*Continued*)

No.	Transform $\bar{f}(p) = \int_0^\infty e^{-pt} f(t)\,dt$	Function $f(t)$
46	$\dfrac{(1-p)^n}{p^{n+\frac{1}{2}}}$	$\dfrac{n!}{(2n)!\sqrt{\pi t}} H_{2n}\sqrt{t}$ where $H_n(t) = e^{t^2}\dfrac{d^n}{dt^n}e^{-t^2}$ is the Hermite polynomial
47	$\dfrac{(1-p)^n}{p^{n+\frac{3}{2}}}$	$-\dfrac{n!}{\sqrt{\pi}\,(2n+1)!} H_{2n+1}(\sqrt{t})$
48	$\dfrac{\sqrt{p+2a}}{\sqrt{p}} - 1$	$ae^{-at}[I_1(at) + I_0(at)]$
49	$\dfrac{1}{\sqrt{p+a}\,\sqrt{p+b}}$	$e^{-\frac{1}{2}(a+b)t} I_0\left(\dfrac{a-b}{2}t\right)$
50	$\dfrac{\Gamma(k)}{(p+a)^k(p+b)^k}$ $\quad k > 0$	$\sqrt{\pi}\left(\dfrac{t}{a-b}\right)^{k-\frac{1}{2}} e^{-\frac{1}{2}(a+b)t} I_{k-\frac{1}{2}}\left(\dfrac{a-b}{2}t\right)$
51	$\dfrac{1}{\sqrt{p+a}\,(p+b)^{\frac{3}{2}}}$	$te^{-\frac{1}{2}(a+b)t}\left[I_0\left(\dfrac{a-b}{2}t\right) + I_1\left(\dfrac{a-b}{2}t\right)\right]$
52	$\dfrac{\sqrt{p+2a} - \sqrt{p}}{\sqrt{p+2a} + \sqrt{p}}$	$\dfrac{1}{t} e^{-at} I_1(at)$
53	$\dfrac{(a-b)^k}{(\sqrt{p+a} + \sqrt{p+b})^{2k}}$ $\quad k > 0$	$\dfrac{k}{t} e^{-\frac{1}{2}(a+b)t} I_k\left(\dfrac{a-b}{2}t\right)$
54	$\dfrac{(\sqrt{p+a} + \sqrt{p})^{-2j}}{\sqrt{p}\,\sqrt{p+a}}$ $\quad j > -1$	$\dfrac{1}{a^j} e^{-\frac{1}{2}at} I_j\left(\dfrac{1}{2}at\right)$
55	$\dfrac{1}{\sqrt{p^2+a^2}}$	$J_0(at)$
56	$\dfrac{(\sqrt{p^2+a^2} - p)^j}{\sqrt{p^2+a^2}}$ $\quad j > 1$	$a^j J_j(at)$
57	$\dfrac{1}{(p^2+a^2)^k}$ $\quad k > 0$	$\dfrac{\sqrt{\pi}}{\Gamma(k)}\left(\dfrac{t}{2a}\right)^{k-\frac{1}{2}} J_{k-\frac{1}{2}}(at)$
58	$(\sqrt{p^2+a^2} - p)^k$ $\quad k > 0$	$\dfrac{ka^k}{t} J_k(at)$
59	$\dfrac{(p - \sqrt{p^2-a^2})^j}{\sqrt{p^2-a^2}}$ $\quad j > -1$	$a^j I_j(at)$

TABLE 8-1. LAPLACE TRANSFORMS (*Continued*)

No.	Transform $\bar{f}(p) = \int_0^\infty e^{-pt}f(t)\,dt$	Function $f(t)$
60	$\dfrac{1}{(p^2 - a^2)^k}$ $\quad k > 0$	$\dfrac{\sqrt{\pi}}{\Gamma(k)}\left(\dfrac{t}{2a}\right)^{k-\frac{1}{2}} I_{k-\frac{1}{2}}(at)$
61	$\dfrac{e^{-kp}}{p}$	$S_k(t) = \begin{cases} 0 & \text{when } 0 < t < k \\ 1 & \text{when } t > k \end{cases}$
62	$\dfrac{e^{-kp}}{p^2}$	0 when $0 < t < k$ $t - k$ when $t > k$
63	$\dfrac{e^{-kp}}{p^j}$ $\quad j > 0$	0 when $0 < t < k$ $\dfrac{(t-k)^{j-1}}{\Gamma(j)}$ when $t > k$
64	$\dfrac{1 - e^{-kp}}{p}$	1 when $0 < t < k$ 0 when $t > k$
65	$\dfrac{1}{p(1 - e^{-kp})} = \dfrac{1 + \coth \frac{1}{2}kp}{2p}$	$S(k,t) = n$ when $(n-1)k < t < nk$ $n = 1, 2, 3, \ldots$
66	$\dfrac{1}{p(e^{kp} - a)}$	0 when $0 < t < k$ $1 + a + a^2 + \cdots + a^{n-1}$ when $nk < t < (n+1)k$ $n = 1, 2, 3, \ldots$
67	$\dfrac{1}{p}\tanh kp$	$M(2k,t) = (-1)^{n-1}$ when $2k(n-1) < t < 2kn$ $n = 1, 2, 3, \ldots$
68	$\dfrac{1}{p(1 + e^{-kp})}$	$\dfrac{1}{2}M(k,t) + \dfrac{1}{2} = \dfrac{1 - (-1)^n}{2}$ when $(n-1)k < t < nk$
69	$\dfrac{1}{p^2}\tanh kp$	$H(2k,t) = \begin{cases} t & \text{when } 0 < t < 2k \\ 4k - t & \text{when } 2k < t < 4k \end{cases}$
70	$\dfrac{1}{p \sinh kp}$	$2S(2k, t + k) - 2 = 2(n-1)$ when $(2n-3)k < t < (2n-1)k$ $t > 0$
71	$\dfrac{1}{p \cosh kp}$	$M(2k, t + 3k) + 1 = 1 + (-1)^n$ when $(2n-3)k < t < (2n-1)k$ $t > 0$
72	$\dfrac{1}{p}\coth kp$	$2S(2k,t) - 1 = 2n - 1$ when $2k(n-1) < t < 2kn$
73	$\dfrac{k}{p^2 + k^2}\coth \dfrac{\pi p}{2k}$	$\lvert \sin kt \rvert$

TABLE 8-1. LAPLACE TRANSFORMS (*Continued*)

No.	Transform $\bar{f}(p) = \int_0^\infty e^{-pt}f(t)\,dt$	Function $f(t)$
74	$\frac{1}{(p^2+1)(1-e^{-\pi p})}$	$\sin t$ when $(2n = 2)\pi < t < (2n-1)\pi$ 0 when $(2n-1)\pi < t < 2n\pi$
75	$\frac{1}{p}e^{-k/p}$	$J_0(2\sqrt{kt})$
76	$\frac{1}{\sqrt{p}}e^{-k/p}$	$\frac{1}{\sqrt{\pi t}}\cos 2\sqrt{kt}$
77	$\frac{1}{\sqrt{p}}e^{k/p}$	$\frac{1}{\sqrt{\pi t}}\cosh 2\sqrt{kt}$
78	$\frac{1}{p^{3/2}}e^{-k/p}$	$\frac{1}{\sqrt{\pi k}}\sin 2\sqrt{kt}$
79	$\frac{1}{p^{3/2}}e^{k/p}$	$\frac{1}{\sqrt{\pi k}}\sinh 2\sqrt{kt}$
80	$\frac{1}{p^j}e^{-k/p} \quad j > 0$	$\left(\frac{t}{k}\right)^{(j-1)/2}J_{j-1}(2\sqrt{kt})$
81	$\frac{1}{p^j}e^{k/p} \quad j > 0$	$\left(\frac{t}{k}\right)^{(j-1)/2}I_{j-1}(2\sqrt{kt})$
82	$e^{-k\sqrt{p}} \quad k > 0$	$\frac{k}{2\sqrt{\pi t^3}}\exp\left(-\frac{k^2}{4t}\right)$
83	$\frac{1}{p}e^{-k\sqrt{p}} \quad k \geqq 0$	$erfc\left(\frac{k}{2\sqrt{t}}\right)$
84	$\frac{1}{\sqrt{p}}e^{-k\sqrt{p}} \quad k \geqq 0$	$\frac{1}{\sqrt{\pi t}}\exp\left(-\frac{k^2}{4t}\right)$
85	$p^{-3/2}e^{-k\sqrt{p}} \quad k \geqq 0$	$2\sqrt{\frac{t}{\pi}}\left[\exp\left(-\frac{k^2}{4t}\right)\right] - k\,erfc\left(\frac{k}{2\sqrt{t}}\right)$
86	$\frac{ae^{-k\sqrt{p}}}{p(a+\sqrt{p})} \quad k \geqq 0$	$(-\exp ak)(\exp a^2t)erfc\left(a\sqrt{t}+\frac{k}{2\sqrt{t}}\right) + erfc\left(\frac{k}{2\sqrt{t}}\right)$
87	$\frac{e^{-k\sqrt{p}}}{\sqrt{p}\,(a+\sqrt{p})}$	$(\exp ak)(\exp a^2t)erfc\left(a\sqrt{t}+\frac{k}{2\sqrt{t}}\right)$
88	$\frac{e^{-k\sqrt{p(p+a)}}}{\sqrt{p(p+a)}}$	0 when $0 < t < k$ $[\exp(-\frac{1}{2}at)]I_0(\frac{1}{2}a\sqrt{t^2-k^2})$ when $t > k$

TABLE 8-1. LAPLACE TRANSFORMS (*Continued*)

No.	Transform $\bar{f}(p) = \int_0^\infty e^{-pt}f(t)\,dt$	Function $f(t)$
89	$\dfrac{e^{-k\sqrt{p^2+a^2}}}{\sqrt{p^2+a^2}}$	0 when $0 < t < k$ $J_0(a\sqrt{t^2-k^2})$ when $t > k$
90	$\dfrac{e^{-k\sqrt{p^2-a^2}}}{\sqrt{p^2-a^2}}$	0 when $0 < t < k$ $I_0(a\sqrt{t^2-k^2})$ when $t > k$
91	$\dfrac{e^{-k(\sqrt{p^2+a^2}-p)}}{\sqrt{p^2+a^2}}$ $k \geqq 0$	$J_0(a\sqrt{t^2+2kt})$
92	$e^{-kp} - e^{-k\sqrt{p^2+a^2}}$	0 when $0 < t < k$ $\dfrac{ak}{\sqrt{t^2-k^2}} J_1(a\sqrt{t^2-k^2})$ when $t > k$
93	$e^{-k\sqrt{p^2-a^2}} - e^{-kp}$	0 when $0 < t < k$ $\dfrac{ak}{\sqrt{t^2-k^2}} I_1(a\sqrt{t^2-k^2})$ when $t > k$
94	$\dfrac{a^j e^{-k\sqrt{p^2+a^2}}}{\sqrt{p^2+a^2}(\sqrt{p^2+a^2}+p)^j}$ $j > -1$	0 when $0 < t < k$ $\left(\dfrac{t-k}{t+k}\right)^{\frac{1}{2}j} J_j(a\sqrt{t^2-k^2})$ when $t > k$
95	$\dfrac{1}{p}\ln p$	$\lambda - \ln t$ $\lambda = -0.5772\cdots$
96	$\dfrac{1}{p^k}\ln p$ $k > 0$	$t^{k-1}\left\{\dfrac{\lambda}{[\Gamma(k)]^2} - \dfrac{\ln t}{\Gamma(k)}\right\}$
97[b]	$\dfrac{\ln p}{p-a}$ $a > 0$	$(\exp at)[\ln a - E_i(-at)]$
98[c]	$\dfrac{\ln p}{p^2+1}$	$\cos t\, Si(t) - \sin t\, Ci(t)$
99	$\dfrac{p\ln p}{p^2+1}$	$-\sin t\, Si(t) - \cos t\, Ci(t)$

[b] $Ei(-t) = -\int_t^\infty \frac{e^{-x}}{x}\,dx$ (for $t > 0$) = exponential integral function and is tabulated in E. Jahnke and F. Emde, "Tables of Functions," Dover Publications, New York, 1943.

[c] $Si(t) = \int_0^t \frac{\sin x}{x}\,dx$ = sine integral function

$Ci(t) = -\int_t^\infty \frac{\cos x}{x}\,dx$ = cosine integral function

TABLE 8-1. LAPLACE TRANSFORMS (*Continued*)

No.	Transform $\bar{f}(p) = \int_0^\infty e^{-pt} f(t)\, dt$	Function $f(t)$
100	$\frac{1}{p} \ln (1 + kp) \qquad k > 0$	$-Ei\left(-\frac{t}{k}\right)$
101	$\ln \frac{p - a}{p - b}$	$\frac{1}{t} (e^{bt} - e^{at})$
102	$\frac{1}{p} \ln (1 + k^2 p^2)$	$-2Ci\left(\frac{t}{k}\right)$
103	$\frac{1}{p} \ln (p^2 + a^2) \qquad a > 0$	$2 \ln a - 2Ci(at)$
104	$\frac{1}{p^2} \ln (p^2 + a^2) \qquad a > 0$	$\frac{2}{a} [at \ln a + \sin at - atCi(at)]$
105	$\ln \frac{p^2 + a^2}{p^2}$	$\frac{2}{t} (1 - \cos at)$
106	$\ln \frac{p^2 - a^2}{p^2}$	$\frac{2}{t} (1 - \cosh at)$
107	$\tan^{-1} \frac{k}{p}$	$\frac{1}{t} \sin kt$
108	$\frac{1}{p} \tan^{-1} \frac{k}{p}$	$Si(kt)$
109	$(\exp k^2 p^2) erfc(kp) \qquad k > 0$	$\frac{1}{k \sqrt{\pi}} \exp\left(-\frac{t^2}{4k^2}\right)$
110	$\frac{1}{p} (\exp k^2 p^2) erfc(kp) \qquad k > 0$	$erf\left(\frac{t}{2k}\right)$
111	$(\exp kp) erfc(\sqrt{kp}) \qquad k > 0$	$\frac{\sqrt{k}}{\pi \sqrt{t}(t + k)}$
112	$\frac{1}{\sqrt{p}} erfc(\sqrt{kp})$	0 when $0 < t < k$ $(\pi t)^{-1/2}$ when $t > k$
113	$\frac{1}{\sqrt{p}} (\exp kp) erfc(\sqrt{kp}) \qquad k > 0$	$\frac{1}{\sqrt{\pi(t + k)}}$
114	$erf\left(\frac{k}{\sqrt{p}}\right)$	$\frac{1}{\pi t} \sin (1k \sqrt{t})$
115	$\frac{1}{\sqrt{p}} \left(\exp \frac{k^2}{p}\right) erfc\left(\frac{k}{\sqrt{p}}\right)$	$\frac{1}{\sqrt{\pi t}} \exp (-2k \sqrt{t})$

TABLE 8-1. LAPLACE TRANSFORMS (*Continued*)

No.	Transform $\bar{f}(p) = \int_0^\infty e^{-pt} f(t)\,dt$	Function $f(t)$
116[d]	$K_0(kp)$	0 when $0 < t < k$ $(t^2 - k^2)^{-1/2}$ when $t > k$
117	$K_0(k\sqrt{p})$	$\frac{1}{2t} \exp\left(-\frac{k^2}{4t}\right)$
118	$\frac{1}{p}(\exp kp) K_1(kp)$	$\frac{1}{k}\sqrt{t(t+2k)}$
119	$\frac{1}{\sqrt{p}} K_1(k\sqrt{p})$	$\frac{1}{k} \exp\left(-\frac{k^2}{4t}\right)$
120	$\frac{1}{\sqrt{p}}\left(\exp\frac{k}{p}\right) K_0\left(\frac{k}{p}\right)$	$\frac{2}{\sqrt{\pi t}} K_0(1\sqrt{2kt})$
121[e]	$\pi[\exp(-kp)] I_0(kp)$	$[t(2k-t)]^{-1/2}$ when $0 < t < 2k$ 0 when $t > 2k$
122	$[\exp(-kp)] I_1(kp)$	$\frac{k-t}{\pi k \sqrt{t(2k-t)}}$ when $0 < t < 2k$ 0 when $t > 2k$

[d] $K_n(x)$ denotes the Bessel function of the second kind for the imaginary argument.
[e] $I_n(x)$ denotes the Bessel function of the first kind for the imaginary argument.

CHAPTER 9

ANALYSIS OF STAGEWISE PROCESSES BY THE CALCULUS OF FINITE DIFFERENCES

9-1. Introduction. Chemical engineers make extensive use of stagewise processes. Typical cases include plate absorption and distillation columns and a series of batch reaction, extraction, or leaching tanks. In each case the operation is characterized by a finite between-stage change in the value of the dependent variable. The finite transition between stages in a stagewise process cannot be treated by the methods of the differential calculus; a new approach is required. The analysis of such situations can be handled by the calculus of finite differences. Tiller and Tour[5] first published examples of the use of the calculus of finite differences in the analysis of chemical engineering stagewise processes. Marshall and Pigford[8] extended the treatment to unsteady-state situations.

Example 9-1. Countercurrent Liquid-Liquid Extraction. As an example of the type of problem under discussion, consider the following: A "batch-continuous" countercurrent liquid-liquid extraction system consists of M separate stages. Each stage consists of a separate tank. Under steady-state operation, L moles of extract solvent and R moles of raffinate solvent are charged to tank n, mixed, allowed to settle, and the raffinate sent to tank $n - 1$ and the extract to tank $n + 1$. This process takes place in each tank. A sketch of the system is shown in Fig. 9-1. The initial

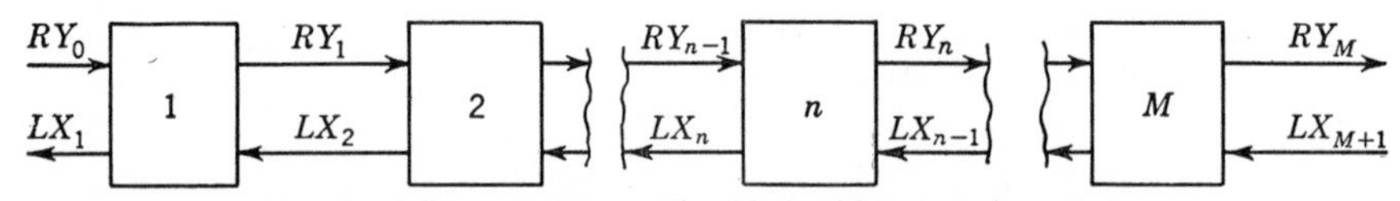

FIG. 9-1. Countercurrent liquid-liquid extraction stages.

raffinate charge to tank 1 on each cycle consists of R moles of solvent with a composition Y_0 (moles of transferable material A per mole of solvent). The initial extract charge to tank M on each cycle consists of L moles of extract solvent with a composition X_{M+1} (moles of transferable material A per mole of solvent). Raffinate and extract are immiscible. Equilibrium is attained in each tank. The equilibrium relation is

$$X_n = kY_n \tag{9-1}$$

It is desired to determine the expressions relating the composition in a given stage to the feed composition for steady-state operation.

Apply a material balance to tank n:

$$\begin{aligned}\text{Input of } A &= RY_{n-1} + LX_{n+1}\\ \text{Output of } A &= RY_n + LX_n\\ \text{Accumulation} &= 0\end{aligned}$$

Consequently,

$$RY_{n-1} + LX_{n+1} = RY_n + LX_n \tag{9-2}$$

When (9-1) is combined with (9-2) there results

$$Y_{n+1} - (\alpha + 1)Y_n + \alpha Y_{n-1} = 0 \tag{9-3}$$

where

$$\alpha = \frac{R}{Lk} \tag{9-4}$$

Equation (9-3) is a *difference equation*. Its solution will give a relation between Y_n, the stage number n, and the inlet compositions to the tank cascade. The *calculus of finite differences* deals with techniques for the solution of difference equations.

The difference equations to be considered here arise from situations in which the dependent variable $y_z = f(z)$ is defined only at *discrete* values of the independent variable z, $z_0, z_1, \ldots, z_n$, which are separated by the *constant* amount $\Delta z = z_{n+1} - z_n$. Replace z by a new independent variable n, defined so that *n will exhibit integer values only and the equidistant increment Δn will be unity:*

$$n = \frac{z_n - z_0}{\Delta z} \qquad n = 0, 1, 2, \ldots \tag{9-5}$$

The dependent variable expression may be written as

$$y_n = \phi(n) = f(z_n) = f(n\,\Delta z + z_0) \tag{9-6}$$

The *differences* of the dependent variable y_n are defined as

First difference: $\Delta^1 y_n = y_{n+1} - y_n$ (9-7)

Second difference:

$$\begin{aligned}\Delta^2 y_n &= \Delta^1 y_{n+1} - \Delta^1 y_n\\ &= (y_{n+2} - y_{n+1}) - (y_{n+1} - y_n)\\ &= y_{n+2} - 2y_{n+1} + y_n\end{aligned} \tag{9-8}$$

kth difference:

$$\Delta^k y_n = \sum_{r=0}^{r=k} \frac{(-1)^r k!}{r!(k-r)!}\, y_{n+k-r} \tag{9-9}$$

In view of Eq. (9-9), a difference equation may be written in either of the alternate forms:

$$F(y_{n+k}, y_{n+k-1}, \ldots, y_{n+1}, y_n, n) = 0 \tag{9-10a}$$

$$\phi(\Delta^k y_n, \Delta^{k-1} y_n, \ldots, \Delta^1 y_n, y_n, n) = 0 \tag{9-10b}$$

The *order* of a difference equation is the difference between the highest and lowest subscripts associated with the dependent variable.

9-2. Solution of Linear Difference Equations. The general linear difference equation of kth order may be written in the form

$$y_{n+k} + N_1 y_{n+k-1} + \cdots + N_{k-1} y_{n+1} + N_k y_n = Q \tag{9-11}$$

$N_1, N_2, \ldots, N_k$ and Q are functions of n or constants. Both the form and solution of (9-11) closely resemble the form and solution of the general linear differential equation (see Sec. 4-20). The complete solution of (9-11) consists of the sum of the solution to the *homogeneous equation, obtained from* (9-11) *by letting Q equal zero, and the particular solution* of the complete equation (9-11):

$$y_n = [c_1 y_1(n) + c_2 y_2(n) + \cdots + c_k y_k(n)] + y_p(n) \tag{9-12}$$

The solution to the homogeneous equation, called the "complementary function," is represented by the terms appearing within the brackets of (9-12), and consists of k linearly independent functions of n, $y_1(n)$, $y_2(n), \ldots, y_k(n)$. Each of these independent functions is multiplied by an arbitrary constant, $c_1, c_2, \ldots, c_k$. All the k arbitrary constants appearing in the complete solution of a kth-order linear difference equation are contained in the complementary function.

The particular solution, denoted in (9-12) by $y_p(n)$, is a specific solution which satisfies (9-11) with Q retained. The particular solution contains no arbitrary constants.

Linear Difference Equation with Constant Coefficients. The general linear difference equation with constant coefficients may be written

$$y_{n+k} + A_1 y_{n+k-1} + \cdots + A_{k-1} y_{n+1} + A_k y_n = Q \tag{9-13}$$

$A_1, A_2, \ldots, A_k$ are constants, and Q is either a constant or a function of n only.

The solution to the *homogeneous equation* corresponding to (9-13),

$$y_{n+k} + A_1 y_{n+k-1} + \cdots + A_{k-1} y_{n+1} + A_k y_n = 0 \tag{9-14}$$

is obtained by assuming each of the linearly independent solutions to be of the form

$$y_n = c\beta^n \tag{9-15}$$

where c is an arbitrary constant and β is a constant whose specific numerical value is to be determined without recourse to the boundary conditions. Substitution of (9-15) in (9-14) gives

$$c\beta^{n+k} + cA_1\beta^{n+k-1} + \cdots + cA_{k-1}\beta^{n+1} + cA_k\beta^n = 0 \tag{9-16}$$

Since $c = 0$ and $\beta = 0$ are trivial cases, (9-16) may be written

$$\beta^k + A_1\beta^{k-1} + \cdots + A_{k-1}\beta + A_k = 0 \tag{9-17}$$

Equation (9-17) is an algebraic equation of the kth degree in β. In general, it will have k roots, $\beta_1, \beta_2, \ldots, \beta_k$. Divide these roots into the following classes:

1. Real and distinct. The partial solution corresponding to the real and distinct (unequal) roots is

$$y_a = c_{a0}\beta_1^n + c_{a1}\beta_2^n + \cdots \tag{9-18a}$$

2. Real and repeated. The partial solution corresponding to a real root repeated $m + 1$ times is

$$y_b = \beta^n(c_{b0} + c_{b1}n + c_{b2}n^2 + \cdots + c_{bm}n^m) \tag{9-18b}$$

3. Complex conjugates. The partial solution corresponding to a root-pair consisting of a complex conjugate, say $r(\cos\theta \pm i \sin\theta)$, is

$$y_c = r^n(c_{c0} \cos\theta n + c_{c1} \sin\theta n) \tag{9-18c}$$

4. Repeated complex conjugates. The partial solution corresponding to a complex-conjugate root *pair* repeated $m + 1$ times is

$$y_d = r^n[(c_{d0} + c_{d1}n + \cdots + c_{dm}n^m) \cos\theta n + (c'_{do} + c'_{d1}n + \cdots + c'_{dm}n^m) \sin\theta n] \tag{9-18d}$$

The *particular solution* of the linear difference equation usually can be obtained by the method of undetermined coefficients which is illustrated below. The more powerful method of variation of parameters is developed in Fort,[2] pages 123 to 125.

Example 9-2. Method of Undetermined Coefficients. Consider the equation

$$y_{n+1} - 2y_n - 8y_{n-1} = 4n^2 + e^n \tag{9-19}$$

The solution to the homogeneous equation is

$$y = c_0(-2)^n + c_1(4)^n \tag{9-20a}$$

It is assumed that the particular solution is of the form

$$y_p = a_0 + a_1n + a_2n^2 + b_0e^n \tag{9-20b}$$

Substitution of (9-20b) in (9-19) gives

$$[a_0 + a_1(n + 1) + a_2(n + 1)^2 + b_0e^{n+1}] - 2(a_0 + a_1n + a_2n^2 + b_0e^n) - 8[a_0 + a_1(n - 1) + a_2(n - 1)^2 + b_0e^{n-1}] = 4n^2 + e^n \tag{9-21}$$

If (9-21) is to be satisfied, the coefficients of like powers of n must be equal. This requires

$$\begin{aligned} -9a_0 + 9a_1 - 7a_2 &= 0 \\ -9a_1 + 18a_2 &= 0 \\ -9a_2 &= 4 \\ b_0\left(e - 2 - \frac{8}{e}\right) &= 1 \end{aligned} \tag{9-22}$$

Consequently,

$$a_0 = -\frac{44}{81} \qquad a_1 = -\frac{8}{9} \qquad a_2 = -\frac{4}{9} \qquad b_0 = \frac{e}{e^2 - 2e - 8}$$

The complete solution of (9-19) is then

$$y = c_0(-2)^n + c_1(+4)^n - \frac{44}{81} - \frac{8}{9}n - \frac{4}{9}n^2 + \frac{e}{e^2 - 2e - 8}e^n \tag{9-20c}$$

The Riccati Difference Equation

$$y_{n+1}y_n + ay_{n+1} + by_n + c = 0 \tag{9-23}$$

is a nonlinear difference equation which can be reduced to a linear equation by transformation of variables. The reduction is accomplished by translation of the y axis by a constant amount δ. Thus, let

$$y_n = u_n + \delta \tag{9-24}$$

Then (9-23) becomes

$$\begin{aligned}(u_{n+1} + \delta)(u_n + \delta) + a(u_{n+1} + \delta) + b(u_n + \delta) + c &= u_{n+1}u_n \\ + (a + \delta)u_{n+1} + (b + \delta)u_n + [\delta^2 + (a + b)\delta + c] &= 0\end{aligned} \tag{9-25}$$

Now let the numerical value of δ be such that

$$\delta^2 + (a + b)\delta + c = 0 \tag{9-26}$$

Equation (9-26) determines the numerical value of δ and reduces Eq. (9-25) to the form

$$u_{n+1}u_n + (a + \delta)u_{n+1} + (b + \delta)u_n = 0 \tag{9-27}$$

Now divide (9-27) by $u_{n+1}u_n$, and let

$$V_n = \frac{1}{u_n} = \frac{1}{y_n - \delta} \tag{9-28}$$

Then there results

$$(b + \delta)V_{n+1} + (a + \delta)V_n + 1 = 0 \tag{9-29}$$

Equation (9-29) is a linear difference equation with constant coefficients and is solvable by the methods described earlier.

More Complete Treatment. The treatment of more complicated difference equations is discussed in references 1, 2, and 3.

9-3. Illustrative Examples

Completion of Example 9-1. The analysis of the countercurrent liquid-liquid extraction system of Example 9-1 led to the linear difference equation

$$Y_{n+1} - (\alpha + 1)Y_n + \alpha Y_{n-1} = 0 \tag{9-3}$$

A trial solution of the form

$$Y = c\beta^n$$

leads to

$$\beta^2 - (\alpha + 1)\beta + \alpha = 0$$

and $\beta_0 = 1$, $\beta_1 = \alpha$. The general solution of (9-3) is then

$$Y_n = c_0 + c_1\alpha^n \tag{9-30}$$

The constants of integration may be evaluated from the boundary conditions

$$\begin{aligned} Y &= Y_0 && \text{at } n = 0 \\ Y &= \frac{X_{M+1}}{k} = Y_{M+1} && \text{at } n = M + 1 \end{aligned} \tag{9-31}$$

The final result is

$$\frac{Y_n - Y_0}{Y_{M+1} - Y_0} = \frac{\alpha^n - 1}{\alpha^{M+1} - 1} \tag{9-32}$$

where

$$\alpha \doteq \frac{R}{Lk} \tag{9-4}$$

The boundary conditions (9-31) are obtained by considering that the raffinate feed (RY_0) to tank 1 comes from a tank below tank 1 and that the extract feed (LX_{M+1}) to tank M comes from a tank above tank M (see Prob. 9-1).

Example 9-3. Stirred-tank Reactor System. A continuous stirred-tank reactor system consists of M tanks in series, as shown in Fig. 9-2. L ft³/hr of solution is fed

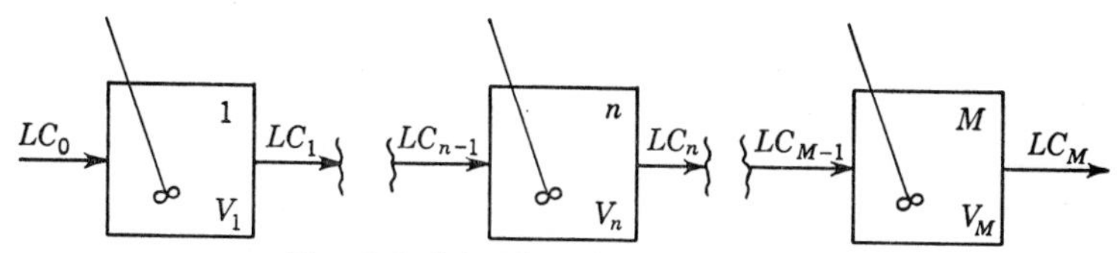

FIG. 9-2. Stirred-tank reactor stages.

to each tank. The feed to the first tank contains a concentration of C_0 moles/ft³ of component A. Each tank is well stirred. Inside the tanks, the irreversible first-order reaction $A \to B$ takes place. The rate at which A reacts is given by the expression

$$-\frac{dW_n}{dt} = (KVC)_n \qquad \text{moles/hr} \tag{9-33}$$

If the reaction-rate constant K is the same for each tank, show that the highest concentration of B which may be produced by the tank system occurs when $V_1 = V_2 = \cdots = V_n$, provided that the total volume $\nu = \sum_{n=1}^{n=m} V_n$ remains constant.

Solution. A material balance written for component A around the $(n + 1)$st tank gives

$$C_{n+1} = \frac{C_n}{K\theta_{n+1} + 1} \tag{9-34}$$

where

$$\theta_n = \frac{V_n}{L} \tag{9-35}$$

Use of (9-34) gives

$$C_1 = \frac{C_0}{K\theta_1 + 1} \tag{9-36}$$

$$C_2 = \frac{C_1}{K\theta_2 + 1} = \frac{C_0}{(K\theta_1 + 1)(K\theta_2 + 1)} \tag{9-37}$$

$$C_M = \frac{C_0}{(K\theta_1 + 1)(K\theta_2 + 1)(K\theta_3 + 1) \cdots (K\theta_M + 1)} \tag{9-38}$$

The highest concentration of B in the fluid leaving the system will occur when C_M, the concentration of A in the fluid leaving the system, is a minimum. The minimum results when

$$dC_M = \sum_{n=1}^{M} \frac{\partial C_M}{\partial \theta_n} d\theta_n = 0 \tag{9-39}$$

$$\frac{\partial C_M}{\partial \theta_n} = \frac{-KC_0}{P(K\theta_n + 1)} \tag{9-40}$$

$$P = (K\theta_1 + 1)(K\theta_2 + 1)(K\theta_3 + 1) \cdots (K\theta_M + 1) \tag{9-41}$$

Combination of (9-39) and (9-40) gives

$$dC_M = \frac{-KC_0}{P} \sum_{n=1}^{M} \frac{d\theta_n}{K\theta_n + 1} = 0 \tag{9-42}$$

Now

$$\sum_{n=1}^{M} \theta_n = \frac{\nu}{L} = \text{const} \tag{9-43}$$

and

$$\sum_{n=1}^{M} d\theta_n = 0 \tag{9-44}$$

From (9-44) it follows that

$$d\theta_M = -\sum_{n=1}^{M-1} d\theta_n \tag{9-45}$$

Combination of (9-42) and (9-45) gives

$$dC_M = \frac{-KC_0}{P} \sum_{n=1}^{M-1} \left[\left(\frac{1}{K\theta_n + 1} - \frac{1}{K\theta_M + 1}\right) d\theta_n\right] = 0 \tag{9-46}$$

For (9-46) to hold identically, the coefficient of each term in the summation must be zero. This condition requires

$$\frac{1}{K\theta_n + 1} - \frac{1}{K\theta_M + 1} = 0 \tag{9-47}$$

or

$$\theta_n = \theta_M \tag{9-48}$$

Since θ_n applies to any tank, the volume of liquid in any tank must be equal to the volume of liquid in every other tank, and the highest concentration of B is obtained when

$$V_1 = V_2 = V_3 = \cdots = V_n \tag{9-49}$$

Example 9-4. Distillation in a Plate Column. A continuous-flow distillation column is fed with a binary mixture of A and B. The relative volatility α of the mixture is constant. The "usual" simplifying assumptions are valid. Consider the plates above the feed plate. Derive the difference equation which relates the liquid composition to the plate number n, if the over-all plate efficiency is 100 per cent. Reduce the equation to a linear one with constant coefficients. Show that the axis translation required involves the points where the "equilibrium" and "operating" lines intersect. Integrate the difference equation and indicate the procedure to be used to evaluate the arbitrary constants in the integrated result.

Solution. Let

x_n = mole fraction of more volatile component in liquid phase on plate n

x_D = mole fraction of more volatile component in product withdrawn from condenser

y_n^* = mole fraction of more volatile component in vapor in equilibrium with liquid of composition x_n

y_n = mole fraction of more volatile component in vapor phase leaving plate n; $y_n = y_n^*$ if the plate efficiency is 100 per cent, as is assumed here

O = mole rate of liquid flow down column above feed plate; assumed constant

V = mole rate of vapor flow up column; assumed constant

D = mole rate of product withdrawal from total condenser

Write a material balance on the more volatile component. The system chosen is shown dotted in Fig. 9-3 and includes the nth plate and the condenser. The result is

$$Vy_{n-1} - Ox_n - Dx_D = 0 \qquad (9\text{-}50)$$

Equation (9-50) is the usual "operating line" on a McCabe-Thiele diagram.

The definition of the relative volatility is

$$\alpha = \frac{y_n^*(1 - x_n)}{(1 - y_n^*)x_n} \qquad (9\text{-}51)$$

which is the equation of the "equilibrium line." Combination of (9-50) and (9-51) gives

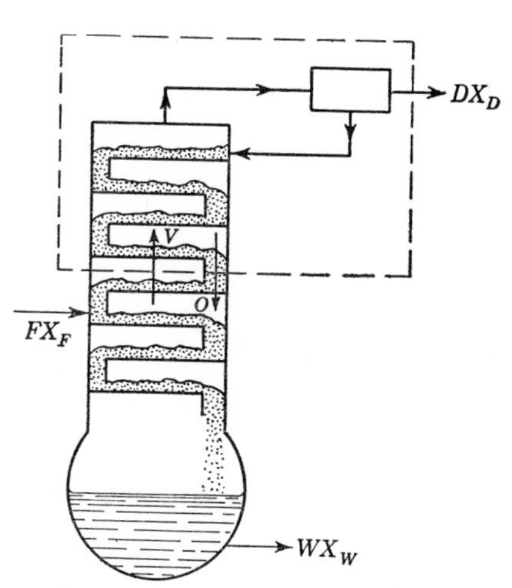

FIG. 9-3. Plate distillation column.

$$x_nx_{n-1} + \frac{x_n}{\alpha - 1} + \left(\frac{Dx_D(\alpha - 1) - \alpha V}{O(\alpha - 1)}\right)x_{n-1} + \frac{Dx_D}{O(\alpha - 1)} = 0 \qquad (9\text{-}52)$$

Let

$$a = \frac{1}{\alpha - 1}$$
$$b = \frac{Dx_D(\alpha - 1) - \alpha V}{O(\alpha - 1)} \qquad (9\text{-}53)$$
$$c = \frac{Dx_D}{O(\alpha - 1)}$$

Then (9-52 becomes

$$x_nx_{n-1} + ax_n + bx_{n-1} + c = 0 \qquad (9\text{-}54)$$

Equation (9-54) is a Riccati equation. With the transformation

$$V_n = \frac{1}{x_n - \delta} \qquad (9\text{-}55)$$

there results

$$(b + \delta)V_n + (a + \delta)V_{n-1} + 1 = 0 \tag{9-56}$$

provided that δ satisfies

$$\delta^2 + (a + b)\delta + c = 0 \tag{9-57a}$$

or

$$\delta = \frac{-(a + b) \pm \sqrt{(a + b)^2 - 4c}}{2} \tag{9-57b}$$

The solution of the linear difference equation (9-56) is

$$V_n = K\left(-\frac{a + \delta}{b + \delta}\right)^n - \frac{1}{(a + \delta) + (b + \delta)} \tag{9-58}$$

where K is an arbitrary constant. Combination of (9-58) and (9-55) gives

$$x_n = \delta + \frac{1}{K\left(-\dfrac{a + \delta}{b + \delta}\right)^n - \dfrac{1}{(a + \delta) + (a + \delta)}} \tag{9-59}$$

where a, b, and c are given by (9-53) and δ by (9-57).

Let the intersection of the equilibrium and operating lines occur at the point x_i, y_i. This point must satisfy both (9-50) and (9-51):

$$Vy_i - Ox_i - Dx_D = 0 \tag{9-60}$$

$$\alpha = \frac{y_i(1 - x_i)}{(1 - y_i)x_i} \tag{9-61}$$

When y_i is eliminated between (9-60) and (9-61), there results

$$x_i^2 + (a + b)x_i + c = 0 \tag{9-62}$$

where a, b, and c are defined by (9-53). Equation (9-62) is identical with Eq. (9-57) which determines δ. Hence

$$\delta = x_i \tag{9-63}$$

and the translation of axes required to linearize the Riccati difference equation obtained in this distillation problem corresponds to taking the new origin of coordinates at the intersection of the equilibrium and operating lines.

The value of the arbitrary constant K appearing in the result (9-59) depends upon conditions below the feed plate. This constant could be evaluated if the feed-plate composition were known.

9-4. Unsteady-state Operation. Chemical engineers frequently operate staged systems in an unsteady-state condition. During the start-up period of such equipment the process conditions change continuously. Similarly, transient conditions occur when the process conditions are adjusted during operation. Information concerning the behavior of the system during the unsteady-state operating periods is of great value to the process engineer. Such factors as the time required to attain substantially steady-state operation and the quality of the product during the transient period often are of decisive importance in the evaluation of alternative processes or of alternative operating procedures.

During the operation of industrial equipment, fluctuations in feedstream composition, energy-supply rates, etc., occur constantly. As a result of these fluctuations, the system continually oscillates about the "steady-state" condition. The design and specification of the instruments and equipment needed to limit and control such oscillations are facilitated if the details of the transient behavior of the system are understood.

The batch operation of such equipment as plate distillation columns involves constantly changing conditions. The design of staged, batch-operated systems demands an unsteady-state analysis.

Analytical solutions which describe the transient behavior of batch systems can be obtained by the application of methods already discussed, provided that the systems are relatively simple. Typical cases are illustrated in the following sections. The formal analysis of more complicated situations requires advanced techniques that cannot be discussed here. However, numerical results for all types of situations ordinarily may be obtained by the procedures described in Chap. 10.

9-5. Illustrative Examples

Example 9-5. Starting a Stirred-tank Reactor. It is proposed to start the stirred-tank reactor system of Example 9-3 by filling each tank with pure solvent and then initiating the flow of L feet^3/hr of solution containing C_0 moles/ft^3 of A (and no B). If the volume of each tank is the same, derive the relation between the exit concentration of B and the time t.

Solution. Denote the concentration of B in tank n at time t by B_n. A material balance written for component B around the $(n + 1)$st tank gives

$$B_{n+1} - B_n - K\theta C_{n+1} = -\theta \frac{\partial B_{n+1}}{\partial t} \tag{9-64}$$

where

$$\theta = \frac{V}{L} \tag{9-65}$$

A material balance written for component A around the $(n + 1)$st tank gives

$$(1 + K\theta)C_{n+1} - C_n = -\theta \frac{\partial C_{n+1}}{\partial t} \tag{9-66}$$

Addition of (9-64) and (9-66) gives

$$F_{n+1} - F_n = -\theta \frac{\partial F_{n+1}}{\partial t} \tag{9-67}$$

where

$$F_n = C_n + B_n \tag{9-68}$$

At time $t = 0$, pure solvent is in the tanks. Hence, the initial boundary conditions are

$$\left.\begin{aligned} F_n &= 0 \\ C_n &= 0 \\ B_n &= 0 \end{aligned}\right\} \quad t = 0, \text{ all } n \qquad \begin{aligned} &(9\text{-}69a) \\ &(9\text{-}69b) \\ &(9\text{-}69c) \end{aligned}$$

The Laplace transform of (9-67) is, following the introduction of (9-69a),

$$(1 + \theta p)\bar{F}_{n+1} - \bar{F}_n = 0 \tag{9-70}$$

Equation (9-70) is a linear difference equation with constant coefficients. Its solution is

$$\bar{F}_n = G\left(\frac{1}{1 + \theta p}\right)^n \tag{9-71}$$

The constant G in (9-71) may be evaluated by means of a material balance written around the first tank, making use of the fact that the liquid entering the tank is of composition $F_0 = C_0$:

$$F_1 - C_0 = -\theta \frac{\partial F_1}{\partial t} \tag{9-72}$$

The Laplace transform of (9-72), following the introduction of the boundary condition at $t = 0$ (9-69a), gives

$$\bar{F}_1 = \frac{C_0}{p(1 + \theta p)} \tag{9-73}$$

Combination of (9-71) and (9-73) gives

$$G = \frac{C_0}{p} \tag{9-74}$$

Combination of (9-71) and (9-74) yields

$$\bar{F}_n = \frac{C_0}{p(1 + \theta p)^n} \tag{9-75}$$

The Laplace transform of the material balance on component B, Eq. (9-64), is, using (9-69c),

$$(1 + \theta p)\bar{B}_{n+1} - \bar{B}_n - K\theta\bar{C}_{n+1} = 0 \tag{9-76}$$

However, the transform of (9-68) coupled with (9-75) gives

$$\bar{C}_{n+1} = \bar{F}_{n+1} - \bar{B}_{n+1} = \frac{C_0}{p(1 + \theta p)^{n+1}} - \bar{B}_{n+1} \tag{9-77}$$

The combination of (9-76) and (9-77) is

$$[1 + (K + p)\theta]\bar{B}_{n+1} - \bar{B}_n = \frac{K\theta C_0}{p(1 + \theta p)^{n+1}} \tag{9-78}$$

The solution of the linear difference equation (9-78) is the sum of the homogeneous and particular solutions. The homogeneous solution is $H/[1 + (K + p)\theta]^n$ and the particular solution is $C_0/p(1 + \theta p)^n$. Hence,

$$\bar{B}_n = \frac{H}{[1 + (K + p)\theta]^n} + \frac{C_0}{p(1 + \theta p)^n} \tag{9-79}$$

The constant H in Eq. (9-79) is evaluated by means of a balance on component B written around the first tank:

$$K\theta C_1 - B_1 = \theta \frac{\partial B_1}{\partial t} \tag{9-80}$$

The transform of (9-80) is

$$(1 + \theta p)\bar{B}_1 = K\theta\bar{C}_1 = K\theta(\bar{F}_1 - \bar{B}_1) \tag{9-81}$$

Combination of (9-81) and (9-73) gives

$$\bar{B}_1 = \frac{1}{1 + (K + p)\theta} \frac{K\theta C_0}{p(1 + \theta p)} \tag{9-82}$$

When (9-82) and (9-79) are joined at $n = 1$,

$$H = -\frac{C_0}{p} \tag{9-83}$$

Combination of (9-79) and (9-83) gives

$$\bar{B}_n = \frac{C_0}{p}\left\{\frac{[1 + (K + p)\theta]^n - (1 + \theta p)^n}{[1 + (K + p)\theta]^n(1 + \theta p)^n}\right\} \tag{9-84}$$

The inversion of (9-84) may be carried out by means of the residue theorem. The poles occur at

$$p = 0 \qquad \text{simple pole}$$

$$p_q = -\frac{1}{\theta} \qquad \text{pole of order } n$$

$$p_i = -\frac{1 + K\theta}{\theta} \qquad \text{pole of order } n$$

The residue at $p = 0$ is obtained by application of (8-84):

$$\rho_0 = \frac{j(0)}{l'(0)} = C_0\left[1 - \frac{1}{(1 + K\theta)^n}\right]$$

The residue at $p = 0$ is the steady-state solution of the problem.

The residue at p_q is obtained by application of (8-85):

$$\rho_q = \exp\left(-\frac{t}{\theta}\right)\sum_{s=1}^{s=m} A_s \frac{t^{s-1}}{(s-1)!}$$

$$A_s = \frac{1}{(n-s)!}\,\phi_q^{n-s}(p_q)$$

Taking into account the fact that the ϕ_q terms are to be evaluated at $p_q = -(1/\theta)$,

$$\phi_q = \frac{C_0}{p(\theta^n)}$$

$$\rho_q = -C_0 \exp\left(-\frac{t}{\theta}\right)\sum_{s=1}^{s=n} \frac{(t/\theta)^{s-1}}{(s-1)!}$$

The residue at p_i also is obtained by application of (8-85). Taking into account the fact that the ϕ_i terms are to be evaluated at $p_i = -[(1 + K\theta)/\theta]$,

$$\phi_i = -\frac{C_0}{p(\theta^n)}$$

$$\rho_i = C_0 \exp\left(-\frac{1 + K\theta}{\theta}t\right)\sum_{s=1}^{s=n} \frac{(t/\theta)^{s-1}}{(1 + K\theta)^{n-s+1}(s-1)!}$$

The complete inversion of (9-84) is then

$$B_n = C_0\left\{1 - \exp\left(-\frac{Lt}{V}\right)\left[\sum_{s=1}^{s=n}\left(\frac{Lt}{V}\right)^{s-1}\frac{1}{(s-1)!}\right] - \left(\frac{L}{L + KV}\right)^n \left[1 - \exp\left(-\frac{L + KV}{V}t\right)\sum_{s=1}^{s=n}\left(\frac{L + KV}{V}t\right)^{s-1}\frac{1}{(s-1)!}\right]\right\} \tag{9-85}$$

Equation (9-85) gives the concentration of the product component leaving the nth reactor at time t.

Mason and Piret[6] have examined a number of other stirred-tank reactor systems involving first-order reactions. They present the results for cases involving unequal tank volumes and unequal reaction-rate constants. The transient behavior of a stirred-tank reactor system involving higher-order reactions generally cannot be analyzed in closed form. A numerical attack suitable for second-order reactions has been given by Acton and Lapidus.[7]

Example 9-6. Rate at Which a Plate Absorber Approaches Steady State. An absorption column contains N actual plates. Each plate holds H_L moles of inert liquid, and the space above each plate holds H_G moles of inert gas. Operation of the column is started by filling each plate with liquid of composition X_s and then feeding L moles/hr of inert solvent of composition X_s to the top of the column and G moles/hr of carrier gas of composition Y_0 to the bottom of the column. All compositions refer to mean values on a plate and are on the basis of moles of transferable component per mole of inert carrier. The plate efficiency E^0 is constant:

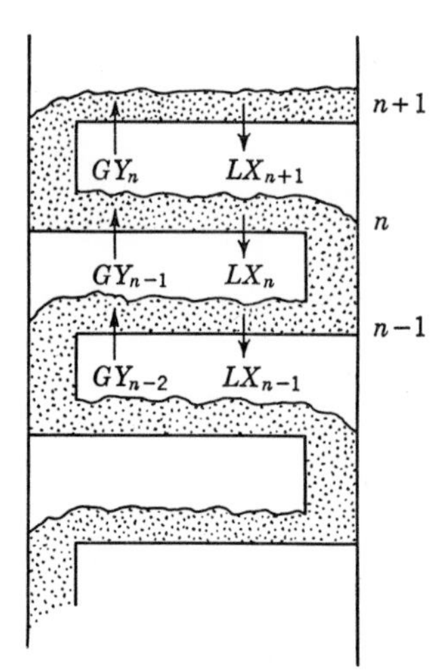

FIG. 9-4. Plate absorber.

$$E^0 = \frac{Y_n - Y_{n-1}}{Y_n^* - Y_{n-1}} \tag{9-86}$$

The equilibrium relation is

$$Y_n^* = mX_n + b \tag{9-87}$$

Determine the relation between the composition of the streams leaving the column and the time since start-up.

Solution. The following nomenclature is used:

E^0 = over-all plate efficiency defined by Eq. (9-86)
G = carrier-gas flow, moles/hr
H_L = carrier liquid holdup on plate, moles
H_G = carrier vapor holdup per plate, moles
L = inert-carrier-solvent flow rate, moles/hr
N = total number of actual plates in column
X_n = moles of transferable component per mole of inert solvent leaving plate n
Y_n = moles of transferable component per mole of carrier gas leaving plate n
Y_n^* = moles of transferable component per mole of carrier gas at equilibrium with solvent of composition X_n [see Eq. (9-87)]

A material balance written around the nth plate (see Fig. 9-4) gives

$$G(Y_{n-1} - Y_n) + L(X_{n+1} - X_n) = H_L \frac{\partial X_n}{\partial t} + H_G \frac{\partial Y_n}{\partial t} \tag{9-88}$$

Substituting the equilibrium condition (9-87) gives

$$G(Y_{n-1} - Y_n) + \frac{L}{m}(Y_{n+1}^* - Y_n^*) = \frac{\partial}{\partial t}\left(\frac{H_L}{m} Y_n^* + H_G Y_n\right) \tag{9-89}$$

Combination of (9-89) and the definition of the plate efficiency (9-86) permits Y^* to be eliminated:

$$G(Y_{n-1} - Y_n) + \frac{L}{mE^0}[(Y_{n+1} - Y_n) + (E^0 - 1)(Y_n - Y_{n-1})] = \frac{\partial}{\partial t}\left\{\frac{H_L}{mE^0}[Y_n + (E^0 - 1)Y_{n-1}] + H_G Y_n\right\} \tag{9-90}$$

With the notation

$$\lambda = \frac{GmE^0}{L} \tag{9-91}$$

$$\alpha = \frac{H_L}{L} \tag{9-92}$$

$$\beta = \frac{H_G mE^0}{L} \tag{9-93}$$

Eq. (9-90) may be written in the form

$$Y_{n+1} - (2 + \lambda - E^0)Y_n + (1 + \lambda - E^0)Y_{n-1} = \frac{\partial}{\partial t}[(\alpha + \beta)Y_n + \alpha(E^0 - 1)Y_{n-1}] \tag{9-94}$$

The applicable boundary conditions are

$$X_n = X_s \qquad Y_n = mX_s + b \equiv Y_s \qquad t = 0 \tag{9-95}$$

$$X_{N+1} = X_s \qquad Y_{n=0} = Y_0 \qquad t > 0 \tag{9-96}$$

The condition $Y_n = Y_s$ given by (9-95) assumes the vapor to be in equilibrium with the solvent at the start of operation and assumes vapor holdup to be small in comparison with liquid holdup.

The Laplace transformation with respect to time of (9-94) is, after introduction of the boundary condition (9-95) and rearrangement,

$$\bar{Y}_{n+1} - [2 + \lambda - E^0 + (\alpha + \beta)p]\bar{Y}_n + [1 + \lambda - E^0 - \alpha(E^0 - 1)p]\bar{Y}_{n-1} + (\beta + \alpha E^0)Y_s = 0 \tag{9-97}$$

The solution of (9-97) is

$$\bar{Y}_n = AZ_1^n + FZ_2^n + \frac{Y_s}{p} \tag{9-98}$$

where

$$Z_1 = \tfrac{1}{2}(c + i\sqrt{4d - c^2}) \tag{9-99a}$$

$$Z_2 = \tfrac{1}{2}(c - i\sqrt{4d - c^2}) \tag{9-99b}$$

$$c = 2 + \lambda - E^0 + (\alpha + \beta)p \tag{9-99c}$$

$$d = 1 + \lambda - E^0 - \alpha(E^0 - 1)p \tag{9-99d}$$

$$i = \sqrt{-1} \tag{9-99e}$$

Note that in the expressions for Z_1 and Z_2 the discriminant in the quadratic formula has been rearranged and the imaginary i introduced. The reason for this departure from the usual procedure shortly will become apparent.

The constants A and F of (9-98) may be evaluated by introducing the boundary conditions (9-96).

The condition $Y = Y_0$ entering the bottom of the column is introduced as follows: Write a material balance around the first plate ($n = 1$). There results

$$G(Y_0 - Y_1) + L(X_2 - X_1) = \frac{\partial}{\partial t}(H_L X_1 + H_G Y_1) \tag{9-100}$$

Replace the X terms in (9-100) by means of (9-86) and (9-87) and make use of the fact that Y_0 is a constant. The (9-100) becomes

$$Y_2 - (2 + \lambda - E^0)Y_1 + (1 + \lambda - E^0)Y_0 = (\alpha + \beta)\frac{\partial Y_1}{\partial t} \tag{9-101}$$

The Laplace transform of (9-101) is

$$\bar{Y}_2 - [2 + \lambda - E^0 + (\alpha + \beta)p]\bar{Y}_1 + \frac{1 + \lambda - E^0}{p} Y_0 + (\alpha + \beta)Y_s = 0 \tag{9-102}$$

Combination of (9-102) and (9-98) gives

$$A(Z_1^2 - cZ_1) + F(Z_2^2 - cZ_2) + \frac{1 + \lambda - E^0}{p}(Y_0 - Y_s) = 0 \tag{9-103}$$

where c is defined by (9-99c).

The condition $X = X_s$ entering the top of the column is introduced by means of a material balance written around the top (Nth) plate:

$$L(X_s - X_N) + G(Y_{N-1} - Y_N) = \frac{\partial}{\partial t}(H_L X_N + H_G Y_N) \tag{9-104}$$

Replace the X terms in (9-104) by means of (9-86) and (9-87), making use of the fact that X_s is a constant and $Y_s = mX_s + b$. Then (9-104) becomes

$$E^0 Y_s - (1 + \lambda)Y_N + (1 + \lambda - E^0)Y_{N-1} = \frac{\partial}{\partial t}[(\alpha + \beta)Y_N + \alpha(E^0 - 1)Y_{N-1}] \tag{9-105}$$

The Laplace transform of (9-105) is

$$\left(\beta + \alpha E^0 + \frac{E^0}{p}\right) Y_s - [1 + \lambda + (\alpha + \beta)p]\bar{Y}_N + d\bar{Y}_{N-1} = 0 \tag{9-106}$$

where d is defined by (9-99d). Combination of (9-98) and (9-106) gives

$$A[dZ_1^{N-1} - (c - 1 + E^0)Z_1^N] + F[dZ_2^{N-1} - (c - 1 + E^0)Z_2^N] = 0 \tag{9-107}$$

Subsequent algebraical manipulation is greatly facilitated by the use of the properties of complex numbers. It is for this reason that Z_1 and Z_2 were written in the form given by (9-99). When written in this way, Z_1 and Z_2 are complex conjugates. With reference to the discussion of Sec. 8-7, let

$$Z_1 = r(\cos\theta + i\sin\theta) \tag{9-108a}$$

$$Z_2 = r(\cos\theta - i\sin\theta) \tag{9-108b}$$

$$r = \sqrt{\frac{c^2}{4} + \frac{4d - c^2}{4}} = \sqrt{d} \tag{9-108c}$$

$$c = 2r\cos\theta = 2\sqrt{d}\cos\theta \tag{9-108d}$$

Now make use of the following relations:

$$Z_1^n = r^n(\cos n\theta + i\sin n\theta) \tag{9-109a}$$

$$Z_2^n = r^n(\cos n\theta - i\sin n\theta) \tag{9-109b}$$

$$\sin(\theta_1 \pm \theta_2) = \sin\theta_1\cos\theta_2 \pm \cos\theta_1\sin\theta_2 \tag{9-109c}$$

$$\cos(\theta_1 \pm \theta_2) = \cos\theta_1\cos\theta_2 \mp \sin\theta_1\sin\theta_2 \tag{9-109d}$$

Equation (9-98) becomes

$$\bar{Y}_n = d^{n/2}[(A + F)\cos n\theta + i(A - F)\sin n\theta] + \frac{Y_s}{p} \tag{9-110}$$

Equation (9-103) becomes

$$A + F = \frac{1}{d}\frac{1 + \lambda - E^0}{p}(Y_0 - Y_s) \tag{9-111}$$

Equation (9-107) becomes

$$d^{n/2}\{(A + F)[(1 - E^0) \cos N\theta - d^{1/2} \cos (N + 1)\theta] + i(A - F) \\ [(1 - E^0) \sin N\theta - d^{1/2} \sin (N + 1)\theta]\} = 0 \tag{9-112}$$

Then, if $d \neq 0$,

$$A - F = i(A + F)\frac{d^{1/2} \cos (N + 1)\theta + (E^0 - 1) \cos N\theta}{d^{1/2} \sin (N + 1)\theta + (E^0 - 1) \sin N\theta} \tag{9-113}$$

Equation (9-110) is then

$$\bar{Y}_n = \frac{1}{p[d^{1/2} \sin (N + 1)\theta + (E^0 - 1) \sin N\theta]}\Big([(1 + \lambda - E^0)(Y_0 - Y_s)d^{(n-2)/2}] \\ \{(\cos n\theta)[d^{1/2} \sin (N + 1)\theta + (E^0 - 1) \sin N\theta] - (\sin n\theta) \\ [d^{1/2} \cos (N + 1)\theta + (E^0 - 1) \cos N\theta]\} \\ + Y_s[d^{1/2} \sin (N + 1)\theta + (E^0 - 1) \sin N\theta]\Big) \tag{9-114}$$

In order to find conditions at the top of the tower, set $n = N$. Then (9-114) becomes

$$\bar{Y}_N = \frac{1}{p[d^{1/2} \sin (N + 1)\theta + (E^0 - 1) \sin N\theta]} \\ \{[(1 + \lambda - E^0)(Y_0 - Y_s)\, d^{(N-1)/2} \sin \theta] + Y_s[d^{1/2} \sin (N + 1)\theta + (E^0 - 1) \sin N\theta]\} \tag{9-115}$$

With the notation

$$j(p) = (1 + \lambda - E^0)(Y_0 - Y_s)\, d^{(N-1)/2} \sin \theta \\ + Y_s[d^{1/2} \sin (N + 1)\theta + (E^0 - 1) \sin N\theta] \tag{9-116}$$

$$l(p) = p[d^{1/2} \sin (N + 1)\theta + (E^0 - 1) \sin N\theta] \tag{9-117}$$

it is seen that the order of $l(p)$ is at least one greater than that of $j(p)$. Consequently, inversion will be attempted by the method of residues. The poles occur at those values of p for which $l(p) = 0$. This requires

$$p = 0 \tag{9-118}$$

$$d^{1/2} \sin (N + 1)\theta + (E^0 - 1) \sin N\theta = 0 \tag{9-119}$$

The residue at the pole at $p = 0$ corresponds to the steady-state solution. Apply (8-84):

$$\rho_0(t) = \frac{j(0)}{l'(0)} \tag{9-120}$$

$$l'(0) = [d^{1/2} \sin (N + 1)\theta + (E^0 - 1) \sin N\theta]_{p=0}$$

From (9-99c) and (9-99d), at $p = 0$,

$$c = 2 + \lambda - E^0$$

$$d = 1 + \lambda - E^0$$

Using (9-99a) and (9-99b) for $p = 0$,

$$Z_1 = 1$$

$$Z_2 = d$$

Combination of (9-109a) and (9-109b) gives, at $p = 0$,

$$\sin N\theta = \frac{1 - d^N}{i2d^{N/2}}$$

$$\sin (N + 1)\theta = \frac{1 - d^{N+1}}{i2d^{(N+1)/2}}$$

$$\sin \theta = \frac{1 - d}{i2d^{1/2}}$$

Hence

$$\frac{j(0)}{l'(0)} = \frac{(Y_0 - Y_s)d^N(1 - d)}{(1 - d^{N+1}) + (E^0 - 1)(1 - d^N)} + Y_s$$
$$= \frac{(Y_0 - Y_s)(E^0 - \lambda)(1 + \lambda - E^0)^N}{E^0 - \lambda(1 + \lambda - E^0)^N} + Y_s$$

The steady-state solution of (9-115) is then

$$\frac{Y_N)_{s.s.} - Y_s}{Y_0 - Y_s} = \frac{(E^0 - \lambda)(1 + \lambda - E^0)^N}{E^0 - \lambda(1 + \lambda - E^0)^N} \tag{9-121}$$

If the plates are perfect ($E^0 = 1$), (9-121) becomes

$$\frac{(Y_N)_{s.s.} - Y_s}{Y_0 - Y_s} = \frac{(1 - \lambda)\lambda^N}{1 - \lambda^{N+1}} \tag{9-122}$$

The inversion leading to the unsteady-state operation of the column is not amenable to a general analytical solution except in special cases. Usually, for $E^0 \neq 1$, the algebra for the general case becomes unmanageable. However, specific situations may always be treated numerically. In the following, a general solution will be obtained for perfect plates ($E^0 = 1$). Then a numerical solution for $E^0 \neq 1$ will be illustrated. One of the special cases for $E^0 \neq 1$ which can be handled in a generalized fashion is given as Prob. 9-7.

Unsteady-state Solution, Perfect Plates ($E^0 = 1$). For this case, Eq. (9-115) becomes

$$\bar{Y}_N = \frac{1}{p \sin (N + 1)\theta} [(Y_0 - Y_s)\lambda d^{(N-2)/2} \sin \theta + Y_s \sin (N + 1)\theta] \tag{9-123}$$

The poles correspond to

$$p = 0$$
$$\sin (N + 1)\theta = 0$$

The steady-state solution obtained from the inversion at the pole $p = 0$ has already been obtained and is given by (9-122). In view of (9-99c and d) and (9-108d),

$$d = \lambda$$
$$c = 1 + \lambda + (\alpha + \beta)p = 2\sqrt{\lambda} \cos \theta \tag{9-124}$$

The condition $\sin (N + 1)\theta = 0$ requires

$$\theta = \frac{k\pi}{N + 1} \qquad \text{where } k \text{ is integer} \tag{9-125}$$

Actually, as will be seen shortly, k only takes on the values 1, 2, 3, . . . , $N + 1$. Denote the values of θ which satisfy (9-125) by θ_k and the corresponding values of p by p_k. Then

$$p_k = \frac{2\sqrt{\lambda} \cos \theta_k - (1 + \lambda)}{\alpha + \beta} \tag{9-126}$$

Examination of (9-125) and (9-126) shows that, corresponding to $k = 0, 1, 2, 3, \ldots, N + 1$, $N + 1$ separate and distinct values of p_k are defined. However, in view of the properties of $\cos \theta$, for $k = N + 2, N + 3, \ldots$, the previously defined values of p_k are repeated. This does not correspond to multiple roots for a given value of p_k; rather for a given value of p_k, several values of θ_k are defined. Hence, the individual p_k's are simple poles, and θ_k is restricted to the range $0 = \theta_k \leqq \pi$ in order to stay on the same branch of the function. Now,

$$\frac{d}{dp}[p \sin (N+1)\theta]_{p_k} = \left[p(N+1) \cos (N+1)\theta \frac{d\theta}{dp}\right]_{p_k}$$
$$= -\frac{[2\sqrt{\lambda} \cos \theta_k - (1+\lambda)](N+1) \cos (N+1)\theta_k}{2\sqrt{\lambda} \sin \theta_k} \tag{9-127}$$

Hence,

$$\Sigma \rho_k = \sum_{k=0}^{k=N+1} \left\{\exp - \left[\frac{1 + \lambda - 2\sqrt{\lambda} \cos k\pi/(N+1)}{\alpha + \beta}\right] t\right\}$$
$$\frac{(Y_0 - Y_s)2\lambda^{(N+1)/2} \sin^2 k\pi/(N+1)}{(-1)^k(N+1)[1 + \lambda - 2\sqrt{\lambda} \cos k\pi/(N+1)]} \tag{9-128}$$

Since $\sin 0 = 0$ and $\sin \pi = 0$, the complete solution for $E^0 = 1$ is

$$\frac{Y_N - Y_s}{Y_0 - Y_s} = \frac{(1-\lambda)\lambda^N}{1 - \lambda^{N+1}}$$
$$+ \sum_{k=1}^{k=N} \left\{\exp - [1 + \lambda - 2\sqrt{\lambda} \cos k\pi/(N+1)] \frac{t}{\alpha + \beta}\right\}$$
$$\frac{(-1)^k 2\lambda^{(N+1)/2} \sin^2 k\pi/(N+1)}{(N+1)[1 + \lambda - 2\sqrt{\lambda} \cos k\pi/(N+1)]} \tag{9-129}$$

where

$$\lambda = \frac{Gm}{L}$$
$$\alpha = \frac{H_L}{L}$$
$$\beta = \frac{H_G m}{L}$$

Unsteady-state Solution, Real Plates. In order to illustrate the method of solution for a tower with plates of finite efficiency, consider a tower with $N = 3$, $E^0 = 1/3$, $\lambda = GmE^0/L = 1.5$, $\alpha = 0.1$, $\beta = 0$. The steady-state part of the solution is given by (9-121). It remains to find the unsteady-state part of the solution. In order to do this, it is necessary to find the poles of (9-115) corresponding to those values of p, p_q, which satisfy

$$d^{1/2} \sin (N+1)\theta + (E^0 - 1) \sin N\theta = 0 \tag{9-130}$$

where

$$d = 1 + \lambda - E^0 - \alpha(E^0 - 1)p \tag{9-99d}$$
$$c = 2 + \lambda - E^0 + (\alpha + \beta)p \tag{9-99c}$$
$$\cos \theta = \frac{c}{2\sqrt{d}} \tag{9-108d}$$

In theory, Eqs. (9-99d and c) and (9-108d) may be used to find the relation between p and θ. Then either p or θ may be replaced in (9-130) and the resulting relation solved for p_q or θ_q. In practice, the algebra involved in such a procedure quickly becomes awkward. A more practical technique involves a trial-and-error method as follows:

1. Guess a value of θ, θ_s, and use this value in (9-130) to find the value of d, d_s, which makes (9-130) vanish.
2. Calculate the value of p, p_s, corresponding to d_s from (9-99d).
3. Using θ_s and d_s, compute the value of c, c_s, which satisfies (9-108d).
4. Using c_s, compute the value of p, p_v, which satisfies (9-99c).
5. Compute $\Delta p = p_s - p_v$, and plot Δp as a function of θ_s, the trial value of θ. The values of p, for which $\Delta p = 0$, are the desired poles.

The labor involved in the above steps can be reduced if the following is kept in mind:

1. $d^{1/2} = r$ and must be positive.
2. p_q must be negative, since the exponential terms in the final inversion must vanish as $t \to \infty$.
3. Organize the calculations in a systematic tabular fashion.

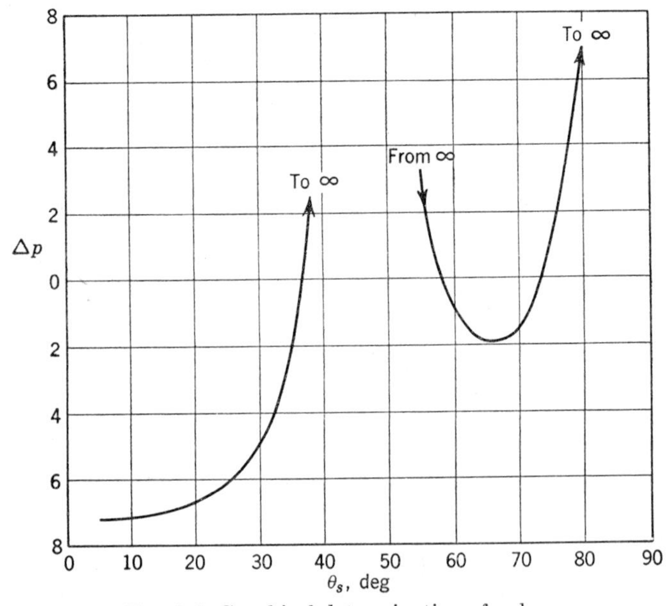

FIG. 9-5. Graphical determination of poles.

The results of the calculation of $\Delta p = p_s - p_v$ are shown plotted as a function of θ_s in Fig. 9-5. The poles correspond to those values of p for which Δp is zero. In Fig. 9-5 only the first quadrant is covered. Calculations for the other three quadrants give different values of θ corresponding to $\Delta p = 0$, as shown in Table 9-1.

For each value of p in Table 9-1, four values of θ were found. However, those values of θ which made $\sqrt{d}$ negative are not permissible. The remaining values of θ form two sets, 0 to 180° and 180 to 360°. Obviously, any number of such sets could be found covering a range of 180°, i.e., 360 to 540°, etc. Any *one* of these sets will result in the same values of the residues of (9-115) as any other one set. Since θ is a multiple-valued function of p, each set corresponds to a separate branch of the function, and only one set should be used in carrying out the inversion of (9-115). In subsequent work here, the set $\theta_1 = 36.5°$, $\theta_2 = 73.8°$, $\theta_3 = 121.6°$ will be used.

TABLE 9-1

p	Δp	$\sqrt{d}$	θ, deg
−13.4	0	1.13	36.5
−13.4	0	−1.13	143.5
−13.4	0	−1.13	216.5
−13.4	0	1.13	323.5
−29.0	0	0.487	73.8
−29.0	0	−0.487	106.2
−29.0	0	−0.487	253.8
−29.0	0	0.487	286.2
−32.4	0	−0.0696	58.4
−32.4	0	0.0696	121.6
−32.4	0	0.0696	238.4
−32.4	0	−0.0696	301.6

Note that although $\theta = 0$ satisfies (9-130), no residue is generated at this point because of the sin θ term which appears in the numerator of (9-115).

With the poles known, the inversion of the unsteady-state portion of (9-115) may be obtained by means of the residue theorem. From the differentiation of the denominator of (9-115), the only term which contributes to the residue is

$$p \frac{d}{dp} [d^{1/2} \sin (N+1)\theta + (E^0 - 1) \sin N\theta]$$

$$= p \left\{ \frac{\sin (N+1)\theta}{2\sqrt{d}} \frac{d(d)}{dp} + [\sqrt{d}\,(N+1) \cos (N+1)\theta + (E^0 - 1)N \cos N\theta] \frac{d\theta}{dp} \right\}$$

$$= p \frac{1}{2 \sin \theta \sqrt{d}} \left\{ [\alpha(1 - E^0) \sin (N+1)\theta \sin \theta] \right.$$

$$+ \frac{\alpha(1 - E^0) \cos \theta}{\sqrt{d}} [\sqrt{d}\,(N+1) \cos (N+1)\theta + (E^0 - 1)N \cos N\theta]$$

$$\left. - (\alpha + \beta)[\sqrt{d}\,(N+1) \cos (N+1)\theta + (E^0 - 1)N \cos N\theta] \right\}$$

$$= \frac{p_j k_j}{2 \sin \theta \sqrt{d}} \tag{9-131}$$

Then, using (9-121) for the steady-state portion,

$$\frac{Y_N - Y_s}{Y_0 - Y_s} = \frac{(E^0 - \lambda)(1 + \lambda - E^0)^N}{E^0 - \lambda(1 + \lambda - E^0)^N}$$

$$+ 2(1 + \lambda - E^0) \sum_j \frac{[\exp (p_j t)](\sqrt{d_j})^N \sin^2 \theta_j}{p_j k_j} \tag{9-132}$$

For $N = 3$, $E^0 = \frac{1}{3}$, $\lambda = 1.5$, $\alpha = 0.1$, $\beta = 0$, Eq. (9-132) becomes

$$\frac{Y_3 - Y_s}{Y_0 - Y_s} = 0.793 - 0.817 \exp - (13.4)t$$

$$+ 2.37 \cdot 10^{-2} \exp - (29.0)t - 1.48 \cdot 10^{-5} \exp - (32.4)t \tag{9-133}$$

Note that at $t = 0$, $Y_3 = Y_s$, which provides a check on the arithmetic operations.

BIBLIOGRAPHY

1. Boole, G.: "Treatise on the Calculus of Finite Differences," Macmillan & Co., Ltd., London, 1880; G. E. Stechert & Co., New York, 1931.
2. Fort, T.: "Finite Differences and Difference Equations in the Real Domain," Oxford University Press, New York, 1948.
3. Milne-Thomson, L. M.: "The Calculus of Finite Differences," Macmillan & Co., Ltd., London, 1933.
4. Kármán, T. v., and M. A. Biot: "Mathematical Methods in Engineering," McGraw-Hill Book Company, Inc., New York, 1940.
5. Tiller, F. M., and R. S. Tour: "Stagewise Operations—Applications of the Calculus of Finite Differences to Chemical Engineering, *Trans. Am. Inst. Chem. Engr.*, vol. 40, pp. 317–32, June, 1944.
6. Mason, D. R., and E. L. Piret: Continuous Flow Stirred Tank Reactor Systems, *Ind. Eng. Chem.*, vol. 42, no. 5, pp. 817–825, 1950.
7. Acton, F. S., and L. Lapidus: Continuous Stirred-tank Reactors, *Ind. Eng. Chem.*, vol. 47, no. 4, pp. 706–710, 1955.
8. Marshall, W. R., Jr. and R. L. Pigford: "The Application of Differential Equations to Chemical Engineering Problems," University of Delaware, Newark, Del., 1947.

PROBLEMS

9-1. By means of material balances written around the first ($n = 1$) and last ($n = M$) tanks, show that the boundary conditions (9-31) of Example 9-1 are correct and that the concept of fictitious tanks at either end is, in this case, legitimate.

9-2. A chemical reaction is carried out in a special tower. The tower contains N bubble cap plates. Each plate contains H moles of liquid consisting largely of an inert nonvolatile solvent and containing a dissolved nonvolatile catalyst. The liquid *does not* flow through the column but is held on the plate to act as a reaction vessel. Pure gas A is fed to the bottom of the column at a rate of G moles/hr. On each plate, some A dissolves and reacts to form the volatile compound B by the reversible reaction

$$A \leftrightarrows B$$

The plates are cooled to a uniform temperature, and the forward reaction takes place at a rate which is proportional to the number of moles of A present and the rate constant K_A; the reverse-reaction rate is proportional to the moles of B and to the rate constant K_B. The B formed is then stripped and carried off by the gas. The over-all plate absorption or desorption efficiencies are

$$E_A = \left(\frac{Y_n - Y_{n-1}}{Y_n^* - Y_{n-1}}\right)_A \qquad \text{for component } A$$

$$E_B = \left(\frac{Y_n - Y_{n-1}}{Y_n^* - Y_{n-1}}\right)_B \qquad \text{for component } B$$

The equilibrium relations are

$$(Y_n^*)_A = m_A(X_n)_A + b$$
$$(Y_n^*)_B = m_B(X_n)_B + c$$

where Y and X denote mole fractions in the gas and liquid, respectively.

Find the integral relation between $(Y_n)_B$ and n at steady state. Discuss the operation of this system.

9-3. A batch distillation column contains M plates, a total condenser, and a still. The column, originally empty, is charged with a binary mixture containing a mole fraction X_0 of the more volatile component. The relative volatility α of the mixture is constant. The column is operated at total reflux until the system is at equilibrium. At this time the fraction f of the original charge is in the still, and the remainder is equally distributed among the M plates and the condenser. Vapor holdup in the system may be neglected. Develop an expression for the composition of the liquid in the condenser as a function of the system parameters. Compute the composition of the liquid in the condenser when $X_0 = 0.5$, $M = 20$, $\alpha = 1.5$, and (a) $f = 1$ and (b) $f = 0.7$, for $E^0 = 1$ and $E^0 = 0.5$. Discuss.

9-4. The recovery of certain valuable constitutents (such as metals, phosphoric acid, etc.) from ores is often accomplished by treating the ore with a reagent which dissolves the valuable constitutent. The treating system frequently consists of a cascade of tank pairs. Each tank pair consists of an agitated tank and an associated clarifier (or thickener). In the agitated tank, reagent and ore are mixed. In the associated clarifier the slurry from the agitated tank is allowed to settle, forming a liquid phase essentially free of solid and a mud phase which consists of a thick slurry of ore residue and liquid phase. Such a system is shown in Fig. 9-6. (Each tank shown represents a tank pair.) In the enriching section, countercurrent action is used to increase the recovery of the constitutent from the ore (particularly when a chemical equlibrium is involved). By the time the mud leaves the enriching section, essentially all the valuable material has gone into solution. However, the liquid associated with the mud contains reagent and ore values. The stripping section is used to recover these values and return them to the system.

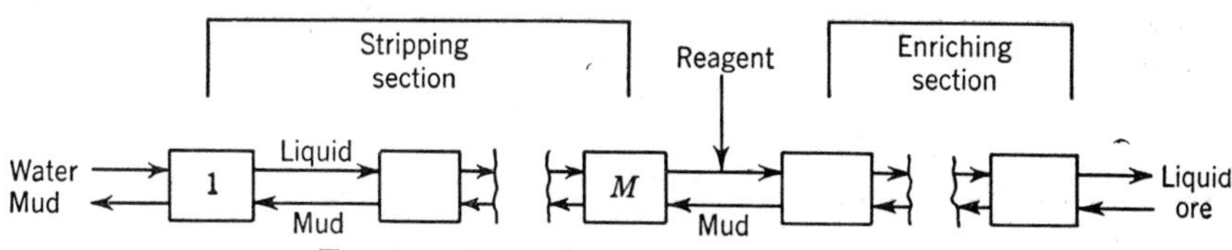

FIG. 9-6. Extraction cascade of Prob. 9-4.

Consider the stripping section. It consists of M stages. The first stage ($n = 1$) is fed with pure water. The last stage ($n = M$) is fed with mud which has associated with it liquid of composition C_{M+1} of recoverable material (principally reagent). It is desired to obtain a relation between the composition of the liquid (C_M) leaving the last stage of the stripping cascade and the number of stages M. The fractional recovery of recoverable material fed to the stripping cascade is also desired. The following assumptions may be used:

1. The mass of mud is not changed by the stripping treatment.
2. Each pound of mud contains f lb of liquid per pound of solids.
3. The liquid associated with the mud leaving a given stage is of the same composition as the clarified liquid leaving the stage.

Data and Nomenclature

C_n = recoverable constituents per pound of water in the liquid leaving the nth stage, lb

C_{M+1} = recoverable constituents per pound of water in the liquid associated with the mud fed to stage M, lb

f = liquid carried by mud per pound of solids, lb

L_n = rate of flow of water from nth stage, lb/hr

L_0 = rate of flow of pure water to first stage of stripping cascade, lb/hr

S = rate of flow of solids associated with mud, lb/hr

9-5. In the stirred-tank reactor problem of Example 9-5, what is the concentration of the component A leaving the nth tank?

9-6. A single stirred-tank reactor of volume ν is filled with pure solvent. A flow of L ft³/hr of solution containing C_0 moles/ft³ of A and B_0 moles/ft³ of B is then initiated. Inside the reactor the *reversible* reaction

$$A \leftrightarrows B$$

takes place. The forward-rate constant is K_1, and the backward-rate constant is K_2. What is the composition of the stream leaving the reactor as a function of time?

9-7. With reference to the plate absorption column of Example 9-6, find the *analytical* expression for the composition of the gas leaving the top plate if $\lambda = E^0 = \frac{1}{2}$, $\beta = 0$. What is the composition of the liquid leaving the tower?

9-8. With reference to the plate absorption column of Example 9-6, what is the composition of the liquid leaving the column if $N = 3$, $E^0 = \frac{1}{3}$, $\lambda = 1.5$, $\alpha = 0.1$, $\beta = 0$?

9-9. Define the approach to equilibrium of a plate absorption column by the expression

$$\frac{Y_N - Y_{Ns.s.}}{Y_0 - Y_s}$$

where Y_N denotes the composition of the vapor leaving the top of the column at time t and $(Y_N)_{s.s.}$ denotes the composition of the vapor leaving the top of the column at steady state ($t \to \infty$). Compare the rate of approach to equilibrium of a real column where $N = 3$, $E^0 = \frac{1}{3}$, $\lambda = 1.5$, $\alpha = 0.1$, $\beta = 0$ with that of an ideal column ($E^0 = 1$) which roughly approximates the real column; i.e., take the ideal column to be $N = (NE^0)_R = 1$, $\lambda = (\lambda/E^0)_R = 4.5$, $\alpha = (\alpha/E^0)_R = 0.3$, $\beta = 0$.

CHAPTER 10

THE NUMERICAL SOLUTION OF PARTIAL DIFFERENTIAL EQUATIONS

10-1. Introduction. Only a small fraction of the partial differential equations which are generated by engineering problems may be solved by formal analytical procedures. More often, an analytical solution of the problem is beyond the power of present methods. When such a situation is encountered, a solution ordinarily may be obtained by numerical techniques. In addition, numerical methods may be used with success by engineers whose formal mathematical background is limited. The following sections develop the numerical methods that have been found most useful in solving the partial differential equations that arise in chemical engineering.

The numerical approach to the solution of a partial differential equation is very similar to the method used in the numerical solution of a total differential equation. The derivatives appearing in the equation are replaced by finite-difference ratios, with the result that the differential equation is replaced by a difference equation. The difference equation then is solved by algebraic or arithmetic procedures.

In a number of cases, the computations needed to obtain an answer by finite-difference methods are less time-consuming than the computations required to evaluate the terms in the infinite series which arise from the known analytical solution of the same problem. Often, then, the numerical solution to a *specific problem* is both shorter and easier than the analytical approach. The specific results of the numerical method are not, however, as readily extended to similar problems as are analytical solutions.

Numerical techniques can handle problems which are too complicated for known analytical methods. In really complex situations, however, it is found that the computations involved are too tedious for hand or ordinary desk computing-machine operations. Fortunately, the calculations required by the numerical methods of solution may be carried out with dispatch by modern high-speed computing machines. Indeed, the very-high-speed digital computers cannot be used unless the partial differential equation is first approximated by a finite difference equation;

i.e., the problem first must be set up in the form needed for the application of a numerical-solution procedure before most high-speed computers can be employed. Consequently, numerical methods can be used to solve problems which are too complicated for either analytical procedures or hand computation if a high-speed machine can be made available.

In the case of partial differential equations, the accuracy of the results obtained by numerical-solution procedures is hard to estimate. It is known, however, that if the numerical method satisfies two criteria, termed "convergence" and "stability," the accuracy is determined by the number of increments employed, and increased accuracy may be obtained at the cost of increased labor. An estimate of the accuracy of the final result is obtained by comparing calculations made using different increment magnitudes. Ordinarily, if the difference between the results of two such calculations is within the desired accuracy, the answers obtained from the calculations using the largest number (i.e., smallest size) of increments may be considered satisfactory.

The convergence criterion deals with the approach of the approximate numerical solution to the exact solution as the number of increments employed becomes infinite in number. Unless the numerical solution converges to the exact solution in the limit when the number of increments approaches infinity, the numerical method is unsatisfactory. A method which converges in the limit to the exact solution is said to fulfill the convergence criterion.

The stability criterion deals with the growth of errors introduced into the calculation. In a practical solution, a finite number of increments must be employed, and only a finite number of significant figures can be carried in the computations. Both of these practical limitations introduce errors into the analysis. These errors are not serious unless they increase in magnitude ("grow") as the solution proceeds. A method which prevents the growth of errors is said to fulfill the stability criterion.

In the case of *linear* partial differential equations with constant coefficients, the conditions to be imposed upon the numerical-solution method in order to satisfy both the convergence and stability criteria can be stated in a precise and convenient manner. For initial-value problems, this is done in Sec. 10-5; for boundary-value problems it is done in Sec. 10-14. A similar treatment of the convergence and stability of nonlinear equations has not yet been formulated. In these situations, the analyst must proceed with caution.

In the following, initial-value problems are considered first. In most chemical-engineering applications an initial-value problem takes the form of an unsteady-state situation in which conditions are everywhere specified at a given instant but only at the boundaries in succeeding instants. Conditions in the interior are to be determined as a function of time. The use of numerical methods in the solution of initial-value problems is illustrated by a number of examples.

The second major application of numerical methods to be considered deals with boundary-value problems. In general, these are steady-state situations in two or more dimensions and are solved numerically by "relaxation" methods.

10-2. Example of a Numerical Solution. In this section the numerical method of solution is illustrated by its application to a specific problem. The problem chosen may be set up directly in a form suitable for numerical solution without recourse to the governing partial differential equation. A more generally applicable formal mathematical technique for reducing a partial differential equation to a form needed for the application of a numerical-solution method is developed in Sec. 10-3.

Example 10-1. Consider a large plane wall of insulating brick 12 in. thick (Fig. 10-1) originally at a uniform temperature of 100°F. Let the temperature of one face

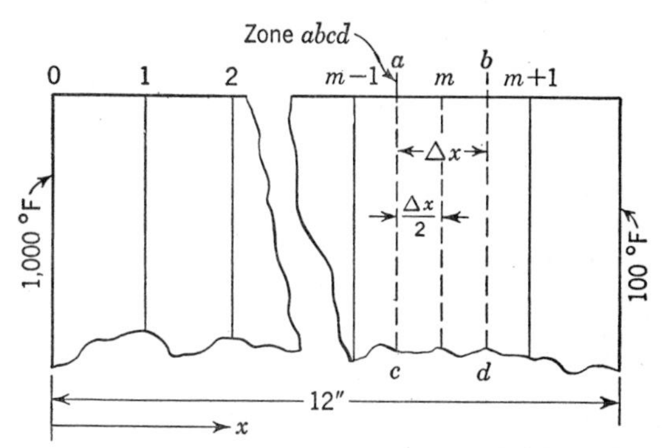

FIG. 10-1. Unsteady-state heating of brick wall.

suddenly be increased to and maintained at 1000°F while the opposite face is maintained at 100°F. What is the temperature distribution in the interior of the brick wall as a function of time? The properties of the brick will be assumed constant at $k = 0.6$ Btu/(hr)(ft^2)(°F/ft); $\rho = 30$ lb/ft^3; $c = 0.23$ Btu/(lb)(°F).

Solution. It is assumed that the wall extends indefinitely in the y and z directions; consequently, the heat flow can be considered to take place in the x direction only. Divide the wall by a number of *equally spaced* planes labeled 0, 1, 2, . . . , $m - 1$, m, $m + 1$, . . . , as shown in Fig. 10-1. Now write a heat balance on the zone $abcd$ which is of thickness Δx and bisected by the plane m. The input of heat into the zone in time Δt is, approximately,

$$\frac{kA(T_{m-1,n} - T_{m,n})}{\Delta x} \Delta t$$

The corresponding output of heat is

$$\frac{kA(T_{m,n} - T_{m+1,n})}{\Delta x} \Delta t$$

and the accumulation (or storage) of heat in the zone in the time Δt is

$$A\ \Delta x \rho c(T_{m,n+1} - T_{m,n})$$

The subscript m denotes the temperature at plane m, and the subscript n denotes the temperature at time t; the subscript $n + 1$ denotes the temperature at time $t + \Delta t$. Using the principle of "input − output = accumulation" and rearranging, the finite-difference heat balance is

$$T_{m+1,n} - 2T_{m,n} + T_{m-1,n} = \frac{\rho c\ \Delta x^2}{k\ \Delta t}(T_{m,n+1} - T_{m,n}) \tag{10-1}$$

It is convenient to denote $(\rho c\ \Delta x^2)/(k\ \Delta t)$ by the symbol M, frequently termed the "modulus":†

$$M = \frac{\rho c\ \Delta x^2}{k\ \Delta t} \tag{10-2}$$

Combining (10-1) and (10-2) gives the working equation

$$T_{m,n+1} = \frac{T_{m-1,n} + (M - 2)T_{m,n} + T_{m+1,n}}{M} \tag{10-3}$$

Equation (10-3) relates in a simple way the temperature at an interior plane m in the solid at time $t + \Delta t$ to the temperatures which existed at the plane m and at distance Δx on either side of the plane at time t. Selection of the value of the modulus M must be made before actual calculation may proceed. Equation (10-3) indicates that the minimum permissible value of M is 2. For $M < 2$ the temperature at plane m would exert a negative influence on the temperature at the same plane one time increment later, a situation which is intolerable on physical grounds. This physical argument is confirmed by the stability and convergence discussion of Sec. 10-5. The rules given there show that for this problem convergence and stability are obtained if $M \geqq 2$.

The numerical calculations are simplified if M is taken to be integer and if the thickness of the wall is an exact multiple of the distance increment Δx. Let $M = 2$ and $\Delta x = 3$ in. $= 0.25$ ft. Then the time increment is fixed by Eq. (10-2):

$$\Delta t = \frac{\rho c\ \Delta x^2}{Mk} = \frac{30 \cdot 0.23(0.25)^2}{2 \cdot 0.6}$$

$$\Delta t = 0.36 \text{ hr}$$

For $M = 2$, Eq. (10-2) becomes

$$T_{m,n+1} = \frac{T_{m-1,n} + T_{m+1,n}}{2}$$

Let $n = 0$ at the instant that the temperature of one face of the wall is suddenly increased, and denote the m value of the hot face by 0. Then the temperature of the plane 3 in. in from the hot face after a time interval of $\Delta t = 0.36$ hr has elapsed is

$$T_{11} = \frac{T_{00} + T_{20}}{2}$$

The next problem to be faced is the value to be assigned to T_{00}. At times infinitesimally less than zero, T_0 is 100°F, whereas at times infinitesimally greater than zero T_0 is prescribed to be 1000°F. The statement of the problem does not permit a unique assignment of the value of T_{00}, but in practice it is found that the best compromise *in the absence of other information* is to use the mean or average value, in this case,

† The concept of a modulus was used by Binder[4] and Schmidt[34] and extended by Dusinberre.[12]

TABLE 10-1
$M = 2 \qquad \Delta x = 0.25$ ft

t, hr	Δt increments	T_0, °F	T_1, °F	T_2, °F	T_3, °F	T_4, °F
0	0	(550)	100	100	100	100
0.36	1	1000	325	100	100	100
0.72	2	1000	550	212.5	100	100
1.08	3	1000	606.25	325	156.25	100
1.44	4	1000	662.5	381.25	212.5	100

$(100 + 1000)/2 = T_{00} = 550$°F. Then

$$T_{11} = \frac{550 + 100}{2} = 325\text{°F}$$

$$T_{21} = \frac{100 + 100}{2} = 100\text{°F} \qquad \text{etc.}$$

Proceeding in this fashion, Table 10-1 was prepared. Note that at the end of the first time increment the surface temperature was raised to the prescribed value of 1000°F. Although a different surface temperature was used in the illustration only

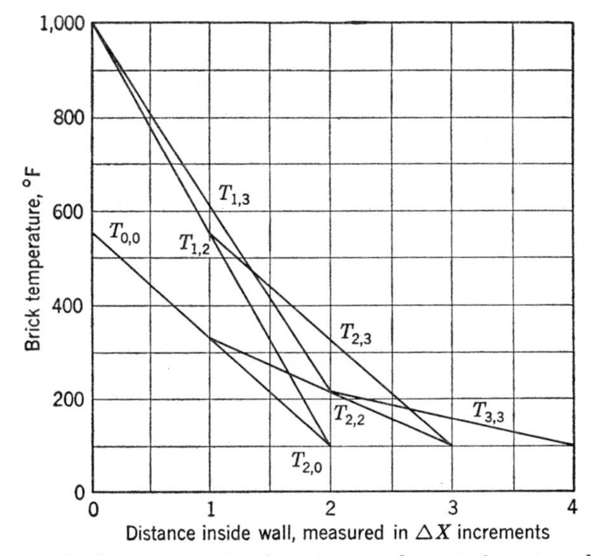

FIG. 10-2. Graphical evaluation of unsteady-state heat transfer.

in the starting procedure, it will be apparent that no complications would be encountered if the surface temperature had been changed at the end of every time interval. *This flexibility is typical of the finite-difference method.*

When $M = 2$, the calculations may also be carried out graphically.† The procedure is illustrated in Fig. 10-2. The ordinate represents temperature, and the abscissa

† Section 10-8 develops a general graphical method.

represents distance from the hot surface. When $M = 2$, the finite-difference relation states that the temperature at any given point at a given time is the arithmetic mean of the temperatures of two points located a distance Δx on each side at a *time* Δt *previously*. The arithmetic mean may be obtained graphically as the intersection of a straight line connecting the points at $-\Delta x$, $+\Delta x$, with the vertical line representing the point in question. Thus for Fig. 10-2, at zero time the temperature at plane 0 is 550°F, and the temperature at plane 2 is 100°F. A straight line connecting these points intersects plane 1 at 325°F. Hence, after one time increment ($t = 0.36$ hr), the temperature at plane 1 is 325°F, and the temperature at planes 2, 3, and 4 remains unchanged at 100°F. At the end of the first time increment the surface temperature is 1000°F, and the temperature of plane 2 is still 100°F. The temperature at plane 1 at the end of the second time increment ($t = 0.72$ hr) is then found by connecting the points 1000°, plane 0, and 100°, plane 2, by a straight line whose intersection at plane 1 is 550°F. The temperature at plane 2 at $t = 0.72$ hr is found by connecting $T_{1,1}(325°)$ with $T_{3,1}(100°)$ and is 212.5°. The remainder of the construction proceeds in a similar fashion.

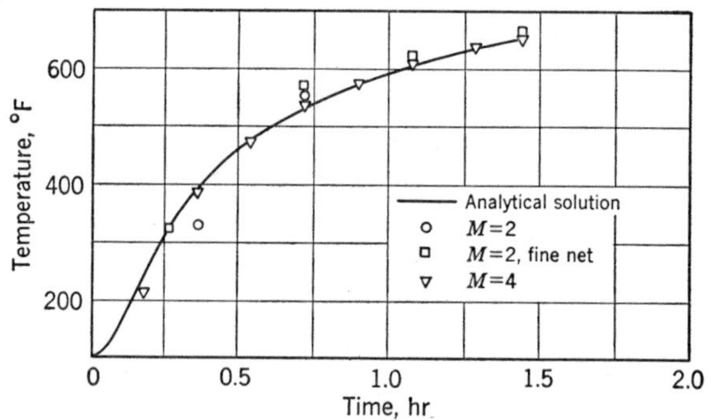

FIG. 10-3. Temperatures at plane 1 of Example 10-1.

For this simple example, an analytical solution may be obtained. The result is

$$\frac{T - T_0}{T_4 - T_0} = \frac{x}{L} + \frac{2}{\pi} \sum_{1}^{\infty} \frac{1}{n} \left(\sin \frac{n\pi x}{L} \right) \exp \left(- \frac{n^2 k \pi^2 t}{\rho c L^2} \right)$$

Temperatures at plane 1 ($x = 0.25$ ft) as a function of time have been calculated by means of the above analytical relation and are compared with the finite-difference solution for $M = 2$ in Fig. 10-3. In the figure the solid line represents the analytical solution; ⊙ points refer to the $M = 2$ solution. Points represented by other symbols refer to solutions obtained by methods to be discussed shortly. Examination of Fig. 10-3 discloses that the simple $M = 2$ solution is most in error at the start of the heating process, i.e., for small values of t. This suggests that a better technique for initiating the finite-difference calculation would result in considerable improvement. Two improved starting techniques will be examined.

The first improvement is of general application and is based on the rather obvious obervation that the use of a smaller time increment at the beginning of the process should improve the calculation. As an example, let $M = 2$ and $\Delta x = 0.125$ ft. Then $\Delta t = 0.09$ hr, one-fourth the previous value. The results shown in Table 10-2 were obtained by the use of the same techniques as before. The calculations were stopped after an elapsed time of 0.36 hr, and the final results were used to start the finite-difference calculation based on $M = 2$, $\Delta x = 0.25$ ft, $\Delta t = 0.36$ hr tabulated in Table 10-3. The calculations of Table 10-3 are plotted as ⊡ points in Fig. 10-3. The comparison shows that the initial use of small increments followed by a transition to a larger increment improves the results obtained near the starting point but slightly reduces the accuracy of the results obtained after the transition to a larger increment.

TABLE 10-2

$M = 2 \qquad \Delta x = 0.125$ ft

t, hr	Δt incre-ments	T_0, °F	$T_{1/2}$, °F	T_1, °F	$T_{3/2}$, °F	T_2, °F	$T_{5/2}$, °F	T_3, °F	$T_{7/2}$, °F	T_4, °F
0	0	(550)	100	100	100	100	100	100	100	100
0.09	1	1000	325	100	100	100	100	100	100	100
0.18	2	1000	550	212.5	100	100	100	100	100	100
0.27	3	1000	606.25	325	156.25	100	100	100	100	100
0.36	4	1000	662.5	381.25	212.5	128.13	100	100	100	100

TABLE 10-3

$M = 2 \qquad \Delta x = 0.25$ ft

Initial points were obtained from the use of the finer increments of Table 10-2.

t, hr	Δt increments	T_0, °F	T_1, °F	T_2, °F	T_3, °F	T_4, °F
0.36	1	1000	381.3	128.1	100	100
0.72	2	1000	564.1	240.7	114.1	100
1.08	3	1000	620.4	339.1	170.4	100
1.44	4	1000	669.6	395.4	219.6	100

The second method of improving the starting conditions, which is due to Fishenden and Saunders,[16] utilizes an analytical solution which is valid for small values of t. Such an analytical solution often may be obtained relatively easily by use of a Laplace-transform solution which is valid for large values of p only. The large p transform solution for the present problem is

$$\frac{T - T_4}{T_0 - T_4} = erfc\left(\frac{c\rho x^2}{4kt}\right)^{1/2} = 1 - erf\left(\frac{c\rho x^2}{4kt}\right)^{1/2}$$

where

$$erfc(Z) = 1 - \frac{2}{\sqrt{\pi}}\int_0^Z e^{-w^2}\,dw$$

Substitution into the above relation gives, for $t = 0.36$ hr, $T_{11} = 386$°F, $T_{21} = 141$°F. These temperatures may be used to start the finite-difference calculation in a manner analogous to that shown in Table 10-3.

The numerical calculations thus far have been limited to values of $M = 2$. In Table 10-4 calculations are shown for $M = 1$, $\Delta x = 0.125$ ft. It will be noted that the calculated temperatures soon oscillate between ridiculously large positive and negative values. This is an unmistakable indication of instability. When a situation such as this arises in practice, the calculation should be abandoned and replaced

TABLE 10-4
$M = 1 \quad \Delta x = 0.125$ ft

t, hr	Δt	T_0	$T_{1/2}$	T_1	$T_{3/2}$	T_2	$T_{5/2}$	T_3	$T_{7/2}$	T_4
0.0	0	(550)	100	100	100	100	100	100	100	100
0.18	1	1000	550	100	100	100	100	100	100	100
0.36	2	1000	550	550	100	100	100	100	100	100
0.54	3	1000	1000	100	550	100	100	100	100	100
0.72	4	1000	100	1450	−350	550	100	100	100	100
0.90	5	1000	2350	−1700	2350	−800	550	100	100	100
1.08	6	1000	−3050	6400	−4850	3700	−1250	550	100	100
1.26	7	1000	10450	−14300	14950	−8800	5500	−1750	550	100

by one using either a different finite difference approximation, a larger modulus, or a combination of both remedies.

Values computed for $M = 4$, $\Delta x = 0.25$ ft are given in Table 10-5 and are compared with the results of previous calculations in Fig. 10-3. These $M = 4$ computations are in remarkable agreement with the analytical curve. In general, it is found that for the *same value of the distance increment*, better accuracy is obtained by the use of a larger modulus. On the other hand, since more time steps are involved, a larger

TABLE 10-5
$M = 4 \quad \Delta x = 0.25$ ft

t, hr	Δt increments	T_0, °F	T_1, °F	T_2, °F	T_3, °F	T_4, °F
0	0	(550)	100	100	100	100
0.18	1	1000	212.5	100	100	100
0.36	2	1000	381.5	128.1	100	100
0.54	3	1000	472.8	184.4	107.0	100
0.72	4	1000	532.5	237.2	124.6	100
0.90	5	1000	575.6	282.9	146.6	100
1.08	6	1000	608.5	322.0	169.0	100
1.26	7	1000	634.8	355.4	190.0	100
1.44	8	1000	656.3	383.9	208.9	100

modulus increases the amount of computational effort. *In a very rough way,* it turns out that the same accuracy is obtained from the same computational effort irrespective of the size of the modulus, *provided that M is sufficiently large to ensure stability and convergence.*†

10-3. Mathematical Formulation of Finite Difference Equations. The reduction of a partial differential equation to a suitable finite difference

† This generalization is clearly too broad and can be used only as a rough rule of thumb. Boelter and Tribus[5] show that for a problem of the type of Example 10-1, with a fixed value of Δx, a choice of M other than 2 requires more labor for a given accuracy, unless M is made greater than 3.55 and sometimes unless $M > 8$.

approximation is most easily effected by means of Taylor's series. As an illustration of the method, consider the partial differential equation which applies to Example 10-1:

$$\frac{\partial^2 T}{\partial x^2} = \frac{c\rho}{k}\frac{\partial T}{\partial t} \tag{10-4}$$

Since $T = f(x,t)$, the $f(x,t)$ may be expanded about t for a fixed value of x:

$$T(x, t + \Delta t) = T(x,t) + \Delta t \frac{\partial T}{\partial t} + \frac{\Delta t^2}{2}\frac{\partial^2 T}{\partial t^2} + \frac{\Delta t^3}{6}\frac{\partial^3 T}{\partial t^3} + \frac{\Delta t^4}{24}\frac{\partial^4 T}{\partial t^4} + \cdots \tag{10-5}$$

As long as Δt is sufficiently small, terms of the order of Δt^2 and higher may be neglected, and a first approximation to $\partial T/\partial t$ is

$$\frac{\partial T}{\partial t} = \frac{T(x, t + \Delta t) - T(x,t)}{\Delta t} = \frac{T_{m,n+1} - T_{m,n}}{\Delta t} \tag{10-6}$$

Two series expansions are needed to obtain the first approximation to $\partial^2 T/\partial x^2$:

$$T(x + \Delta x, t) = T(x,t) + \Delta x \frac{\partial T}{\partial x} + \frac{\Delta x^2}{2}\frac{\partial^2 T}{\partial x^2} + \frac{\Delta x^3}{6}\frac{\partial^3 T}{\partial x^3} + \frac{\Delta x^4}{24}\frac{\partial^4 T}{\partial x^4} + \cdots \tag{10-7a}$$

$$T(x - \Delta x, t) = T(x,t) - \Delta x \frac{\partial T}{\partial x} + \frac{\Delta x^2}{2}\frac{\partial^2 T}{\partial x^2} - \frac{\Delta x^3}{6}\frac{\partial^3 T}{\partial x^3} + \frac{\Delta x^4}{24}\frac{\partial^4 T}{\partial x^4} - \cdots \tag{10-7b}$$

If Eqs. (10-7a,b) are added and terms of the order of Δx^4 neglected, the first approximation to $\partial^2 T/\partial x^2$ is found to be

$$\frac{\partial^2 T}{\partial x^2} = \frac{T(x + \Delta x, t) - 2T(x,t) + T(x - \Delta x, t)}{\Delta x^2} = \frac{T_{m+1,n} - 2T_{m,n} + T_{m-1,n}}{\Delta x^2} \tag{10-8}$$

When the finite difference approximations (10-6) and (10-8) are used to replace the partial derivatives in (10-4) and the definition of the modulus (10-2) is introduced, there results

$$T_{m,n+1} = \frac{T_{m-1,n} + (M - 2)T_{m,n} + T_{m+1,n}}{M} \tag{10-3}$$

which is the same as the result found earlier in Sec. 10-2 by means of elementary heat-flow considerations. The advantage of the Taylor's-series approach resides in its applicability to a wide variety of problems and in the manifold variations it permits. As examples of its application, the following additional cases are considered.

It is desired to determine a better approximation to $\partial T/\partial t$ than that given by (10-6). Expand $T(x, t + 2\,\Delta t)$ to obtain

$$T(x, t + 2\,\Delta t) = T(x,t) + 2\,\Delta t \frac{\partial T}{\partial t} + 2\,\Delta t^2 \frac{\partial^2 T}{\partial t^2} + \frac{4}{3}\Delta t^3 \frac{\partial^3 T}{\partial t^3} + \cdots \qquad (10\text{-}9)$$

Then combine (10-9), (10-8), and (10-5), neglecting terms of the order of Δt^3 and higher, to obtain

$$\frac{\partial T}{\partial t} = \frac{4T_{m,n+1} - 3T_{m,n} - T_{m,n+2}}{2\,\Delta t} \qquad (10\text{-}10)$$

Equation (10-10) is a better approximation to $\partial T/\partial t$ than (10-6). Note, however, that it involves temperatures at three time increments. Consequently, its use will require iteration procedures in which $T_{m,n+2}$ is "guessed" in the first trial calculations and subsequently replaced by the better value obtained after completion of the first trial in the second round of calculation. Iteration is awkward, but this penalty is sometimes accepted in situations which would otherwise require very small increments. Ordinarily, for the development of numerical techniques to be used with computers, a noniterative procedure is preferred.

It is desired to determine the finite difference approximation of $\partial^2 T/(\partial t)(\partial x)$. Use of Taylor's series† gives

$$T_{m+1,n+1} = T_{m,n} + \Delta x \frac{\partial T}{\partial x} + \Delta t \frac{\partial T}{\partial t} + \frac{1}{2}\left(\Delta x^2 \frac{\partial^2 T}{\partial x^2} + 2\,\Delta x\,\Delta t \frac{\partial^2 T}{\partial t\,\partial x} + \Delta t^2 \frac{\partial^2 T}{\partial t^2}\right) + \cdots \qquad (10\text{-}11)$$

If the terms $\partial T/\partial x$, $\partial^2 T/\partial x^2$, $\partial T/\partial t$, and $\partial^2 T/\partial t^2$ in Eq. (10-11) are replaced by their respective finite difference approximations [for example, $\partial T/\partial t$ by (10-6) and $\partial^2 T/\partial x^2$ by (10-8)], there results

$$\frac{\partial^2 T}{\partial t\,\partial x} = \frac{2T_{m+1,n+1} - 3T_{m+1,n} - 3T_{m,n+1} + 2T_{m,n} - T_{m-1,n} - T_{m,n-1}}{2\,\Delta t\,\Delta x} \qquad (10\text{-}12)$$

† In the general form,

$$f(x,y) = f(x_0,y_0) + \left[(x - x_0)\left(\frac{\partial f}{\partial x}\right)_{0,0} + (y - y_0)\left(\frac{\partial f}{\partial y}\right)_{0,0}\right] + \frac{1}{2!}\left[(x - x_0)^2\left(\frac{\partial^2 f}{\partial x^2}\right)_{0,0} + 2(x - x_0)(y - y_0)\left(\frac{\partial^2 f}{\partial x\,\partial y}\right)_{0,0} + (y - y_0)^2\left(\frac{\partial^2 f}{\partial y^2}\right)_{0,0}\right] + \cdots + \frac{1}{n!}\left[(x - x_0)\frac{\partial}{\partial x} + (y - y_0)\frac{\partial}{\partial y}\right]^n (f)_{0,0} + \cdots$$

Similar methods may be used to find the finite difference approximations to higher derivatives and alternative approximations to given derivatives.

10-4. Boundary Conditions for Initial-value Problems. In the heat-flow problem of Sec. 10-2, the values of the dependent variable (temperature) were specified throughout the wall at $t = 0$ and at the two surfaces of the wall for all later times. These boundary conditions are sufficient to determine the subsequent temperature history of the wall, but they are not the only type of boundary conditions that may be specified. An equally valid set of boundary conditions would result if the surface-temperature specification were replaced by the derivative at the surface of the temperature normal to the surface. As will be seen shortly, this would be accomplished if the heat flux at the surface or, alternatively, the surrounding temperature and heat-transfer coefficient were specified. In more general problems, it may be necessary to specify both the value of the dependent variable and its derivative or, in other cases, the values of higher derivatives. In the application of finite-difference methods, it is frequently necessary to replace a derivative specified at a boundary by a suitable finite difference approximation; the objective of this section is the development of the necessary methods.

Suppose that the heat flux F [in suitable units of, say, Btu/(hr)(ft²)] had been specified at the hot surface of the wall of Example 10-1 instead of the surface temperature. The interior temperatures are desired as a function of time and position. Equation (10-4), and its finite-difference equivalent (10-3) still apply. In order to use (10-3), however, a way must be found to determine the surface temperature T_0 from the specified heat flux F. If F is considered *positive when heat flows from the surroundings to the wall,* the surface condition is

$$F = -k\left(\frac{\partial T}{\partial x}\right)_s \tag{10-13†}$$

† The derivation of this relation is instructive. Consider an element of area δS at the surface. Construct an element of volume in the solid wall of area δS and of infinitesimal length ϵ and write an unsteady-state heat balance:

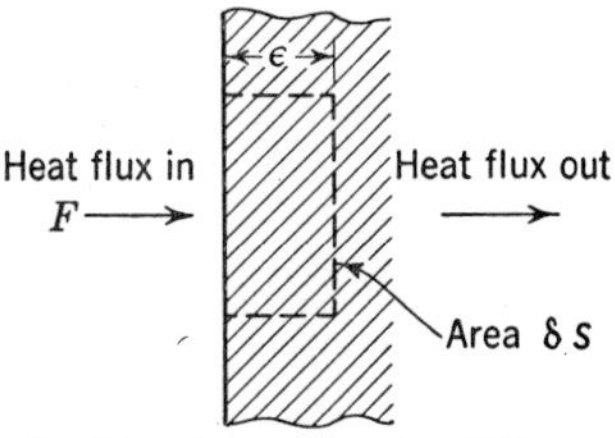

$$\text{Input} = F\,\delta S\,dt \qquad \text{Output} = -\left[k\left(\frac{\partial T}{\partial x}\right)_{x=\epsilon}\right]\delta S\,dt$$

$$\text{Accumulation} = \frac{\partial}{\partial t}\left[\rho c T\,\delta S\,\epsilon\right]dt$$

and
$$F = -k\left(\frac{\partial T}{\partial x}\right)_{x=\epsilon} + \epsilon\rho c\,\frac{\partial T}{\partial t}$$

Surface boundary condition

Now let $\epsilon \rightarrow 0$, whereupon the term $\epsilon\rho c(\partial T/\partial t)$ vanishes and $(-\partial T/\partial x)_{x=\epsilon}$ becomes $(-\partial T/\partial x)_s$. Hence, at the surface

$$F = -k\left(\frac{\partial T}{\partial x}\right)_s$$

The first-order finite difference approximation to (10-13) is

$$F_n = \frac{k(T_{0,n} - T_{1,n})}{\Delta x} \tag{10-14}$$

when F_n is the flux at time $n\ \Delta t$. Rearrangement of Eq. (10-14) gives

$$T_{0,n} = T_{1,n} + \frac{\Delta x}{k} F_n \tag{10-15}$$

which relates the surface temperature to the interior temperature and the surface-temperature gradient (through F_n) at the *same instant.* At the start of the process, use of Eq. (10-15) at $t = 0$ leads to a dilemma of the same sort as that discussed in Sec. 10-2, namely, at $t = 0$ the surface temperature is given both by the specification of the initial temperature distribution and by (10-15), and in general they are different. In the absence of other information, the use of the mean of the two values is recommended.

As an illustration of the use of a boundary condition of a type covered by (10-15), the following problem is solved.

Example 10-2. Drying of a Slab. A large flat slab of clay is to be dried. The slab is to be dried from both sides, is 4.0 cm thick, and has an initial uniform water concentration (C_i) of 0.500 g of water per cubic centimeter of stock. The movement of water within the clay occurs by diffusion; the diffusion constant is $D = 0.25$ cm²/hr. It is known that, with the proposed drying conditions, the drying will occur in the "constant-rate period" at a rate of 0.1 g/(hr)(cm²) of water as long as the surface-moisture concentration remains above 0.22 g of water per cubic centimeter of stock. It is desired to predict the duration of the constant-rate period, the amount of water evaporated, and the distribution of water within the clay at the end of the constant-rate period.

The solution of the problem is obtained by calculating the time required for the surface concentration to fall to 0.22 and by calculating the corresponding water concentrations within the clay. A water balance applied to a differential element of slab volume gives the following equations:

$$\frac{\partial^2 C}{\partial x^2} = \frac{1}{D}\frac{\partial C}{\partial t} \tag{10-16}$$

with the boundary conditions

$$F = -D\left(\frac{\partial C}{\partial x}\right)_s = -0.1 \qquad \text{at } x = 0 \tag{10-17a}$$

$$F = -D\left(\frac{\partial C}{\partial x}\right)_s = +0.1 \qquad \text{at } x = L \tag{10-17b}$$

Equations (10-16) and (10-17) may be identified with Eqs. (10-4) and (10-13). With the notation

$$M = \frac{\Delta x^2}{D\ \Delta t} \tag{10-18}$$

the finite difference approximation to (10-16) is essentially identical to (10-3):

$$C_{m,n+1} = \frac{C_{m+1,n} + (M - 2)C_{m,n} + C_{m-1,n}}{M} \tag{10-19}$$

and the finite-difference boundary condition is essentially identical to (10-15):

$$C_{0,n} = C_{1,n} + \frac{\Delta x}{D} F = C_{1,n} - \frac{0.1\ \Delta x}{D} \tag{10-20}$$

with a similar relation for the opposite face of the slab. In this problem, however, the facts that the initial concentration is symmetrically distributed about the mid-point of the slab *and* the boundary conditions at the two surfaces are identical *both* can be used to simplify the calculation. Under these special circumstances, the concentration distribution is symmetrical about the center line of the slab, and consequently the numerical calculations may be restricted to one-half the slab. Let $M = 2$ and $\Delta x = 0.25$ cm. Then $\Delta t = (0.25)^2/(2 \cdot 0.25) = 0.125$ hr and

$$C_{0,n} = C_{1,n} - 0.100 \tag{10-21}$$

$$C_{m,n+1} = \frac{C_{m+1,n} + C_{m-1,n}}{2} \tag{10-22}$$

At $t = 0$ ($n = 0$), $C_{0,0}$ is 0.500, according to the stated initial conditions, whereas Eq. (10-21) yields 0.400. Straddle these values by starting with their mean: $C_{00} = 0.450$. *At subsequent times,* however, use the value specified by Eq. (10-21). The numerical calculations are given in Table 10-6. A surface concentration of 0.22 is reached between 1.000 and 1.125 hr, interpolation indicates 1.042 hr, and the corresponding concentration distribution is shown in the row marked "interpolated." The amount of water evaporated up to this point is $0.1 \cdot 1.042 = 0.104$ g of water *per square centimeter of surface area.* Notice that in Table 10-6 the concentrations at the middle of the slab were obtained by application of (10-22), with $C_{m+1,n}$ taken equal to $C_{m-1,n}$ as a result of the symmetry. Also note that more figures have been carried in the calculations than are justified by the original data. This was done to minimize "round-off" errors. In a final tabulation of the results of the calculations only three significant figures are justified.

In other problems, the flux F at the boundary may not be specified directly. As examples of such cases, situations are now considered in which (1) it is desired to calculate the flux from interior conditions, (2) a transfer coefficient is specified, (3) a solid-solid boundary is present, and (4) an insulated boundary is present.

Occasionally, interior and surface conditions may be known or determined by the problem and the flux F at the surface desired. A typical case in point is Example 10-1, where the slab temperatures can be calculated directly without knowledge of the surface flux. If the surface flux, i.e., the heat-transfer rate, is desired, it may be calculated by means of Eq. (10-14).

The surface flux is directly related to the surface-transfer coefficient. For example, when convection controls, the heat-transfer rate is conventionally written in the form

$$h(T_{a,n} - T_{0,n}) = -k \left(\frac{\partial T}{\partial x}\right)_s = F_n \tag{10-23}$$

where $T_{a,n}$ is the ambient temperature and h is the heat-transfer coeffi-

TABLE 10-6

$M = 2 \qquad \Delta x = 0.25$ cm

Time increment	Elapsed time, hr	C_0 $x = 0$	C_1 $x = 0.25$	C_2 $x = 0.5$	C_3 $x = 0.75$	C_4 $x = 1.0$	C_5 $x = 1.25$	C_6 $x = 1.5$	C_7 $x = 1.75$	C_8 $x = 2.0$ mid-plane	C_9 $x = 2.25$
0	0	(0.450)	0.500	0.500	0.500	0.500	0.500	0.500	0.500	0.500	0.500
1	0.125	0.375	0.475	0.500	0.500	0.500	0.500	0.500	0.500	0.500	0.500
2	0.250	0.3375	0.4375	0.4875	0.500	0.500	0.500	0.500	0.500	0.500	0.500
3	0.375	0.3125	0.4125	0.4688	0.4938	0.500	0.500	0.500	0.500	0.500	0.500
4	0.500	0.2907	0.3907	0.4532	0.4844	0.4969	0.500	0.500	0.500	0.500	0.500
5	0.625	0.2720	0.3720	0.4376	0.4751	0.4922	0.4985	0.500	0.500	0.500	0.500
6	0.750	0.2548	0.3548	0.4236	0.4649	0.4868	0.4961	0.4993	0.500	0.500	0.500
7	0.875	0.2392	0.3392	0.4099	0.4552	0.4805	0.4931	0.4981	0.4997	0.500	0.4997
8	1.000	0.2246	0.3246	0.3972	0.4452	0.4942	0.4893	0.4964	0.4991	0.4997	0.4991
9	1.125	0.2109	0.3109	0.3849	0.4357	0.4673	0.4853	0.4942	0.4981	0.4991	0.4981
(Interpolated)	1.042	0.220	0.320	0.392	0.442	0.472	0.488	0.496	0.499	0.500	0.499

cient. Combination of (10-23) and (10-14) yields

$$T_{0,n} = \frac{NT_{a,n} + T_{1,n}}{N + 1} \tag{10-24}$$

where N, *the surface modulus*, is given by

$$N = \frac{h\,\Delta x}{k} \tag{10-25}$$

When the surface coefficient and the ambient conditions are known, Eq. (10-24) or a corresponding relation may be used to calculate the boundary conditions. As before, in starting the calculation, the use of an average of the value of T_0 initially specified and that given by (10-24) is recommended.

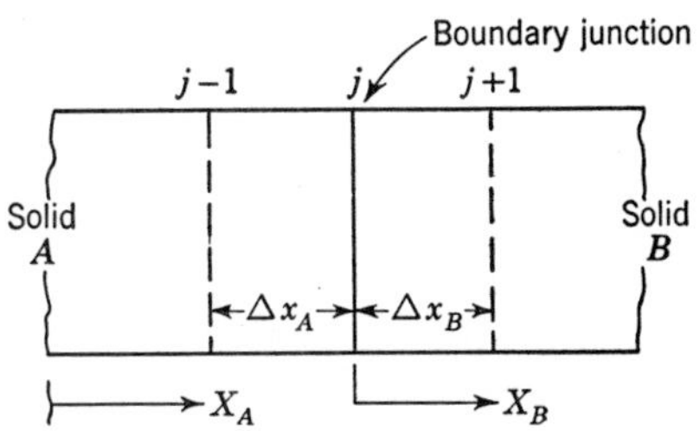

FIG. 10-4. Junction of two solids.

A solid-solid boundary in a heat-flow problem is a specific case of a general boundary condition in which the flux across the boundary is continuous. The same principles as those used in the case of heat flow may be applied to other problems. Consider a solid A which is in perfect thermal contact with solid B at the plane j of Fig. 10-4. An analysis analogous to that used to obtain Eq. (10-13) leads to

$$-F_A = -k_A\left(\frac{\partial T}{\partial x_A}\right)_j = -k_B\left(\frac{\partial T}{\partial x_B}\right)_j = F_B \tag{10-26}$$

When the partial derivatives are replaced by finite differences, there results

$$\frac{k_A}{\Delta x_A}(T_{j-1} - T_j) = \frac{k_B}{\Delta x_B}(T_j - T_{j+1}) \tag{10-27}$$

Equation (10-27) may be used to connect the solutions in the two regions at the *same instant of time*. In proceeding from solid A to solid B by means of (10-27), Δt_A *must equal* Δt_B, but the moduli M_A and M_B or the distance increments Δx_A and Δx_B may be different.

At an isolated boundary, the net flux F will be zero. In this case, Eq. (10-15) may be used with $F_n = 0$.

10-5. Convergence and Stability of Numerical Solutions of Initial-value Problems. The question of the convergence and stability of the numerical methods used to approximate the solution of a partial differential equation is of more than academic importance. Unless these criteria are fulfilled, the engineer cannot trust the results of his calculations.

It will be recalled that convergence implies that the finite difference approximation will reduce to the exact solution when the size of the increments employed is made infinitesimally small. Stability implies that errors associated with the use of increments of finite size, numerical mistakes, or round-off errors will not grow as the calculation proceeds. When a finite-difference procedure is both convergent and stable, a comparison of calculations made using two different increment intervals is generally a good indication of the reliability that may be assigned to the results.

Unfortunately, analytical tests for convergence and stability that are valid for the wide variety of problems that arise in practice are not available. However, tests have been developed which apply to the finite-difference methods used to solve *linear* partial differential equations with *constant* coefficients. These criteria are summarized below:

1. The existence of a solution of the corresponding finite difference approximation is, in general, assured if the exact *initial*-value problem possesses a solution. In most engineering situations, the existence of a physical problem generally implies that a solution exists, although it is not always easy to determine the relevant boundary conditions.

2. Available evidence† indicates that a finite-difference method which satisfies the criteria required for stability also will prove to be a convergent solution. As yet unexamined cases may exist which do not conform to the above statement. Furthermore, convergence *does not* guarantee stability.

3*a*. *Sufficient* conditions for stability of the finite difference equation

$$T_{m,n+1} = c_1 T_{m+1,n} + c_2 T_{m,n} + c_3 T_{m-1,n} \qquad (10\text{-}28a)‡$$

$$1 \leqq m \leqq L - 1 \qquad n \geqq 1§$$

with the boundary conditions

$$\left.\begin{aligned} T_{0,n} &= h_1 T_{1,n} \\ T_{L,n} &= h_2 T_{L-1,n} \end{aligned}\right\} n \geqq 1 \qquad (10\text{-}28b)$$

$$T_{m,0} = f(m) \qquad 0 \leqq m \leqq L$$

require

$$c_1 \geqq 0 \qquad c_2 \geqq 0 \qquad c_3 \geqq 0$$

$$c_1 + c_2 + c_3 \leqq 1 \qquad (10\text{-}29)$$

$$0 \leqq h_1 \leqq 1 \qquad 0 \leqq h_2 \leqq 1$$

b. The five-point difference equation

$$T_{m+1,n} - 2aT_{m,n} + b^2 T_{m-1,n} = cT_{m,n+1} + dT_{m,n-1} \qquad (10\text{-}30a)$$

† See references 9, 17, 21, and 27.

‡ In the following stability criteria T represents any applicable dependent variable and is not restricted to temperature.

§ L denotes the last distance increment.

where a, b, and c are *real constants* with the boundary conditions

$$\begin{aligned} T_{0,n} &= h_1 T_{1,n} + \mu(n) \qquad \text{at } m = 0 \\ T_{L,n} &= h_2 T_{L-1,n} + v(n) \qquad \text{at } m = L \end{aligned} \tag{10-30b†}$$

which satisfy

$$\begin{aligned} 0 &\leqq h_1 \leqq \frac{1}{|b|} \\ 0 &\leqq h_2 \leqq |b| \end{aligned} \tag{10-31‡}$$

and additional arbitrary initial conditions at $n = 0$ and $n = 1$ (or merely at $n = 0$ if $d = 0$), is stable *if, and only if,* the following conditions are fulfilled:

i. The roots of the equation

$$R^2 + \frac{2}{c}(a - b \cos \alpha)R + d = 0 \tag{10-32}$$

cannot exceed unity in absolute value for any real value of α.

ii. Neither $R = +1$ nor $R = -1$ may be a *repeated* root of (10-32).

If condition i is satisfied but ii violated, the difference equation (10-30) may be linearly unstable; i.e., errors may grow at a rate which is directly proportional to the number of elapsed time increments.

c. The nine-point difference equation

$$\begin{aligned} (a_3 T_{m+1,n+1} + b_3 T_{m,n+1} + c_3 T_{m-1,n+1}) + (a_2 T_{m+1,n} + b_2 T_{m,n} + c_2 T_{m-1,n}) \\ + (a_1 T_{m+1,n-1} + b_1 T_{m,n-1} + c_1 T_{m-1,n-1}) = 0 \end{aligned} \tag{10-33a}$$

where the coefficients a, b, c are constants and meet the requirement that corresponding a's and c's are in a constant ratio

$$c_1 = p^2 a_1 \qquad c_2 = p^2 a_2 \qquad c_3 = p^2 a_3 \tag{10-33b}$$

where *p is real,* with the boundary conditions

$$\begin{aligned} T_{0,n} &= h_1 T_{1,n} + \mu(n) \qquad \text{at } m = 0 \\ T_{L,n} &= h_2 T_{L-1,n} + v(n) \qquad \text{at } m = L \end{aligned} \tag{10-33c}$$

† h_1, h_2, $\mu(n)$, and $v(n)$ may be constants or functions of time. They may not, however, be direct functions of interior conditions.

‡ If $h_1 = 0$, the less stringent condition

$$0 \leqq h_2 \leqq \frac{L|b|}{L-1}$$

applies. If $h_2 = 0$,

$$0 \leqq h_1 \leqq \frac{L}{(L-1)|b|}$$

applies.

which satisfy

$$0 \leqq h_1 \leqq \frac{1}{|p|} \qquad 0 \leqq h_2 \leqq |p| \tag{10-34}†$$

and additional arbitrary initial conditions at two consecutive values of n (or merely at one value of n if $a_1 = b_1 = c_1 = 0$), is stable *if, and only if,* the following conditions are fulfilled:

i. The roots of the equation

$$(b_3 + 2pa_3 \cos \alpha)R^2 + (b_2 + 2pa_2 \cos \alpha)R + (b_1 + 2pa_1 \cos \alpha) = 0 \tag{10-35}$$

cannot exceed unity in absolute value for any real value of α.

ii. Neither $R = +1$ nor $R = -1$ may be a *repeated* root of (10-35).

If condition i is satisfied but ii is violated, the difference equation (10-33) may be linearly unstable.

The conditions to be applied to the "root" equations (10-32) and (10-35) are frequently simplified if the condition that neither root of the equation

$$R^2 + AR + B = 0 \tag{10-36}$$

can exceed unity in absolute value is expressed as

$$|A| \leqq B + 1 \leqq 2 \tag{10-37}$$

(provided that A and B are real).

The convergence criteria are demonstrated by comparison of the exact solutions of the governing partial differential equation and of the finite difference "equivalent." The stability criteria are obtained as a result of an examination of the form of the exact solution of the finite difference equation when it is regarded as an error-propagating equation. The derivations of these criteria are given in references 9, 17, 21, and 27.

The stated stability criteria apply only to difference equations with *constant coefficients.* However, if the coefficients obtained in an actual problem depend upon position or time (m and/or n), it may prove possible to break up the equation into a series of equations, each equation being applied to a limited number of space and/or time intervals. In the system of equations, the truly variable coefficients of any given equation may be replaced by suitable average values. The stability

† If $h_1 = 0$, the less stringent condition

$$0 \leqq h_2 \leqq \frac{L|p|}{L - 1}$$

applies. If $h_2 = 0$,

$$0 \leqq h_1 \leqq \frac{L}{(L - 1)|p|}$$

applies.

criteria may then be applied to each equation in turn. If stability may be obtained in each case, the system of equations may be used to solve the problem. Alternatively, the most stringent of the stability requirements so determined may be imposed on the original variable coefficient equation and the calculation carried out in the manner originally intended. This "average-coefficient" procedure is a reasonable one and usually is reliable, *but it cannot be guaranteed to be valid.*

Unless the boundary conditions of the problem to be solved are covered by the boundary conditions associated with the stability criteria, the tests do not apply. However, certain heuristic procedures are frequently helpful. If the finite difference equation does not meet the stability criteria when no consideration is given to the actual boundary conditions, it should not be used. If the boundary conditions which apply to the actual problem are not covered by the listed stability criteria but all other requirements are satisfied, the proposed procedure *may* or *may not* be stable. If a calculation is to be carried out under such circumstances, it is recommended that more stringent conditions be applied than would be indicated by the listed criteria. It must be recognized that this procedure does not guarantee stability.

As an example of the use of the stability criteria, the finite difference equation

$$T_{m,n+1} = \frac{T_{m+1,n} + (M-2)T_{m,n} + T_{m-1,n}}{M} \tag{10-3}$$

with the boundary conditions

$$\begin{aligned} T_{0,n} &= \frac{N_0 T_{a,n} + T_{1,n}}{N_0 + 1} \\ T_{L,n} &= \frac{T_L T_{b,n} + T_{L-1,n}}{N_L + 1} \end{aligned} \tag{10-24}$$

will be examined. If Eq. (10-3) is compared with the five-point stability-criteria difference equation (10-30a), it is found that $a = (2 - M)/2$, $b^2 = 1$, $c = M$, $d = 0$. The "root" equation (10-32) is then

$$R^2 + \frac{2}{M}\left(\frac{2-M}{2} - \cos\alpha\right) R = 0$$

and

$$R = 0 \qquad R = \frac{2}{M}\left(\cos\alpha - \frac{2-M}{2}\right)$$

The $R = 0$ root satisfies the requirement $-1 \leqq R \leqq 1$. When applied to the second root, this condition demands

$$-1 \leqq \frac{2}{M}\left(\cos\alpha + \frac{M-2}{2}\right) \leqq 1$$

or

$$-1 \leqq 1 - \frac{2}{M}(1 - \cos\alpha) \leqq 1$$

For any positive value of M, the right-hand inequality is satisfied. The left-hand inequality gives

$$M \geqq 1 - \cos \alpha$$

If the above condition is to hold for *all* real values of α, then $M \geqq 2$ is required. The value $M = 2$ is permitted even though for $\alpha = \pi$ it would lead to $R = -1$, since this would not be a *repeated* root. The difference equation (10-3) can then be made to satisfy the stability requirements. However, the particular boundary conditions (10-24) must be considered. A comparison of (10-24) with (10-30*b*) shows that

$$h_1 = \frac{1}{N_0 + 1} \qquad \mu(n) = \frac{N_0 T_{a,n}}{N_0 + 1}$$

$$h_2 = \frac{1}{N_L + 1} \qquad v(n) = \frac{N_L T_{b,n}}{N_L + 1}$$

The conditions (10-31) then require

$$0 \leqq \frac{1}{N_0 + 1} \leqq 1$$

$$0 \leqq \frac{1}{N_L + 1} \leqq 1$$

These conditions are satisfied for all positive values of N.

In a practical problem, the restrictions on M imposed by the stability requirements of the finite difference approximations used in previous illustrations may prove inconvenient. In such cases some workers prefer to adopt a different finite difference formulation which involves less stringent stability restrictions. One such formulation, proposed by von Neumann[27] and used by Hartree,[15,20] Crank and Nicolson,[10] and others, may be obtained as follows:

In the derivation of the finite difference approximation to the heat-flow equation given in Sec. 10-2, it will be recalled that the rate at which heat crossed the boundary of the volume element was taken to be the rate at the time $n\,\Delta t$. It was then assumed that this rate could be used during the time interval between $n\,\Delta t$ and $(n + 1)\,\Delta t$. Clearly, this is an approximation, since the actual rate of heat flow changes during the time interval. A better value would be the *average* of the heat-flow rate at the two time extremes, $n\,\Delta t$ and $(n + 1)\,\Delta t$. In Fig. 10-1, the heat-flow rate across the plane ac at time $n\,\Delta t$ is

$$\frac{kA(T_{m-1,n} - T_{m,n})}{\Delta x}$$

and the rate at time $(n + 1)\,\Delta t$ is

$$\frac{kA(T_{m-1,n+1} - T_{m,n+1})}{\Delta x}$$

The average rate during the time interval $n\,\Delta t \rightarrow (n+1)\,\Delta t$ is

$$\frac{kA}{2\,\Delta x}\,(T_{m-1,n} - T_{m,n} + T_{m-1,n+1} - T_{m,n+1})$$

When the same type of average is applied to the surface bd, the average rate is found to be

$$\frac{kA}{2\,\Delta x}\,(T_{m,n} - T_{m+1,n} + T_{m,n+1} - T_{m+1,n+1})$$

The improved finite difference approximation is then

$$\frac{2\rho c\,\Delta x^2}{k\,\Delta t}\,(T_{m,n+1} - T_{m,n}) = (T_{m+1,n+1} - 2T_{m,n+1} + T_{m-1,n+1}) + (T_{m+1,n} - 2T_{m,n} + T_{m-1,n}) \qquad (10\text{-}38a)$$

The same result may be obtained by a formal mathematical procedure. The Taylor-series expansion used in Sec. 10-3 to obtain $(\partial^2 T/\partial x^2)_{m,n}$ gave a value at the time $n\,\Delta t$. A similar expansion about the temperature $T_{m,n+1}$ gives the value of $(\partial^2 T/\partial x^2)_{m,n+1}$ at the time $(n+1)\,\Delta t$:

$$\left(\frac{\partial^2 T}{\partial x^2}\right)_{m,n+1} \cong \frac{T_{m+1,n+1} - 2T_{m,n+1} + T_{m-1,n+1}}{\Delta x^2} \qquad (10\text{-}39)$$

Substitution of the finite-difference equivalents in the expression

$$\frac{\partial^2 T}{\partial x^2} \cong \frac{1}{2}\left[\left(\frac{\partial^2 T}{\partial x^2}\right)_{m,n} + \left(\frac{\partial^2 T}{\partial x^2}\right)_{m,n+1}\right]$$

and the approximation for $\partial T/\partial t$ given by (10-6) in

$$\frac{\partial^2 T}{\partial x^2} = \frac{c\rho}{k}\frac{\partial T}{\partial t} \qquad (10\text{-}4)$$

gives (10-38). In order to determine the stability criteria to be applied to (10-38), it is rearranged in a form which can be compared directly with the nine-point difference equation (10-33a):

$$[T_{m+1,n+1} - 2(M+1)T_{m,n+1} + T_{m-1,n+1}] + [T_{m+1,n} + 2(M-1)T_{m,n} + T_{m-1,n}] = 0 \qquad (10\text{-}38b)$$

Then

$$a_1 = b_1 = c_1 = 0 \qquad a_2 = c_2 = a_3 = c_3 = 1$$

$$b_2 = 2(M-1) \qquad b_3 = -2(M+1)$$

The condition (10-33b) is satisfied by $p = 1$. The "root" equation (10-35) is

$$[2\cos\alpha - 2(M+1)]R^2 + [2\cos\alpha + 2(M-1)]R = 0$$

and

$$R = 0$$

$$R = -\frac{\cos\alpha + (M-1)}{\cos\alpha - (M+1)} = -\left[1 + \frac{2M}{\cos\alpha - (M+1)}\right]$$

The condition $|R| \leqq 1$ is satisfied for *all positive* values of M, and consequently the finite difference approximation (10-38) is strongly stable. It is for this reason that some workers prefer an equation of the form of (10-38), despite the fact that its solution ordinarily involves iteration.

The analytical convergence and stability criteria given are limited to linear equations with constant coefficients and particular boundary conditions. They do not cover the wide variety of situations encountered by the engineer. Some heuristic methods which increase the scope of the criteria already have been suggested. In other new situations the following points may prove helpful:

1. Employ criteria which are more stringent than those demanded by the linear difference equation which is most similar to the case at hand. In addition, apply physical principles to estimate reasonable values of the moduli.

2. Carefully examine the solution for evidence of oscillation or unreasonable physical results. If such anomalies occur, change the modulus, the increments, or the form of the finite difference approximation. Note that oscillations may not be apparent until late in the calculations.

3. Repeat part of the calculation, using more increments or larger modulus or a better finite difference approximation, such as an iteration method. Compare the results of the independent methods.

10-6. Solution by Iteration. In Sec. 10-5 it was shown that in many cases the most stable form of the finite difference approximation to a partial differential equation involved an iterative method of solution. The procedure to be followed in such a case is illustrated in this section by means of an example.

Example 10-3. The temperature distribution in the brick wall of Example 10-1 is to be obtained by the finite difference approximation (10-38) derived in Sec. 10-5.

Solution. Let $M = (\rho c\, \Delta x^2)/(k\, \Delta t) = 2$ and $\Delta x = 0.25$ ft. Then

$$\Delta t = \frac{30 \cdot 0.23(0.25)^2}{2 \cdot 0.6} = 0.36 \text{ hr}$$

Equation (10-38) then becomes

$$T_{m,n+1} = \tfrac{1}{6}[(T_{m+1,n} + 2T_{m,n} + T_{m-1,n}) + (T_{m+1,n+1} + T_{m-1,n+1})] \qquad (10\text{-}40)$$

This reduces to the series of linear equations

$$\begin{aligned} T_{1,n+1} &= \tfrac{1}{6}[(T_{0,n} + 2T_{1,n} + T_{2,n}) + (T_{0,n+1} + T_{2,n+1})] \\ T_{2,n+1} &= \tfrac{1}{6}[(T_{1,n} + 2T_{2,n} + T_{3,n}) + (T_{1,n+1} + T_{3,n+1})] \\ T_{3,n+1} &= \tfrac{1}{6}[(T_{2,n} + T_{3,n} + T_{4,n}) + (T_{2,n+1} + T_{4,n+1})] \end{aligned} \qquad (10\text{-}41)$$

In the stated problem, the boundary conditions are $T_{m,0} = 100°\text{F}$, $T_{0,n} = 1000°\text{F}$, $T_{4,n} = 100°\text{F}$. Study of the three linear algebraic equations comprising (10-41) will show that they may be solved to give $T_{1,n+1}$, $T_{2,n+1}$, $T_{3,n+1}$, *explicitly* in terms of the temperatures at the boundaries at time $n\,\Delta t$. However, when many distance increments are involved or in more complicated problems, the algebraical solution becomes excessively tedious or impossible, and the following iterative method is used (see also Table 10-7).

1. The temperatures T_2, T_3 at time $n = 1\ \Delta t$ are *guessed*. In the present example, $T_{2,1}$ is guessed to be 220°F, $T_{3,1}$ is guessed to be 120°F.

2. The temperature $T_{1,1}$ is then calculated; here $T_{1,1} = \frac{1}{6}(1000 + 200 + 100 + 1000 + 220) = 420$°F.

3. The temperature $T_{2,1}$ is then calculated, using the value $T_{1,1}$ obtained from 2 and the guessed value of $T_{3,1}$: $T_{2,1} = \frac{1}{6}(100 + 200 + 100 + 420 + 120) = 156$°F.

4. The temperature $T_{3,1}$ is then calculated, using the value $T_{2,1}$ obtained from 3: $T_{3,1} = \frac{1}{6}(100 + 200 + 100 + 156 + 100) = 109.5$°F.

5. The above process is repeated until convergence is obtained.

The calculations for the first time step are shown in Table 10-7. Note that the surface temperature at zero time was taken to be the full 1000°F. *The iterative method, in contrast to the "step-ahead" technique, works best when the full initial conditions are applied immediately.* The temperature at $x = 0.25$, $t = 0.36$ of 409°F given by the iteration method with $M = 2$ may be compared with the values shown in Fig. 10-3. The analytical solution gives 385°F, the corresponding step-ahead solution with $M = 2$ and $\Delta x = 0.25$ ft gives 325°F; with $M = 4$ and $\Delta x = 0.25$ ft, the step-ahead method gives 381.5°F. In this case, better results for the same effort ($M = 4$) are obtained with the noniterative method. The advantage of the iterative procedure is the additional stability safety factor it provides in complex situations.

TABLE 10-7. ITERATION METHOD
$M = 2 \qquad \Delta x = 0.25$ ft

t, hr	Δt increment	T_0, °F	T_1, °F	T_2, °F	T_3, °F	T_4, °F
0	0	1000	100	100	100	100
0.36	1	1000	$(420)_1$	$(220)_1$	$(120)_1$	100
0.36	1	1000	$(409)_2$	$(156)_2$	$(109.5)_2$	100
0.36	1	1000	$(409)_3$	$(153)_3$	$(109)_3$	100

10-7. Equations with Variable Coefficients.

The finite-difference method may be used to obtain numerical solutions of more complicated equations than those previously considered. For example, an equation of the form

$$\frac{\partial T}{\partial t} = A(T,x,t)\,\frac{\partial^2 T}{\partial x^2} + B(T,x,t)\,\frac{\partial T}{\partial x} + C(T,x,t) \tag{10-42}$$

when A, B, and C denote functions of T, x, and t, may be reduced to the finite difference equation

$$\frac{T_{m,n+1} - T_{m,n}}{\Delta t} = A(T_{m,n}, m\,\Delta x, n\,\Delta t)\,\frac{T_{m+1,n} - 2T_{m,n} + T_{m-1,n}}{\Delta x^2} + B(T_{m,n}, m\,\Delta x, n\,\Delta t)\,\frac{T_{m+1,n} - T_{m,n}}{\Delta x} + C(T_{m,n}, m\,\Delta x, n\,\Delta t) \tag{10-43}$$

by the methods of Sec. 10-3. The calculation of $T_{m,n+1}$ may be carried out in a straightforward manner by use of (10-43) and the stipulated boundary conditions. Alternatively, better approximations may be used, giving rise to equations that are solved by iterative procedures.

Certain specialized forms of Eq. (10-42) are of frequent occurrence. For example, the equation for one-dimensional radial heat flow in a cylindrical rod of constant thermal properties is

$$\frac{\partial T}{\partial t} = \frac{k}{c\rho}\left(\frac{\partial^2 T}{\partial r^2} + \frac{1}{r}\frac{\partial T}{\partial r}\right) \tag{10-44}$$

This is conveniently handled as follows: Consider a cylinder of radius R. Divide the radius into a number of equally spaced distance increments each of length Δr. Let $r = m\,\Delta r$. The finite difference equivalent of (10-44) is obtained as

$$T_{m,n+1} = \frac{(1 + 1/m)T_{m+1,n} + (M - 2 - 1/m)T_{m,n} + T_{m-1,n}}{M} \tag{10-45}$$

where

$$M = \frac{\rho c\,\Delta r^2}{k\,\Delta t} \tag{10-46}$$

Equation (10-45) becomes indeterminate at the center of the cylinder ($r = m = 0$). However, at $r = 0$, $\partial T/\partial r = 0$ because of symmetry, and $\lim_{r\to 0} (1/r)(\partial T/\partial r) = (\partial^2 T/\partial r^2)_{r=0}$. Then, at $r = 0$,

$$\left(\frac{\partial^2 T}{\partial r^2} + \frac{1}{r}\frac{\partial T}{\partial r}\right)_{r=0} = 2\,\frac{\partial^2 T}{\partial r^2} = 4\,\frac{T_{1,n} - T_{0,n}}{\Delta r^2} \tag{10-47}$$

when the symmetry is invoked. The finite difference equation to be used at $r = 0$ is then

$$T_{0,n+1} = \frac{4T_{1,n} + (M - 4)T_{0,n}}{M} \tag{10-48}$$

Spherical symmetry may be handled in the same manner. For example, the basic heat-flow equation is

$$\frac{\partial T}{\partial t} = \frac{k}{\rho c}\left(\frac{\partial^2 T}{\partial r^2} + \frac{2}{r}\frac{\partial T}{\partial r}\right) \tag{10-49}$$

With equal distance increments the finite difference approximations become

$$T_{m,n+1} = \frac{(1 + 2/m)T_{m+1,n} + [M - 2(1 + 1/m)]T_{m,n} + T_{m-1,n}}{M} \tag{10-50}$$

and

$$T_{0,n+1} = \frac{6T_{1,n} + (M - 6)T_{0,n}}{M} \tag{10-51}$$

10-8. Graphical Methods. In Sec. 10-2 a graphical method for the solution of the simple one-dimensional heat-conduction equation was discussed. It is possible to apply graphical techniques to more complicated situations; the method will be developed in this section. Although graphical procedures often are very convenient, they ordinarily demand a modulus of 2 and may prove to be unstable.

The general graphical method is due to Schmidt.[35] It enables the finite-difference solution of an equation of the general form

$$\frac{\partial T}{\partial t} = a_0 P(x) \left[\frac{\partial^2 T}{\partial x^2} + Q(x) \frac{\partial T}{\partial x} \right] \tag{10-52}$$

to be evaluated graphically. In Eq. (10-52), $P(x)$ and $Q(x)$ are functions of x or constants. This form includes such cases as radial or spherical heat flow in one dimension, heat and mass transfer to flowing fluids, etc. In the following, the method of setting up the graphical procedure will be outlined. Then the conditions under which the method is valid will be demonstrated. Finally, the problem of heat transfer to a flowing fluid will be set up and solved as an illustration of the use of the method.

The procedure is as follows:

1. The x interval is divided into a number of increments in accordance with the relation

$$x_{m+1} - x_m = \epsilon \sqrt{P(x_m)} \tag{10-53}$$

ϵ is a small *constant* number. $P(x)$ is the function appearing in Eq. (10-52). In general, the several increments in x will not be equal to one another.

2. The time increment is determined by the requirement

$$M = \frac{\epsilon^2}{a_0 \, \Delta t} = 2 \tag{10-54}$$

3. A new distance coordinate is introduced by means of the transformation

$$\xi = \int [\exp \int - Q(x) \, dx] \, dx = f(x) \tag{10-55}$$

As a result of the double integration required to determine $f(x)$, two arbitrary constants appear. The calculator may assign any desired values to these constants.

4. Increments in ξ corresponding to the increments in x described under 1 are determined according to the relation

$$\xi_{m+1} - \xi_m = f(x_{m+1}) - f(x_m) \tag{10-56}$$

This may be accomplished conveniently by plotting ξ as a function of x and determining the increments of ξ graphically, as shown in Fig. 10-5.

5. Starting values of the dependent variable (T) are plotted vs. ξ, as shown in Fig. 10-5. The value of the dependent variable $T_{m,n+1}$ at the end of the next time increment is then found as the intersection of a straight line connecting the values of $T_{m-1,n}$ and $T_{m+1,n}$ with the vertical line at ξ_m. In Fig. 10-5, point A represents $T_{m-1,n}$, point B represents $T_{m,n}$, and point C represents $T_{m+1,n}$. Then point D, the intersection of the straight line connecting points A and C with the vertical line through ξ_m, represents $T_{m,n+1}$. The remainder of the values of the dependent variable at the time $(n + 1)\,\Delta t$ are then found in the same manner. The cycle is continued in the same way. Thus, despite the complication of the original equation (10-52), the graphical method finally reduces to the same procedure as that used in the graphical evaluation of the simple heat-conduction problem illustrated in Sec. 10-2.

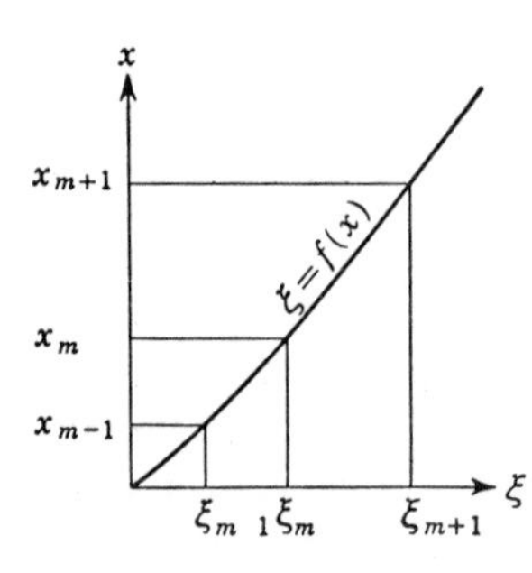

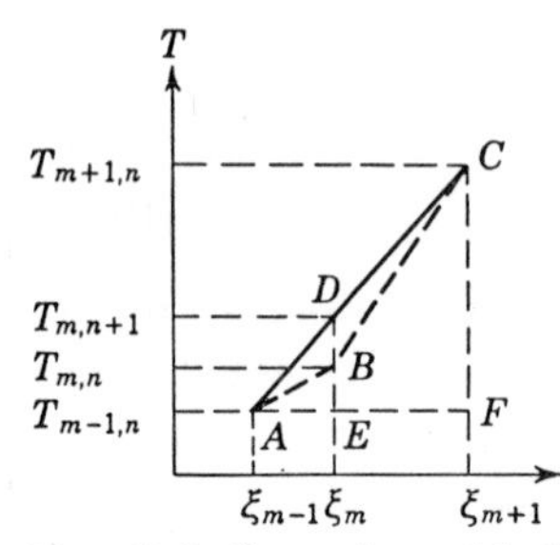

FIG. 10-5. General graphical procedure.

The validity of the method is established in the following way: With reference to Fig. 10-5, the following relations are consequences of the geometrical construction:

$$\frac{ED}{FC} = \frac{AE}{AF}$$

but

$$\begin{aligned} ED &= BE + BD \\ BE &= T_{m,n} - T_{m-1,n} \\ FC &= T_{m+1,n} - T_{m-1,n} \\ AE &= \xi_m - \xi_{m-1} \\ AF &= \xi_{m+1} - \xi_{m-1} \\ &= (\xi_{m+1} - \xi_m) + (\xi_m - \xi_{m-1}) \end{aligned}$$

Consequently, the length BD is

$$BD = \frac{(\xi_m - \xi_{m-1})(T_{m+1,n} - T_{m,n}) - (\xi_{m+1} - \xi_m)(T_{m,n} - T_{m-1,n})}{(\xi_{m+1} - \xi_m) + (\xi_m - \xi_{m-1})} \tag{10-57}$$

It is now necessary to show that it is possible to interpret BD as

$$BD = T_{m,n+1} - T_{m,n}$$

by use of suitable increment spacings and a distorted coordinate system.

First, consider the finite difference approximation to (10-52) when *unequal* increments in Δx are used. Define the increments in x by the relation

$$x_{m+1} - x_m = \epsilon\phi(x_m) \tag{10-58}$$

where ϵ is a *small* constant number and ϕ is a distorting function as yet undetermined. Then

$$x_m - x_{m-1} = \epsilon\phi(x_{m-1}) \tag{10-59}$$

A Taylor's-series expansion about $\phi(x_m)$ gives

$$\phi(x_{m-1}) \cong \frac{\phi(x_m)}{1 + \epsilon\phi'(x_m)} \cong \phi(x_m)[1 - \epsilon\phi'(x_m)] \tag{10-60}$$

when ϵ is small and terms of order ϵ^2 and higher are neglected. The prime (′) symbol denotes differentiation. Equation (10-59) then may be replaced by

$$x_m - x_{m-1} = [1 - \epsilon\phi'(x_m)]\epsilon\phi(x_m) \tag{10-61}$$

Taylor's-series expansions of $T_{m-1,n}$ and $T_{m+1,n}$ about $T_{m,n}$, together with (10-58) and (10-61), give

$$\left(\frac{\partial^2 T}{\partial x^2}\right)_{m,n} \cong \frac{2}{\epsilon^2[\phi(x_m)]^2[2 - \epsilon\phi'(x_m)]}\left[(T_{m+1,n} - T_{m,n}) - \frac{T_{m,n} - T_{m-1,n}}{1 - \epsilon\phi'(x_m)}\right] \tag{10-62}$$

$$\frac{\partial T}{\partial x} \cong \frac{1}{\epsilon\phi(x_m)[2 - \epsilon\phi'(x_m)]}\left\{(T_{m+1,n} - T_{m,n}) + \frac{T_{m,n} - T_{m-1,n}}{[1 - \epsilon\phi'(x_m)]^2}\right\} \tag{10-63}$$

Use of (10-62) and (10-63) gives as the finite difference approximation to (10-52)

$$T_{m,n+1} - T_{m,n} = \left\{\frac{a_0\,\Delta t}{\epsilon^2}\frac{2P(x_m)}{[\phi(x_m)]^2[2 - \epsilon\phi'(x_m)]}\right\} \left((T_{m+1,n} - T_{m,n}) - \left[\frac{T_{m,n} - T_{m-1,n}}{1 - \epsilon\phi'(x_m)}\right] + \frac{Q(x_m)\epsilon\phi(x_m)}{2}\left\{(T_{m+1,n} - T_{m,n}) + \frac{T_{m,n} - T_{m-1,n}}{[1 - \epsilon\phi'(x_m)]^2}\right\}\right) \tag{10-64}$$

Now consider the expression for BD, (10-57). The new coordinate is defined in terms of the distorting function $f(x)$ as

$$\xi = f(x) \tag{10-65}$$

Then, strictly,

$$\xi_{m+1} - \xi_m = f(x_{m+1}) - f(x_m) \tag{10-66}$$

As an approximation, Taylor's-series expansion is used to give

$$\xi_{m+1} - \xi_m \cong \epsilon\phi(x_m)f'(x_m) + \frac{\epsilon^2[\phi(x_m)]^2 f''(x_m)}{2} \tag{10-67}$$

and

$$\xi_m - \xi_{m-1} \cong \epsilon\phi(x_m)f'(x_m)[1 - \epsilon\phi'(x_m)] - \frac{\epsilon^2[\phi(x_m)]^2 f''(x_m)[1 - \epsilon\phi'(x_m)]^2}{2} \tag{10-68}$$

Use of (10-67) and (10-68) in the expression (10-57) gives

$$BD \cong \frac{1 - \epsilon\phi'(x_m)}{[2 - \epsilon\phi'(x_m)] + \dfrac{\epsilon\phi(x_m)f''(x_m)}{f'(x_m)}\left\{\epsilon\phi'(x_m) - \dfrac{\epsilon^2}{2}[\phi'(x_m)]^2\right\}}$$
$$\left((T_{m+1,n} - T_{m,n}) - \frac{T_{m,n} - T_{m-1,n}}{1 - \epsilon\varphi'(x_m)}\right.$$
$$- \frac{f''(x_m)}{f'(x_m)}\{\epsilon\phi(x_m)[1 - \epsilon\phi'(x_m)]\}$$
$$\left.\left\{(T_{m+1} - T_{m,n}) + \frac{T_{m,n} - T_{m-1,n}}{[1 - \epsilon\phi'(x_m)]^2}\right\}\right) \tag{10-69}$$

Comparison of (10-64), the finite difference approximation to (10-52), with (10-69) shows that BD may be taken equal to $T_{m,n+1} - T_{m,n}$ if

$$\frac{\epsilon^2}{a_0\,\Delta t} = 2 \tag{10-54}$$

$$\epsilon\phi(x_m) = \epsilon\sqrt{P(x_m)} = x_{m+1} - x_m \tag{10-53}$$

$$\frac{f''(x)}{f'(x)} = -Q(x) \tag{10-70}$$

and certain terms, small if ϵ is small, are neglected. Direct integration of (10-70) gives

$$\xi = \int[\exp\int - Q(x)\,dx]\,dx = f(x) \tag{10-55}$$

The fact that only the derivatives of $f(x)$ are involved in (10-70) permits the integration constants arising in (10-55) to be specified arbitrarily.

It is to be emphasized that the graphical method will fail if the finite difference approximations used fail to prove stable for $M = 2$.

Example 10-4. The use of the graphical method will be illustrated by application to a problem of heat transfer to a fluid in laminar motion. A fluid is flowing at a steady rate through a round pipe of radius R. The parabolic laminar velocity profile has been established. The fluid, originally at a uniform temperature T_i, suddenly enters a heated section where the pipe wall is at the new uniform temperature T_w. It is desired to predict the temperature history of the fluid as it passes through the heated section. It will be assumed that the established parabolic velocity profile remains unchanged, and conduction along the x direction may be neglected.

The basic differential equation is obtained by setting up an energy balance on a stationary annular volume in the fluid of width dr and length dx.

$$\text{Input} = 2\pi r\,dr\,\rho VcT - k\,2\pi r\,dx\,\frac{\partial T}{\partial r}$$

$$\text{Output} = 2\pi r\,dr\,\rho Vc\left(T + \frac{\partial T}{\partial x}dx\right) - k\,2\pi\,dx\left[r\frac{\partial T}{\partial r} + \frac{\partial}{\partial r}\left(r\frac{\partial T}{\partial r}\right)dr\right]$$

$$\text{Accumulation} = 0 \qquad \text{steady state}$$

Consequently,

$$\frac{\partial T}{\partial x} = \frac{k}{V\rho c}\left(\frac{\partial^2 T}{\partial r^2} + \frac{1}{r}\frac{\partial T}{\partial r}\right) \tag{10-71}$$

But
$$V = 2\bar{V}\left[1 - \left(\frac{r}{R}\right)^2\right] \tag{10-72}$$

where V is the velocity at radius r and $\bar{V}$ is the mean velocity over the cross section πR^2.

For convenience, let

$$y = \frac{r}{R} \tag{10-73}$$

$$\text{Re} = \frac{2R\rho\bar{V}}{\mu} \tag{10-74}$$

$$\text{Pr} = \frac{c\mu}{k} \tag{10-75}$$

$$a_0 = \frac{k}{2R\rho\bar{V}c} = \frac{1}{\text{Re Pr}} \tag{10-76}$$

$$t = \frac{x}{R} \tag{10-77}$$

$$\theta = \frac{T - T_i}{T_w - T_i} \tag{10-78}$$

Then (10-71) becomes

$$\frac{\partial\theta}{\partial t} = \frac{a_0}{1 - y^2}\left(\frac{\partial^2\theta}{\partial y^2} + \frac{1}{y}\frac{\partial\theta}{\partial y}\right) \tag{10-79}$$

with the boundary conditions

$$\left.\begin{aligned} \theta &= 0 \quad \text{at } x = 0, \text{ all } y \\ \theta &= 1 \quad \text{at } x = \infty, \text{ all } y \\ \theta &= 1 \quad \text{at } y = 1, \quad x \geqq 0 \\ \frac{\partial\theta}{\partial y} &= 0 \quad \text{at } y = 0, \quad x \geqq 0 \end{aligned}\right\} \tag{10-80}$$

Comparison of (10-79) and (10-52), together with the conditions (10-53) to (10-55), results in the transformation expressions

$$y_{m+1} - y_m = \epsilon\sqrt{\frac{1}{1 - y_m{}^2}} \tag{10-81}$$

$$\Delta t = \frac{\epsilon^2}{2a_0} = \frac{\epsilon^2 \text{ Re Pr}}{2} \tag{10-82}$$

$$\begin{aligned} \xi &= \int \left(\exp \int -\frac{dy}{y}\right) dy \\ &= b_1 \ln y + b_2 \end{aligned} \tag{10-83}$$

where b_1 and b_2 are arbitrary constants. Equation (10-81) is used to determine the increments in y. y ranges from 0 to 1, giving a total interval of 1. Since the boundary conditions are given at both $y = 0$ $(-\partial\theta/\partial y = 0)$ and $y = 1$ $(\theta = 1)$, it is desirable (although not absolutely necessary) to divide the total interval in such a way that the initial and final values of y_m are 0 and 1, respectively. This requires a certain amount of trial and error. First, the desired number of increments is tentatively estimated. In this illustrative example, 5 will be chosen, although in practice a larger number would be preferable. The necessary value of ϵ is then estimated and the values of y_m calculated by means of (10-81). For example, first try $\epsilon = 0.2$. The values of y_m are then obtained (Table 10-8). At $m = 5$, $y_5 = 1.294$ for $\epsilon = 0.2$, which exceeds the desired value of 1. By trial it is found that for $\epsilon = 0.174$, $y_5 = 1.004$, which is near enough to 1.00. Consequently $\epsilon = 0.174$, and the corresponding y_m values are as shown in Table 10-8. The value of $\Delta t = \Delta x/R$ *imposed on the*

TABLE 10-8

m	y_m for $\epsilon = 0.2$	y_m for $\epsilon = 0.174$
0	0.000	0.000
1	0.200	0.174
2	0.404	0.351
3	0.623	0.537
4	0.878	0.744
5	1.294	1.004

solution is now found from (10-82) for the particular values of Re and Pr which apply to the problem at hand.

Equation (10-83) is used to find the values of ξ_m corresponding to y_m. Since y varies from 0 to 1, ln y will be negative. For convenience, then take $b_1 = -10$ and

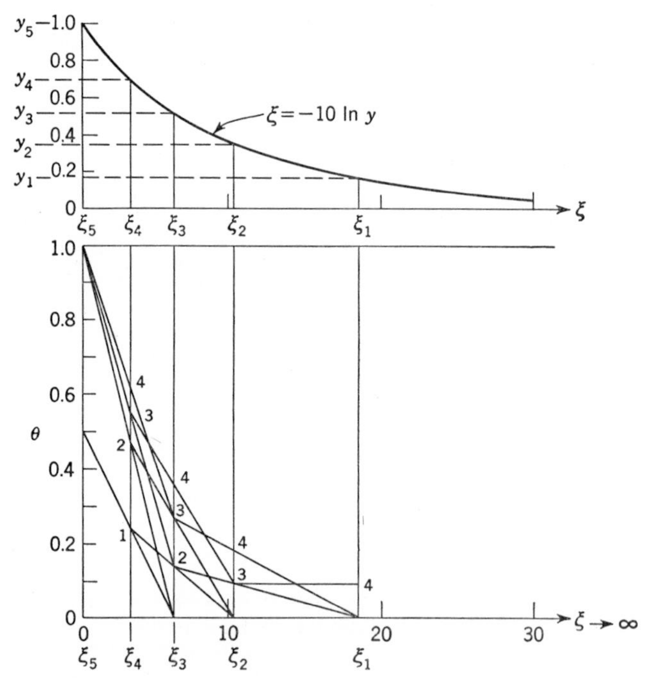

FIG. 10-6. Graphical solution of the problem of heat transfer to a fluid in laminar motion.

$b_2 = 0$. ξ_m is then calculated as $\xi_m = -10 \ln y_m$ or, alternatively, ξ_m is plotted as a function of y_m, as shown in Fig. 10-6. The increments in ξ, $\xi_{m+1} - \xi_m$ are then the increments to be used in the graphical calculation, as shown in Fig. 10-6. At $n = 0$, $\theta_0 = \theta_1 = \cdots = \theta_4 = 0$. θ_5 is taken to be the average $(1 + 0)/2 = 0.5$, since the graphical technique is a step-ahead method. $\theta_{4,1}$ is then found as the intersection of the straight line connecting $\theta_{5,0}$ and $\theta_{3,0}$ with the vertical line at ξ_4. No other temperatures change during the first increment. After the first increment, θ_5 is taken to be its stipulated value, $\theta_5 = 1$, and the construction proceeds. No difficulty is encountered until the fourth increment at ξ_1. $\theta_{2,3}$ is known, and $\theta_{0,3}$ is still 0. However, ξ_0 is located an infinite distance from ξ_1 (or ξ_2) and cannot be plotted. An infinite increment implies a zero slope; consequently, $\theta_{1,4}$ is taken equal to $\theta_{2,3}$, that is, is obtained as the intersection of a horizontal line from $\theta_{2,3}$ with the vertical line at ξ_1. Because of the symmetry about $y = 0$ (ξ_0), $\theta_{1,n} = \theta_{-1,n}$ and $\theta_{0,n+1} = \theta_{1,n}$.

In the example just considered, the boundary condition specified the value of the dependent variable at the boundary. If the slope $\partial T/\partial x$ is specified instead, the procedure is modified slightly. Thus, suppose that the boundary condition is given in the form

$$\left(\frac{\partial T}{\partial x}\right)_s = F \tag{10-84}$$

Since

$$\xi = f(x) \tag{10-55}$$

$$\left(\frac{\partial T}{\partial x}\right)_s = \left(\frac{\partial T}{\partial \xi}\right)_s f'(x_s) \cong \frac{T_{s+1} - T_s}{\xi_{s+1} - \xi_s} f'(x_s) = F \tag{10-85}$$

or

$$\frac{T_{s+1} - T_s}{\xi_{s+1} - \xi_s} = \frac{F}{f'(x_s)} \tag{10-86}$$

Consequently, the intersection of a line of slope $F/f'(x_s)$ through T_{s+1} with the vertical line at ξ_s will determine T_s. This is illustrated in Fig. 10-7.

Graphical methods may be applied to more complicated cases than these just considered. An illustration of such a situation is given in Example 10-6. References 25 and 26 develop modified forms of the graphical procedure outlined above.

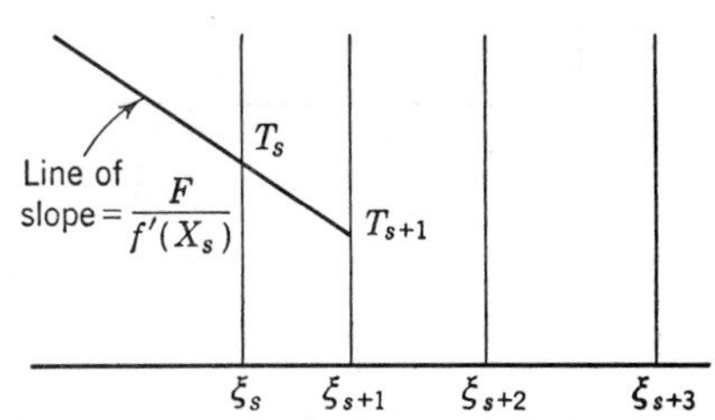

FIG. 10-7. Boundary conditions for graphical method.

10-9. Three or More Independent Variables. The previous discussion has been limited to cases involving only two independent variables. Problems involving three or more independent variables may be handled by analogous methods, although the calculations become more tedious. For example, consider the simple unsteady-state heat-conduction equation in two linear dimensions:

$$\frac{\partial T}{\partial t} = \frac{k}{\rho c}\left(\frac{\partial^2 T}{\partial x^2} + \frac{\partial^2 T}{\partial y^2}\right) \tag{10-87}$$

Divide the solid into a series of rectangles, each of length Δx and of width Δy. Let $m\,\Delta x$ denote the x distance of a point in the solid from a suitable reference y axis, $p\,\Delta y$ denote the y distance of the same point from a suitable reference x axis, and $n\,\Delta t$ denote the elapsed time since the start of the process. The finite difference equivalent of (10-87) is then

$$T_{m,p,n+1} - T_{m,p,n} = \frac{k\,\Delta t}{\rho c}\left(\frac{T_{m+1,p,n} - 2T_{m,p,n} + T_{m-1,p,n}}{\Delta x^2} + \frac{T_{m,p+1,n} - 2T_{m,p,n} + T_{m,p-1,n}}{\Delta y^2}\right) \tag{10-88}$$

Again, the concept of a modulus may be employed:

$$M = \frac{\rho c\,\Delta x^2}{k\,\Delta t} \tag{10-89}$$

$$P = \frac{\rho c\,\Delta y^2}{k\,\Delta t} \tag{10-90}$$

Δx and Δy need not be equal, but Δt must be independent of the position in space. Equation (10-88) with the moduli M and P is similar to the finite difference approximation to the one-distance coordinate case, and the solution of (10-88) proceeds in the same way. The boundary conditions are somewhat more troublesome, however, and deserve discussion.

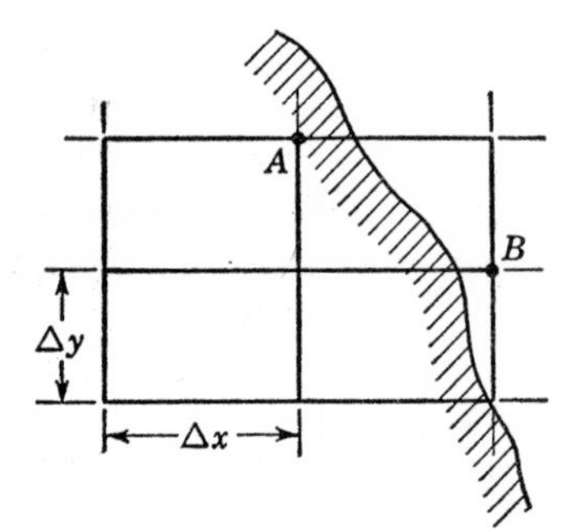

FIG. 10-8. Irregular boundary.

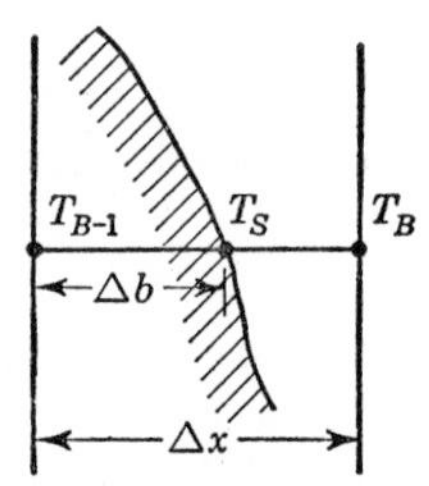

FIG. 10-9. Boundary conditions at irregular boundaries.

When a physical problem involves an irregular boundary such as that shown in Fig. 10-8, it is frequently convenient to use rectangular coordinates even though the resulting net points do not coincide with the boundary. When this occurs, the actual boundary is, in effect, replaced by the net points nearest the boundary on *either side* of the boundary. Thus, in Fig. 10-8, the net points labeled A and B would be used. As a very crude approximation, the values of the dependent variable at points A and B might be taken equal to the nearest surface values. However, this is not necessary, and a better procedure may be developed.

Suppose that the value of the dependent variable is specified along the irregular boundary. The value of the dependent variable at a net point near the boundary is found by interpolation or extrapolation.

Using Fig. 10-9, suppose that T_s is specified and T_B is desired. Linear extrapolation gives

$$\frac{T_s - T_{B-1}}{\Delta b} \cong \frac{T_B - T_{B-1}}{\Delta x} \tag{10-91}$$

which may be used to relate T_B to T_s and T_{B-1}.

In other cases, the derivative of the dependent variable normal to the boundary at the boundary, $(\partial T/\partial n)_b$, may be specified along the boundary. Then, with reference to Fig. 10-10,

$$\left(\frac{\partial T}{\partial n}\right)_b \cong \frac{T_B - T_i}{\Delta s} \tag{10-92}$$

$$\frac{T_2 - T_1}{\Delta x} \cong \frac{T_i - T_1}{\Delta i} \tag{10-93}$$

and combining (10-92) and (10-93) gives

$$T_B = T_1 + \frac{\Delta i}{\Delta x}(T_2 - T_1) + \Delta s \left(\frac{\partial T}{\partial n}\right)_b \tag{10-94}$$

The above treatment of the boundary conditions is, of course, approximate. It becomes more precise as the distance increments Δx and Δy are reduced in size. Additional improvement results from the use of nonlinear interpolation or extrapolation, but the added complication is seldom justified. In the case of an *insulated* boundary, a special treatment is recommended. Thus, although Eq. (10-94) formally may be

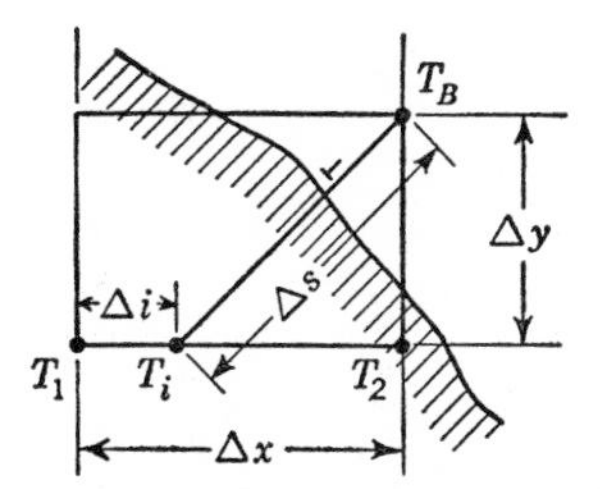

FIG. 10-10. Boundary conditions at irregular boundaries.

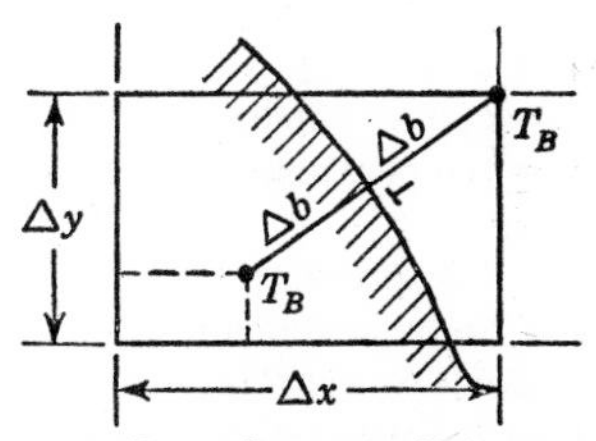

FIG. 10-11. Boundary condition at irregular insulated boundary.

applied with $(\partial T/\partial n)_b = 0$ to the case of an insulated boundary, such a procedure fails to recognize the fact that with *finite intervals* the gradient parallel to the boundary need not be zero in the *two-* or *three-linear-dimension* case. This objection is eliminated if the *image-point* procedure is utilized. It is applied as follows: In Fig. 10-11, the value of the dependent variable T_B situated a distance Δb outside the true insulated boundary along the normal to the boundary is taken *equal* to the value of the dependent variable symmetrically located a distance Δb inside the insulated boundary along the same normal. By linear interpolation, T_B

may then be related to the values of the dependent variable at surrounding net points.

10-10. Illustrative Examples. In this section three examples of the application of the finite-difference method to chemical-engineering situations are worked out in detail in order further to illustrate the versatility of the method.

Example 10-5. Regenerative Heat Exchanger. A regenerative heat exchanger is under consideration for use in an air liquefaction system. The regenerator consists of a cylindrical column packed with copper spheres. It is desired to calculate the temperature of the air and the spheres in the exchanger as a function of position and time when the exchanger is operated as follows:

1. Initially, the spheres and air inside the exchanger are at $-180°F$ (280°R).
2. Suddenly "warm" air at $-110°F$ (350°R), a pressure of 50 atm, and a density of 23 lb/ft³ is pumped into the top of the vertical exchanger at a steady mass velocity (G_0) of 2,000 lb/(ft²)(hr).

The following data and nomenclature are to be used:

a = area of sphere heat-transfer surface per cubic foot of gross exchanger volume, 39.3 ft²/ft³
c_p = specific heat of air at constant pressure, 0.23 Btu/(lb)(°R)
c_B = specific heat of spheres, 0.067 Btu/(lb)(°R)
f = volumetric fraction of voids in exchanger, ft³ of gaseous air/ft³ of gross exchanger volume, 0.345 ft³/ft³
h = convection heat-transfer coefficient between spheres and air, 20 Btu/(hr)(ft²)(°R)
G = mass flow rate of air, lb/(ft²)(hr)
L = packed depth of exchanger, 4.68 ft
t = time, hr
T = gas temperature, °R
ρ_A = air density, lb/ft³
ρ_B = copper-sphere density, 555 lb/ft³
θ = temperature of spheres, °R

Subscripts

m = distance increment
n = time increment

Permissible Simplifications

1. Thermal properties are constant.
2. Radial temperature gradient within a given copper sphere is insignificant.
3. Transfer of heat by sphere-to-sphere contact or by conduction within air is insignificant.
4. Air properties (velocity, density, temperature, etc.) are not a function of *radial* position.
5. Air density is inversely proportional to absolute temperature (approximately true because of the variation in the compressibility factor in this range).
6. Air-pressure drop may be neglected.
7. Any *change* in the mass holdup of air in any section may be neglected.
8. The exchanger walls are adiabatic and of negligible heat capacity.

Solution. The exchanger height is divided into a number of equally spaced intervals, each of length Δz, as shown in Fig. 10-12. Consider the volume enclosed by the m and $m + 1$ planes at time $n \Delta t$.

1. A finite-difference energy balance is written on the *air* contained in the volume:

$$G_{m,n}c_pT_{m,n} - G_{m+1,n}c_pT_{m+1,n} - h\,\Delta z a(T_{m+\frac{1}{2},n} - \theta_{m+\frac{1}{2},n}) = \frac{\Delta z f c_p}{\Delta t}[(\rho_A T)_{m+\frac{1}{2},n+1} - (\rho_A T)_{m+\frac{1}{2},n}] \quad (10\text{-}95)$$

Since ρ_A varies as $1/T$ and the pressure is essentially constant, the right-hand side of Eq. (10-95) is zero. In addition, if the change in the air holdup is neglected,

$$G_{m+1} = G_m = G_0 \quad (10\text{-}96)$$

With the notation

$$M = \frac{G_0 c_p}{ha\,\Delta z} \quad (10\text{-}97)$$

Eq. (10-95) becomes

$$T_{m+1,n} - T_{m,n} = \frac{1}{M}(\theta_{m+\frac{1}{2},n} - T_{m+\frac{1}{2},n}) \quad (10\text{-}98)$$

2. A finite-difference energy balance is written on *both* the air and copper spheres contained in the volume:

$$G_0c_p(T_{m,n} - T_{m+1,n}) = \frac{\Delta z(1-f)\rho_B c_B}{\Delta t}(\theta_{m+\frac{1}{2},n+1} - \theta_{m+\frac{1}{2},n}) \quad (10\text{-}99)$$

With the notation

$$N = \frac{\Delta z(1-f)\rho_B c_B}{G_0 c_p\,\Delta t} \quad (10\text{-}100)$$

Eq. (10-99) becomes

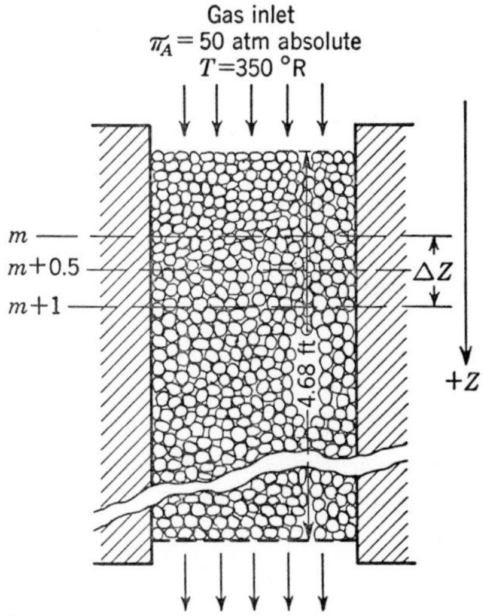

FIG. 10-12. Regenerative heat exchanger.

$$\theta_{m+\frac{1}{2},n+1} - \theta_{m+\frac{1}{2},n} = \frac{1}{N}(T_{m,n} - T_{m+1,n}) \quad (10\text{-}101)$$

Equation (10-98) contains the term $T_{m+\frac{1}{2},n}$. As an approximation, it is replaced by

$$T_{m+\frac{1}{2},n} = \frac{T_{m+1,n} + T_{m,n}}{2} \quad (10\text{-}102)$$

Equation (10-98) then is replaced by Eq. (10-103) below:

$$T_{m+1,n} = \frac{1}{1+2M}[2\theta_{m+\frac{1}{2},n} + (2M-1)T_{m,n}] \quad (10\text{-}103)$$

Combining (10-101) and (10-103) gives

$$\theta_{m+\frac{1}{2},n+1} = \frac{1}{2}\left[\left(\frac{2}{N} + 1 - 2M\right)T_{m,n} + \left(1 + 2M - \frac{2}{N}\right)T_{m+1,n}\right] \quad (10\text{-}104)$$

Equations (10-103) and (10-104) are the working equations. In order to select suitable values for the moduli M and N, physical considerations are invoked. Examination of Eq. (10-103) discloses that a negative effect of $T_{m,n}$ on $T_{m+1,n}$ is avoided if

$$M \geqq \tfrac{1}{2} \quad (10\text{-}105)$$

Suitable N values may be estimated if $T_{m+1,n}$ is eliminated between Eqs. (10-103) and (10-104):

$$\theta_{m+\frac{1}{2},n+1} = \frac{[N(1+2M) - 2]\theta_{m+\frac{1}{2},n} + 2T_{m,n}}{N(1+2M)} \tag{10-106}$$

Consequently,

$$N \geqq \frac{2}{1+2M} \tag{10-107}$$

The value selected for M fixes Δz; the value selected for N then fixes Δt when Δz is fixed. Let $M = N = 1$, which satisfies (10-105) and (10-107). Then

$$\Delta z = \frac{1 \cdot 2000 \cdot 0.23}{20 \cdot 39.3} = 0.585 \text{ ft}$$

$$\Delta t = \frac{0.586(1 - 0.345)555 \cdot 0.067}{2000 \cdot 0.23} = 0.0310 \text{ hr}$$

$$\frac{L_e}{\Delta z} = \frac{4.68}{0.585} = 8.0 \text{ distance increments}$$

Equation (10-103) becomes

$$T_{m+1,n} = -\tfrac{1}{3}(2\theta_{m+\frac{1}{2},n} + T_{m,n}) \tag{10-108}$$

and (10-104) becomes

$$\theta_{m+\frac{1}{2},n+1} = -\tfrac{1}{2}(T_{m,n} + T_{m+1,n}) \tag{10-109}$$

At $n = 0$, $\theta_{m,0} = 280°\text{R}$ *for all m*

$T_{m,0} = 280°\text{R}$ *for m > 0*

At $m = 0$, $T_{0,n} = 350°\text{R}$ for all m

However, the fact that a "step-ahead" finite-difference procedure has been set up makes it desirable to take

$$T_{0,0} = \frac{350 + 280}{2} = 315°\text{R}$$

but to take $T_{0,n} = 350°\text{R}$ for all subsequent times ($n > 0$). Some calculating labor may be saved if it is noted that Eqs. (10-108) and (10-109) are unchanged if 280°R is subtracted from both sides, i.e., if the new variables†

$$\begin{aligned} \theta' &= \theta - 280 \\ T' &= T - 280 \end{aligned} \tag{10-110}$$

are used. Then as boundary conditions, $\theta'_{m,0} = 0$, $T'_{0,0} = 35$, $T'_{0,n} = 70$ ($n > 0$), and $T'_{m,0} = 0$ ($m > 0$). The calculation is started using Eq. (10-109):

$$\theta'_{\frac{1}{2},1} = \tfrac{1}{2}(35 + 0) = 17.5 \qquad m = \tfrac{1}{2},\ n = 1$$

$$\theta'_{\frac{3}{2},1} = \theta'_{\frac{5}{2},1} = \theta'_{m+\frac{1}{2},1} = \tfrac{1}{2}(0 + 0) = 0 \qquad \text{etc.}$$

After the sphere temperatures at all bed locations ($m + \frac{1}{2}$ values) have been determined for $n = 1$, Eq. (10-108) is used to determine air temperatures at $n = 1$:

$$T'_{1,1} = \tfrac{1}{3}(2 \cdot 17.5 + 70) = 35$$

$$T'_{2,1} = \tfrac{1}{3}(2 \cdot 0 + 35) = 11.67 \qquad \text{etc.}$$

The remainder of the calculations are tabulated in Table 10-9.

The analytical solution of this problem has been obtained by Anzelius[2] and by Schumann.[36] The result is an infinite series of Bessel functions. Numerical evaluation of the infinite series[36] for $T_{5,5}$ gives 315°R, which is to be compared with the result of 280.0 + 35.0 = 315.0°R from Table 10-9.

† If desired, all variables may be made dimensionless.

Table 10-9. Numerical Solution of Example 10-5

$M = N = 1 \qquad \Delta z = 0.585$ ft $\qquad \Delta t = 0.0310$ hr

Elapsed t, hr	n	T'_0	$\theta'_{0.5}$	T'_1	$\theta'_{1.5}$	T'_2	$\theta'_{2.5}$	T'_3	$\theta'_{3.5}$	T'_4	$\theta'_{4.5}$	T'_5	$\theta'_{5.5}$	T'_6	$\theta'_{6.5}$	T'_7	$\theta'_{7.5}$	T'_8
0	0	(35)	0	0	0	0	0	0	0	0	0	0	0	0	0	0	0	0
0.0310	1	70	17.5	35.0	0	11.7	0	3.89	0	1.30	0	0.43	0	0.14	0	0.50	0	0.02
0.0620	2	70	52.5	58.3	23.3	35.0	7.78	16.9	2.60	7.35	0.86	3.02	0.23	1.20	0.10	0.46	0.04	0.18
0.0930	3	70	64.2	66.1	46.7	53.1	25.8	35.0	12.1	19.7	5.19	10.0	2.11	4.75	0.83	2.14	0.32	0.93
0.1240	4	70	68.1	68.7	59.6	62.4	44.1	50.2	27.4	35.0	14.9	21.6	7.39	12.1	3.40	6.33	1.54	3.13
0.1550	5	70	69.4	69.6	65.5	66.9	56.3	59.8	42.6	48.3	28.3	35.0	16.9	22.9	9.23	13.8	4.83	7.81
0.1860	6	70	69.8	69.9	68.2	68.8	63.3	65.2	54.1	57.8	41.6	47.0	28.9	34.9	18.3	23.9	10.8	15.2

Although the foregoing analysis dealt specifically with a regenerative heater, practically the same treatment may be used to handle absorption, ion exchange, and similar processes.

Example 10-6. Batch Distillation with Holdup. During a batch distillation the tower conditions vary with time. The calculation of the product composition should include the effect of holdup within the column, plate efficiency, unequal heats of vaporization, and similar complications. In the general case, these factors render an analytical solution impossible, but a finite-difference method may always be applied.

Consider the batch distillation of a binary mixture of A and B in a column containing P actual plates and the still. It is desired to predict the product composition

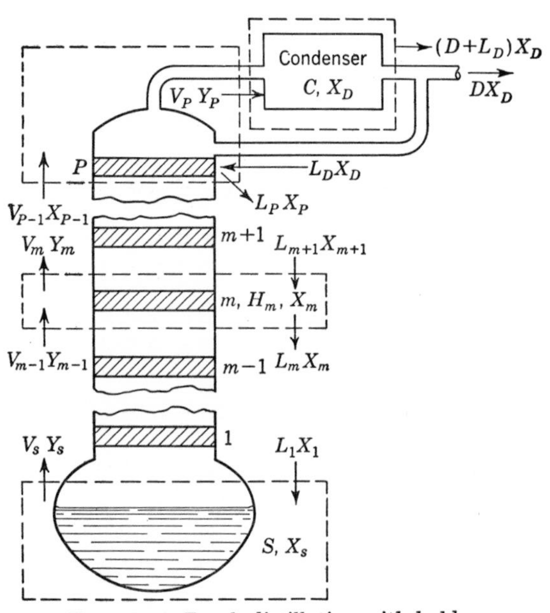

FIG. 10-13. Batch distillation with holdup.

during the course of the distillation. With reference to Fig. 10-13 and with the notation defined below, the material balances may be written.

Symbols Used in Example 10-6

C = condenser contents at any time, moles
D = product rate, moles/hr
E^0 = Murphree plate efficiency
F = fraction of original still charge removed as product
H = liquid holdup per plate, moles
L = liquid overflow rate from plate, moles/hr
P = number of actual plates
R = reflux ratio, L/D
S = still contents at any time, moles
t = time, hr
V = vapor rate, moles/hr
X = mole fraction of more volatile component in liquid

y = mole fraction of more volatile component in vapor
y^* = mole fraction of more volatile component in vapor that is in equilibrium with liquid

Subscripts

D = product
m = plate location
P = top plate
n = number of ΔF increments
s = still

The material balances are as follows:

For the still,

$$L_1 - V_s = \frac{\partial S}{\partial t} \tag{10-111a}$$

$$L_1 X_1 - V_s y_s = \frac{\partial}{\partial t}(S X_s) \tag{10-111b}$$

For plate m,

$$(L_{m+1} - L_m) - (V_m - V_{m-1}) = \frac{\partial}{\partial t}(H_m) \tag{10-112a}$$

$$(L_{m+1} X_{m+1} - L_m X_m) - (V_m y_m - V_{m-1} y_{m-1}) = \frac{\partial}{\partial t}(H_m X_m) \tag{10-112b}$$

For the condenser,

$$V_P - (D + L_D) = \frac{\partial C}{\partial t} \tag{10-113a}$$

$$V_P y_P - (D + L_D) X_D = \frac{\partial}{\partial t}(C X_D) \tag{10-113b}$$

when it is assumed that the material in the still, on each plate, and in the condenser is of uniform composition, and vapor holdup is neglected. Note that plate 1 is fed by the still, and plate P is fed by the condenser. In the general case, the vapor and downflow rates and the molal holdup will be functions of time and position. They may be related to one another by means of energy balances and the relation between the partial molal volume of each component and the composition and temperature. In this illustration, these complications will be side-stepped, and the following idealizations invoked:

1. Condenser holdup is neglected.
2. The molal plate holdup and plate efficiency are constant and are the same for each plate.
3. The molal overflow and vaporization rates are independent of plate location.

The material-balance equations (10-111a) and (10-113a) then become

$$\frac{dS}{dt} = L - V = -D \tag{10-114}$$

It is convenient to replace time by defining the new variable F, the fraction of the original still charge distilled from the system,

$$F = \frac{S_0 - S}{S_0} \tag{10-115}$$

Then

$$dt = \frac{S_0\, dF}{D} \tag{10-116}$$

and Eqs. (10-111) to (10-113) become

$$\frac{dX_s}{dF} = \frac{1}{1 - F_n} [R(X_{1,n} - X_{s,n}) - (R + 1)(y_{s,n} - X_{s,n})] \tag{10-117a}$$

$$\frac{dX_m}{dF} = \frac{S_0}{H} [R(X_{m+1,n} - X_{m,n}) - (R + 1)(y_{m,n} - y_{m-1,n})] \tag{10-117b}$$

$$\frac{dX_P}{dF} = \frac{S_0}{H} [R(y_{P,n} - X_{P,n}) - (R + 1)(y_{P,n} - y_{P-1,n})] \tag{10-117c}$$

where

$$y_P = X_D \qquad y_0 = y_s \tag{10-117d}$$

Additional information is supplied by the charge composition, the vapor-liquid equilibrium relation

$$y_m^* = \phi(X_m) \tag{10-118}$$

by the plate-efficiency relation

$$y_{m-1} - y_m = E_m^0(y_{m-1} - y_m^*) \tag{10-119}$$

and by specification of the manner in which either R or X_D is to vary with F.

The finite difference equivalents to (10-117a, b, c) are obtained by replacing the derivatives by finite differences. The resulting equations are

$$X_{s,n+1} = \frac{R + 1}{(1/\Delta F) - n} \left[\frac{R}{R + 1} (X_{1,n} - X_{s,n}) - y_{s,n} - X_{s,n}) \right] + X_{s,n} \tag{10-117e}$$

$$X_{m,n+1} = \frac{\Delta F S_0(R + 1)}{H} \left[\frac{R}{R + 1} (X_{m+1,n} - X_{m,n}) - (y_{m,n} - y_{m-1,n}) \right] + X_{m,n} \tag{10-117f}$$

$$X_{P,n+1} = \frac{\Delta F S_0(R + 1)}{H} \left[\frac{R}{R + 1} (y_{P,n} - X_{P,n}) - (y_{P,n} - y_{P-1,n}) \right] + X_{P,n} \tag{10-117g}$$

$$y_P = X_D \qquad y_0 = y_s \tag{10-117d}$$

The selection of a suitable value of ΔF must be based upon physical grounds. It seems reasonable to suppose that the method will be stable if the moles of a given component carried by any stream entering or leaving the plate during the increment ΔF are less than the holdup of that component on the plate. On this basis, the following conservative criterion is selected:

$$\frac{HX_m}{\Delta F S_0(R + 1)y_m} \geqq 1 \tag{10-120}$$

The application of the finite difference equations is best illustrated by a specific case. It is desired to separate an equimolal mixture of A and B in a batch column containing three plates and a still. The still efficiency is 1.0; the plate efficiency $E^0 = 0.5$. The relative volatility of the system is $\alpha = 2$. The holdup for each plate is 5 moles; the condenser holdup is negligible. The still is to be charged *through the reflux* line with 115 moles of liquid at its boiling point. A reflux ratio of 4 is to be used throughout the distillation. The "usual" simplifying assumptions apply. What is the composition of the product as a function of the fraction distilled?

Because of the manner of charging, initially all plates are of uniform composition ($x = 0.5$), $S_0 = 100$, and $H = 5$. The increment in F, ΔF, is taken to be 0.005, which just meets the criterion (10-120). The working equations are then

$$X_{s,n+1} = \frac{5}{200 - n} [\tfrac{4}{5}(X_{1,n} - X_{s,n}) - (y_{s,n} - X_{s,n})] + X_{s,n} \qquad (10\text{-}121a)$$

$$X_{m,n+1} = \tfrac{1}{2} [\tfrac{4}{5}(X_{m+1,n} - X_{m,n}) - (y_{m,n} - y_{m-1,n})] + X_{m,n} \qquad (10\text{-}121b)$$

$$X_{P,n+1} = \tfrac{1}{2}[\tfrac{4}{5}(y_{P,n} - X_{P,n}) - (y_{P,n} - y_{P-1,n})] + X_{P,n} \qquad (10\text{-}121c)$$

$$y_m^* = \frac{\alpha X_m}{1 + (\alpha - 1)X_m} = \left(\frac{2X_m}{1 + X_m}\right)_n \qquad (10\text{-}122)$$

$$y_s = y_s^* \qquad (10\text{-}123a)$$

$$(y_{m-1} - y_m)_n = 0.5(y_{m-1} - y_m^*)_n \qquad \text{for } m > 0 \qquad (10\text{-}123b)$$

The calculations then proceed as follows:

1. A table with rows and columns similar to Table 10-10 is prepared.
2. The starting conditions ($n = 0$) are inserted in the table.
3. Equations (10-121*a, b, c*) are used in turn to find $X_{s1}, X_{11}, \ldots, X_{P1}$.
4. The equilibrium relation (10-122) is used to calculate y^* at $n = 1$.
5. The plate-efficiency relation (10-123*a*) or (10-123*b*) is used to calculate $y_s, y_{11}, \ldots, y_{P1}$.
6. The process is repeated for $n = 2$ and subsequent intervals.
7. After several intervals the calculations are checked by material balances applied to the column.

The numerical results of the calculations carried out for the above example are shown in Table 10-10. A material balance at the end of five increments shows 57.1

TABLE 10-10. BATCH DISTILLATION
$\Delta F = 0.005$

n	F	X_s	y_s	X_1	y_1^*	y_1	X_2	y_2^*	y_2	$X_3 = X_P$	y_P^*	$y_P = X_D$
0	0	0.5	0.667	0.5	0.667	0.667	0.5	0.667	0.667	0.5	0.667	0.667
1	0.005	0.496	0.664	0.5	0.667	0.665	0.5	0.667	0.666	0.567	0.724	0.695
2	0.010	0.492	0.660	0.499	0.666	0.662	0.526	0.690	0.676	0.604	0.754	0.715
3	0.015	0.488	0.657	0.508	0.670	0.664	0.550	0.710	0.687	0.629	0.772	0.730
4	0.020	0.484	0.653	0.523	0.686	0.670	0.570	0.726	0.698	0.666	0.800	0.749
5	0.025	0.480	0.649	0.533	0.695	0.672	0.595	0.746	0.709	0.674	0.805	0.757

moles of light component in the column and distillate. In the original charge, 57.5 moles were present. An error of this magnitude is acceptable.

The stability criterion (10-120) adopted is probably unduly conservative. Calculations carried out using $\Delta F = 0.01$ gave results in substantial agreement with those for $\Delta F = 0.005$. On the other hand, if ΔF is taken to be 0.1, the distillate leaving the top plate at the end of the first increment is found to *exceed* a mole fraction of unity! If it were desired to carry out rough calculations with increments of this magnitude, the effect of plate holdup should be ignored.

References 30 and 31 discuss the use of finite-difference methods in cases where the "usual" simplifying assumptions are not valid. Multicomponent systems may be handled by straightforward modifications of the binary equation development.

Example 10-7. The Design of a Fixed-bed Catalytic Reactor. In this example the finite-difference method is applied to the design of a fixed-bed catalytic reactor in which the temperature and composition of the reactants are a function of both radial position and reactor height. Simultaneous radial and height composition variation occurs when the reaction rate is a function of temperature and the reactor is heated or cooled. The importance of the radial gradients is indicated by consideration of an exothermic reaction occurring in a fixed-bed cylindrical reactor, cooled at the wall. Because of higher temperatures, the reactants are depleted, and the products are produced faster near the center than in regions close to the wall. The result is a radial transport of the reactants toward the center and of the products toward the wall. In this way, the important central reaction zone is continuously fed with reactants which keep the reaction going, and it is depleted of products which tend to retard the reaction. Neglecting lateral diffusion may therefore result in seriously underestimating the over-all conversion.

The following treatment was first reported by Baron:[3] Consider a cylindrical reactor packed with pellets whose effective diameter d is much smaller than the reactor diameter D. For simplicity, it is assumed that the mean molecular weight and molal specific heat of the system are constant. These conditions are approximately true if the disappearance of 1 mole of reactants is followed by the appearance of 1 mole of products, or if inerts are present in large amounts, or if the conversion is low and if the temperature variation of the specific heat is moderate. If necessary, these restrictions may be removed at the expense of additional complication. It is assumed that the necessary kinetic data have been obtained in laboratory studies† and that for the particular catalyst and flow conditions to be employed the rate at which component i reacts is known as a function of temperature and composition in, say, the form

$$R_i = \phi(T,p,y_A,y_B, \ldots ,y_i) \tag{10-124}$$

= moles of component i undergoing reaction in unit time per unit mass of catalyst

The symbols are defined as follows:

Nomenclature for Example 10-7

$a, b, \ldots, s$ = stoichiometric factors
c_p = molal specific heat of fluid
d = effective catalyst pellet diameter
D = reactor diameter
E = eddy diffusivity
f = fractional conversion of limiting component
K = proportionality factor
N = radial mass transfer rate
p = fluid pressure
q = radial heat-transfer rate
r = radius (local)
r' = dimensionless radius, $r' = r/D$
R = reaction rate, moles/(unit time) (unit mass of catalyst)
T = temperature
V = superficial fluid velocity
y = mole fraction
z = axial coordinate
z' = dimensionless coordinate, $z' = z/D$
ρ = fluid *molal* density
ρ_c = catalyst mass density
λ = molal heat of reaction

† Using, for example, the "differential-reactor" method of Hougen, Watson, and coworkers.[23,37]

Subscripts

A = component A
i = component i

Superscripts

0 = conditions at the entrance to the reactor
′ (prime) = a dimensionless coordinate

In addition, an expression is needed for the rate at which heat and mass are transported radially. The radial transport is due to two factors: molecular transport and the bulk transport which results from the deflection of the bulk velocity as the fluid flows around the catalyst particle. Under normal operating conditions, the molecular-transport term is insignificant in comparison with the bulk radial transport and may be neglected. An approximate expression for the bulk-radial-transport term may be developed from a consideration of the motion of the fluid in the presence of the catalyst granules. Only the results will be given here; the derivation is given in references 3 and 11. It is found that the radial transport rate is given by expressions of the form

$$(N_i)_{\text{radial}} = -E\frac{\partial y_i}{\partial r} \tag{10-125}$$

= moles of i per unit area per unit time transferred radially

$$(q)_{\text{radial}} = -Ec_p\frac{\partial T}{\partial r} \tag{10-126}$$

= heat per unit area per unit time transferred radially

where

$$E = K\rho V \tag{10-127}$$

K is primarily a function of the packing characteristics. In certain flow ranges it is also a function of the viscosity and, according to Danckwerts and Sugden,[11] a weak function of the velocity. Hacker,[18] however, found K to be independent of velocity.† Despite the lack of information concerning the exact variation of K with packing characteristics and flow conditions, Eq. (10-127) may be used with K regarded as a constant if K has been measured under conditions which substantially duplicate the proposed reactor conditions.

With the above assumptions and with the restriction that the rate at which heat is transferred across a given particle, between particles by direct contact, and by radiation is negligible and that longitudinal conduction is insignificant, a differential energy balance may be written on a volume of width dr and length dz:

$$\text{Input} = (2\pi r\,dr)\rho Vc_pT - 2\pi r\,dz\,K\rho Vc_p\frac{\partial T}{\partial r}$$

$$\text{Output} = \text{input} + (2\pi r\,dr)\frac{\partial}{\partial z}(\rho Vc_pT)\,dz - 2\pi\,dz\,K\left(\frac{\partial}{\partial r}r\rho Vc_p\frac{\partial T}{\partial r}\right)dr - 2\pi r\,dr\,dz\,\rho_c\lambda_AR_A$$

$$\text{Accumulation} = 0$$

Hence,

$$\frac{\partial}{\partial z}(\rho Vc_pT) = \frac{K}{r}\frac{\partial}{\partial r}\left(r\rho Vc_p\frac{\partial T}{\partial r}\right) + \rho_c\lambda_AR_A \tag{10-128}$$

Proceeding in the same fashion, a material balance on component i is obtained in the form

$$\frac{\partial}{\partial z}(\rho Vy_i) = \frac{K}{r}\frac{\partial}{\partial r}\left(r\rho V\frac{\partial y_i}{\partial r}\right) - \rho_cR_i \tag{10-129}$$

and similar expressions for all other components.

† For further discussion and data, see also Coberly and Marshall,[8] Hougen and Piret,[22] and Smith.[6,19]

The solution of Eqs. (10-128) and (10-129) is greatly simplified if ρV may be taken to be constant. The experimental measurements of Coberly and Marshall[8] indicate that this assumption is valid, but Hall and Smith[19] conclude from heat-transfer measurements that some radial variation exists. In the following, ρV will be considered constant.

The reaction-rate terms R_i are functions of the local composition as well as of the temperature. Consequently, the solution of the general problem involves the integration of the system of equations comprising the energy balance (10-128) and a material-balance equation corresponding to (10-129) for each component but one. In the present case, however, considerable simplification is afforded by the assumptions that the mean molecular weight and the molal mass velocity ρV are constant. It is shown below that under these circumstances the reactants are always present in stoichiometric amounts; i.e., the mole fractions are the same as those that are obtained from stoichiometric calculations based on a batch reaction. Consequently, only one differential equation of the type (10-129) need be integrated together with (10-128), the remainder of the mole fractions being determined by stoichiometry. The proof is as follows:

Consider the reaction

$$aA + bB + \cdots + sS + \cdots \rightleftharpoons cC + dD + \cdots + pP \tag{10-130}$$

If the rate of reaction of component A is R_A, then the rate of reaction of B is $R_A(b/a)$, of S is $R_A(s/a)$, and of any component I

$$R_i = R_A \frac{i}{a} \tag{10-131}$$

with a suitable sign associated with i: + if component I reacts with A; − if component I is formed from A. In view of (10-131), (10-129) may be written as

$$\frac{\partial y_i}{\partial z} = \frac{K}{r}\frac{\partial}{\partial r}\left(r\frac{\partial y_i}{\partial r}\right) - \frac{\rho_c i R_A}{\rho V a} \tag{10-132a}$$

and a similar expression for component A:

$$\frac{\partial y_A}{\partial z} = \frac{K}{r}\frac{\partial}{\partial r}\left(r - \frac{\partial y_i}{\partial r}\right) - \frac{\rho_c R_A}{\rho V} \tag{10-132b}$$

If R_A is eliminated between (10-132a) and (10-132b), there results

$$\frac{\partial[y_i - (i/a)y_A]}{\partial z} = \frac{K}{r}\frac{\partial}{\partial r}\left[r\frac{\partial}{\partial r}\left(y_i - \frac{i}{a}y_A\right)\right] \tag{10-133}$$

Equation (10-133) applies to the catalytic reactor. The boundary conditions are

$$y_i - \frac{i}{a}y_A = y_i^0 - \frac{i}{a}y_A^0 \qquad \text{at } z = 0 \tag{10-134}$$

$$\frac{\partial y_i}{\partial r} = \frac{\partial y_A}{\partial r} = 0 \qquad \text{at } r = 0 \text{ and } r = \frac{D}{2}$$

The solution of (10-133) with the boundary conditions (10-134) is

$$y_i - \frac{i}{a}y_A = y_i^0 - \frac{i}{a}y_A^0 \tag{10-135}$$

Equation (10-135) enables the mole fraction of any constituent to be calculated if the mole fraction of one component is known. The situation is then completely analogous to a batch reaction. It is convenient to let component A be the limiting reactant, i.e., that component which disappears first as the reaction proceeds.

The equations to be solved are then

$$\frac{\partial T}{\partial z} = \frac{K}{r}\frac{\partial}{\partial r}\left(r\frac{\partial T}{\partial r}\right) + \frac{\rho_c \lambda_A R_A}{\rho V c_p} \tag{10-136}$$

$$\frac{\partial y_A}{\partial z} = \frac{K}{r}\frac{\partial}{\partial r}\left(r\frac{\partial y_A}{\partial r}\right) - \frac{\rho_c R_A}{\rho V} \tag{10-137}$$

$$y_i = y_i^0 + \frac{i}{a}(y_A - y_A^0) \tag{10-138}$$

together with the known reaction-rate expression

$$R_A = \phi(T,p,y_A,y_B, \ . \ . \ . \ ,y_i) \tag{10-139}$$

and the specification of the temperature at the reactor wall (or of the temperature gradient at the reactor wall). Generally it may be assumed that the pressure is constant within the reactor. If not, an additional equation relating the pressure to the flow variables is needed.

The solution of the system of Eqs. (10-136) to (10-139) is facilitated by the introduction of new variables. Let f denote the fractional conversion of component A:

$$f = 1 - \frac{y_A}{y_A^0} \tag{10-140}$$

Define the new length parameters

$$r' = \frac{r}{D} \tag{10-141}$$

$$z' = \frac{z}{D} \tag{10-142}$$

Combination of (10-138) to (10-140) gives

$$R_A = \gamma(T,p,f) \tag{10-143}$$

Introduction of the new variables into (10-136) and (10-137) results in the equations

$$\frac{\partial T}{\partial z'} = \frac{K}{D}\left(\frac{\partial^2 T}{\partial r'^2} + \frac{1}{r'}\frac{\partial T}{\partial r'}\right) + \frac{R_A \lambda_A D \rho_c}{\rho V c_p} \tag{10-144}$$

$$\frac{\partial f}{\partial z'} = \frac{K}{D}\left(\frac{\partial^2 f}{\partial r'^2} + \frac{1}{r'}\frac{\partial f}{\partial r'}\right) + \frac{R_A D \rho_c}{\rho V y_A^0} \tag{10-145}$$

The simultaneous solution of (10-143) to (10-145) (and, if necessary, an additional pressure relation) may be accomplished by finite-difference methods when the applicable boundary conditions are specified. Baron[3] has proposed a graphical solution that is frequently convenient. It proceeds as follows:

If the pressure may be regarded as constant throughout the reactor, Eq. (10-143) may be used to calculate the reaction rate R_A for values of T and f in the range of interest. Then, for specific values of T and f, the terms $R_A \lambda_A D \rho_c / \rho V c_p$ and $R_A D \rho_c / \rho V y_A^0$ are known.

Equations (10-144) and (10-145) differ from the general equation (10-52) for which a graphical solution was developed in Sec. 10-8 only as a result of the known reaction-rate terms. For an equation of the form of (10-52), the distance BD of Fig. 10-5 represented the increment $(T_{m,n+1} - T_{m,n})$. For an equation of the form of (10-144) it is evident that the distance BD equals

$$BD \cong (T_{m,n+1} - T_{m,n}) - \frac{R_A \lambda_A D \rho_c \, \Delta z'}{\rho V c_p} \tag{10-146}$$

Consequently, if the graphical procedure of Sec. 10-8 is followed, the temperature

$T_{m,n+1}$ is equal to the sum of the distance BD and the value of $(R_A\lambda_A D\rho_c\,\Delta z')/\rho V c_p$ evaluated at $T_{m,n}$ and $f_{m,n}$. Similar statements apply to the fractional conversion. The working diagram is then laid out on this basis. Using the conditions (10-53) to (10-56), there results

$$\epsilon = \Delta z' \tag{10-147}$$

$$\frac{\Delta r'^2}{KD\,\Delta z'} = 2 = \frac{D\,(\Delta r')^2}{K\,\Delta z'} \tag{10-148}$$

$$\xi = \int \left(\exp \int -\frac{dr'}{r'} \right) dr' = \log r' \tag{10-149}$$

$$\xi_{m+1} - \xi_m = \log\,(m+1)\,\Delta r' - \log m\,\Delta r' \tag{10-150}$$

These equations determine the relation between $\Delta r'$ and $\Delta z'$ and the coordinate distortion. Then, with reference to Fig. 10-14, the problem is laid out in the following manner:

Quadrant I: The fractional conversion is plotted vs. ξ.

Quadrant II: The temperature is plotted vs. ξ.

Quadrant III: The temperature is plotted vs. the heat release term $(R_A\lambda_A D\rho_c\,\Delta z')/\rho V c_p$ for several fixed values of the fractional conversion. For example, the curve marked f_1 applies to the fixed value f_1.

Quadrant IV: The fractional conversion is plotted vs. the reaction-rate term $(R_A D\rho_c\,\Delta z')/\rho V y_A{}^0$ for several fixed values of the temperature.

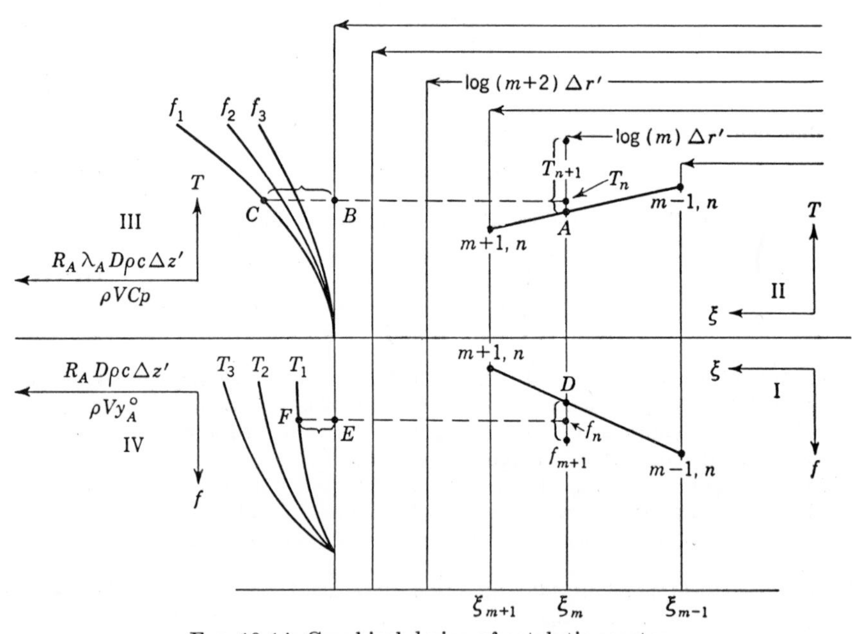

Fig. 10-14. Graphical design of catalytic reactor.

With the interior temperatures and conversions known at a distance $n\,\Delta z'$ for all r', the temperature at slice m at a distance $(n+1)\,\Delta z'$ is found by first connecting the temperatures at distance n at slices $m-1$ and $m+1$ by a straight line. The intersection of this line with the vertical line at ξ_m (point A) is then located. Now a horizontal line from $T_{m,n}$ is drawn to the left until it intersects the curve in quadrant

III corresponding to the fractional conversion $f_{m,n}$. The bracketed horizontal distance BC is then added to A, giving the temperature $T_{m,n+1}$. An analogous procedure is followed in quadrants I and IV to determine $f_{m,n+1}$. The boundary conditions are introduced in the manner outlined in Sec. 10-8. Modifications of this graphical procedure may be applied to other problems.

The above graphical method is not necessarily stable under all conditions, since it demands a modulus of 2. The chief factor affecting the stability would appear to be the form of the reaction-rate expression (10-143). If it is strongly temperature-dependent (for example, a strong exponential temperature dependence), the graphical results should be scrutinized carefully. Baron[3] tested the graphical solution by calculation of the conversion of SO_2 to SO_3, using the rate data of Hall and Smith.[19] He predicted a conversion of 25 per cent; the experimentally observed result was 30 per cent. The agreement is reasonable when the assumptions involved in the analysis are taken into consideration. However, the experimental data of Hall and Smith show an essentially *linear* relation between the reaction rate and temperature for the $SO_2 \rightarrow SO_3$ reaction. Consequently, the use of the graphical method to calculate the conversion of SO_2 does not represent a severe test of the procedure.

Crank and Nicolson[10] have studied the application of finite-difference procedures to the calculation of a chemical reaction proceeding in a solid. The reaction rate was taken to be an exponential function of the temperature (Arrhenius type of temperature dependence). Crank and Nicolson used an iteration procedure which was a modification of the method discussed in Sec. 10-6 and which was presumably stable. They found that, despite the use of iteration, the calculated results behaved erratically when the reaction rate was strongly temperature-dependent unless a large number of increments were employed.

The literature contains numerous other examples of the application of finite-difference methods to initial-value problems. Some of them employ alternative procedures, particularly at the boundaries, which in specific instances are better approximations than those discussed in this chapter. In addition to the references already specifically mentioned, the following may prove of interest. The book "Numerical Analysis of Heat Flow" by Dusinberre[12] considers a large number of problems in detail. References 4, 5, 25, and 34 examine graphical heat-flow solutions. Hartree[20] combines the finite-difference method with the use of the differential analyzer.

10-11. Boundary-value Problems. The solution of a boundary-value problem can often be considered to represent the value approached by the solution of a corresponding initial-value problem after an infinite amount of time. Thus, the solution of the two-dimensional steady-state heat-conduction equation

$$\frac{\partial^2 T}{\partial x^2} + \frac{\partial^2 T}{\partial y^2} = 0 \tag{10-151}$$

may be obtained from the solution of the unsteady-state equation

$$\frac{\partial T}{\partial t} = \frac{k}{\rho c}\left(\frac{\partial^2 T}{\partial x^2} + \frac{\partial^2 T}{\partial y^2}\right) \tag{10-87}$$

by permitting the time variable to approach infinity. This fact suggests a method for the solution of a boundary-value problem: Set it up as an initial-value problem, apply finite-difference methods, and extend the calculations far into the time domain. Unfortunately, this process is also time-consuming and not suitable for general application. However, it does embody the principle underlying a more practical approach—the *relaxation method.* The relaxation method is a finite-difference technique that combines great flexibility with a systematic procedure for obtaining a convergent solution to a boundary-value problem.

Consider the finite difference approximation to (10-151):

$$\frac{T_{m+1,p} - 2T_{m,p} + T_{m-1,p}}{\Delta x^2} + \frac{T_{m,p+1} - 2T_{m,p} + T_{m,p-1}}{\Delta y^2} = 0 \qquad (10\text{-}152)$$

Let

$$\Delta x = \Delta y \qquad (10\text{-}153)$$

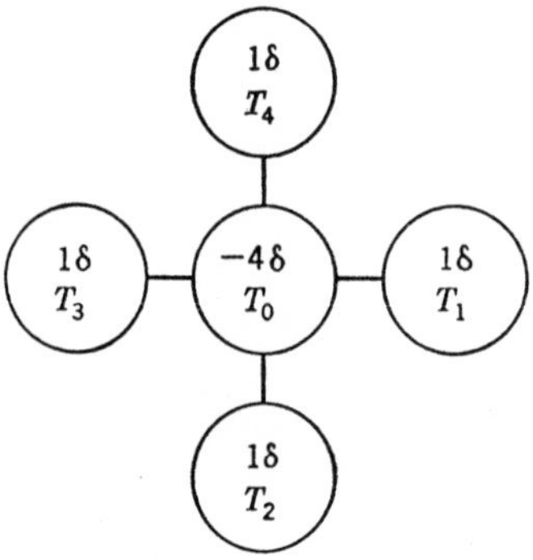

and, with reference to Fig. 10-15, let

$$\begin{aligned} T_{m,p} &= T_0 \\ T_{m+1,p} &= T_1 \\ T_{m,p-1} &= T_2 \\ T_{m-1,p} &= T_3 \\ T_{m,p-1} &= T_4 \end{aligned} \qquad (10\text{-}154)$$

FIG. 10-15. Relaxation pattern.

Then Eq. (10-152) becomes

$$T_1 + T_2 + T_3 + T_4 - 4T_0 = 0 \qquad (10\text{-}155)$$

Equation (10-155) states that at steady state the temperature at a point is equal to the *arithmetic mean* of the temperatures at surrounding points. Now suppose that the temperatures at all interior points in a steady-state heat-conduction problem are guessed and the residual R defined by the equation

$$T_1 + T_2 + T_3 + T_4 - 4T_0 = R_0 \qquad (10\text{-}156)$$

is calculated for each point. In general, the residuals at some points will be different from zero, showing that the guessed temperature distribution was not correct. Indeed, it can be shown that the residual is directly proportional to the rate at which the temperature at a given point changes with time in the corresponding unsteady-state case. The magnitude of a given residual then is a measure of the departure of the temperature at that point from the steady-state condition.

In general, the temperature at the point with the largest residual (in absolute magnitude) is most in error. Clearly then, the temperature at this point should be modified first, and the modification should be made so that the residual at that point is decreased in absolute magnitude.

This is the principle of the relaxation method: The solution is guessed, the residuals are calculated, and the temperatures associated with the largest residuals in turn are changed until all residuals are essentially zero. There are, of course, certain "tricks of the trade" that facilitate the process. These are now considered.

Examination of Eq. (10-156) discloses that the residual at point 0 is most strongly dependent upon T_0 itself. Consequently, if T_0 is changed by an amount δ but all other temperatures remain fixed, the residual at 0 is *decreased* an amount 4δ, whereas the residuals at points 1, 2, 3, and 4 are *increased* by δ only. The "relaxation pattern" is that shown in Fig. 10-15. It is this "leverage" that makes the relaxation method converge quickly. *In setting up a finite difference approximation for use with the relaxation method, the analyst deliberately should attempt to introduce leverage of this sort.*

The relaxation method proceeds by attacking that point whose residual is of the greatest numerical value by modifying the estimate of the associated value T_0 in such a way that the magnitude of the maximum residual is decreased. A straightforward attack would involve the complete liquidation of the largest residual R_0 by adding to T_0 exactly one-fourth of R_0 before proceeding to the next point. However, this is seldom the best procedure (except near the end of a nearly completed calculation), since the algebraic reduction of a residual at a given point is accompanied by an algebraic increase in the residuals at the four neighboring points. It is more fruitful to either "overrelax" or "underrelax." Overrelaxation, i.e., to add *more* than one-fourth R_0 to T_0 so that the residual changes sign, is recommended when the largest residuals surrounding T_0 are of the same sign as R_0, and underrelaxation is recommended when the opposite situation exists. In this way, the residuals are made to alternate in sign from point to point, a condition which usually leads to rapid convergence.

The relaxation pattern, such as the one shown in Fig. 10-15, can be used to calculate the residuals at different net points during the course of the relaxation operations. Consequently, it is not necessary to apply Eq. (10-156) at each step, and considerable labor is avoided.

The above points are best illustrated by a specific example.

Example 10-8. The inner walls of a square flue are maintained at 700°F while the outer walls are maintained at 100°F. The walls are 2 ft thick, and the inside perimeter is 24 ft. The wall thermal properties are independent of temperature; the thermal conductivity is constant at a value of 0.6 Btu/(hr)(ft²)(°F/ft). Calculate the interior temperature distribution and the heat-transfer rate under steady-state conditions.

The partial differential equation (10-151) applies to this problem. Then, with $\Delta x = \Delta y$, the relaxation equation (10-156) and the relaxation pattern of Fig. (10-15) also apply. It is best to begin with a coarse net. After the coarse-net temperature distribution has been established, the calculation is repeated, using a finer net until finally the desired detail is achieved.

For the coarse net, let $\Delta x = \Delta y = 1$ ft. The resulting net pattern is shown in Fig. 10-16. Because of the symmetry, only one-eighth of the wall need be considered. However, note that, for the same reason, the relaxation pattern of A will be doubly affected by B and the relaxation pattern of E will be doubly affected by D.

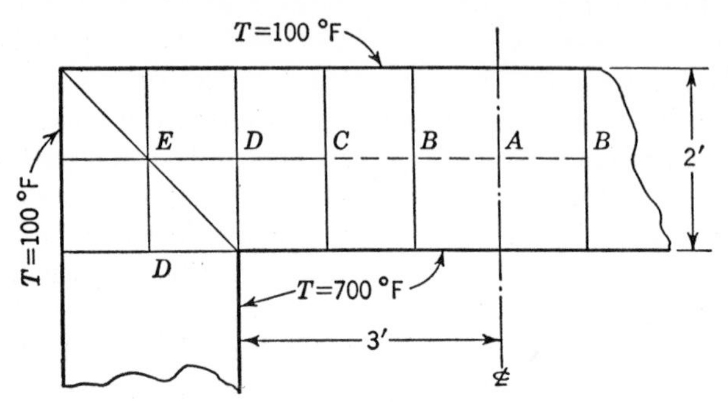

FIG. 10-16. Square flue, coarse net.

In actual practice, the successive arithmetic operations are usually carried out on an enlarged diagram similar to Fig. 10-16. The fixed boundary values are entered in ink and interior values and residuals in pencil. As the calculations proceed, old interior values are erased and replaced by new ones. In a lengthy calculation, the residuals periodically are checked by means of Eq. (10-156) in order to avoid prolonged propagation of numerical errors.

TABLE 10-11. RELAXATION CALCULATIONS

Step	T_A	R_A (2 ←)	T_B	R_B	T_C	R_C	T_D	R_D	T_E	R_E (2 →)
0	400	0	400	0	400	0	400	0	400	−600
1	400	0	400	0	400	0	400	−175	225*	100
2	400	0	400	0	400	−50	350*	25	225	0
3	400	0	400	−15	385	10	350	10	225	0
4	400	−10	395*	5	385	5	350	10	225	0
5	398.5*	−4	395	3.5	385	8.5	353.5*	−4	225	7
6	398.5	−1	396.5*	−2.5	385	10	353.5	−2	227*	−1
7	398.5	−1	396.5	0	387.5*	0	353.5	0.5	227	−1

In this example, Table 10-11 has been prepared in order to illustrate the detailed operational steps. The arrows shown in the table serve as reminders that certain points are to be weighted twice. Asterisks denote the point at which the temperature was changed. It is desirable to start the calculation with the best possible guess as to the interior temperatures. Here, however, the uniform value of 400°F is assigned to all interior points. The residuals are then calculated, giving the results shown for step 0. Only R_E (-600) differs from zero, and it is to be attacked first. In step 1, T_E is reduced 175° to 225°F. This overrelaxes point E, making $R_E = 100$ *and*

$R_D = -175$. The other residuals are unchanged. In step 2, point D is overrelaxed when $-50°$ is added to T_D. This changes R_C by -50 and R_E by -100. In step 5, points A and D are changed simultaneously in order to hasten convergence. At the end of step 7, the maximum residual is unity, indicating that the interior temperatures differ from those that would be obtained in a completely relaxed pattern by the order of ± 0.25°F. This *does not* mean that the results differ from the *true* values by this amount, however. In general, the discrepancy is greater than this. Values which more closely approach the true temperatures require the use of a finer net.

In this example, it is clear that the region of greatest uncertainty lies between points E and C. As a good approximation, the temperatures at points C, B, and A may be regarded as "exact," and the finer net is imposed only between E and C, as shown in Fig. 10-17. In this region $\Delta x = \Delta y = 0.5$ ft. The temperatures along the right-hand boundary containing point C are found by linear interpolation and are *regarded*

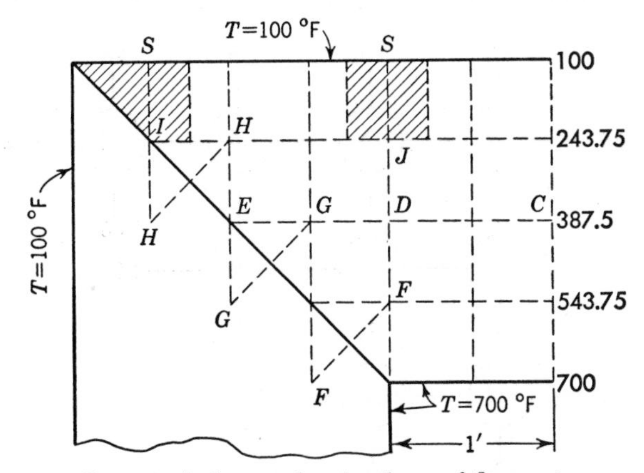

FIG. 10-17. Square flue, local use of finer net.

as *exact*. The necessary symmetry points F, G, and H are also shown. *Starting* values for the new interior points arising from the finer net are obtained by linear interpolation. The relaxation process is now applied and the temperatures corresponding to a "relaxed" fine net calculated. If deemed necessary, the process may be repeated, using a still finer net.

After the interior temperatures have been caculated, the rate of heat transfer may be determined. The local heat flux at the surface is

$$\frac{q}{A} = -k\left(\frac{\partial T}{\partial n}\right)_b \tag{10-157}$$

If the approximate boundary condition (10-92) is used, (10-157) becomes (when referred to a typical boundary point)

$$\frac{q}{A} = \frac{k(T_j - T_s)}{\Delta y} \tag{10-158}$$

The local heat flux given by (10-158) is applied to a surface area extending a distance $\Delta x/2$ on either side of a boundary net line. Thus, the heat flow per unit depth from the shaded area surrounding the line JS of Fig. 10-17 is taken to be

$$q = \frac{k(T_j - T_s)\,(\Delta x)}{\Delta y} \tag{10-159}$$

The sum of all such terms is then the surface heat flux. This is a straightforward procedure except at a corner. A simple approximate rule may be used at corners: The effective lane area is taken to be the mean of the lane areas at either end of the net line. Thus, for the shaded area at the corner of Fig. 10-17, the lane area at point I is $\Delta x/2$; at point S the lane area is $3\ \Delta x/2$. The effective lane area is then $\frac{1}{2}\ \Delta x(\frac{3}{2} + \frac{1}{2}) = \Delta x$, and the heat flow from the corner is approximately $k(T_I - T_s)\ \Delta x/\Delta y$.

If the finer net calculations of Example 10-7 are carried out in detail, it will be noticed that the algebraic sum of the residuals is not changed by the relaxation process unless the relaxed point is adjacent to a boundary. Because of this, it is desirable to transfer the interior residuals to points near the boundary as soon as possible.

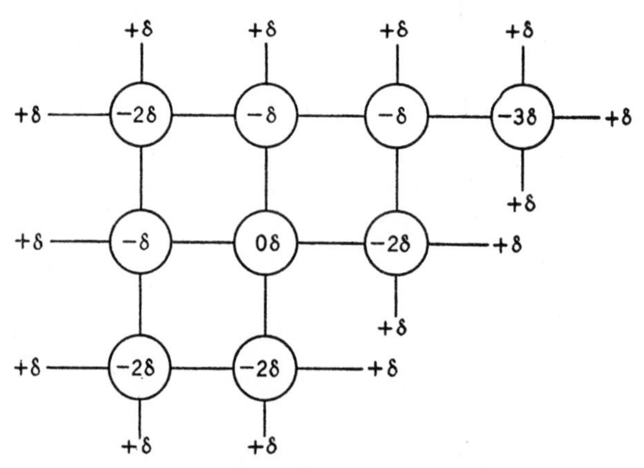

FIG. 10-18. Block-relaxation pattern.

When the residuals in a given region are largely of the same sign, it is frequently helpful simultaneously to alter all the temperatures in this region by the same amount. This technique is termed "block relaxation." With the relaxation pattern of Fig. 10-15, it may be shown that if all the entries in the block of Fig. 10-18 are increased by the amount δ, the residuals at these points and at points immediately adjacent to the block are changed as indicated. If a net line which extends from a point in a block to a point *outside* the block is termed a "free line," the block-relaxation rule for the point-relaxation pattern of Fig. 10-15 is as follows:

If each entry in the block is increased by δ, the residual at any point in the block is decreased by δ times the number of free lines leading from that point, and the residual at a point immediately adjacent to the block is increased by δ times the number of free lines which terminate at that point.

If the boundary points are fixed and all interior points simultaneously are increased by an amount δ, the residual at any point is decreased by δ

times the number of lines joining that point to the boundary. The net decrease in the over-all residual is then δ times the total number of lines N joining interior to boundary points. Block relaxation of the entire net is particularly helpful at the start of the relaxation process. The interior points are guessed and the net over-all residual, ΣR, calculated. The estimate at each interior point is then increased by the amount $\Sigma R/N$. In this way, the net over-all residual is reduced to approximately zero.

10-12. Treatment of Boundary Conditions. In Sec. 10-11, the values of the dependent variable were assumed to be specified at net points coincident with the boundary. This is not always the case. In some problems the derivative of the dependent variable normal to the boundary may be specified at the boundary; in problems involving an irregular boundary, the net points may not coincide with the boundary. Simple formal approximation methods for handling these situations were developed in Sec. 10-9. In this section, these relations will be used to generate relaxation patterns at net points in the immediate vicinity of the boundary. It is to be noted that the boundary-condition approximations used in this chapter are, in general, the simplest that can be developed. Better approximations can be derived that in specific situations give better results when a specified net spacing is employed. Examples may be found in articles by Southwell[32,33] and Emmons.[14] The simpler approximations are, however, more easily applied and become more exact as the net is refined.

Consider first the case where the boundary temperatures are specified but where all the net points do not coincide with the boundary. This situation is illustrated in Fig. 10-9, and the relation between values at net points near the boundary and the boundary temperature is given by Eq. (10-91). The easiest relaxation procedure is generally as follows: Estimate all interior temperatures and calculate *net* boundary temperatures by means of (10-91). Regard the net boundary temperatures as exact and relax interior points, using a coarse net. Recalculate the net boundary temperatures, using Eq. (10-91) and the relaxed interior values, and repeat the relaxation process, using a finer net. This process is continued until convergence is obtained with a net of the desired spacing. In practice, it is found that convergence is rapid.

If the normal derivatives are specified along the boundary, the problem is treated in a similar fashion. Thus, with reference to Fig. 10-10, the net boundary temperature T_B is related to the interior temperatures and to the normal derivative by Eq. (10-94). The interior temperatures are estimated and the net boundary temperatures calculated by means of (10-94). The process then proceeds exactly as before.

10-13. Other Applications. The previous discussion has considered the application of relaxation methods to boundary-value problems which satisfy Laplace's equation. The method has been applied with success to a number of involved problems governed by one or more nonlinear partial differential equations. The two books by Southwell[32,33] deal with a variety of such problems. Two cases of particular interest to chemical engineers will be considered here.

Poisson's equation

$$\frac{\partial^2 T}{\partial x^2} + \frac{\partial^2 T}{\partial y^2} + f(x,y) = 0 \tag{10-160}$$

applies to the case of steady-state conduction with internal-heat generation. When the derivatives are replaced by finite differences and Δx is taken equal to Δy, the residual at an interior point 0 is given by the expression

$$R_0 = T_1 + T_2 + T_3 + T_4 - 4T_0 + \Delta x^2 f(x_0,y_0) \tag{10-161}$$

If Eq. (10-161) is compared with the Laplace-equation relaxation relation (10-155), it will be noted that they differ only as a result of the *known* quantity $\Delta x^2 f(x_0,y_0)$. The relaxation *patterns* are identical in the two cases. Consequently, if the term $\Delta x^2 f(x_0,y_0)$ is included in the calculation of the *initial* residuals at all net points, the subsequent relaxation steps are identical to those used in the solution of Laplace's equation.

A specialization of a partial differential equation of the form

$$\frac{\partial}{\partial x}\left(P\frac{\partial T}{\partial x}\right) + \frac{\partial}{\partial y}\left(Q\frac{\partial T}{\partial y}\right) = F \tag{10-162}$$

is of frequent occurrence. The functions P, Q, and F may involve x, y, T, $\partial T/\partial x$, and $\partial T/\partial y$. In the general case, a relaxation pattern for (10-162) may be derived. Expand (10-162) with the result

$$P\frac{\partial^2 T}{\partial x^2} + Q\frac{\partial^2 T}{\partial y^2} = F - \frac{\partial P}{\partial x}\frac{\partial T}{\partial x} - \frac{\partial Q}{\partial y}\frac{\partial T}{\partial y} \tag{10-163}$$

Let

$$f = \frac{Q}{P} \tag{10-164a}$$

and

$$G = \frac{1}{P}\left(F - \frac{\partial P}{\partial x}\frac{\partial T}{\partial x} - \frac{\partial Q}{\partial y}\frac{\partial T}{\partial y}\right) \tag{10-164b}$$

Equation (10-162) then becomes

$$\frac{\partial^2 T}{\partial x^2} + f\frac{\partial^2 T}{\partial y^2} = G \tag{10-165}$$

The finite difference equivalent of (10-165) is

$$\frac{T_1 - 2T_0 + T_3}{\Delta x^2} + f_0\frac{T_2 - 2T_0 + T_4}{\Delta y^2} = G_0 \tag{10-166}$$

f_0 and G_0 are the values of f and G at the point 0. If $\Delta x = \Delta y$, the residual form of (10-166) is

$$R_0 = T_1 + f_0T_2 + T_3 + f_0T_4 - 2(1 + f_0)T_0 - \Delta x^2 G_0 \quad (10\text{-}167)$$

The relaxation pattern corresponding to Eq. (10-167) is shown in Fig. 10-19. Note that the values of f at point 2 and at point 4 are involved. If f varies from point to point in the net, the numerical coefficients associated with the relaxation pattern will vary from point to point. In carrying out the relaxation process, values of T are estimated. The corresponding values of f and G are then calculated. If the quantities $\partial T/\partial x$ or $\partial T/\partial y$ are involved, these terms are evaluated by suitable finite difference equivalents. The residuals at each net point are then calculated by means of Eq. (10-167) and the relaxation pattern *partially* liquidated. Revised values of f and G are then calculated at each net point and the system further relaxed. This process is continued until convergence is obtained.

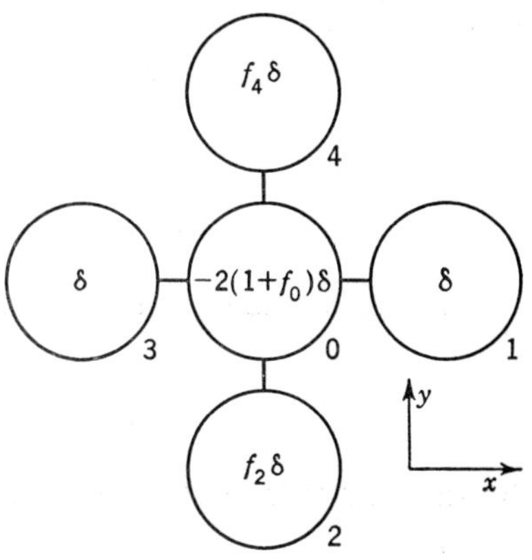

FIG. 10-19. Relaxation pattern.

If the terms involving T and/or its derivatives which appear in G are linear, it may prove simpler to replace these derivatives by finite differences and formulate an alternative finite difference equivalent to Eq. (10-165). The relaxation pattern will then differ from that shown in Fig. 10-19.

10-14. Convergence of the Relaxation Method. Experience has shown that when the physical problem possesses a unique solution and when the exact *boundary-value* equation has been correctly formulated, the finite difference equation can be solved by *some type* of relaxation process. In addition, as the net is continually refined, the solution of the approximate problem approaches the solution of the exact problem. Formal proof of the above statement is limited to the case of boundary-value problems governed by linear, self-adjoint elliptic equations in which the dependent variable is specified along the boundary. The specific relaxation process which leads to a convergent solution cannot be delineated in the general case; it must be found by experiment.

The question of stability does not arise in a boundary-value problem; only the problem of convergence need be considered.

10-15. Conformal Mapping. Two-dimensional boundary-value problems governed by Laplace's equation (or simple modifications thereof) but involving irregular boundaries often may be transformed into a problem with regular boundaries by means of conformal-mapping methods. The validity of the conformal-mapping procedure is a consequence of the properties of analytic functions of a complex variable. Standard references such as Churchill[7] discuss the procedure in detail. Attention here will be confined to the case when part or all of the boundary is made up of nearly circular curves. It will be shown that the circular boundary may be replaced by a straight boundary by means of a coordinate transformation.

Let r and θ denote the cylindrical polar coordinates of a point x, y in the x-y plane. Define a new coordinate system by the relations

$$\begin{aligned} u &= \ln \frac{r}{a} \\ v &= \theta \\ a &= \text{arbitrary constant} \end{aligned} \tag{10-168}$$

Now replace the rectangular coordinates x and y in Laplace's equation

$$\frac{\partial^2 T}{\partial x^2} + \frac{\partial^2 T}{\partial y^2} = 0 \tag{10-151}$$

by the new coordinates u and v. The result is

$$\frac{\partial^2 T}{\partial u^2} + \frac{\partial^2 T}{\partial v^2} = 0 \tag{10-169}$$

i.e., the form of Laplace's equation is not changed by the coordinate transformation. If θ is restricted to the range $0 \leqq \theta < 2\pi$, it is found that a region in the x, y plane bounded by two concentric circles of radii a and b, with centers at $x = 0$, $y = 0$, becomes a rectangular area in the u, v plane enclosing the region

$$0 \leqq v < 2\pi \qquad 0 \leqq u \leqq \ln \frac{b}{a}$$

The circle $r = a$ becomes the line $u = 0$; the circle $r = b$ becomes the line $u = \ln b/a$. If the center is at the origin, arcs of circles become segments of lines parallel to the v axis; segments of radial lines become segments of lines parallel to the u axis. These concepts are illustrated in Fig. 10-20.

Since Laplace's equation remains unchanged in the new coordinate system, the relaxation process using a square (or rectangular) net may be applied to the transformed problem in the usual manner. After a solution has been obtained in the u, v plane, the inverse transformation will relocate the solution in the x, y plane. Clearly, if values of T are specified along a boundary in the x, y plane, these same values apply to the transformed points in the u, v plane. If the derivatives normal to the boundary are specified in the x, y plane, the relations

$$\frac{\partial T}{\partial r} = \frac{1}{ae^u}\frac{\partial T}{\partial u} \qquad \frac{1}{r}\frac{\partial T}{\partial \theta} = \frac{1}{ae^u}\frac{\partial T}{\partial v} \tag{10-170}$$

may be used to determine the normal derivatives in the u, v plane.

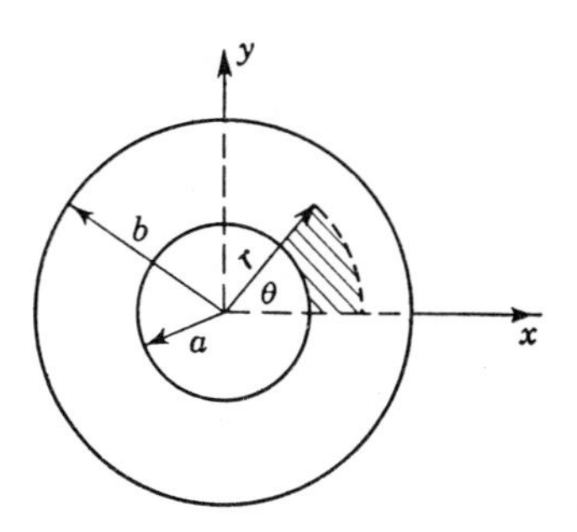

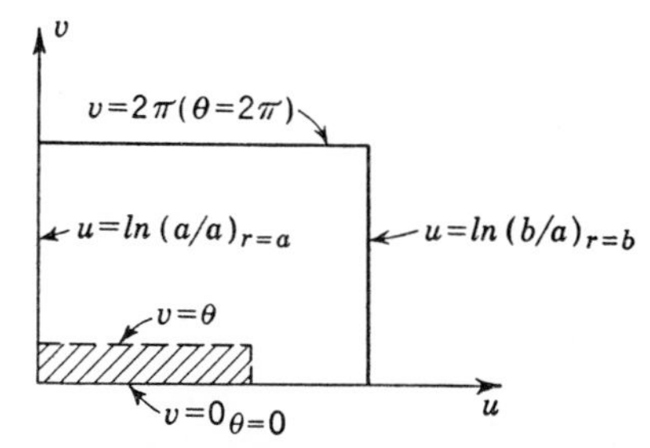

FIG. 10-20. Conformal mapping.

In terms of u and v, Poisson's equation

$$\frac{\partial^2 T}{\partial x^2} + \frac{\partial^2 T}{\partial y^2} + f(x,y) = 0 \tag{10-160}$$

becomes

$$\frac{\partial^2 T}{\partial u^2} + \frac{\partial^2 T}{\partial v^2} + a^2e^{2u}F(u,v) \tag{10-171}$$

where

$$F(u,v) = f(ae^u \cos v,\ ae^u \sin v) \tag{10-172}$$

BIBLIOGRAPHY

1. Householder, A. S.: "Principles of Numerical Analysis," McGraw-Hill Book Company, Inc., New York, 1953.
2. Anzelius, A.: *Z. angew. Math. u. Mech.*, vol. 6, pp. 291–294, 1926.
3. Baron, T.: *Chem. Eng. Progr.*, vol. 48, no. 3, pp. 118–124, 1952.
4. Binder, L.: dissertation, Munich, 1911.
5. Boelter, L. M. K., and M. Tribus: "Numerical Methods of Analysis in Engineering," arranged by L. E. Grinter, The Macmillan Company, New York, 1949.

6. Bunnel, D. G., H. B. Irvin, R. W. Olson, and J. M. Smith: *Ind. Eng. Chem.*, vol. 41, p. 1977, 1949.
7. Churchill, R. V.: "Introduction to Complex Variables and Applications," McGraw-Hill Book Company, Inc., New York, 1948.
8. Coberly, C. A., and W. R. Marshall, Jr.: *Chem. Eng. Progr.*, vol. 47, no. 3, p. 147, 1951.
9. Courant, R., K. Friedricks, and H. Lewy: "Uber die partiellen Differenzengleichungen der Mathematischen Physik," *Math. Ann.*, vol. 100, pp. 32–74, 1928.
10. Crank, J., and P. Nicolson: *Proc. Cambridge Phil. Soc.*, vol. 43, pp. 50–67, 1947.
11. Danckwerts, P. V., and A. C. Sugden: S.M. thesis, Massachusetts Institute of Technology, 1947.
12. Dusinberre, G. M.: "Numerical Analysis of Heat Flow," McGraw-Hill Book Company, Inc., New York, 1949.
13. Emmons, H. W.: *Quart. Appl. Math.*, vol. 2, no. 3, pp. 173–195, 1940.
14. Emmons, H. W.: *Trans. ASME*, vol. 65, p. 607, 1943.
15. Eyres, N. R., D. R. Hartree, J. Ingham, R. Jackson, R. J. Sargant, and J. B. Wagstaff: *Trans. Roy. Soc. (London)*, vol. 240, no. 813, pp. 1–57, 1946.
16. Fishenden, M., and O. A. Saunders: "The Calculation of Heat Transmission," H. M. Stationery Office, London, 1932.
17. Fowler, C. M.: *Quart. Appl. Math.*, vol. 3, no. 4, pp. 361–376, 1946.
18. Hacker, D.: B.S. thesis, University of Illinois, 1949.
19. Hall, R. E., and J. M. Smith: *Chem. Eng. Progr.*, vol. 45, p. 459, 1949.
20. Hartree, D. R., and J. R. Womersley: *Proc. Roy. Soc. (London)*, vol. A210, p. 353, 1937.
21. Hildebrand, F. B.: "Methods of Applied Mathematics," Prentice-Hall, Inc., Englewood Clifis, N.J., 1952.
22. Hougen, J. O., and E. L. Piret: *Chem. Eng. Progr.*, vol. 47, no. 6, p. 295, 1951.
23. Hougen, O. A., and K. M. Watson: *Ind. Eng. Chem.*, vol. 35, p. 529, 1943.
24. Levy, H., and E. A. Baggott: "Numerical Studies in Differential Equations," C. A. Watts & Co., Ltd., London, 1934.
25. Mersman, W. A., W. P. Berggren, and L. M. K. Boelter: The Conduction of Heat in Composite Infinite Solids, *Univ. Calif. Publs. Eng.*, vol. 5, no. 1, pp. 1–22, 1942.
26. Nessi, A., and L. Nissolle: "Méthodes graphiques pour l'étude des installations de chauffage," Dunod, Paris, 1929.
27. O'Brien, G. G., M. A. Hyman, and S. Kaplan: *J. Math. Phys.*, vol. 29, pp. 223–251, 1951.
28. Richardson, L. F.: *Trans. Roy. Soc. (London)*, vol. A210, p. 307, 1910.
29. Richardson, L. F.: *Trans. Roy. Soc. (London)*, vol. A226, p. 299, 1927.
30. Rose, A., and R. C. Johnson: *Chem. Eng. Progr.*, vol. 49, pp. 15–21, 1953.
31. Rose, A., and T. J. Williams: *Ind. Eng. Chem.*, vol. 42, p. 2494, 1950.
32. Southwell, R. V.: "Relaxation Methods in Engineering Science," Oxford University Press, New York, 1940.
33. Southwell, R. V.: "Relaxation Methods in Theoretical Physics," Oxford University Press, New York, 1946.
34. Schmidt, E.: "Foeppls Festschrift," Springer-Verlag OHG, Berlin, 1924.
35. Schmidt, E.: Forschung auf dem Gebiete des Ingenieurewesen, vol. 13, no. 5, pp. 177–185, 1942.
36. Schumann, T. E. W.: *J. Franklin Inst.*, vol. 208, pp. 405–416, 1929.
37. Watson, K. M., and R. H. Dodd: *Trans. Am. Inst. Chem. Engrs.*, vol. 42, p. 263, 1946.

PROBLEMS

10-1. A composite slab consists of 9 in. of *A* (magnesite brick) and 8.70 in. of *B* (alumina brick), in good thermal contact. The initial distribution of temperature is specified below. Suddenly the exposed side of *A* is brought to and maintained at 100°F, while the exposed side of *B* is suddenly brought to and maintained at 0°F at the same time. Find the temperature distribution (*a*) at the end of 3.05 hr and (*b*) when the steady state is reached (at $\theta = \infty$).

NOTES: The initial distribution of temperature in *A* is given by the equation

$$T = 22.2x - 2.46x^2$$

where x is distance (inches) from the bare face of *A*, towards *B*, and T is temperature in °C.

The initial distribution of temperature in *B* is given by

$$T = 10x - 90$$

The following tables show T at various values of x:

x in A..	0	1	2	3	4	4.5	5	6	7	8	9
T......	...	19.78	34.6	44.5	49.4	50.0	49.4	44.5	34.6	19.78	0

x in B.........	9	10	12	14	16	17.7
T.............	0	10	30	50	70	87

DATA:

	A	*B*
Thermal conductivity k.........	2.15	0.58
Density ρ, lb/ft³...............	159	90.0
Specific heat c_p.................	0.22	0.20
Thermal diffusivity α...........	0.0615	0.0322

10-2. A ceramic company is manufacturing large slabs 4 in. thick. They are carefully dried in a tunnel drier, leaving the drier at 300°F, and are sent directly to a small ring furnace for kilning. The present cycle of operations is represented by the following conditions:

Cycle	Time, min	T_{av} (gas), °F	$(h_c + h_r)_{av}$, Btu/(hr)(°F)(ft²)
1	13	600	15
2	13	800	19
3	13	1000	24

These firing conditions produce a satisfactory product although the minimum temperature produced in the slab is barely enough to result in the desired sintering.

Because of a rush order, it is decided to fire the slabs at 1000°F directly following the drying operation in order to speed up production. The plant manager desires a preliminary estimate of the percentage decrease in firing time that could be introduced by this method and still maintain a satisfactory product.

DATA (standard English engineering units):

$$k = 0.6 \qquad \rho = 100 \text{ lb/ft}^3 \qquad c = 0.25$$

10-3. An airplane has been in flight at constant air speed for a considerable time at a high altitude and then descends to a low altitude in a period of 5 min. It is desired to calculate the surface temperatures on *both* sides of the windows during the descent period. The windows are constructed of a laminated safety glass, having an over-all thickness of 0.529 in., as follows: 0.250 in. of glass inside, 0.154 in. of plastic, and 0.125 in. of glass outside.

The coefficient of heat transfer h on the inside may be taken as constant at 4 Btu/(hr)(ft²)(°F), while that on the outside varies linearly with time from 10 to 25, during the 5-min descent. The air temperature inside the cabin is constant at 70°F, while the temperature of the outside air depends on the altitude and varies linearly with time from −40°F to +60°F during the 5-min period of descent.

DATA:

	Glass	Plastic
k, Btu/(hr)(ft)(°F)........	0.40	0.075
ρ, lb/ft³..................	150	80
c_p, Btu/(lb)(°F)...........	0.24	0.50
$k/\rho c_p$, ft²/hr..............	0.0111	0.001875

$h_i = 4$ Btu/(hr)(ft²)(°F) $\qquad t_{ai} = 70°\text{F}$

$h_0 = 10 + 3\theta_{\text{min}}$, Btu/(hr)(ft²)(°F) $\qquad t_{a0} = -40 + 20\theta_{\text{min}}$

10-4. In the reduction of the partial differential equation

$$\frac{\partial^2 T}{\partial x^2} = \alpha \frac{\partial T}{\partial t}$$

to a finite difference equation, the substitutions

$$\frac{\partial T}{\partial t} = \frac{T_{m,n+1} - T_{m,n-1}}{2\,\Delta t}$$

$$\frac{\partial^2 T}{\partial x^2} = \frac{T_{m+1,n} - 2T_{m,n} + T_{m-1,n}}{\Delta x^2}$$

were made. Under what conditions is the resulting difference equation stable?

10-5. A solid undergoes a "spontaneous" exothermic chemical reaction at a rate determined by the equation

$$\frac{\partial w}{\partial t} = -wAe^{-B/T}$$

A large block of this solid, initially uniformly at such a low temperature T_i that the reaction rate is negligible, is exposed to surroundings at a high temperature T_A in the following way: Four of the faces are insulated, and the other two parallel faces, separated by the distance L, are exposed to the hot surroundings at time $t = 0$. If the following assumptions are permissible, devise a *finite-difference* numerical method for calculating the solid temperature T as a function of time t and position x. Show *exactly* how you would proceed to carry out the calculations and indicate where and how your method might lead to serious inaccuracies. Indicate the methods that might be used to improve the accuracy of your technique *other than using smaller increments.*

DATA AND ASSUMPTIONS:

1. Thermal conductivity k, specific heat c, and density ρ are always constant, *even after complete reaction has occurred.*

2. The heat-transfer coefficient between the solid face and the medium at T_A is constant at a value h.

3. The reaction releases q units of energy per pound of material reacted, regardless of the temperature level.

w = pounds reactant per unit volume
A = const
B = const
T = temperature
t = time

10-6. A steel billet 24 in. in diameter is to be reheated before rolling. It is placed in a soaking pit which is heated by combustion gases which are at a temperature of 1800°F. The walls of the soaking pit are at a temperature of 1500°F. The billet was at a uniform temperature of 80°F when introduced into the pit. The heat-transfer coefficient due to conduction and convection from the gas to the billet is 5 Btu/(hr)(ft^2)(°F). The thermal conductivity of the billet is 20 Btu/(hr)(ft^2)(°F/ft); the surface emissivity of the billet is 0.9; the soaking pit walls may be considered black. What is the temperature at the center of the billet at the end of 1 hr's heating if end effects are neglected?

10-7. When the surrounding gas is at a pressure of 1 atm, limestone decomposes at an appreciable rate only at temperatures of 910°C and higher. At 910°C (1672°F) and ordinary heat-supply rates, the decomposition is essentially isothermal, and the decomposition rate is controlled by the heat-addition rate.

The flat surfaces of a large slab of limestone 2 in. thick are suddenly brought to and maintained at a temperature of 1010°C (1852°F). The initial temperature of the entire slab was 125°C (257°F). It is desired to predict the time required to obtain complete decomposition of the slab. The following data and simplifying assumptions may be invoked:

1. *Properties of Limestone*

   Density, 103 lb/ft^3
   Specific heat, 0.335 Btu/(lb)(°F)
   Thermal conductivity, 0.54 Btu/(hr)(ft^2)(°F/ft)

2. *Decomposition Reaction at* 910°C

   $CaCO_3$ + impurities → CaO + $CO_{2(g)}$ + impurities
   ΔH = 1640 Btu per pound of CO_2 produced
   Pounds of CO_2 evolved per pound of limestone at complete decomposition = 0.44

3. *Decomposition Products*

   CO_2
   Specific heat, 0.305 Btu/(lb)(°F)
   CaO + impurities
   Density, 58 lb/ft^3
   Specific heat, 0.281 Btu/(lb)(°F)
   Thermal conductivity, 0.32 Btu/(hr)(ft^2)(°F/ft)

4. *Assumptions*

a. Linear dimensions of slab remain unchanged.
b. Holdup of CO_2 may be neglected.
c. CO_2 is in thermal equilibrium with solid which it contacts.
d. Edge effects may be neglected.
e. Thermal properties are independent of temperature.

10-8. In the design of a high-pressure steam boiler it is proposed to use alloy steel tubes in the boiling section. These tubes will be 2 in. ID by 3 in. OD, with water at 3,000 psi boiling inside the tubes. In one section of the furnace these tubes will be embedded in the refractory wall and hence will receive heat from one side only (see

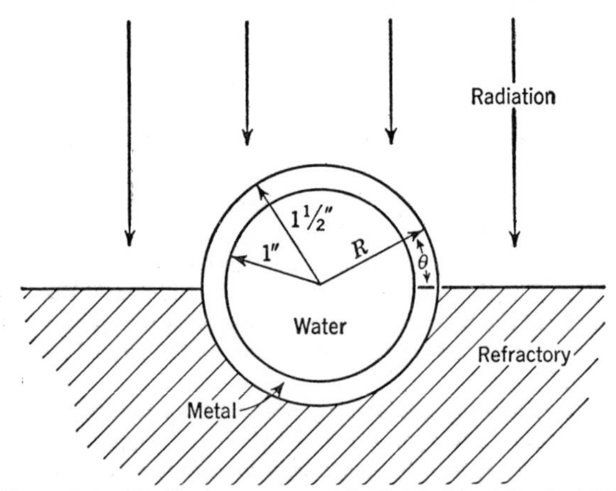

FIG. 10-21. Radiation to steam-boiler tube of Prob. 10-8.

Fig. 10-21). Under the proposed operating conditions, the heat-flow rate will be substantially independent of the tube-wall temperature. Because of the geometry of the system, the radiation density will not be uniform but is given by the expression

$$dq = QLR_0 \sin \theta \, d\theta$$

where Q = Btu/(hr)(ft² of *projected* tube area) = 121,000
L = tube length, ft
R_0 = tube radius, ft
θ = polar angle (see Fig. 10-21)

The metal-wall resistance is so large in comparison with the water-side heat-transfer coefficient that the temperature of the inner wall of the tube is essentially the boiling temperature of the water (695°F). The thermal conductivity of the tubes is 31 Btu/(hr)(ft²)(°F/ft).

In order to evaluate the feasibility of this design, it is desired to calculate the tube-wall temperatures. Such information will permit thermal stresses to be assessed.

INDEX